First published 1990 by CRC Press
Taylor & Francis Group
6000 Broken Sound Parkway NW, Suite 300
Boca Raton, FL 33487-2742

Reissued 2018 by CRC Press

Library of Congress Cataloging-in-Publication Data

Ghannoum, Mahmoud A. (Mahmoud Afif)
Candida adherence to epithelial cells / authors, Mahmoud A. Ghannoum and Samir S. Radwan
p. cm.
Includes bibliographical references.
ISBN 0-8493-5979-1
1. Candidiasis—Pathogenicity. 2. Candida—Adhesion.
3. Epithelial cells. I. Radwan, Samir S. (Samir Sayed) II. Title.
(DNLM: 1. Candida-physiology. 2. Candida Albicans—pathogenicity.
3. Cell Adhesion. QW 180.5.D38 G411c)
QR201.C27G43 1990
616'.015—dc20 89-22307

A Library of Congress record exists under LC control number: 89022307

ISBN 13: 978-1-315-89131-6 (hbk)
ISBN 13: 978-1-351-07041-6 (ebk)

Visit the Taylor & Francis Web site at http://www.taylorandfrancis.com and the
CRC Press Web site at http://www.crcpress.com

PREFACE

Infections caused by the pathogenic yeast *C. albicans* and to a lesser extent related members of the *Candida* genus are among the most prevalent of all opportunistic infections and without doubt represent the most important of the mycotic infections. In recent years Candidosis has become of even greater significance in view of the advances in surgical and therapeutic modalities for the treatment of a variety of medical conditions (e.g., cancer, immunocompromised or physiologically compromised patients) some of which may predispose a host to infection. Of particular relevance is the emergence of acquired immune deficiency syndrome (AIDS), AIDS-related complex (ARC) and the widespread use of immuno-suppressive therapy.

The capacity of a microorganism to cause disease is dependent upon a number of determinants related both to the pathogen and the host. One of these determinants is adhesion, which represents the first intimate contact between pathogen and host. *Candida* adhesion to epithelial cells is an essential prerequisite for establishment of infection. This fact is reflected in extensive research on the phenomenon of adherence. Improved understanding of this process will certainly be of major clinical significance. It may, for example, lead to the design of prophylactic agents for the control and the prevention of candidosis. This would improve on the present situation in which antimycotic therapy is the main approach. Antimycotic agents often cause toxicity, especially when combined with aggressive therapy for primary disease treatment. As yet no book has been devoted to this subject although there have been several works on bacterial adhesion.

This book gives a comprehensive account of the literature in the area of candidal adherence to surfaces. At the same time, we have provided background material to allow the reader to become familiar with *Candida albicans* and its relation to its host. In writing the book we have attempted to make each chapter as self-contained as possible, so that it is not necessary for the reader to start at the beginning of the book and work his/her way through. Naturally, some areas of repetition were unavoidable but justifiable.

The table of contents is detailed at the beginning of each chapter and is, therefore, not outlined in this preface. However, we would direct the attention of the reader to Chapters 6 and 7, in which our own research data is heavily drawn upon. These chapters examine two areas which we feel should be addressed by future workers.

We have used the terms "adherence" and "adhesion" interchangeably. There are those who feel that the word "adhesion" is more appropriate, but this convention has not been generally adopted.

We hope that the book will prove to be of value both to research workers interested in the pathogenicity of *Candida* in particular and to students of mycology or microbiology in general.

We would like to thank those research workers whose kind permission allowed us to reproduce their material. In addition we would like to thank the following people for their help: Michael Robinson and Peter Whittaker who read and revised carefully the whole manuscript; Vivienne Armer and Kenneth Moore who read parts of the manuscript, the members of our laboratory (Ashraf, Moeen, Iman, and Imad) who helped in many ways, Rima Yassin who carried out literature searches, S. Rajendra who was responsible for the photography, Sanjeev Reddy for typing the manuscript; and last but not least, Khaled Abu Elteen, without whose help this book would not have been possible.

We wish to thank Kuwait University, Research Management Unit, and the Kuwait Foundation for the Advancement of Science for their financial support for certain aspects of this work (Projects No. S0038, S0040 87/27/07 and S0039).

One of us (M.A.G.) would like to express his thanks to Marie and the children (Afif and Emma) for their patience and support throughout.

Mahmoud Afif Ghannoum
Samir Sayed Ahmad Radwan

THE AUTHORS

Mahmoud Afif Ghannoum, Ph.D., is Associate Professor of Microbiology in the Department of Botany and Microbiology, Faculty of Science, at Kuwait University of Kuwait, Arabian Gulf.

Dr. Ghannoum received his B.Sc. degree in Biology-Chemistry from the American University of Beirut, Lebanon in 1973. He obtained his M.Sc. (Medicinal Chemistry) and Ph.D. (Medicinal and Pharmaceutical Microbiology) degrees in 1974 and 1978, respectively, from the Department of Chemistry, Loughborough University of Technology, Loughborough, United Kingdom. He served as an Assistant Professor at the Higher Institute of Technology (Malta) from 1978 to 1980, as an Assistant Professor at Kuwait University Faculty of Science, Department of Botany and Microbiology from 1980 to 1985, and as an Associate Professor at the same Department since 1985.

Dr. Ghannoum is a member of the Society for General Microbiology, U.K., and a member of the International Society for Human and Animal Mycology, and was previously a member of the Institute of Biology, Society for Applied Bacteriology, Royal Microscopical Society and Institute of Medical Laboratory Sciences.

Dr. Ghannoum has been the recipient of research grants from the Research Management Unit, Kuwait University, Kuwait Foundation for the Advancement of Science and Environment Protection Council, Kuwait. He has collaborated with research workers from University of California, Davis, Loughborough University of Technology, U.K., and National University, Maynooth, Ireland. He has received the Abdul Hameed Shawman Prize for Young Arab Scientists in Biological Sciences for 1987.

Dr. Ghannoum is the author of more than 55 papers and has been a co-author of two books. His current research interests relate to *Candida* and Candidamycosis.

Samir Sayed Ahmad Radwan, Dr.rer.nat., is Professor of Microbiology in the Faculty of Science, The University of Kuwait.

Dr. Radwan received his B.Sc. and M.Sc. degrees from Ain Shams University, Cairo, Egypt in 1958 and 1961, respectively. He obtained his Dr. rer. nat. from the University of Muenster, FRG, in 1964. After doing postdoctoral work at the Federal Center of Lipid Research in Muenster, FRG, Dr. Radwan was appointed lecturer of Microbiology in the Faculty of Science, Ain Shams University, Cairo, Egypt, in 1965. He became an Associate Professor of Microbiology in 1971 and Professor of Microbiology in 1976.

Dr. Radwan is a member of the Deutsche Gesellschaft fuer Fettforschung and the International Association for Plant Tissue Culture.

He was the recipient of the Alexander van Humboldt Fellowship (1972-1974), he was awarded the State Prize of Egypt in biological sciences in 1980 and the Prize of the Kuwait Foundation for the Advancement of Science in Biological Science also in 1980. His name is included in the International Who's Who of Intellectuals, 8th Edition (International Biographical Centre), the Men of Achievement Honours List (IBC), and Edition II of Five Thousand Personalities of the World (The American Biographical Institute).

Dr. Radwan has published more than 67 original papers and review articles. His current research interest is in microbial lipids and their biological significance.

TABLE OF CONTENTS

Chapter 1

MICROFLORA OF THE HUMAN BODY

TABLE OF CONTENTS

I. INTRODUCTION

Since this volume deals with *Candida* adherence to epithelial cells, it is important that an introductory chapter be devoted to the normal flora of the body. This is particularly important because in natural environments (e.g., oral cavity, gut, vagina) a mixed flora is present. These microorganisms certainly compete for available adherence sites on the host epithelium or may interact directly with the yeast, or both. In this chapter we shall describe the indigenous normal flora with regard to its distribution, entrance into, and exit from the body. In addition, the relation of these microorganisms to the adherence of *Candida* will be discussed. It is not intended to give a detailed account of the normal flora, as this subject has already been adequately covered by earlier authors and is beyond the scope of this volume.

II. DISTRIBUTION OF MICROORGANISMS IN THE HUMAN BODY

A. THE SKIN

The human skin is directly exposed to air and consequently to airborne microorganisms. However, the skin surface is not a suitable habitat for microbial activity. Although dead cells and skin secretions may provide adequate nutrients for commensal microorganisms, the moisture content is normally too low and thus represents a limiting factor for the growth of microorganisms. Only the relatively moist areas, e.g., the scalp, sweat glands, and underarm regions, may be relatively rich in microorganisms. It is important to note that the human skin has an acid pH value of between 4 and 6.[1] This is another factor which limits the number of bacteria capable of propagation. Two types of skin microorganisms are recognized — transient and resident species. The transients come into contact with the skin but do not contribute to any activity and soon die. The residents, on the other hand, can actively propagate on human skin. The latter comprise only a few groups of bacteria, viz. *Staphylococcus* spp. and coryneform bacteria. Generally, Gram-negative bacteria are rare on human skin, but can flourish if the Gram-positive residents are eliminated, e.g., by antibiotic treatment. Skin residents hydrolyze glycerides of the skin releasing the constituent fatty acids which inhibit other microorganisms. For more detail of the skin microflora the reader is referred to pertinent reviews (e.g., Skinner and Carr[2]).

The effects of host factors on the normal skin microflora have been studied.[3,4] For example,it has frequently been demonstrated that the identities of microorganisms may be related to race.[5,6] The skin microflora also varies according to age;[7] this is a reflection of changes in conditions on the skin commensurate with aging. Sex determines the identity of the skin microorganisms only in the young adults.[8] Another important factor is diabetes, which is known to be associated with an increase in the level of glucose on the skin;[9,10] this situation may allow for surface and systemic fungal infections leading to diseases such as candidosis.[11,12]

Several benefits have been attributed to the normal skin microflora.[13] As mentioned above, several bacteria are capable of liberating fatty acids from skin glycerides;[14] these fatty acids are known to inhibit *Staphylococcus aureus* and *Streptococcus* spp.[15-19] The beneficial effect of the resident skin bacteria is apparent from results, indicating that the application of antibacterial drugs mainly active against *Streptococcus* spp. and coryneform bacteria usually favor the growth of Gram-negative bacteria.[20] Similar findings were reported in studies of neonates.[21-23]

Selwyn[24] described the carriage of antibiotic-producing organisms on human skin and showed how such organisms may protect the carrier against colonization of potential pathogens such as *Staphylococcus aureus*. Selwyn and his colleagues showed that these antibiotic producers can restrict the growth of susceptible strains *in vitro*.[25-27] Hsu and Wiseman[28] reported that 4.9% of 1065 isolates of *S. aureus* and 8.5% of 387 coagulase-negative cocci produced antibiotics active against the Oxford strain of *S. aureus*. Noble and Willie[29] studied the production of antibiotics active against *S. aureus* from healthy skin and in the nose and skin lesions. The authors found that approximatly 5% of these sites yielded such organisms, but did not obtain any evidence indicating that the presence of producer organisms in skin lesions had a protective effect against colonization by *S. aureus*.[29]

B. THE MOUTH AND DIGESTIVE TRACT

The human mouth is inhabited by many microorganisms, particularly anaerobic bacteria [30-35] Such organisms occur mainly on and among the teeth[36] and can also colonize the epithelium of the oral cavity. Oral microorganisms cover their nutrient requirements mainly from food materials embedded between the teeth. Consequently, the identities of the microbial inhabitants vary with the diet in the infant and adult phases. Saliva also provides some of the nutrient requirements of the oral microflora.

Saliva consists of inorganic substances, viz. chlorides, bicarbonates, phosphates, sodium, calcium, potassium, and trace elements, as well as organic compounds such as proteins and mucoproteins, with small amounts of carbohydrates.[37] Several antimicrobial substances occur in saliva, e.g., the enzymes lysozyme and lactoperoxidase.[38-40] Babies less than 1 year old have in the oral cavity mainly anaerobic bacteria such as *Streptococcus* spp. and *lactobacillus* spp. in addition to a few aerobes.[41,42] With the onset of teething, a dramatic alteration in the mouth microflora occurs, mainly involving a further shift in the balance towards the anaerobes. Dental plaque consists of bacterial cells enclosed within a polysaccharide matrix. Such bacteria belong predominantly to the genus *Leptotrichia* (or *Fusobacterium);* the cells being filamentous and anaerobic.[43-46] In addition, *Streptococcus* spp. coryneform bacteria and others may be present. The process by which bacteria colonize the smooth tooth surface is now established. Initially, a thin film of acidic salivary glycoproteins becomes attached to the tooth and provides the bacterial colonization site.[47-52] At first *Streptococcus* spp. grow producing a relatively dense bacterial film. At later stages this film becomes colonized by the filamentous bacteria. Mouth bacteria, comprising the plaque, play an essential role in the development of dental caries. Tooth decay occurs only in crevices where food debris can accumulate. When easily fermentable substrates, such as sugars, become available, bacteria ferment these sugars to lactic acid which decalcifies the tooth surface leading to its deterioration. Bacterial proteolytic enzymes then decompose the matrix of the tooth enamel and cells penetrate deeper into this tissue. The benefits which the normal mouth microflora provide to man are demonstrated by the fact that patients treated with antibiotics which inhibit *Streptococcus* spp. in the oral cavity may permit growth of *Candida* or Gram-negative rod-like bacteria.[53,54]

The digestive tract consists of the esophagus, stomach, small, and large intestine, and usually contains in total more than 10^{14} bacterial cells. The stratified squamous epithelium of the esophagus is colonized by *Lactobacillus* spp.[55,56] Being highly acidic, with a pH value of about 2, the stomach represents a natural barrier preventing the passage of microorganisms to the intestine. Nevertheless, the stomach wall is normally colonized by *Lactobacillus* spp. and *Streptococcus* spp. which are known to tolerate high acidity.[57,58] The duodenum, the first part of the small intestine, is also acidic and therefore its microflora resembles that of the stomach. The ileum, in contrast, is more alkaline and contains considerably more bacteria in the lumen mixed with the digesting material. The large intestine is the richest part in microorganisms of the whole digestive tract. Bacteria which occur in its lumen are mainly facultative to anaerobic. Genera such as *Escherichia, Bacteroides, Clostridium, Streptococcus, Bifido bacterium, Klebsiella,* and others are normally found in the colon.[59-68] The intestinal microorganisms perform several metabolic activities, some of which may provide important benefits to man. These activities include vitamin synthesis and nitrogen fixation, as well as production of gas, odor, and organic acids.[69] It has also been suggested that bacteria may be active in promoting the muscular contraction of the intestine.[70] The beneficial role of the intestinal microflora is demonstrated by the fact that patients taking antibiotics orally usually suffer from shifts in the microbial equilibrium.[71-73] Under such conditions, antibiotic-resistant strains of *Staphylococcus, Proteus,* as well as *Candida albicans,* which are normally suppressed by the indigenous intestinal microorganisms, may grow actively, causing problems (see Chapter 2). Factors affecting the intestinal microflora have been reviewed.[69,74,75]

Several reports[74,76,77] have appeared concerning the fecal microflora of man. It has long been known that microbial cells may make up about 40% of the total volume of feces.[78] Such microorganisms pass through dehydrating conditions in the colon and rectum. Therefore, feces represent microbiologically a rather unique habitat. The anaerobic bacteria predominate. The following genera can be expected to be found in human feces: *Lactobacillus, Streptococcus, Peptococcus, Bifidobacterium, Clostridium, Eubacterium, Propionobacterium, Bacteroides, Escherichia,* and others. In addition, yeasts such as *Candida* spp. and *Pityrosporum* spp. may be found.[77,79-89]

C. THE VAGINA

The normal vaginal microflora has been described in several publications.[90-99] Bacteria have been isolated from the vagina of women at various stages of the menstrual cycle, as well as of pregnant women. Several genera, viz. *Corynebacterium, Lactobacillus, Staphylococcus, Streptococcus, Micrococcus, Bacteroides,* and others have been identified. Yeasts such as *Candida albicans* and *C. glabrata* have also repeatedly been recorded in the vagina. De Louvois et al.[93] isolated 19 genera and groups of microorganisms from the lower genital tract of 280 unselected pregnant women. Coryneform bacteria, *Lactobacillus* spp., and *Staphylococcus epidermidis* were abundant in most of the cases examined (83.4 to 66.1%). *Streptococcus* spp. were encountered frequently and the majority of the β-hemolytic streptococci were assigned to Lancefield B. The percentage of *Escherichia coli* and yeasts (which includes *C. albicans, C. glabrata, and Trichomonas vaginalis)* in the posterior fornix was high.[93]

Corbishley[94] studied the microbial flora of the vagina and cervix qualitatively and semiquantitatively in 40 women attending an intrauterine contraceptive device clinic. *Lactobacillus* spp. were detected in 61% of vaginal and in 53% in cervical specimens; fecal bacteria, including anaerobes, were even more frequently found. Ohm and Galask[92] studied the aerobic and anaerobic flora of 100 women prior to their undergoing hysterectomy and found that anaerobes predominated. The anaerobic Gram-positive cocci were the most common isolates, followed by anaerobic Gram-positive rods and anaerobic Gram-negative cocci. *C. glabrata* and *C. albicans* were isolated from 4% of the cultures.[92]

D. THE RESPIRATORY TRACT

Staphylococcus aureus normally predominates in the nose, whereas *S. albus* and *Corynebacterium* spp. occur in limited numbers. Various research groups have demonstrated that American negroids are less commonly nasal carriers of *S. aureus* than American caucasoids.[2] Several other species belonging to the genera *Streptococcus* and *Neisseria* were recorded in the nasopharynx occasionally together with pathogens like *Haemophilus influenza, Neisseria meningitidis,* and *Klebsiella pneumoniae.* Under normal conditions the trachea, bronchi, and pulmonary tissues are devoid of microorganisms. Microorganisms that may reach these regions are trapped on the mucous surfaces and swept upwards by the cilia of the epithelial tissue to be swallowed or removed by expectoration.[4,55,100-102]

Table 1 presents information about the normal microbial flora of the human body.

III. HOW MICROORGANISMS ENTER INTO THE HUMAN BODY

The human body is directly exposed to microorganisms externally through the skin and internally through the epithelial tissue lining the digestive tract, the respiratory system, and the urinogenital organs. The intact skin and epithelial tissues under normal conditions are quite efficient as insulators from the external environment. At the same time there continuously occurs through such anatomical barriers some material exchange between the body and environment. The anatomical barriers are frequently exposed to mechanical damage by a variety of factors, thus providing entry sites for microorganisms (see Chapter 2). It has been found that the internal barriers of the human body, i.e., the epithelial tissues, provide more such sites than the skin does. Hence, the body is provided internally, especially in the conjunctiva and respiratory tract, with efficient cleaning systems and/or antimicrobial devices. Such tissue can be colonized or penetrated by microorganisms only after the latter have firmly attached themselves to the surface cells. Most of the infectious diseases of man occur through the respiratory and digestive tracts.

Infectious microorganisms are either provided with specific mechanisms of attachment to the surface cells, or if not, they can enter the body only through some mechanical damage of the barrier, e.g., a wound. In this context, it is to be noted that some bacteria, e.g., *Vibrio cholerae* do not need to overcome any anatomical barrier in order to cause the disease. The cells remain outside in the digestive tract and secrete toxic substances which either act locally or after being absorbed.

A. ENTRY THROUGH THE SKIN

Microorganisms can infect the body through breaks in the skin, or alternatively through the agency of insect and animal bites. Skin breaks normally arise in a conventional process of vaccination, e.g., against smallpox. Contaminated needles are also responsible for infections through the skin. Insect bites may transmit the causative agents of typhus plague *(Rickettsia prowazeki)*, malaria *(Plasmodium* spp.), etc. and animal bites can transmit those agents causing rabies.

B. ENTRY THROUGH THE INTESTINAL TRACT

Microorganisms in the intestinal tract have to withstand the effect of the food-flow. Therefore, not only the indigenous microflora but also the pathogenic microorganisms that infect the body through the intestinal epithelium are frequently provided with some mechanism of attachment to the epithelial cells.[103,105] Otherwise such cells could easily be washed away by the food-flow. In spite of such mechanisms of attachment, considerable numbers of microorganisms are continuously removed with the feces.

Attachment sites for pathogenic microorganisms occur not only in the intestinal tract but also in the respiratory and genitourinary tracts. There is a lot of information on such sites in experimental animals such as rats (see Bitton and Marshall[106]). A few microorganisms are adherent to the small intestinal mucosa of both mice and man, but there are no pallisaded masses of bacteria comparable to those found in the mouse stomach.[56,107-109] In many cases it has been shown that there are specific sites in the epithelium where the microbial cells become attached.[110-113] In man there are several pieces of evidence for the specificity in adherence. (1) The pathogenic cell does not show an equal affinity to all epithelial cells, but rather prefers those where subsequent penetration normally occurs. (2) The specificity in adherence between bacterial cells and epithelial cells has frequently been demonstrated *in vitro*. On the molecular level there is evidence indicating that there are complementary macromolecules on the pathogen cell and the epithelial cell; such molecules interact, resulting in the specific adherence.[114-120] For example, it is well known that a few strains of *Escherichia coli* can colonize the small intestine causing diarrhea.[121,122] These enteropathogenic strains are characterized by producing an acidic polysaccharide capsule known as K antigen which is responsible for the specific adherence to intestinal mucosal surfaces.[111,123-127]

After attachment, in most cases, pathogens penetrate the epithelial layer; some travel away to various organs of the body whereas others remain within the submucosa and propogate.[113] Penetration usually takes place through lesions in the mucous membrane. For a successful penetration the pathogen does not necessarily need to be motile. The normal microflora play a decisive role in the processes of attachment and penetration. Apparently, the pathogen must

TABLE 1
Normal Microbial Flora of the Human Body

Microbial group	Resident microorganism	Anatomical site: Skin	Gastrointestinal tract: Oral	Small intestine	Large intestine	Feces	Respiratory tract	Genitourinary tract	Vagina
Bacteria									
	Clostridium sp.		–	–	+			+	+
	Bacteriodes sp.					+		+	
	Enteric bacteria	+	+						
	Proteus sp.	+	–	+	+			+	
	Salmonella		–	–	+				
	Shigella		–	–	+/–			+	
	Bifidobacterium sp.					+		+	
	Escherichia sp.		–	+	+	+			
	Eubacterium sp.		–	+/–	+	+		+	
	Fusobacterium sp.		+	–	+	+	+	+	+
	Neisseria sp.		+	–	–		+		
	Actinomyces sp.		+	–	+/–				
	Corynobacterium		+	–	–		+		
	Lactobacillus sp.		+	+	+	+			+[a]
	Branhamella sp.		+	–	+		+		
	Gardnerella vaginalis							+	
	Haemophillus sp.						+		
	Staphylococcus sp.	+	+	–	+		+	+	
	S. aureus	+					+		
	S. epidermidis	+					+		
	Streptococcus sp.	+	+	+	+	+	+	+	+[b]
	Peptostreptococcus sp.				+	+	+		
	Borrelia sp.		+	–	–		+		
	Treponema sp.		+	–	–		+		
	Klebsiella					+			
	Diphtheroids	+	+	+	–		+	+	+[b]

TABLE 1 (continued)
Normal Microbial Flora of the Human Body

		Anatomical site							
			Gastrointestinal tract						
Microbial group	Resident microorganism	Skin	Oral	Small intestine	Large intestine	Feces	Respiratory tract	Genitourinary tract	Vagina
Yeast									
	Candida sp.		+	+	+				+
	C. albicans	+[c]					+	+	+
	C. parapsilosis	+							
	Pityrosporum sp.	+				+			
Fungi									
	Aspergillus sp.						+		
Virus									
	Adenoviruses						+		
	Coxsackieviruses						+		
	Epstein-Barr virus						+		
	Herpes-Simplex						+		
	Influenza viruses						+		
	Rhinoviruses						+		
Flagellated protozoa									
	Giardia		–	+/–	–				
	Trichomonas vaginalis					+			

[a]During child-bearing period.
[b]Before puberty.
[c]Average frequency of isolation <10.

compete with the indigenous microorganisms for nutrients and space at the adherence sites.[128,129] Several other factors may affect the penetration process. One of these factors is the presence or absence in mucous of secretory antibodies that may act against infection. Another factor is the ability of certain pathogens to produce enzymes that facilitate their entrance.[106,130] The ability to survive in the presence of acids, bile, and proteolytic enzymes is another important factor because pathogens usually become subjected to one or the other of these substances in the human body.[131-134]

C. ENTRY THROUGH THE RESPIRATORY TRACT

We have already mentioned that the respiratory system is normally provided with an efficient system for removing inhaled particles and airborne microorganisms.[135,136] However, some of the minute particles with a diameter less than 5μm may succeed in reaching the alveoli.[137] Normally these particles are removed after they have been phagocytosed by macrophages in the alveoli. However, some pathogens that are provided with a special mechanism for adherence to the epithelial cells may succeed in infection. For example, the influenza virus is provided on its surface with hemoagglutinin which reacts specifically with sialic acid moieties of the epithelial glycoproteins leading to firm attachment.[113] Subsequently, the attached virus can infect the cell. Causative agents of pneumonia and whooping cough have projections on their surfaces which mediate the cell adhesion to sites on the epithelium rich in specific compounds such as neuraminic acid. On the other hand, lesions may arise in the epithelium of the respiratory tract through the activities of certain viruses and bacteria. Such lesions are apparently portals for infection by pathogens.[135]

Many factors may affect the mucociliary defenses of the respiratory tract, thus affecting the probability of infection. Such factors include, e.g., cigarette smoking and air pollutants. Certain microorganisms are able to inhibit the activity of epithelial cilia. Examples are *Bordetella pertussis* and *Haemophilus influenzae*. Other microorganisms, such as *Mycobacterium tuberculosis*, may survive for short or long periods after being phagocytosed.[136]

D. ENTRY THROUGH THE URINOGENITAL TRACT

The urinary tract is provided with an excellent cleaning mechanism; it becomes repeatedly flushed at short intervals with urine, which is normally sterile. Hence, bacteria must be provided with highly efficient adhesion mechanisms in order to become established on the epithelium.[138] The causative agent of gonorrhoea is provided with such an efficient mechanism. The cells adhere through their numerous fimbriae to the epithelial cells.[129,139] In males the bladder is normally sterile, whereas the urethra may contain bacteria in the terminal end. Since the urethra in the female is only about 5 cm long, it is much more easily and frequently invaded by microorganisms than in the male. Such invasion may be enhanced by sexual intercourse, which in turn may also lead to infection of female bladder. Urine itself can cover the nutrient requirements of many microorganisms. Therefore, factors leading to incomplete emptying, such as stones, enlarged prostate, or unusual deformation of the bladder, normally lead to urinary infections.[140-143]

In contrast to the urinary tract, the vaginal epithelium is easily colonized by microorganisms because it is not provided with any cleaning mechanism. However, some lactic acid bacteria normally inhabiting the vagina produce lactic acid from glycogen, thus lowering the pH value to about 5.0 (see Chapter 5).[144-146] This acidity appears to be the major factor acting against infection. In girls before puberty and women after the menopause there is no glycogen in the vaginal epithelium. Hence, the pH value remains high and bacteria such as *Streptococcus* spp. and *Staphylococcus* spp. can easily colonize the vagina.[147-151]

IV. ENCOUNTER OF THE PARASITE WITH THE HOST CELLS

After the parasite has succeeded in attachment to a site in the human body, it begins to establish itself. A successful infection is not possible unless a relatively large number of pathogenic cells are available. Therefore, the initial stage of establishment usually involves active microbial propagation at the adherence site. Subepithelial invasion usually follows.[113]

A. THE ADHERENCE SITES

The molecular bases for the adherence process apparently vary according to the nature of the pathogen.[152] It is not intended to give a review of this subject here. For this, the reader may refer to some excellent works.[153,154] Although the present volume emphasizes the adhesion of *Candida albicans*, it may be useful to refer to the adherence of other microorganisms. In the following discussion an attempt is made to give a brief description of the surface components believed to be involved in the attachment of various microorganisms to surfaces. Such components mediate adhesion even against any electrical repulsion between the microorganism and the site of adherence. Microbial surface structures concerned with adherence are apparently the cell walls, capsules, and cell appendages. Structures like capsules may bring about attachment by simply acting like a "glue". However, the bases of adherence are certainly much more complex than this, especially when we consider that the nature of the substratum is equally important in this phenomenon. It has been repeatedly demonstrated that different mucosal surfaces sometimes have a high degree of selectivity for bacteria.[155]

1. Viruses

Bacteriophages establish bonding with components of bacterial surfaces. The T-coliphage is the best studied of these viruses. It adheres by the tail fibers to specific receptors in the surface of *E. coli*.[156-163] The nature of such receptors has been reviewed.[164] It is believed that almost every structure exposed or originating on the bacterial cell surface can act as phage receptor.[106,165,166] Chemical substances which can be involved in the reception process are the lipopolysaccharide and lipoprotein components of the wall of Gram-negative bacteria, the peptidoglycans of Gram-positive bacteria, teichoic acids, proteins, and carbohydrates. Some phages attach to the sex pili of the host bacterial cells through unknown mechanisms.[167-171]

There is evidence to suggest that some animal viruses are not provided with surface compounds that specifically attach to surface components of the host cells.[172] Several viruses can however, agglutinate red blood corpuscles, i.e., they have specific combining groups. Some animal viruses seem to attach specifically to a side chain of a mucoprotein in the cell surface consisting of N-acetyl neuraminic acid. The adherence of such viruses is blocked by neuraminidase. With other animal viruses the attachment sites are composed of glycolipo-proteins, and for still others, e.g., the polio virus, the site is a lipoprotein in the plasma membrane.[173,174] The attachment site for the avian tumor virus to fibroblast cells of chicken embryo appears to consist of glycoproteins. This site may be solubilized by trypsin and the solubilized material has been shown to inhibit the infectivity of this virus *in vitro*.[175]

2. Bacteria

Bacteria vary among themselves in the nature of the cell surface components involved in adherence to their hosts. A group of bacteria adhere through special cell appendages such as fimbriae, or even holdfasts. Other bacteria do not have such appendages and appear to adhere through particular polymers at their surfaces. Fimbriae are numerous cell appendages on certain bacterial species and are much shorter than flagella; they are not involved in cell motility.[176] Like flagella, however, fimbriae are composed of proteins.[177] Their functions are still under active investigation, yet there is now reason to believe that they may be involved in the adherence of bacterial cells to various surfaces.[178] They probably do so by first overcoming the initial electrostatic repulsion barrier between the bacterial cell and the adherence substrate. Several microscopic observations have revealed that such appendages may mediate the adherence of enteric bacteria to fungi as well as to plant and animal cells.[178] Fimbriae may also bring about the attachment of individual bacterial cells in the same culture, producing various aggregates.

Thus, pellicles on the surface of liquid media produced by highly aerobic bacteria such as *Pseudomonas* spp. and *Rhizobium* spp. which ensure adequate cell aeration, arise due to the attachment of the cell fimbriae. Similarly, the star-like aggregations in liquid cultures of certain bacteria are mediated by clusters of polarly inserted fimbriae.[179-184] Many pathogenic bacteria are known to adhere to epithelial cells through their fimbriae, e.g., *Gonococcus* spp. and *Streptococcus*. spp.[185-190] Some aquatic bacteria, e.g., the genus *Caulobacter*, produce stalks that are mere extensions of the whole cells, ending in holdfasts by which the cells attach to animate or inanimate substrates.[191] In the life cycle of this bacterium a mono-flagellated swarmer cell is produced by budding. It swims about and secretes some adhesive material at its flagellar end. The flagellum is lost, the stalk develops, and the holdfast adheres to the adhesive material. Up to 100 (but normally only a few) cells may adhere through their holdfast, giving the characteristic rosettes. The adherence substance consists of polysaccharides.[192-195]

Many species of bacteria adhere nonspecifically to various substrates, apparently without the involvement of any microscopic cell appendage. This attachment to suitable substrate surfaces is mediated via adhesive macromolecules in the cell envelope. In some cases, such macromolecules may be seen in electron micrographs as their fibrils. Chemically they consist of polysaccharides and/or glycoproteins and they affect nonspecific adhesion simply by forming cell-to-cell bridges. Inanimate objects, such as glass slides coated with such polymers, trap bacteria from which such polymers have been obtained.[196-201] It is also known that bacteria attach specifically to certain sites of the body.[112,202-204] Such bacteria may be either indigenous or simply allochthonous. Only in a very few cases, however, could the molecular bases of such adhesion be understood. Such cases include the attachment of *Streptococcus* mutants to salivary mucinous glycoprotein absorbed to teeth or to artificial surfaces and the attachment of *S. mibis*, *S. sanguis*, and *S. salivarius* to glycoproteins of the buccal epithelial cells.[155,205-208]

3. Fungi

Attachment of fungal hyphae is interesting not only from the medical viewpoint but also with regard to biodegradation problems caused by these microorganisms. Fungi can colonize and degrade even objects submerged in sea water.[209,210] It is well known that filamentous fungi and yeasts may produce some adhesive polysaccharides which may play a role in their attachment.[211] This process can be mediated in some other fungi by glycoproteins.[212] There is a large gap in information regarding the adherence mechanism of pathogenic fungi to human cells.

4. Protozoa

The mechanisms through which protozoa can adhere to surfaces are of particular interest in view of the fact that these microorganisms normally lack cell walls and other envelopes.[213] Except for a few specific cases, the molecular bases of protozoal adherence are still far from clear. However, it is generally believed that protozoa attach themselves to animate and inanimate substrates by one of three mechanisms: (1) by suction mechanism, (2) by nesting into a suitable pocket, and (3) through the secretion of some adhesive substances.[214-217]

B. MICROBIAL PROPAGATION

As mentioned, the first step for a parasite to establish itself after attachment is the active propagation at the site of adherence. This propagation leads in many cases to the spreading of infection on the epithelium.[112] There is evidence suggesting that the rate of multiplication of bacteria colonizing mucosal surfaces is low. Experimentally derived estimates of the growth rate in the intestinal tract of various rodents and bovine rumens indicate that the microbial biomass doubles two to three times per day. This is also true of the oral microflora.[112,218-220] The importance of active microbial propagation for a successful infection is apparent when we consider that the immune system of the body needs a period of time in order to be fully mobilized. In other words, a pathogen capable of quick propagation may complete the infection process without the interference of the immune system. However, chemical defense mechanisms other than immunity may arise during the propagation of certain parasites, especially viruses. Such mechanisms may affect the microbial propagation. It is well known that human epithelial cells respond after, at most, a few hours to almost all viral infections by the production of low molecular weight proteins called interferons. These compounds are directly coded for by the infected cells and can protect uninfected cells so that the infection process is slowed or even blocked.[221]

Factors which may affect the rate of microbial propagation on the adherence site have not been adequately studied. One such factor is how wet the infection surface is. Apparently, invading microorganisms can propagate much more easily and quickly on wet surfaces than on dry. Thus, local infections of the skin usually take much longer to establish themselves than respiratory or intestinal infections.[112]

C. MICROBIAL PENETRATION

It is important to note that several microbial infections normally remain confined to the epithelial surfaces.[113] The best known of these are infections caused by influenza virus, trachoma-chlamydias, *Staphylococcus* spp., *Salmonella* spp., *Streptococcus* spp., *Corynebacterium diphtheriae*, and *Candida albicans*. In most of these cases the pathogens cannot overcome the defense forces of the host cells. As a rule, many of the Gram-negative bacteria remain confined to the epithelial surfaces.[106,154,222,223] These include the coliform bacteria and *Pseudomonas aeruginosa*. However, under certain conditions, e.g., in immunosuppressed patients, such bacteria may give rise to systemic infections.[224-227] In a few cases Gram-negative bacteria enter the epithelium, yet remain close to the surface. This is the case in *Shigella* dysentery and salmonellosis. In fewer cases, Gram-negative bacteria may invade the epithelium causing systemic infection, e.g., *Salmonella typhi* and *S. paratyphi*. Dermatophytic fungi also remain confined to the skin in the dead keratinized surface layers. Why such fungi cannot penetrate deeper into the body cells is not yet known. Probably the living cells contain factors which inhibit their growth.

A large number of microorganisms can enter the epithelial layers and even establish systemic infections.[112,113] These include measles, smallpox and other viruses, psittacosis-chlamydias, *Mycobacterium tuberculosis*, *Treponema pallidum*, *Bacillus anthracis*, the causative agents of Q-fever and typhus, fungi causing histoplasmosis and blastomycosis, as well as protozoa causing malaria, toxoplasmosis, and dysentery. Such microorganisms enter the blood or lymph, become contained within leukocytes, and ultimately infect susceptible cells in various body tissues or organs.[135] Some systemic infectants, however, can directly propagate in the blood or lymph, i.e., not necessarily intracellularly.[221] Such latter infectants become more exposed to the body defense forces than the intracellular infectants. Since the present volume is mainly concerned with the adherence of *C. albicans*, i.e., with the initial step of infection, it is not intended to give details here of the microbial subepithelial invasion nor of the body defense mechanisms.[228-231] Readers interested in these subjects are referred to other works.[106,153,154]

V. HOW MICROORGANISMS LEAVE THE HUMAN BODY

Microorganisms are normally shed from various body surfaces and may be transmitted in part to other persons.

A. EXIT FROM THE SKIN

Skin scales carrying microorganisms are continuously shed by the body. One of the most important potential pathogens continuously shed from the nose is *Staphylococcus aureus*. This process is the basis for cross infection in hospitals by the bacterium.[39,232,233]

B. EXIT FROM THE INTESTINAL TRACT

Inhabitants of the intestinal tract are shed in feces. These include those shed initially into the bile, such as the causative agent of typhoid.[86,234,235] Intestinal infections usually result in feces becoming more fluid with more frequent discharge. This is one of the means by which the body rids itself of intestinal infectants. Pathogenic microorganisms in feces persist for varying periods of time. Long-persisting microorganisms are those capable of producing endospores such as *Clostridium* spp. and *Bacillus* spp. or cysts such as *Entamoeba histolytica*. Occasionally drinking water may become infected with fecal microorganisms which are then transmitted to healthy humans. This is how the causative agent of cholera epidemics is transmitted.[74,135]

C. EXIT FROM THE RESPIRATORY TRACT

Microorganisms of the respiratory tract are shed together with aerosols normally produced during breathing and speech. Especially large amounts of aerosols are associated with singing and particularly with coughing and sneezing.[236,237] Aerosols normally carry nonpathogenic commensal microorganisms. When a person is diseased they can transmit pathogenic microorganisms such as *Streptococcus* spp., the causative agents of meningitis, tuberculosis, and the common cold. It should be mentioned in this context that pathogenic microorganisms of the respiratory tract can reach the saliva. In this case saliva may represent a vehicle of microbial transmission, especially during activities such as kissing and spitting.[135,238-240]

D. EXIT FROM THE URINOGENITAL TRACT

Under normal conditions urine is sterile. However, a pathogenic microorganism that can reach the urine can potentially infect the kidney and bladder. Pathogens in urine usually reach sewage and, in a similar manner to fecal pathogens, can contaminate drinking water, thus representing a serious hazard to man. Even animal pathogens such as *Leptospira* which normally infects rats, can infect man in this manner. Many pathogens are also transmitted during sexual intercourse and these are among the most common ones.[135]

VI. RELATION OF NORMAL FLORA TO *CANDIDA* ADHERENCE

The role of normal flora in the control of infection by *Candida* is well known. Evidence to support this came mainly from studies using antibiotic-treated or germ-free animals.[241-245] These studies established that suppression of the normal flora predisposes the host to infection by opportunistic microorganisms including *Candida*.[246] Interference with the adherence of *Candida* is one of the mechanisms proposed for the control of yeast infection by indigenous bacterial flora. (For a review of the mechanisms of candidosis control see Chapter 2). This point received considerable attention, and some reports show that certain commensal bacteria enhance candidal adherence, while others block it. Other bacteria have no significant effect whatsoever. The suggestion that the indigenous bacterial flora interferes with the adherence of *C. albicans* came first from Liljemark and Gibbons.[242] These authors showed that cells of *C. albicans* attach in lower numbers to epithelial cells from the tongue or cheek of conventional rats rather than from germ-free rats. In addition, mixed human salivary bacteria and strains of *Streptococcus salivarius* and *S. mitior* were shown to suppress mouth and gut colonization by this yeast in gnotobiotic mice.[242] Sobel et al.[247] reported a reduction in adherence of *C. albicans* to vaginal cells by pre-incubating the epithelial cells with two vaginal isolates of *Lactobacillus* and washing free of nonadherent bacteria before incubating with the yeast. Samaranayake and MacFarlane,[249] using an *in vitro* technique to investigate the effect of oral commensal bacteria on the adherence of *C. albicans* to HeLa cells, showed that *S. salivarius* and *S. mitior*, but not *S. mutans* reduced candidal adhesion to HeLa cells.[248] Makrides and MacFarlane,[249] using the same technique, showed that cell suspensions and cell suspension supernatants of *E. coli* and *Klebsiella aerogenes* significantly enhanced candidal adherence to HeLa cells. In contrast, *Staphylococcus aureus*, *Streptococcus mitior*, and a strain of *S. sanguis* significantly reduced adherence. Cell suspensions and cell suspension supernatants of *S. milleri* and a second strain of *S. sanguis* had no significant effect on candidal adherence.[249] Makrides and MacFarlane[250] showed that pre-incubation of HeLa cell monolayers with *E. coli* cell suspensions led to a

TABLE 2
Effect of Bacteria on Yeast Attachment to Epithelial Cells (EC)

Experiment[a]	No. of *C.albicans* attached per EC [mean ± SE]	Adherence (%)	*P* Value[b]
Yeast (control)	1.7 ± 0.2346	100	Control
Yeast and piliated *K. pneumoniae*	4.5 ± 0.6691	264[b]	<0.0005
Yeast and non-piliated *K. pneumoniae*	1.8 ± 0.3778	117[b]	NS
Yeast (control)	2.6 ± 0.5523	100	Control
Yeast and *E.coli*	9.2 ± 1.0834	353[b]	<0.0005

[a] *lKebsiella pneumonia* and *E. coli* attachment studies were performed on different days and enumerated by direct counting.

[b] Compared with the yeast (control) group. NS = not significant.

Abstracted from Centeno, A.C., Davis, C.P., Cohen, M.S., and Warren, M.M., *Infect. Immun.*, 39, 1354, 1983. With permission.

significant increase in the adherence of *C. albicans* to these epithelial cells. Centeno et al.[251], however, showed that pre-incubation with piliated, but not with nonpiliated, strains of *E. coli* and *K. pneumonia* enhanced the subsequent attachment of *C. albicans* to the epithelial cells (Table 2).[251] Evidently, the strain of *E. coli* used by Makrides and MacFarlane[250] had large numbers of fimbriae on the cell surface. Acrylic denture strips precoated with *S. salivarius* cells had a significantly lower number of *Candida* cells attached to them than the noncoated strips.[252]

The mechanisms by which the indigenous microflora interferes with the adherence of *C. albicans* to epithelial cells have been investigated. Kennedy and Volz[241] suggested that the indigenous microflora reduce the mucosal association of *C. albicans* by forming a dense layer of bacteria in the mucus gel bathing these epithelial layers, surpassing yeast cells for adhesion sites (epithelial receptors). Another possibility is the production of inhibitory substances, such as volatile fatty acids, which modify *Candida* adhesin(s) or mucosal receptor(s), or both, thus making the occurrence of the yeast-mucosal association difficult.[241] Differences observed in the capabilities of various members of the indigenous microflora which affect adherence of *Candida* may be explained on the basis of differences in their abilities to adhere to mucosal surfaces. Thus, *S. salivarius* and *S. mitior*, which have been shown to adhere well to oral mucosal surfaces,[202,203,253] were able to suppress *C. albicans* in the oral cavity, whereas *S. mutans*, which is known to attach poorly to oral epithelial cells,[202,253] had little effect on yeast colonization.[242]

The mechanisms proposed for the enhancement of *Candida* adherence, following pretreatment with *E. coli*, suggest the presence of mannose-like receptors on both the epithelial and yeast cell surfaces, with type 1 fimbriae of *E. coli* acting as a bridging mechanism between these two cells.[250,251] This hypothesis is supported by scanning electron microscopy studies (Figure 1).

Marrie and Costerton[254] examining (by electron microscopy) scrapings of *C. albicans* plaques from patients with oral candidosis, observed adhesion of bacteria *(E. coli, Bacteroides melaninogenicus,* and *Staphylococcus saprophyticus)* to *Candida* cells. This association is apparently mediated by a ruthenium red positive matrix, suggesting that adherence is mediated through the presence of a polysaccharide matrix which occupies the zone of adhesion.

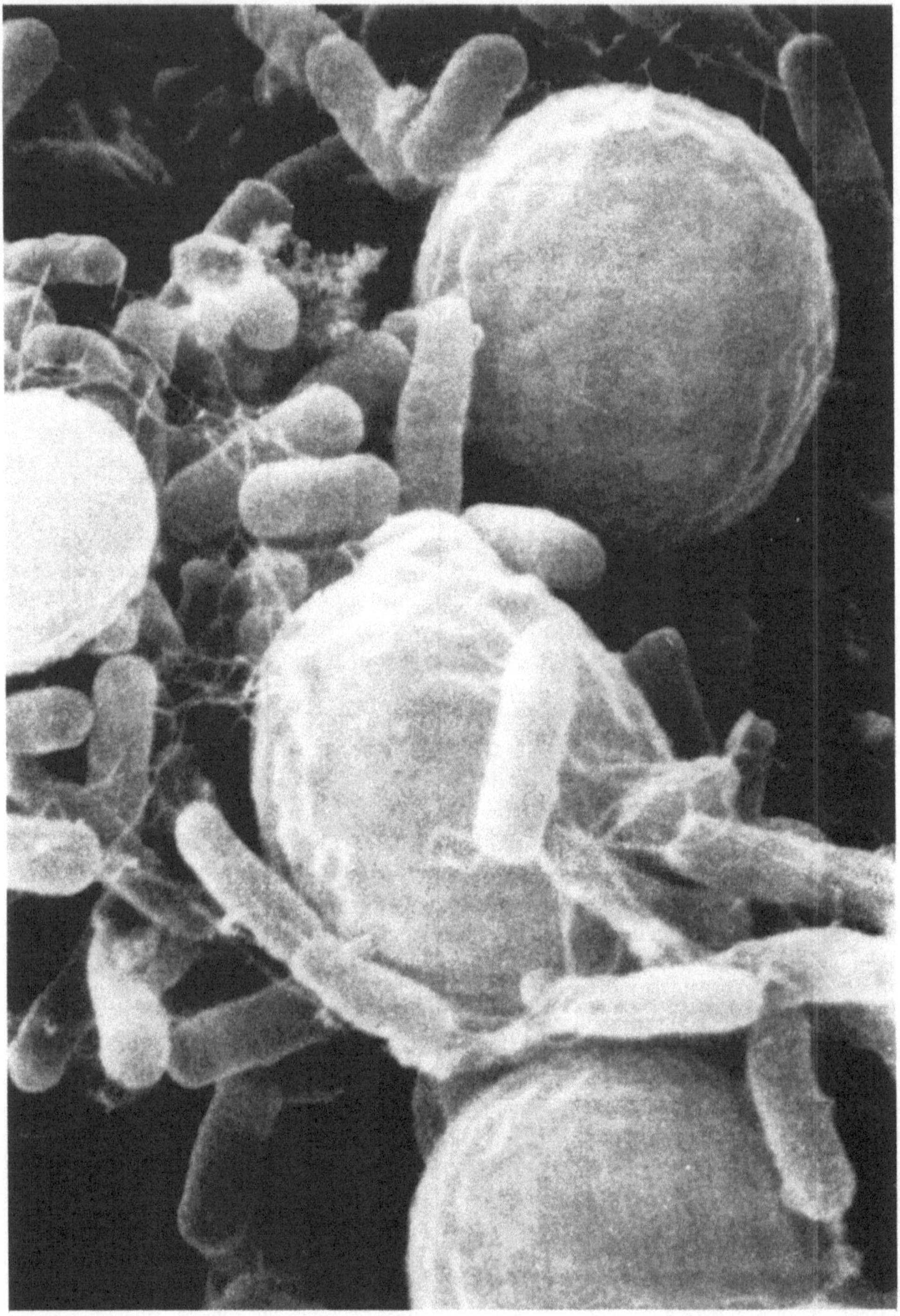

FIGURE 1. Scanning electron micrograph showing the adherence of *C. albicans* MRL 3553 to HeLa cells which had been pretreated with *E. coli* NCTC10418. (Magnification × 23,000.) (From Makarides, H.C. and MacFarlane, T.W., *Microbios*, 38, 177, 1983. With permission.)

REFERENCES

1. **Noble, W.C.,** Observation on the surface flora of the skin and on the skin pH, *Br. J. Dermatol.*, 80, 279, 1968.
2. **Skinner, F.A. and Carr, J.G.,** *The Normal Microbial Flora of Man*, Academic Press, London, 1974.
3. **Noble, W.C. and Somerville, D.A.,** Host factors and their effect on the normal flora, in *Microbiology of Human Skin*, Noble W.C. and Somerville, D.A., Eds., Saunders, London, 1974, 284.
4. **Evans, H.E., Akpata, S.O., and Baki, A.,** Factors influencing the establishment of the neonatal bacterial flora. I. The role of host factors, *Arch. Environ. Health*, 21, 514, 1970.
5. **Findlay, G.M. and Abraham, C.,** The incidents of staphylococci in the nose and on the skin of Africans and Europeans in West Africa, *J. R. Army Med. Corps*, 87, 272, 1946.
6. **Millian, S.J., Baldwin, J.N., Rheins, M.S., and Weiser, H.H.,** Studies on the incidence of coagulase positive staphylococci in a normal unconfined population, *Am. J. Public Health*, 50, 791, 1960.
7. **Somervilla, D.A.,** The Normal Flora of the Skin in Different Age Groups, Ph.D. Thesis, University of Otago, N.Z., 1966.
8. **Somerville, D.A. and Noble, W.C.,** Microcolony size of microbes on human skin, *J. Med. Microbiol.*, 6, 323, 1973.
9. **Fusaro, R.M. and Goetz, F.C.,** Common cutaneous manifestations and problems of diabetes mellitus, *Postgrad. Med.*, 49, 84, 1971.
10. **Williams, R.E.O.,** Healthy carriage of *Staphylococcus auerus:* its prevalence and importance, *Bacteriol. Rev.*, 27, 56, 1963.
11. **Baker, R.D.,** Diabetes and Mucromycosis, *Diabetes*, 9, 143, 1960.
12. **Landau, J.W. and Newcomer, V.D.,** Acute cerebral phycomycosis (Mucromycosis). Report of a patient successfully treated with amphotericin B and cycloheximide and review of pertinent literature, *J. Pediatr.*, 61, 363, 1962.
13. **Williams, R.E.O.,** Benefit and mischief from commensal bacteria, *J. Clin. Pathol.*, 26, 811, 1973.
14. **Reisner, R.M. and Puhvel, M.,** Lipolytic activity of *Staphylococcus albus*, *J. Invest. Dermatol.*, 53, 1, 1969.
15. **Ricketts, C.R., Squire, J.R., and Topley, E.E.,** Human skin lipids with particular reference to the self-sterilizing power of the skin, *Clin. Sci.*, 10, 89, 1951.
16. **Lacey, R.W.,** *In vivo* effect of acetone, alcohol and soap on behaviour of *Staphylococcus aureus*, *Br. J. Exp. Pathol.*, 49, 209, 1968.
17. **Aly, R., Maibach, H.E., Shinefield, H.R., and Strauss, W.G.,** Survival of pathogenic microorganisms on human skin, *J. Invest. Dermatol.*, 58, 205, 1972.
18. **Pillsbury, D.M. and Rebell, G.,** The bacterial flora of the skin, *J. Invest. Dermatol.*, 18, 173, 1952.
19. **Aly, R., Maibach, H.I., Rahman, R., Shinefield, H.R., and Mandel, A.D.,** Correlation of human *in vivo* and *in vitro* cutaneous antimicrobial factors, *J. Infect. Dis.*, 131, 579, 1975.
20. **Shehadeh, N.H. and Kligman, A.M.,** The effect of topical anitbacterial agents on the bacterial flora of the axalli, *J. Invest. Dermatol.*, 40, 61, 1963.
21. **Evans, Z.A., Rendtorff, R.C., and Rosenberg, E.W.,** Efficacy of dermal antisepsis, *N. Engl. J. Med.*, 284, 675, 1971.
22. **Forfar, J.O., Gould, J.C., and MacCabe, A.F.,** Effect of hexachlorophane on incidence of staphylococcal and gram-negative infection in the newborn, *Lancet*, 2, 177, 1968.
23. **Fulton, J.E., McGinley, K., Leyden, J., and Marples, R.R.,** Gram-negative folliculitis in acne vulgaris, *Arch. Dermatol.*, 98, 349, 1968.
24. **Selwyn, S.,** Natural antibiosis among skin bacteria as a primary defence against infection, *Br. J. Dermatol.*, 93, 487, 1975.
25. **Marsh, P.D. and Selwyn, S.,** Studies on antagonism between human skin bacteria, *J. Med. Microbiol.*, 10, 161, 1977.
26. **Marsh, P.D. and Selwyn, S.,** Continuous-culture studies of interactions among human skin-commensal bacteria, *J. Med. Microbiol.*, 10, 261, 1977.
27. **Milyani, R.M. and Selwyn, S.,** Quantitative studies on competitive activities of skin bacteria growing on solid media, *J. Med. Microbiol.*, 11, 379, 1978.
28. **Hsu, C.-Y. and Wiseman, G.M.,** Antibacterial substances from staphylococci, *Can. J. Microbiol.*, 13, 947, 1967.
29. **Noble, W.C. and Willie, J.A.,** Carriage of inhibitor-producing organisms on human skin, *J. Med. Microbiol.*, 13, 329, 1980.
30. **Carlsson, J.,** Dental plaque as a source of salivary streptococci, *Odontol. Revy*, 18, 173, 1967.
31. **Carlsson, J.,** Presence of various types of non-haemolytic streptococci in dental plaque and in other sites of the oral cavity in man, *Odontol. Revy*, 18, 55, 1967.
32. **Gibbons, R.J., Socransky, S.S., Sawyer, S., Kapsimalis, B., and MacDonald, J.B.,** The microbiota of the gingival crevice area of man. II. The predominant cultivable organisms, *Arch. Oral Biol.*, 8, 281, 1963.

33. **Gordon, D.F. and Gibbons, R.J.,** Studies of the predominant cultivable microorganisms from human tongue, *Arch. Oral Biol.*, 11, 627, 1966.
34. **Gordon D.F. and Jong, B.B.,** Indigenous flora from human saliva, *Appl. Microbiol.*, 16, 428, 1968.
35. **Socransky, S.S. and Maganiello, A.D.,** The oral microbiota of man from birth to senility, *J. Periodontol.*, 42, 485, 1971.
36. **Gibbons, R.J., Socransky, S.S., de Araujo, W.C., and Van Houte, J.,** Studies on the predominant cultivable microflora of dental plaque, *Arch. Oral Biol.*, 9, 365, 1964.
37. **Oppenheim, F.G.,** Preliminary observations on the presence and origin of serum albumin in human saliva, *Helv. Odontol. Acta*, 14, 10, 1970.
38. **Hoffman, H.,** Oral microbiology, *Adv. Appl. Microbiol.*, 8, 195, 1966.
39. **Holmberg, K. and Hallader, H.O.,** Production of bactericidal concentrations of hydrogen peroxide by *Streptococcus sanguis*, *Arch. Oral Biol.*, 18, 423, 1973.
40. **Kelstrup, J. and Gibbons, R.J.,** Bacteriocins from human and rodent streptococci, *Arch. Oral Biol.*, 14, 251, 1969.
41. **Carlsson, J., Grahnen, H., Jonsson, G., and Wikner, S.,** Establishment of *Streptococcus sanguis* in the mouths of infants, *Arch. Oral Biol.*, 15, 1143, 1970.
42. **McCarthy, C., Snyder, M.L., and Parker, R.B.,** The indigenous oral flora of man. I. The new born to the 1 year old infant, *Arch. Oral Biol.*, 10, 61, 1965.
43. **Bowden, G.H. and Hardie, J.M.,** Anaerobic organisms from the human mouth, in *Isolation of Anaerobes*, Shapton, D.A. and Board, R.G., Eds., Soc. Appl. Bact. Tech. Ser. No. 5, Academic Press, London, 1971.
44. **Loesche, W.J., Socransky, S.S., and Gibbons, R.J.,** *Bacteroides oralis*: proposed new species isolated from the oral cavity of man, *J. Bacteriol.*, 88, 1329, 1964.
45. **Hadi, A.W. and Russell, C.,** Quantitative estimations of fusiforms in saliva from normal individuals and cases of acute ulcerative gingivitis, *Arch. Oral Biol.*, 13, 1371, 1968.
46. **Hadi, A.W. and Russell, C.,** Fusiforms in gingival material, *Br. Dent. J.*, 126, 82, 1969.
47. **Newman, H.N.,** Zone demarcation of organic films present on human enamel surfaces *in vivo*, *Br. Dent. J.*, 134, 273, 1973.
48. **Newman, H.N.,** Organic films on enamel surfaces. I. Vestigial enamel organ. *Br. Dent. J.*, 135, 64, 1973.
49. **Newman, H.N.,** Clinical significance of microorganisms and organic films on human enamel, *Proc. R. Soc. Med.*, 65, 908, 1972.
50. **Armstrong, W.G.,** The composition of the organic films formed on human teeth, *Caries Res.*, 1, 89, 1967.
51. **Leach, S.A., Critchley, P., Kolendo, A.B., and Saxton, C.A.,** Salivary glycoproteins as components of the enamel integuments, *Caries Res.*, 1, 104, 1967.
52. **Leach, S.A. and Melville, T.H.,** Investigation of some human oral organisms capable of releasing the carbohydrates from salivary glycoproteins, *Arch. Oral Biol.*, 15, 87, 1970.
53. **Sprunt, K. and Redman, W.,** Evidence suggesting importance of role of interbacterial inhibition in maintaining balance of normal flora, *Ann. Intern. Med.*, 68, 579, 1968.
54. **Geddes, D.A.M. and Jenkins, G.A.,** Intrinsic and extrinsic factors influencing the flora of the mouth, in *The Normal Microbial Flora of Man*, Skinner, F.A. and Carr, J.G., Eds., Soc. Appl. Bact. Symp. Ser. No. 3, Academic Press, London, 1974, 87.
55. **Marples, R.R., Fulton, J.E., Leyden, J., and McGinley, K.J.,** Effect of antibiotics on the nasal flora in acne patients, *Arch. Dermatol.*, 99, 647, 1969.
56. **Savage, D.C., Dubos, R., and Schaedler, R.W.,** The gastrointestinal epithelium and its autochihonous bacterial flora, *J. Exp. Med.*, 127, 67, 1968.
57. **Drasar, B.S., Shiner, M., and McLeod, G.M.,** Studies on the intestinal flora. I. The bacterial flora of the gastrointestinal tract in healthy and achlorhydric persons, *Gastroenterology*, 56, 71, 1969.
58. **Franklin, M.A. and Skoryna, S.G.,** Studies on natural gastric flora. I. Bacterial flora of fasting human subjects, *Can. Med. Assoc. J.*, 95, 1349, 1966.
59. **French, J.M.,** Problems raised by the treatment of steatorrhoea with antibacterial drugs, *Postgrad. Med. J.*, 37, 259, 1961.
60. **Bornside, G.H. and Cohn, I.,** The normal microbial flora: comparative bacterial flora of animals and man, *Am. J. Dig. Dis.*, 10, 844, 1965.
61. **Kalser, M.H., Cohen, R., Arteaga, I., Yawn, E., Mayoral, L., Hoffert, W.R., and Frazier, D.,** Normal viral and bacterial flora of the human small and large intestine, *N. Engl. J. Med.*, 274, 500 and 558, 1966.
62. **Donaldson, R.M., Jr.,** Normal bacterial populations of the intestine and their relation to intestinal function, *N. Engl. J. Med.*, 270, 938, 1964.
63. **Donaldson, R.M., Jr.,** Small bowel bacterial over growth, *Adv. Internal Med.*, 16, 191, 1970.
64. **Gorbach, S.L.,** Intestinal microflora, *Gastroenterology*, 60, 110, 1971.
65. **Gorbach, S.L., Plaut, A.G., Nahas, L., Weinstein, L., Spanknebel, G., and Levitan, R.,** Studies of intestinal microflora. II. Microorganisms of the small intestine and their relations to oral and fecal flora, *Gastroenterology*, 53, 856, 1967.

66. **Gorbach, S.L., Nahas, L., Lerner, P.I., and Weinstein, L.**, Studies of intestinal microflora. I. Effect of diet, age and periodic sampling on numbers of fecal microorganisms in man, *Gastroenterology*, 53, 845, 1967.
67. **Gorbach, S.L., Nahas, L., Plaut, A.G., Weinstein, L., Levitan, R., and Patterson, J.F.**, Studies of intestinal microflora. IV. The micro flora of ileostomy effluent: a unique microbial ecology, *Gastroenterology*, 53, 874, 1967.
68. **Macy, J.M. and Probst, I.**, The biology of gastrointestinal bacteroides, *Annu. Rev. Microbiol.*, 33, 561, 1979.
69. **Drasar, B.S. and Barrow, P.A.**, *Intestinal Microbiology*, Van Nostrand Reinhold, London, 1985.
70. **Savage, D.C. and McAllister, J.S.**, Cecal enlargement and microbial flora in suckling mice given antibacterial drugs, *Infect. Immun.*, 3, 342, 1971.
71. **Finegold, S.M.**, Interaction of antimicrobial therapy and intestinal flora, *Am. J. Clin. Nutr.*, 23, 1466, 1970.
72. **Finegold, S.M., Posnich, D.J., Miller, L.G., and Hewitt, W.L.**, The effect of various antibacterial compounds on the normal fecal flora, *Ernaehrungsforschung*, 10, 316, 1965.
73. **Azar, H. and Drapanas, T.**, Relationship of antibiotics to wound infection and enterocolitis in colon surgery, *Am. J. Surg.*, 115, 209, 1968.
74. **Savage, D.C.**, Microbial ecology of the gastrointestinal tract, *Annu. Rev. Microbiol.*, 31, 107, 1977.
75. **Drasar, B.S. and Hill, M.J.**, *Human Intestinal Flora*, Academic Press, London, 1974.
76. **Moore, W.E.C. and Holdeman, L.V.**, Human fecal flora: the normal flora of 20 Japanese-Hawaiians, *Appl. Microbiol.*, 27, 961, 1974.
77. **Holdeman, L.V., Good, I.J., and Moore, W.E.C.**, Human fecal flora: variation in bacterial composition within individuals and a possible effect of emotional stress, *Appl. Environ. Microbiol.*, 31, 359, 1976.
78. **Moore, W.E.C. and Holdeman, L.V.**, Discussion of current bacteriological investigation of the relationships between intestinal flora, diet and colon cancer, *Cancer Res.*, 35, 3418, 1975.
79. **Finegold, S.M., Flora, D.J., Attebery, H.R., and Sutter, V.L.**, Fecal bacteriology of colonic polyp patients and control patients, *Cancer Res.*, 35, 3407, 1975.
80. **Holdeman, L.V. and Moore, W.E.C.**, New genus, *Coprococcus*, twelve new species and amended description of four previously described species of bacteria from human faeces, *Int. J. Syst. Bacteriol.*, 24, 260, 1974.
81. **Gossling, J. and Moore, W.E.C.**, Gemmiger-formicilis, N/Gen, N-Sp, an anaerobic budding bacterium from intestines, *Int. J. Syst. Bacteriol.*, 25, 202, 1975.
82. **Mallory, A., Savage, D., Kern, F., Jr., and Smith, J.G.**, Patterns of bile acids and microflora in the human small intestine. II. Microflora, *Gastroenterology*, 64, 34, 1973.
83. **Mata, L.J., Carrillo, C., and Villatoro, E.**, Fecal microflora in healthy persons in a preindustrial region, *Appl. Microbiol.*, 17, 596, 1969.
84. **Cooke, E.M.**, A quantitative comparison of the faecal flora of patients with ulcerative colitis and that of normal persons, *J. Pathol. Bacteriol.*, 9, 439, 1967.
85. **Levison, M.E. and Kaye, D.**, Fecal flora in man: effect of cathartic, *J. Infect. Dis.*, 119, 591, 1969.
86. **Smith, H.W. and Crabb, W.E.**, The faecal bacterial flora of animals and man: its development in the young, *J. Pathol. Bacteriol.*, 82, 53, 1961.
87. **Speck, R.S., Calloway, D.H., and Hadley, W.K.**, Human fecal flora under controlled diet intake, *Am. J. Clin. Nutr.*, 23, 1488, 1970.
88. **Akama, K. and Otani, S.**, *Clostridium perfringens* as the flora in the intestine of healthy persons, *Jpn. J. Med. Sci. Biol.*, 23, 161, 1970.
89. **Attebery, H.R., Sutter, V.L., and Finegold, S.M.**, Effect of a partially chemically defined diet on normal human fecal flora, *Am. J. Clin. Nutr.*, 25, 1391, 1972.
90. **Slotnick, I.J., Stelluto, M., and Prystowsky, H.**, Microbiology of female genital tract. III. Comparative investigation of the cervical flora of parturients receiving either rectal or vaginal examinations, *Am. J. Obstet. Gynecol.*, 85, 519, 1963.
91. **Neary, M.P., Allen, J.A., Okubadejo, O.A., and Payne, D.J.H.**, Preoperative vaginal bacteria and post-operative infection in gynaecological patients, *Lancet*, 8, 1291, 1973.
92. **Ohm, M.J. and Galask, R.P.**, Bacterial flora of the cervix from 100 prehysterectomy patients, *Am. J. Obstet. Gynecol.*, 122, 683, 1975.
93. **De Louvois, J., Hurley, R., and Stanley, V.C.**, Microbial flora of the lower genital tract during pregnancy: relationship to morbidity, *J. Clin. Pathol.*, 28, 731, 1975.
94. **Corbishley, C.M.**, Microbial flora of the vagina and cervix, *J. Clin. Pathol.*, 30, 745, 1977.
95. **Goldacre, M.J., Watt, B., Loudon, N., Milne, L.J.R., Loudon, J.D.O., and Vessey, M.P.**, Vaginal microbial flora in normal young women, *Br. Med. J.*, 1, 1450, 1979.
96. **Gordon, A.M., Hughes, H.E., and Barr, G.T.D.**, Bacterial flora in abnormalities of the female genital tract, *J. Clin. Pathol.*, 19, 429, 1966.
97. **Slotnick, I.J., Hildebrandt, R.J., and Prystowsky, H.**, Microbiology of the femal genital tract. IV. Cervical and vaginal flora during pregnancy, *Obstet. Gynecol.*, 21, 312, 1963.
98. **Slotnick, I.J., Hildebrandt, R.J., Walsh, H., and Prystowsky, H.**, Microbiology of the female genital tract. V. Cervical flora in patients with premature rupture of membranes, *Obstet. Gynecol.*, 24, 93, 1964.

99. **Morris, C.A. and Morris, D.F.,** Normal vaginal microbiology of women of childbearing age in relation to the use of oral contraceptives and vaginal tampons, *J. Clin. Pathol.*, 20, 636, 1967.
100. **Gould, J.C. and McKillop, E.J.,** The carriage of *Staphylococcus pyogenes var aureus* in the human nose, *J. Hyg. (Cambridge)*, 52, 304, 1954.
101. **O'Grady, F. and Wittstadt, F.B.,** Nasal carriage of *Staphylococcus pyogenes*. II. Bacterial ecology of the nose, *Am. J. Hyg.*, 77, 187, 1963.
102. **Davis, N.A. and Davis, G.H.,** Ecology of nasal *Staphylococci*, *J. Bacteriol.*, 89, 1163, 1965.
103. **Freter, R., Brickner, H., Botney, M., Cleven, D., and Aranki, A.,** Mechanisms that control bacterial populations in continuous-flow culture models of mouse large intestinal flora, *Infect. Immun.*, 39, 676, 1983.
104. **Berg, R.D.,** Mechanisms confining indigenous bacteria to the gastrointestinal tract, *Am. J. Clin. Nutr.*, 33, 2472, 1980.
105. **Freter, R., Stauffer, E., Cleven, D., Holdeman, L.V., and Moore, W.E.C.,** Continuous-flow cultures as *in vitro* models of the ecology of large intestinal flora, *Infect. Immun.*, 39, 666, 1983.
106. **Bitton, G. and Marshall, K.C.,** *Adsorption of Micro-Organisms to Surfaces*, John Wiley & Sons, New York, 1980.
107. **Tannock, G.W. and Smith, J.M.B.,** The microflora of the pig stomach and its possible relationship to ulceration of the pars oesophagea, *J. Comp. Pathol.*, 80, 359, 1970.
108. **Plaut, A.G., Gorbach, S.L., Nahas, L., Weinstein, L., Spanknebel, G., and Levitan, R.,** Studies of intestinal microflora. III. The microbial flora of human small intestine mucosa and fluids, *Gastroenterology*, 53, 868, 1967.
109. **Takeuchi, A., Sprinz, H., La Brec, E.H., and Formol, S.B.,** Experimental bacillary dysentery. An electron microscopic study of the response of the intestinal mucosa to bacterial invasion, *Am. J. Pathol.*, 47, 1011, 1965.
110. **Arbuckle, J.B.R.,** The location of *Escherichia coli* in the pig intestine, *J. Med. Microbiolo.*, 3, 333, 1970.
111. **Smith, H. and Pearce, J.H.,** Microbial pathogenicity in man and animals, *22nd Symp. Soc. General Microbiol.*, Cambridge University Press, Cambridge, 1972.
112. **Gibbons, R.J. and van Houte, J.,** Bacterial adherence in oral microbial ecology, *Annu. Rev. Microbiol.*, 29, 19, 1975.
113. **Smith, H.,** Microbial surfaces in relation to pathogenicity, *Bacteriol. Rev.*, 41, 475, 1977.
114. **Ofek, I., Beachey, E.H., and Sharon, N.,** Surface sugars of animal cells as determinates of recognition in bacterial adherence, *Trends Biochem. Sci.*, 3, 159, 1978.
115. **Ofek, I., Mirelman, D., and Sharon, N.,** Adherence of *Escherichia coli* to human mucosal cells mediated by mannose receptors, *Nature*, 265, 623, 1977.
116. **Ofek, I. and Beachey, E.H.,** Mannose binding and epithelial cell adherence of *Escherichia coli*, *Infect. Immun.*, 22, 247, 1978.
117. **Neeser, J.R., Koellreutter, B., and Wuersch, P.,** Oligomannoside-type glycopeptides inhibiting adhesion of *Escherichia coli* strains mediated by type 1 pili: preparation of potent inhibitors from plant glycoproteins, *Infect. Immun.*, 52, 428, 1986.
118. **Stanislawski, L., Simpson, W.A., Hasty, D., Sharon, N., Beachey, E.H., and Ofek, I.,** Role of fibronectin in attachment of *Streptococcus pyogenes* and *Escherichia coli* to human cell lines and isolated oral epithelial cells, *Infect. Immun.*, 48, 257, 1985.
119. **Firon, N., Ashkenazi, S., Mirelman, D., Ofek, I., and Sharon, N.,** Aromatic alpha-glycosides of mannose are powerful inhibitors of the adherence of type 1 fimbriated *Escherichia coli* to yeast and intestinal epithelial cells, *Infect. Immun.*, 55, 472, 1987.
120. **Beachey, E.H. and Simpson, W.A.,** The adherence of group A streptococci to oropharyngeal cells: the lipoteichoic adhesion and fibronectin receptor, *Infection*, 10, 107, 1982.
121. **Ahren, C.M. and Svennerholdm, A.M.,** Experimental enterotoxin-induced *Escherichia coli* diarrhea and protection induced by previous infection with bacteria of the same adhesin or enterotoxin type, *Infect. Immun.*, 50, 255, 1985.
122. **Dupont, H.L., Formal, S.B., Hornick, R.B., Snyder, M.J., Libonati, J.P., Sheahan, D.G., La Brec, E.H., and Kalas, J.P.,** Pathogenesis of *Escherichia coli* diarrhea, *N. Engl. J. Med.*, 285, 1, 1971.
123. **Arbuckle, J.B.R.,** Observation on the association of pathogenic *Escherichia coli* with small intestinal villi of pigs, *Res. Vet. Sci.*, 20, 233, 1976.
124. **Hohmann, A. and Wilson, M.R.,** Adherence of enteropathogenic *Escherichia coli* to intestinal epithelium *in vivo*, *Infect. Immun.*, 12, 866, 1975.
125. **Nagy, B., Moon, H.W., and Isaacson, R.E.,** Colonization of porcine small intestine by *Escherichia coli*: ileal colonization and adhesion by pig enteropathogens that lack K88 antigen and by some acapsular mutants, *Infect. Immun.*, 13, 1214, 1976.
126. **Riising, H.J., Svendsen, J., and Larsen, J.L.,** Occurrence of K-88 negative *Escherichia coli* serotypes in pig with post weaning diarrhoea, *Acta Pathol. Microbiol. Scand. B*, 83, 63, 1975.
127. **Rutter, J.M., Jones, G.W., Brown, G.T.H., Burrows, M.R., and Luther, P.D.,** Anti-bacterial activity in colostrum and milk associated with protection of piglets against enteric disease caused by K88-positive *Escherichia coli*, *Infect. Immun.*, 13, 667, 1976.

128. **Schlessinger, D.**, *Microbiology—1975*, American Society for Microbiology, Washington, D.C., 1975.
129. **Mardh, P.A. and Westron, L.**, Adherence of bacteria to vaginal epithelial cells, *Infect. Immun.*, 13, 661, 1976.
130. **Linstedt, G., Linstedt, S., and Gustafsson, B.E.**, Mucus in intestinal content of germfree rats, *J. Exp. Med.*, 121, 201, 1965.
131. **Florey, H.W.**, The relative amounts of lysozyme present in the tissues of some animals, *Br. J. Exp. Pathol.*, 11, 251, 1930.
132. **Percy-Robb, I.W. and Collee, J.G.**, Bile acids: a pH-dependent antibacterial system in the gut, *Br. Med. M.*, 3, 813, 1972.
133. **Shearman, D.J.C., Parkin, D.M., and McClelland, D.B.L.**, The demonstration and function of antibodies in the gastrointestinal tract, *Gut*, 13, 483, 1972.
134. **Floch, M.H., Gershengoren, W., Diamond, S., and Hersch, T.**, Cholic acid inhibition of intestinal bacteria, *Am. J. Clin. Nutr.*, 23, 8, 1970.
135. **Mims, C.A.**, *The Pathogenesis of Infectious Disease*, Academic Press, London, 1977.
136. **Lidwell, O.M.**, Aerial dispersal of microorganisms from the human respiratory tract, in *The Normal Microbial Flora of Man*, Skinner, F.A. and Carr, J.G., Eds., Academic Press, London, 1974, 135.
137. **Duguid, J.P.**, The size and duration of air carriage of respiratory droplets and droplet nuclei, *J. Hyg. (Cambridge)*, 44, 471, 1946.
138. **Gibbons, R.J., Spirell, D.M., and Skobe, Z.**, Selective adherence as a determinant of the host tropisms of certain indigenous and pathogenic bacteria, *Infect. Immun.*, 13, 238, 1976.
139. **Swenson, J.**, Studies on gonococcus infection. IV. Pili: their role in attachment of gonococci to tissue culture cells, *J. Exp. Med.*, 137, 571, 1973.
140. **Svanborg-Eden, C., Hanson, L.A., Jodal, U., Lindberg, U., and Akerlund, A.S.**, Variable adherence to normal human urinary tract epithelial cells of *Escherichia coli* strains associated with various forms of urinary tract infections, *Lancet*, 2, 490, 1976.
141. **Svanborg-Eden, C.**, Attachment of *Escherichia coli* to human urinary tract epithelial cells. An *in vitro* test system applied in the study of urinary tract infections, *Scand. J. Infect. Dis. Suppl. 15*, 1978.
142. **Kallenins, G. and Winberg, J.**, Bacterial adherence to periurethral cells in girls prone to urinary tract infections, *Lancet*, 2, 540, 1978.
143. **Kallenins, G., Mellby, R., and Winberg, J.**, *in vitro* adhesion of uropathogenic *Escherichia coli* to human periurethral cells, *Infect. Immun.*, 28, 972, 1980.
144. **Davis, E.M. and Pearl, S.A.**, Biology of the human vagina in pregnancy, *Am. J. Obstet. Gynecol.*, 35, 77, 1938.
145. **Wylie, G. and Henderson, A.**, Identity and glycogen-fermenting ability of lactobacilli isolated from the vagina of pregnant women, *J. Med. Microbiol.*, 21, 363, 1969.
146. **Rogosa, M. and Sharpe, M.E.**, Species differentiation of human vaginal lactobacilli, *J. Gen. Microbiol.*, 23, 197, 1960.
147. **Finch, R.G., French, G.L., and Phillips, I.**, Group B Streptococci in the female genital tract, *Br. Med. J.*, 1, 1245, 1976.
148. **Cohen, C.**, Influence of pH on vaginal discharge, *Br. J. Vener. Dis.*, 45, 241, 1969.
149. **Levison, M.E., Corman, L.C., Carrington, E.R., and Kaye, D.**, Quantitative microflora of the vagina, *Am. J. Obstet. Gynecol.*, 127, 80, 1977.
150. **Botta, G.A.**, Hormonal and type-dependent adhesion of group B streptococci to human vaginal cells, *Infect. Immun.*, 25, 1084, 1979.
151. **Sobel, J.D., Schneider, J., Kaye, D., and Levison, M.E.**, Adherence of bacteria to vaginal epithelial cells at various times in the menstrual cycle, *Infect. Immun.*, 32, 194, 1981.
152. **Ward, J.B. and Berkeley, R.C.W.**, The microbial cell surface and adhesion, in *Microbial Adhesion to Surfaces*, Berkeley, R.C.W., Lynch, J.M., Melling, J., Rutter, P.R., and Vincent, B., Eds., Ellis Harwood, Chichester, 1980, 47.
153. **Berkeley, R.C.W., Lynch, J.M., Melling, J., Rutter, P.R., and Vincent B.**, *Microbial Adhesion to Surfaces*, Ellis Harwood, Chichester, 1980.
154. **Savage, D.C. and Fletcher, M.**, *Bacterial Adhesion*, Plenum, New York, 1985.
155. **Gibbons, R.J.**, Adherence of bacteria to host tissue, in *Microbiology—1977*, Schlessinger, D., Ed., American Society for Microbiology, Washington, D.C., 1977, 395.
156. **Schiffenbauer, M. and Stotzky, G.**, Adsorption of coliphages by clay minerals and microbial cells, Abst. Annu. Meet. Am. Soc. Microbiol., 167, 1978.
157. **Berg, G., Bodily, H.L., Lennette, E.H., Melnick, J.L., and Metcalf, T.G.**, *Viruses in Water*, American Public Health Assoc., Washington, D.C., 1976.
158. **Bystricky, V., Stotzky, G., and Schiffenbauer, M.**, Electron microscopy of coliphages adsorbed to clay minerals: application of the critical point drying method, *Can. J. Microbiol.*, 21, 1278, 1975.
159. **Babich, H. and Stotzky, G.**, Clay minerals reduce inactivation rates of bacteriophages in lake water, *Water Res.*, 14, 185, 1980.
160. **Kohn, A. and Fuchs, P.**, Initial effects of viral infection in bacterial and animal host cells, *Adv. Virus Res.*, 18, 159, 1973.

161. **Wright, A. and Barzilia, N.,** Isolation and characterization of non-converting mutants of bacteriophage e^{14}, *J. Bacteriol.*, 105, 937, 1971.
162. **Bayer, M.E.,** Adsorption of bacteriophages to adhesions between wall and membrane of *Escherichia coli, J. Virol.*, 2, 346, 1968.
163. **Simon, L.D. and Anderson, T.F.,** The infection of *Escherichia coli* by T_2 and T_4 bacteriophages as seen in the electron microscope. II. Structure and function of the base plate, *Virology*, 32, 279, 1967.
164. **Lindberg, A.A.,** Bacteriophage receptors, *Annu. Rev. Microbiol.*, 27, 205, 1973.
165. **Fawcett, D.W.,** Surface specializations of absorbing cells, *J. Histochem. Cytochem.*, 13, 75, 1965.
166. **Marvine, D.A. and Hohn, B.,** Filamentous bacterial viruses, *Bacteriol. Rev.*, 33, 172, 1969.
167. **Wolpert, J.S. and Albersheim, P.,** Host-symbiont interactions. I. The lectins of legumes interact with the O-antigen-containing lipopolysaccharide of their symbiont *rhizobia, Biochem. Biophys. Res. Commun.*, 70 729, 1976.
168. **Freimer, N.B., Ögmundsdottir, H.M., Blackwell, C.C., Sutherland, I.W., Graham, L., and Weir, D.M.,** The role of cell wall carbohydrates in binding of microorganisms to mouse peritoneal exudate macrophages, *Acta Pathol. Microbiol. Scand. B*, 86, 53, 1978.
169. **Prehm, P., Jann, B., Hann, K., Schmidt, G., and Stirm, S.,** On bacteriophage T_3 and T_4 receptor region within the cell wall lipopolysaccharide of *Escherichia coli* B, *J. Mol. Biol.*, 101, 277, 1976.
170. **Roberts, J.W. and Steitz, J.E.A.,** Reconstitution of infective bacteriophage R17, *Proc. Natl., Acad. Sci. U.S.A.*, 58, 1416, 1967.
171. **Archibal, A.R.,** Phage-receptors in gram-positive bacteria, in *Virus Receptors, Receptors and Recognition*, Series B, Vol. 7, Part I., *Bacterial Viruses*, Randall, L. and Philipson, L., Eds., Chapman and Hall, London, 1980.
172. **Newton, A.,** Adsorption and penetration of viruses, in *20th Symp. Soc. Gen. Microbiol.*, Charles, H.A. and Knight, J.G., Eds.,Cambridge University Press, Cambridge, U.K., 1970.
173. **Holland, J.J.,** Enterovirus entrance into specific host cells, and subsequent alterations of cell protein and nucleic acid synthesis, *Bacteriol. Rev.*, 28, 3, 1964.
174. **Holland, J.J. and McLaren, L.C.,** The location and nature of enterovirus receptors in susceptible cells, *J. Exp. Med.*, 114, 161, 1961.
175. **Moldow, C.F., Volberding, P., McGrath, M., and Lee, J.J.,** Avian tumor-virus, interactions with chicken fibroblast membranes-partial characterization of initial attachment site activity, *J. Gen. Virol.*, 37, 385, 1977.
176. **Ottow, J.C.G.,** Ecology, physiology, and genetics of fimbriae and pili. Review, *Annu. Rev. Microbiol.*, 29, 79, 1975.
177. **Brinton, C.C.,** Structure, function, synthesis and genetic control of bacterial pili and a molecular model for DNA and RNA transport in gram-negative bacteria, *Trans. N.Y. Acad. Sci.*, 27, 1003, 1965.
178. **Old, D.C.,** Inhibition of the interaction between fimbrial haemagglutinins and erythrocytes by D-mannose and other carbohydrates, *J. Gen. Microbiol.*, 71, 149, 1972.
179. **Ahrens, R., Moll, G., and Rheinheimer, G.,** Function of fimbriae in peculiar star formation of *Agrobacterium luteum, Arch. Microbiol.*, 63, 321, 1968.
180. **Bradley, D.E.,** Shortening of *Pseudomonas aeruginosa* pili after RNA-phage adsorption, *J. Gen. Microbiol.*, 72, 303, 1972.
181. **Mayer, F.,** Contraction of fimbria of starforming soil bacterium *Pseudomonas echinoides, Arch. Microbiol.*, 76, 166, 1971.
182. **Corpe, W.A., Matsuuchi, L., and Armbruster, B.,** Secretion of adhesive polymers and attachment of marine bacteria to surfaces, in *Proc. 3rd Int. Biodegradation Symp.*, Sharpley, J.M. and Kaplan, A.M., Eds., Applied Science, London, 1975, 433.
183. **Ridgeway, H.F., Wagner, R.M., Dawsey, W.T., and Lewin, R.A.,** Fine structure of the cell envelope layers of *Flexibacter polymorphus, Can. J. Microbiol.*, 21, 1733, 1975.
184. **Leak, L.V.,** Fine structure of mucilaginous sheath of *Anabaena sp., J. Ultrastruct. Res.*, 21, 61, 1967.
185. **Swanson, J.,** Adherence of gonococci, in *Microbiology—1977*, Schlessinger, D., Ed., American Society for Microbiology, Washington, D.C., 1977, 422.
186. **Swanson, J., Stephen, M.D., Kraus, J., and Gotschlich, E.C.,** Studies on gonococcus infection. I. Pili and zones of adhesion: their relation to gonococcal growth patterns, *J. Exp. Med.*, 134, 886, 1971.
187. **Gibbons, R.J., van Houte, J., and Liljemark, W.F.,** Parameters that effect the adherence of *Streptococcus salivarius* to oral epithelial surfaces, *J. Dent. Res.*, 51, 424, 1972.
188. **Swanson, J., Hus, K.C., and Gotschlich, E.C.,** Electron microscopic studies on streptococci, *J. Exp. Med.*, 130, 1063, 1969.
189. **Ellen, R.P. and Gibbons, R.J.,** M protein-associated adherence of *Streptococcus pyogenes* to epithelial surfaces: prerequisite for virulence, *Infect. Immun.*, 5, 826, 1972.
190. **Beachey, E.H. and Ofek, I.,** Epithelial cell binding of group A streptococci by lipoteichoic acid on fimbriae denuded of M protein, *J. Exp. Med.*, 143, 759, 1976.
191. **Shapiro, L. and Maizel, J.V., Jr.,** Synthesis and structure of *Caulobacter crescentus* flagella, *J. Bacteriol.*, 113, 478, 1973.

192. **Corpe, W.A.,** Peryphytic marine bacteria on the formation of microbial film on solid surfaces, in *Effect of the Ocean Environment on Microbial Activities,* University Park Press, Baltimore, 1974, 397.
193. **Whittenbury, R. and Dow, C.S.,** Morphogenesis and differentiation in *Rhodomicrobium vannielii* and other budding and prosthecate bacteria, *Bacteriol Rev.,* 28, 231, 1964.
194. **Staley, J.T.,** Incidence of prosthecate bacteria in a polluted stream, *Appl. Microbiol.,* 22, 496, 1971.
195. **Stove, J.L. and Stanier, R.Y.,** Cellular differentiation in stacked bacteria, *Nature,* 196, 1189, 1962.
196. **Corpe, W.A.,** An acid polysaccharide produced by a primary film-forming marine bacterium, *Dev. Ind. Microbiol.,* 11, 402, 1970.
197. **Fletcher, M. and Floodgate, G.D.,** An electron microscopic demonstration of an acidic polysaccharide involved in the adhesion of marine bacterium to solid surfaces, *J. Gen. Microbiol.,* 74, 325, 1973.
198. **Busch, P.L. and Stumm, W.,** Chemical interaction in the aggregation of bacteria: bioflocculation in waste treatment, *Environ. Sci. Technol.,* 2, 49, 1968.
199. **Friedman, B.A., Dugan, P.R., Pfister, R.M., and Remsen, C.C.,** Finestructure and composition of zoogloeal matrix surrounding *Zoogloea ramigera, J. Bacteriol.,* 96, 2144, 1968.
200. **Friedman, B.A., Dugan, P.R., Pfister, R.M., and Remsen, C.C.,** Structure of exocellular polymers and their relationship to bacterial flocculation, *J. Bacteriol.,* 98, 1328, 1969.
201. **Jones, H.C., Roth, I.L., and Sanders, W.M.,** Electron microscopic study of a slime layer, *J. Bacteriol.,* 99, 316, 1969.
202. **Gibbons, R.J. and van Houte, J.,** Selective bacterial adherence to oral epithelial surfaces and its role as an ecological determinant, *Infect. Immun.,* 3, 567, 1971.
203. **van Houte, J., Gibbons, R.J., and Pulkkinen, A.,** Adherence as an ecological determinant for streptococci in human mouth, *Arch. Oral Biol.,* 16, 1131, 1971.
204. **Savage, D.C.,** Electron microscopy of bacteria adherent epithelial in the intestinal canal, in *Microbiology—1977,* Schlessinger, D., Ed., American Society for Microbiology, Washington, D.C., 1977, 422.
205. **Ericson, T. and Magnusson, I.,** Affinity for hydroxyapatite of salivary substances inducing aggregation of oral streptococci, *Caries Res.,* 10, 8, 1976.
206. **Gibbons, R.J. and Fitzgerald, R.J.,** Dextran-induced agglutination of *Streptococcus mutans* and its potential role in formation of microbial dental plaques, *J. Bacteriol.,* 98, 341, 1969.
207. **Kelstrup, J. and Funder-Nielsen, T.D.,** Adhesion of dextran to *Streptococcus mutans, J. Gen. Microbiol.,* 81, 485, 1974.
208. **Germaine, G.R. and Schachtele, C.F.,** *Streptococcus mutans* dextran sucrase: mode of interaction with high-molecular-weight dextran and role in cellular aggregation, *Infect. Immun.,* 13, 365, 1976.
209. **Jones, E.B.G. and Le Campion-Alsumard, T.,** Biodeterioration of polyurethane by marine fungi, *Int. Biodeterior. Bull.,* 6, 119, 1970.
210. **Wilcox, W.W.,** Anatomical changes in wood cell walls attached by fungi and bacteria, *Bot. Rev.,* 36, 1, 1970.
211. **Martin, S.M. and Adams, G.A.,** A survey of fungal polysaccharides, *Can. J. Microbiol.,* 2, 715, 1956.
212. **Crandall, M.A. and Brock, T.D.,** Molecular basis of mating in the yeast *Hansenula wingei, Bacteriol. Rev.,* 32, 139, 1968.
213. **Sladeckova, A.,** Limnological investigation methods for the periphyton (AUFWUCHS), *Bot. Rev.,* 28, 286, 1962.
214. **Hascall, G.K. and Rudzinska, M.A.,** Metamorphosis in *Tokophrya infusionum;* an electron microscope study, *J. Protozool.,* 17, 311, 1970.
215. **Hascall, G.K.,** The stalk of the suctorian, *Tokophrya infusionum:* histochemistry, biochemistry and physiology, *J. Protozool.,* 20, 701, 1973.
216. **Holberton, D.V.,** Attachment of *Giardia* — a hydrodynamic model based on flagellar activity, *J. Exp. Biol.,* 60, 207, 1974.
217. **Bauchop, T. and Clarke, R.T.J.,** Attachment of the ciliate *Epidinium crawley* to plant fragments in the sheep rumen, *Appl. Environ. Microbiol.,* 32, 417, 1976.
218. **El-Shazly, K. and Hungate, R.E.,** Fermentation capacity as a measure of net growth of rumen microorganisms, *Appl. Microbiol.,* 13, 62, 1965.
219. **Gibbons, R.J. and Kapsimalis, B.,** Estimates of overall rate of growth of intestinal microflora of hamsters, guinea pigs and mice, *J. Bacteriol.,* 93, 510, 1967.
220. **Gibbons, R.J.,** Bacteriology of dental caries, *J. Dent. Res.,* 43, 1021, 1964.
221. **McNabb, P.C. and Tomasi, T.B.,** Host defense mechanisms at mucosal surfaces, *Annu. Rev. Microbiol.,* 35, 477, 1981.
222. **Beachey, E.H., Eisenstein, B.I., and Ofek, I.,** *Bacterial Adherence in Infectious Diseases,* Upjohn Company, Michigan, 1982.
223. **Elliott, K.,** *Adhesion and Microorganism Pathogenicity,* Ciba Found. Symp. 80, Pitman Medical, London, 1981.
224. **Wolfson, J.S., Sober, A.J., and Rubin, R.H.,** Dermatologic manifestation of infections in immunocompromised patients, *Medicine,* 64, 115, 1985.

225. **Zinner, S.H. and McCabe, W.R.,** Gram-negative rod bacteremia in cancer patients. A review with emphasis on the antibody response, *Eur. J. Cancer,* 11, 39, 1975.
226. **Schimpff, S.C., Greene, W.H., Young, U.M., and Wiernik, P.H.,** Significance of *Pseudomonas aeruginosa* in the patient with leukemia of lymphoma. *J. Infect. Dis.,* 130, 524, 1974.
227. **Schimpff, S.C.,** Diagnosis of infection in patients with cancer, *Eur. J. Cancer,* 11, 29, 1975.
228. **Kimura, L.H. and Pearsall, N.H.,** Adherence of *Candida albicans* to human buccal epithelial cells, *Infect. Immun.,* 21, 64, 1978.
229. **King, R.D., Lee, J.C., and Morris, A.L.,** Adherence of *Candida albicans* and other *Candida* species to mucosal epithelial cells, *Infect. Immun.,* 27, 667, 1980.
230. **Lee, J.C. and King, R.D.,** Characterization of *Candida albicans* adherence to human vaginal epithelial cells *in vitro, Infect. Immun.,* 41, 1024, 1983.
231. **Sandin, R.L., Rogers, A.L., Patterson, R.J., and Beneke, E.S.,** Evidence for mannose-mediated adherence of *Candida albicans* to human buccal cells *in vitro, Infect. Immun.,* 35, 79, 1982.
232. **Rebell, G., Pillsbury, D.M., Dest-Phalle, M., and Ginsburg, D.,** Factors affecting the rapid disappearance of bacteria placed on the normal skin, *J. Invest. Dermatol.,* 14, 247, 1950.
233. **Sprunt, K. and Redman, W.,** Evidence suggesting importance of role of interbacterial inhibition in maintaining balance of normal flora, *Ann. Intern. Med.,* 68, 579, 1968.
234. **Budd, W.,** *Typhoid Fever: Its Nature, Mode of Spreading and Prevention,* Longmans Green, London, 1973.
235. **Hornick, R.B., Greisman, S.E., Woodward, T.E., Dupont, H.L., Dawkins, A.T., and Snyder, M.J.,** Typhoid fever: pathogenesis and immunologic control, *N. Engl. J. Med.,* 283, 686, 1970.
236. **Bourdillon, R.B. and Lidwell, O.M.,** Sneezing and the spread of infection, *Lancet,* 2, 365, 1941.
237. **Buckland, F.E. and Tyrell, D.A.V.,** Experiments on the spread of colds. Laboratory studies on the dispersal of nasal secretion, *J. Hyg. (Cambridge),* 62, 365, 1964.
338. **Hambreus, A.,** Dispersal and transfer of *Staphylococcus aureus* in an isolation ward for burned patients, *J. Hyg. (Cambridge),* 71, 787, 1973.
239. **Hamburger, M. and Robertson, O.H.,** Expulsion of group A haemolytic streptococci in droplets and droplet nuclei by sneezing, coughing and talking, *Am. J. Med.,* 4, 690, 1948.
240. **Riley, R.J., Mills, C.C., O'Grady, F., Sultan, L.V., Wittstaat, F., and Shiopori, D.N.,** Infectiousness of air from tuberculosis ward, *Am. Rev. Respir. Dis.,* 85, 511, 1962.
241. **Kennedy, M.J. and Volz, P.A.,** Ecology of *Candida albicans* gut colonization: inhibition of *Candida* adhesion, colonization and dissemination from the gastrointestinal tract by bacterial antagonism, *Infect. Immun.,* 49, 654, 1985.
242. **Liljemark, W.F. and Gibbons, R.J.,** Suppression of *Candida albicans* by human oral streptococci in gnotobiotic mice, *Infect. Immun.,* 8, 846, 1973.
243. **Balish, E. and Phillips, A.W.,** Growth, morphogenesis, and virulence of *Candida albicans* after oral inoculation in the germ-free and conventional chick, *J. Bacteriol.,* 91, 1736, 1966.
244. **Seelig, M.S.,** Mechanisms by which antibiotics increase the incidence and severity of candidiasis and alter the immunological defenses, *Bacteriol. Rev.,* 30, 442, 1966.
245. **Shastry, J.C.M., Ramachandran, K., Mohapatra, L.N., and Tandon, B.N.,** A study of *Candida* in throat swabs and gastrointestinal tract of the patients on broad spectrum antibiotic or steroid treatment, *Indian J. Med. Res.,* 57, 133, 1969.
246. **Ekenna, O. and Sherertz, R.J.,** Factors affecting colonization and dissemination of *Candida albicans* from the gastrointestinal tract of mice, *Infect. Immun.,* 55, 1558, 1987.
247. **Sobel, J.D., Myers, P.G., Kaye, D., and Levison, M.E.,** Adherence of *Candida albicans* to human vaginal and buccal epithelial cells, *J. Infect. Dis.,* 143, 76, 1981.
248. **Samaranayake, L.P. and MacFarlane, T.W.,** Factors affecting the *in vitro* adherence of the fungal oral pathogen *Candida albicans,* to epithelial cells of human origin, *Arch. Oral Biol.,* 27, 869, 1982.
249. **Makrides, H.C. and MacFarlane, T.W.,** Effect of commensal bacteria on the adherence of *Candida albicans* to epithelial cells *in vitro, Microbios Lett.,* 21, 55, 1982.
250. **Makrides, H.C. and MacFarlane, T.W.,** An investigation of the factors involved in increased adherence of *C. albicans* to epithelial cells mediated by *E.coli, Microbios,* 38, 177, 1983.
251. **Conteno, A., Davis, C.P., Cohen, M.S., and Warren, M.M.,** Modulation of *Candida albicans* attachment to human epithelial cells by bacteria and carbohydrates, *Infect. Immun.,* 39, 1354, 1983.
252. **Samaranayake, L.P., McCourtie, J., and MacFarlane, T.W.,** Factors affecting the *in vitro* adherence of *Candida albicans* to acrylic surfaces, *Arch. Oral Biol.,* 25, 611, 1980.
253. **Liljemark, W.F. and Gibbons, R.J.,** Proportional distribution and relative adherence of *Streptococcus miteor (mitis)* on various surfaces in the human oral cavity, *Infect. Immun.,* 6, 852, 1972.
254. **Marrie, T.J. and Costerton, J.W.,** The ultrastructure of *Candida albicans* infections, *Can. J. Microbiol.,* 27, 1156, 1981.

Chapter 2

CANDIDA AND THE HUMAN BODY

TABLE OF CONTENTS

I. INTRODUCTION

A. PATHOGENIC SPECIES

Of the 196 described species of *Candida* (including *Torulopsis),*[1] only ten, are isolated with any frequency from medical specimens. Starting with the most pathogenic, these comprise: *Candida albicans, C. tropicalis, C. stellatoidea, C. glabrata* (synonym: *Torulopsis glabrata), C. krusei, C. parapsilosis, C. pseudotropicalis,* (a current name *C. kefyr), C. guilliermondii, C. viswanathii,* and *C. lusitaniae.*[2] The first two species, i.e., *C. albicans* and *C. tropicalis* account for between 70 and 97% of isolated yeast species.[3-7] However, *C. glabrata* has been cultured frequently and its incidence of infection has progressively increased.[8] In one report, *C. glabrata* was rated second in its frequency of isolation to *C. albicans,*[4] and in another it was reported to be the most frequently isolated organism followed by *C. albicans.*[9] Whether this high recovery of *C. glabrata* represents transient self-limiting infections of low virulence or a potentially systemic infection and requiring attention as a possible source of infection is an important question.[8,10] Other species of *Candida* that have been unequivocally demonstrated in human infections include: *C. brumpttii, C. claussenii, C. intermedia, C. lambica, C. macedoniensis, C. robusta, C. norvegensis, C. zeylanoides, C. catenula, C. ravautii, C. lipolytica, and C. famata (Torulopsis candida).*[11-16]

In addition to these the following species have been isolated from man but their association with infection is not yet established: *C. aaseri, C. conglobata, C. curvata, C. langeronii, C. melibiosia, C. membranaefaciens, C. obtusa, C. rugosa, C. silvae, C. tenuis,* and *C. utilis.*[2]

It is our belief that the list of *Candida* species incriminated in infection will continue to grow with the increased use of aggressive chemotherapy and/or radiotherapy and as new agents and surgical modalities are used. This view is supported by experimental evidence, using animal models, which shows that immunosuppresion predisposes the host to invasion by ordinary harmless yeasts.[17]

B. TAXONOMIC CONSIDERATIONS

Candida is a heterogeneous genus presently classified within the family Cryptococcaceae, Fungi Imperfecti (Deuteromycetes). Summarizing the characteristics of yeast genera Barnett et al.[18] describe the genus *Candida* as *"Vegetative Cells* Budding Cells; sometimes pseudo- and septate hyphae. *Sexual reproduction* None. *Physiological Fermentation* +/–; nitrate +/–; urease +/–. *Comments*... This genus is to be restricted to asexual yeasts with ascomycetous characteristics, such as a two-layered cell wall DBB- (Diazonium blue B reaction) and urease –; ... asexual yeasts with basidiomycetous characteristics will be accommodated in the genera *Cryptococcus* or *Rhodotorula."*

In addition to these characteristics, the following were tabulated by Kreger-van Rij:[19] inositol assimilation positive or negative. Inositol-positive strains form pseudomycelia. Coenzyme Q-6, Q-7, Q-8, Q-9, and Q-10.

The genus *Torulopsis* is included in the genus *Candida* since the characteristics of the latter species described above include species with and without hyphae; there is presently no practical means of distinguishing the many other species of the two genera.[20,21]

It is not our intention to give a detailed review of the taxonomy and classification of the genus *Candida,* the origin of its generic name, nor the criteria used for differentiation of species within this genus. The taxonomy and systematics has been dealt with comprehensively in several references.[18-24] Winner and Hurley[25] and Odds[12] give the historical account of the origin of the generic name and discuss various characteristics used to establish the species within this genus. There are many reports describing conventional methods for specimen collection, identification,[26-28] and for susceptibility testing;[29-32] other references describe and evaluate commercially available identification systems.[37] Table 1 lists the most important characteristics of various *Candida* species which have been isolated from human sources.

II. CANDIDA, A NORMAL INHABITANT OF THE HUMAN BODY

Yeast and yeast-like fungi are regularly and universally part of the normal flora of skin surfaces, buccal and vaginal mucosa, and the intestinal tract. In contrast, mycelial fungi do not establish a permanent "normal flora" on human body surfaces.[11]

A couple of generalizations can be made in relation to *Candida* as a part of the normal human flora:

1. Although the literature on yeast carriage in human is voluminous, reports dealing with the carriage of *Candida* in normal healthy subjects are few. The vast majority of these investigations are concerned with patient populations suffering from various ailments (conditions which predisposes them to a higher incidence of fungal colonization).
2. The most commonly isolated *Candida* species from various sites in healthy humans is *C. albicans* followed by *C. glabrata*. The only exception is the skin, where *C. albicans* is rarely isolated as a normal inhabitant.

A. *CANDIDA* FLORA OF THE ORAL CAVITY

Yeasts are part of the normal microbial flora which man has contacted prior to birth.[38] *Candida* represents the most common genus occurring in the oral cavity, and *C. albicans* (both yeast and/or mycelial forms)[39] the most often recovered species.[40-42]

Other *Candida* species occurring in smaller percentage include *C. glabrata*,[9,43] *C. tropicalis, C. stellatoidea,* and *C. pseudotropicalis*.[44] Also *C. brumptii, C. lipolytica var. lipolytica, C. parapsilosis,* and *C. zeylanoides* have been isolated.[45] The yeast count in positive mouths ranges from 10 to 10,000; in 90.3% the count is 1000 or less.[44] In children the titer of *Candida* carriers was between 1400 and 4000 colonies per ml.[46] Using imprint culture, Arendorf and Walker[39] showed that the count in dentate healthy subjects did not exceed 30 colonies per ml. This value increased to a maximum of 49 colonies per ml in complete denture wearers. These authors considered that colony counts in excess of these values implied a candidal infection. The frequency of yeast recovery from clinically normal mouths of healthy adults (in the majority of investigations) ranges from 10 to 50% with some reports bringing it up to 80%.[44,47]

This disparity may reflect differences in sampling techniques, e.g., throat swab, mouth wash, impression cultures, epithelial smears, or imprint cultures, in the sites sampled and the selection of subjects.[12,39,48] However, these studies agree unanimously that fungi are frequent inhabitants of the oral cavity. Various factors, including dental hygiene, affect the microbial flora of the mouth. *Candida* carriage varies with age, sex, diet, seasonal variation, time of day, denture wearing, the use of dummies in infants, etc., but not necessarily in that order of importance. Many other predisposing factors will be reviewed below in the section "*Candida* as an opportunist", because those factors will predispose an individual to infection rather than increase colonization.

Although age has been considered to be vital in determining the frequency of *Candida* carriage in normal subjects,[46] a review of the literature available does not support this. The frequency of *Candida* in the adult population varies from 13 to 50% depending on the method of sampling.[48,49] Similar variation (10 to 45%) in the frequency of *Candida* in the oral cavity was reported for children.[41,42,50] A more meaningful comparison between age groups could be obtained if adult and child populations were examined simultaneously employing the same methodology and careful selection of subjects. It will be possible then to determine whether age plays a role in increasing the frequency of *Candida* carriage or not. In a study by Gergely and Uri[51] positive cultures of fungi were more frequent in females than in males. In another study by Berdicevsky et al.,[46] no significant differences in the prevalence of *Candida* carriers or in titer were found between boys and girls. The authors found no correlation between yeast carriage and the concentration of salivary electrolytes or salivary IgA. However, they reported a significantly higher IgA concentration in children 6 to 12 years old when compared with children between 3 and 5.5 years.

TABLE 1
Characteristics of Various *Candida* Species Isolated From Human Sources

	C. albicans[a]	*C. catenulata*[b]	*C. famata*	*C. glabrata*	*C. guilliermondii*	*C. intermedia*	*C. krusei*	*C. lambica*	*C. lipolytica*	*C. lusitaniae*	*C. macedoniensis*[c]	*C. norvegensis*	*C. parapsilosis*	*C. robusta*[d]	*C. tropicalis*	*C. viswanathii*	*C. zeylanoides*
ABILITY TO FERMENT																	
D-Glucose	+	v	v	+	+	+	+	+	−	+	+	+	+	+	+	+	v
D-Galactose	v	v	−	−	v	v	−	−		v	+	−	v	v	+	+	−
Maltose	+	v	−	−	−	v	−	−		v	−	−	v	v	+	+	−
Meα-D-glucoside	v	−		−		v							v		v	v	−
Sucrose	v	−	v	−	+	+	−	−		+	+	−	v	v	v	v	−
α,α′-Trehalose	v	−	v	v		v						−	v	−	v	v	v
Melobiose	−	−		−		−							−		−	−	−
Lactose	−	−	−	−	−	−	−	−		−	v	−	−	−	−	−	−
Cellobiose	−	−		−		v							−	−	−	−	−
Melezitose	−	−		−		v							v		v	−	−
Raffinose	−	−	v	−	+	v				−	v	−	−		−	−	−
Inulin	−	−		−		−							−		−	−	−
Starch	v	−		−		−							−		v	D	−
ABILITY TO ASSIMILATE AS SOLE SOURCE OF CARBON																	
D-Galactose	+	+	+	−	+	+	−	−	−	v	+	−	+	v	+	+	v
L-Sorbose	v	−	v	−	+	+	v	−			−	−	v		v	−	v
D-Glucosamine	v	v	v	−		v	−						v		v	−	v
D-Ribose	v	v	v	−	v	v	−	−	−	+	v	−	v	−	v	−	−
D-Xylose	+	v	+	−	+	+	−	+	−	+	v	v	+	−	+	+	−
L-Arabinose	v	−	+	−	+	v	−	−	−	v	v	−	+	−	−	D	−
D-Arabinose	v	−	v	−	+	v	−	−			−	−	−		−	−	−
L-Rhamnose	−	−	v	−	v	v	−	−	−	+	−	−	−	−	−	−	−
Sucrose	v	−	+	−	+	+	−	−	−	+	+	−	+	v	v	+	−
Maltose	+	v	+	−	+	+	−	−	−	+	−	−	+	v	+	+	−
α,α′-Trehalose	v	v	+	v	+	+	−	−	−	+	v	−	+	v	+	+	+

Meα-D-glucoside	v	–		–		v							+		v	+	–
Cellobiose	–	–	+	–	+	+	–	–	–	+	v	+	–	–	v	+	v
Salicin	–	–	+	–	+	+	–	–			v	v	–		v	v	v
Arbutin	–	–	+	–	+	v	–	–	–		+		–		v	+	v
Melibiose	–	–	v	–	+	–	–	–	–	–	–	–	–	–	–	–	–
Lactose	–	–	v	–	–	+	–	–	–	–	+	–	–	–	–	–	–
Raffinose	–	–	+	–	+	+	–	–	–	–	+	–	–	v	–	–	–
Melezitose	v	–	+	–	+	+	–	–			–	–	+		v	+	–
Inulin	–	–	v	–	+	v	–	–			+	–	–		–	–	–
Starch	+	v	v	–	–	v	–	–	–	–	–	–	–	v	+	+	–
Glycerol	v	+	+	v	+	v	+	+			v		+		v	+	+
Erythritol	–	–	v	–	–	–	–	–	+	–	–	–	–	–	–	–	–
Ribitol	v	v	+	–	+	v	–	–	–	+	–	–	v	–	v	+	v
Xylitol	+	v		–		v							v		v	v	–
L-Arabinitol	–	–		–		–							v		–	–	–
D-Glucitol	v	v	+	–	+	+	–	–			v	–	+		+	+	+
D-Mannitol	+	+	+	–	+	+	–	–	+	+	v	–	+	v	+	+	+
Galactitol	–	–	v	–	+	v	–	–			–		–		–	–	–
myo-Inositol	–	–	–	–	+	–	–	–	–	–	–	–	–	–	–	–	–
D-Glucono-1,5-lactone	v	v		v		v							+		+	+	v
2-Keto-D-gluconate	+	v		v		+							+		+	+	+
5-Keto-D-gluconate	+			–									v		+		v
D-Gluconate	v	v		+		v							v		v	+	v
D-Glucuronate	–	–		–		–							–		–	–	–
DL-Lactate	+	+	+	v	v	–	+	+			+	+	–		v	–	–
Succinate	+	+	+	–	+	+	+	+	+	+	v	+	+	–	+	+	+
Citrate	+	+	v	–	+	+	v	+	+	v	v	+	+	–	v	+	+
Methanol	–	–		–		–							–		–	–	–
Ethanol	+	+		v		+							+		+	+	v
ABILITY TO ASSIMILATE AS SOLE SOURCE OF NITROGEN																	
Nitrate	–	–	–	–	–	–	–	–	–	–	–	–	–	–	–	–	–
Nitrite	–	–		–		–							–		–	–	v
Ethylamine	+	v	+	–		+	+						+	–	+	+	v
L-Lysine	+	+		v		+							+		+	+	v
Cadaverine	+	+		–		+							+	–	+	+	v
Creatine	–	–	v	–		–	–						–		–	–	–
Creatinine	–	–	v	–		–	–						–		–	–	v

TABLE 1 (continued)
Characteristics of Various *Candida* Species Isolated From Human Sources

	C. albicans[a]	*C. catenulata*[b]	*C. famata*	*C. glabrata*	*C. guilliermondii*	*C. intermedia*	*C. krusei*	*C. lambica*	*C. lipolytica*	*C. lusitaniae*	*C. macedoniensis*[c]	*C. norvegensis*	*C. parapsilosis*	*C. robusta*[d]	*C. tropicalis*	*C. viswanathii*	*C. zeylanoides*
ABILITY TO ASSIMILATE AS SOLE SOURCE OF CARBON AND NITROGEN																	
Alanine	w			−	+		−				+		+		+		
Arginine	+			−	+		−				+		+		+		
Aspartate	+			−	+		−				+		+		+		
Cysteine	−			−	−		−				−		−		−		
Glutamate	+			w	+		+				+		+		+		
Glycine	w			−	+		−				−		−		−		
Histidine	−			−	−		−				−		−		−		
Methionine	−			−	−		−				−		−		−		
Valine	+			+	+		+				+		+		+		
IN AEROBIC GROWTH TESTS D-GLUCOSE IS SOLE SOURCE OF CARBON																	
Growth w/o Vitamins	v	−	v	−	−	v	+	−	−	+	−	−	−	v	v	−	v
Growth w/o *myo*-Inositol	+	+	+	+	+	+	+	+		+	+	+	+		+	+	+
Growth w/o Pantothenate	+	+	+	+	+	+	+	+		+	+	+	+		+	+	+
Growth w/o Biotin	v	−	−	v	−	v	+	+		+	+	+	−		v	−	v
Growth w/o Thiamin	+	−	+	−	+	+	+	−		+	+	−	+		+	+	+
Growth w/o Biotin and Thiamin	v	−	−	−	−	v	+	−		+	+	−	−		v	−	v
Growth w/o Pyridoxine	+	+	+	−	+	+	+	+		+	+	−	+		+	+	+
Growth w/o Niacin	+	+	+	−	+	+	+	+		+	−	+	+		+	+	+
Growth w/o Folic acid	+	+	+	+	+	+	+	+		+	+	+	+		+	+	+
Growth w/o PABA	+	+	+	+	+	+	+	+		+	+	+	+		+	+	+
Growth at 25°C	+	+		+		+							+		+	+	+
Growth at 30°C	+	+		+		+							+		+	+	+
Growth at 35°C	+	+	+	+		v		+					+		+	+	v
Growth at 37°C	+	+	v	+	+	−	+	+	−	+	+	+	+	v	+	+	v
Growth at 42°C	+	−		+	+	−						+	v		+	v	−

0.01% Cyclohex.[e] growth	+	+		−		v							v	−	+	+	+
0.1% Cyclochex. growth	+	v		−		−							−		+	+	+
50% D-Glucose growth	v	v		v		v	−	−	−	−			+	v	v	+	v
60% D-Glucose growth	−	−		−		−							−		−	−	v
ADDITIONAL TESTS																	
Starch formation	−	−		−		−							−		−	−	−
Acetic acid production	−	−		−		−							−		−	−	−
Urea hydrolysis	−	−	−	−	−	−	v	−	+				−		−	−	−
Diazonium Blue B reaction	−	−	−	−	−	−	−	−	−	−	−	−	−		−	−	−
Mole % guanine + cytosine	37.1	54.5	36.9	(f)	(g)	44.4	39.6	(h)	49.6	(i)	41.3		40.5	39.5	(j)	46.3	57.6
Coenzyme Q number		9	9	6	9	9	7		9	8	6	7	9		9		9
True mycelium	v	−	−	−	−	−	−	−	+	−	−	−	−		v	−	−
Pseudomycelium	+	+	−	−	+	+	+	+	+	+	+	+	+	v	+	+	+
Presence of capsule	−	−	−	−	−	−	−	−	−	−	−	−	−	−	−	−	−
Pellicle in broth after 3 d.	−	v	v	−	v	+	+	+	v	−	−	+	−	−	v	−	−
Clamydospores	+	−	−	−	−	−	−	−	−	−	−	−	−	−	v	−	−
Germ-tube	+		−	−	−		−		−	−	−		−	−	−		

Codes for responses to tests:
+ positive, − negative, D positive response delayed >7 d, w weak positive response, v variable response.

[a] Isolated as *C. stellatoidea* and *C. claussenii*.
[b] Isolated as *C. brumptii* and *C. ravautii*.
[c] Isolated as *C. kefyr* and *C. pseudotropicalis*.
[d] Isolated as *Saccharomyces cerevisiae*.
[e] Cycloheximide.
[f] 39.6—40.2
[g] 44.1—44.4
[h] 42.4—43.2
[i] 43.4—46.3
[j] 35.9—36.1

Daily variations in the amount and species of fungi in the oral cavity have been found in several studies. These variations are caused by factors such as food ingestion, inactivity during sleep, and brushing the teeth.[52] According to Williamson,[53] daily variation was lowest in the morning. Gergely and Uri[54] studied the day-by-day variation in the mycotic flora of the mouth. None of the ten subjects they studied gave consistent negative results throughout the study period, and fungi were isolated from subjects with open cavities almost uniformly throughout the 8 d study period. Seasonal variation was also significant; during summer 15 to 25% more positive cultures were obtained.[51]

Denture wearing is a further factor responsible for the increased frequency of isolation of yeasts, particularly of *C. albicans* from the mouth.[45,55-57] Overnight wearing of dentures is associated with increased density and denture colonization (Figure 1).[39] The frequency of *Candida* increases from 44% in non-denture wearers to 85% in denture wearers.[45,55] Nutrition was one of the factors implicated in the pathogenesis of oral candidosis.[12] In particular, diets rich in carbohydrates have been identified by many workers as creating an environment conducive to the growth and multiplication of *Candida* species.[58-61] The mechanisms by which dietary carbohydrates promote candidal growth have been reviewed by Samaranayake and MacFarlane[62] who proposed a unifying hypothesis explaining the exacerbation of mucosal candidosis observed in environments replete with carbohydrates. A relationship between the use of dummy sucking and oral colonization and infection with *Candida* was reported.[63,64] These workers suggested that a greater prevalence of *Candida* in the mouths of children with dummies indicate either an absolute increase in prevalence of colonization, or possibly, denser colonization. They concluded that such foreign bodies in the mouth predispose an infant to candidal infection.

The tongue and, additionally in denture wearers, the fitting surface of maxillary prostheses, seem to be the primary oral reservoir from which the rest of the mouth (particularly the palate and cheeks) becomes secondarily colonized.[34]

B. *CANDIDA* FLORA OF THE VAGINA

It is well known that yeasts may be present in the vagina in the absence of clinical symptoms.[65-67] Some controversy, however, still exists regarding this commensal presence of yeast.[68] Women become colonized with yeasts for various physiological reasons (e.g., menstrual cycle, hormonal status) and remain so for a long period of time. Vaginitis is usually initiated by an isolate which has been present as a normal inhabitant.[68] Harris et al.[69] established a correlation between the incidence of *C. albicans* in the maternal vagina at term and its incidence in the mouth of the newborn child. A similar correlation has been stressed by other workers (for review see Winner and Hurley[25]). Although the vaginal flora has been studied in selected patients, e.g., in women with vaginal discomfort or discharge,[70-72] in women attending gyneco-obstetrics clinics,[73-75] in patients before gynecological operations,[76,77] and in pregnant women,[78,79] only a few studies have been undertaken on normal women.[80-84] Yeasts are usually recovered from the vaginae of women of child-bearing age.[12] A maximum recovery rate is reported in women up to around 40 years old, with a few isolates in older subjects.[85-88] This may be explained on the basis that women of child-bearing age have more maternally derived hormones.

Candida is the most frequently isolated genus, with *C. albicans* predominating over other species, followed by *C. glabrata*.[84,89] Other species isolated from the lower genital tract of unselected pregnant women are *Rhodotorula rubra*, *C. krusei*, *Saccharomycopsis lipolytica*, *C. castellii*, *Saccharomyces cerevisiae*, *S. uvarum*, and *Pichia norvegensis*.[84] However, other *Candida* species have been recovered from the human vagina, albeit in small percentages.[2] According to Hurley et al.[84] *C. albicans*, *C. krusei*, and *C. glabrata* are pathogenic yeasts; the other species are almost certainly commensals or transient organisms. However, Mizuno[90] suggested that *C. albicans* may be present in the vagina as commensal, and that its presence does not necessarily indicate morbidity.

The frequency of isolation of yeasts reported by various reviewers ranges between 5 to 16%

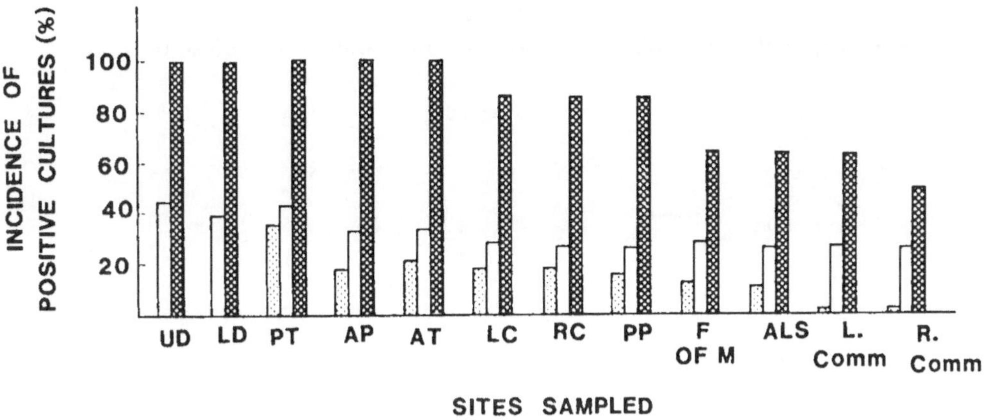

FIGURE 1. Frequency of detection of candida in healthysubjects and denture stomatitis patients. UD, upper denture; LD, lower denture; PT, posterior tongue; AP, anterior palate; AT, anterior tongue; LC, left cheek; RC, right cheek; PP, posterior palate; F OF M, floor of mouth, ALS, anterior labial sulcus; L. Comm., left commissure; and R. Comm., right commissure. , Healthy dentate; , complete denture wearers; and , denture stomatitis (From Arendorf, T.M. and Walker, D.H., *Br. Dent. J.*, 147, 267, 1979. With permission.)

in non-pregnant, healthy women.[12,25,47] However, these values increase to between 7.8 to 67.8% both during pregnancy and in the presence of local morbidity.[2,91] Yeasts are isolated more frequently in spring than in winter.[84] It is believed that the vaginal epithelium, under the influence of estrogenic hormones, contains glycogen, and this glycogen is fermented by lactobacilli (Doderlein's bacillus) to give lactic acid.[92,93] This creates unusually acid conditions in the human vagina (and that of the monkey), the pH being approximately 4.5. A favorable environmental condition is thus created for yeast colonization and growth. (For a review on the environmental conditions prevailing in the healthy vagina, see Ryley.[68])

C. *CANDIDA* FLORA OF THE GASTROINTESTINAL TRACT

Candida spp. are normal inhabitants of the gastrointestinal (GI) tract of man, animals, and birds.[25,94-96] The GI tract is suggested as the main reservoir for *C. albicans* and *C. glabrata,* and perhaps of some other *Candida* spp.[57,96] The most commonly isolated *Candida* species from the GI tract of normal healthy carriers is *C. albicans,* followed by *C. glabrata. C. albicans* represents between 50 to 75% of the yeast present.[11,96-98]

Other *Candida* species isolated from healthy subjects, albeit in small numbers, include: *C. guilliermondi, C. krusei, C. parapsilosis, C. tropicalis, Rhodotorula minuta, R. pilimanae, R. rubra, Geotrichum candidum,* and *Trichosporon beigelii* (Table 2).[11,99] Studying the GI flora in man is difficult, because obtaining reliable samples is a problem.[100] Therefore, the best quantitative studies of the flora are restricted to those in which the flora of fecal samples has been characterized; it is assumed that this material is representative of the colon content.[101] Even when the fecal flora is adopted as a measure of the normal tract flora, difficulty still exists if comparisons are made between various investigators. There are a number of reasons for this, e.g., the difference in the nature of fecal specimen used for yeast isolation (whether it is stool or swab [anal or rectal]), the different methods employed by various investigators to obtain their results, and differences in the age groups selected. Taking these points into consideration the percentage of yeast carriage in the feces of normal healthy subjects is reported to vary from 10 to 68% while the percentage with *C. albicans* varied between 12.2 and 70%.[95,12,42,47,102]

The concentration of *Candida* spp. varies considerably in different parts of the GI tract. In one study, using a special double-lumen tube and fluoroscopic placement for aspiration of the gut surfaces, *C. albicans* was the fungus most frequently recovered in high concentration in all

TABLE 2
Percentage of Yeast Isolated from the Normal Adult Gastrointestinal Tract

Yeast	Throat	Jejunum	Ileum	Colon
C. albicans	30	54	55	65
C. glabrata		4		
C. guilliermondii		4	5	
C. krusei		4		22
C. parapsilosis	4	12		
C. stellata			5	
C. tropicalis		4		
Cryptococcus laurentii			5	
Rhodotorula minuta			5	
R. pilimanae				6
R. rubra		4		

Abstracted from Cohen, A. et al., *N. Engl. J. Med.*, 280, 638, 1969.

the segments of the GI tract.[99] Both the frequency of isolation and the concentration of *C. albicans* increased progressively from the oropharynx to the colon: 30% in the oropharynx, 54% in the jejunum, 55% in the ileum and 65% in the fecal specimens. In the same study it was shown that the mycoflora of the jejunum and ileum apparently is relatively stable in the normal adult regardless of seasonal changes.[99] Despite the high concentration of *Candida* species, infection does not occur under normal circumstances due to the defense mechanisms of the GI tract. A variety of mechanisms have evolved at mucosal surfaces to prevent invasion or damage by pathogens. These include both immune and non-immune defenses. Predominant among immune mechanisms is the occurrence of secretory IgA, which is secreted into the fluids bathing the mucus membranes. Non-immune mechanisms serving host defense at the mucosal level in the GI tract include peristalsis, an indigenous bacterial flora, and the secretion of numerous biologically active substances such as gastric acids, antibacterial substances, and others.[102] (For review see McNabb and Tomasi[103]).

The association between indigenous yeasts and particular areas of the GI tract in mice has been studied by Savage and Dubos.[104] Using special histological techniques, they found that yeast form layers on the epithelium of the glandular mucosa of the stomachs of both mice and rats. These layers differ in several respects from those formed in the stomachs of mice and rats by *Lactobacillus* spp. and anaerobic *Streptococcus* spp.[105-107] The bacterial layers occur on the stratified squamous epithelium of the nonglandular mucosa of the stomach, whereas the yeast layers are confined to the secreting epithelium of the glandular mucosa.

D. *CANDIDA* FLORA OF THE SKIN

While healthy oral, GI, and vaginal mucosa can support a substantial but innocuous saprophytic population of *Candida* species, the same is not true of healthy skin. *Candida* species are generally not considered to comprise a major portion of the normal skin flora, being recovered from a maximum of 5% of random skin cultures.[108] The majority of earlier investigators failed to recover *C. albicans* from healthy skin, e.g., Drouhet[109] was unable to find *C. albicans* in any of 2000 samples he screened. Similarly Benham and Hopkins[110] failed to isolate it from the 100 normal subjects they investigated. Croft and Black,[114] Downing et al.,[111] MacKinnon,[112] and diMenna[113] were equally unsuccessful. However, other workers have isolated *C. albicans,* albeit in a small percentage of the populations they studied (between 0.12 to 2.0%).[115-117] It is possible also to isolate yeasts, presumably in the course of transient carriage, from the skin of those who attend patients with skin disorders and in some other individuals as contaminants.[118,119] Marples[121-123] reported that nonlipophilic yeasts, for example *C. glabrata,* and nonpathogenic species of *Candida*, frequently form sparse populations in the interdigital spaces of the foot, and in tropical countries these yeasts may be inhabitants of the general skin surface. This is associated with the increase in environmental temperature and humidity in these areas. There may be several factors involved in this inability of *Candida* to establish itself on intact, dry skin, including the lack of suitable nutrients, suitable environmental conditions, particularly of humidity,[122] and the presence of inhibitory substances, such as lactic acid, fatty acids, or inhibitor-producing organisms on the human skin.[124-127]

The rare occurrence of yeast on the skin of healthy subjects may be contrasted to its frequent occurrence, and not infrequent pathogenicity, on the skin of those whose susceptibility is heightened for one reason or another.[2] Patients' studies reveal that, irrespective of the type of illness, yeasts are more frequently found on the skins of sick people than on those of healthy subjects.[7] Considering that skin is rarely colonized by yeast, the appearance of local skin lesions, due to the pathogenicity of *Candida*, could indicate widespread infection. It has been proposed that such skin lesions could assist in establishing the diagnosis of disseminated candidosis.[108,128] However, this has been disputed as skin lesions are seldom present in patients who do not have readily detectable septicemia. Another cause of the low frequency of yeasts on healthy skin could be the lack of satisfactory methods for sampling yeast carriage. Barnetson and Milne,[129] using the adhesive-tape technique improved the recovery of yeast on comparison with the wet-swabbing technique. Although a dramatic increase in the recovery of yeast from normal healthy skin is not expected, an improved sampling technique could raise the frequency of *Candida* isolation. The imprint technique proposed by Arendorf and Walker[39] is a possibility that should be assessed for the isolation of yeast from healthy skin.

The most commonly isolated yeast species from healthy skin include: *C. glabrata, C. parapsilosis, C. guilliermondii, C. pseudotropicalis, C. tropicalis, C. zeylanoides, C. famata, Debaryomyces hansenii, D. kloeckeri, Rhodotorula* spp., *Cladosporium* spp., *Trichosporon cutaneum, Cryptococcus* spp. (other than *Crypt. neoformans), Saccharomyces* spp., and *Geotrichum candidum*. (Compiled by Noble and Somerville[124] from References 113,121,130-132).

Candida albicans is more commonly found on the skin in adults over 65 years of age than in young people. Marples and Somerville[133] recorded this yeast on the normal skin of 1.4% of students aged 17 to 22 years but from as many as 27% of geriatric subjects. Further reports support this data and also show a higher rate of recovery from females than from males.[134] The high incidence of yeasts in the elderly could be due to some unmanifested underlying factors in this age group.

Candida species exist, in a precarious microbial balance in the various sites in healthy man (Table 3). When this balance is disrupted, *Candida* proliferates within the normal microbial flora and eventually invades tissues. A number of conditions have been reported to favor the predominance of *Candida* organisms within the normal microbial flora. These factors and the mechanism(s) predisposing to *Candida* infection will be discussed in the following sections.

III. *CANDIDA* — AN OPPORTUNIST

Host-parasite relationships are determined by a delicate balance between the pathogenicity of the organism and the strength of the host defenses. The disturbance of the body defenses has increased greatly with the use of antibiotics, immunosuppressives drugs, hyperalimentation fluids, polyethylene catheters, pressure monitoring devices, heroin abuse, organ transplantation, abdominal surgery, and prosthetic cardiac valves. Numerous reviews have related the introduction of these various surgical and therapeutic modalities to the incidence within the medical community of a relatively new spectrum of infectious diseases.[135-143] The majority of these infections are caused by opportunistic organisms. These are intrinsically of low pathogenicity and are capable of causing disease only in a host whose resistance is lowered.

TABLE 3
Yeast Species Isolated from Various Sites in Healthy Man

	Oral cavity	Vagina	Gastrointestinal tract	Skin
Species isolated	*C. albicans, C. glabrata, C. tropicalis, C. stellatoidea, C. pseudotropicalis, C. brumptii, C. lipolytica, C. parapsilosis,* and *C. zeylanoides*	*C. albicans, C. glabrata, Rhodotorula rubra, C. krusei, Saccharomycopsis lipolytica, C. castelli, Saccharomyces cervisiae, S. uvarum,* and *Pichia norvegensis*	*C. albicans, C. glabrata, C. guilliermondii, C. krusei, C. parapsilosis, C. tropicalis, Rhodotorula minuta, R. pilimanae, R. rubra, Geotrichum candidum,* and *Trichosporon beigelii*	*C. glabrata, C. parapsilosis, C. albicans, C. guilliermondii, C. pseudotropicalis, C. tropicalis, C. zeylanoides, T. famata, Debaryomyces hansenii, D. kloeckeri, Rhodotorula* spp., *Trichosporon* spp., *Crytococcus* spp., *Saccharomyces* spp., and *Geochum candidum*
Frequency of yeast	10 to 50 % up to 80%	5 to 16 % in non-pregnant healthy women	10 to 68%	Up to 50%
Factors affecting the normal flora	Dental hygiene, age, sex, diet, seasonal variation, time of day, denture wearing, and use of dummies	Age, menstrual cycle, hormonal status, and seasonal variation	Age, variation according to different parts of the gastrointestinal tract	Nature of the skin, lack of nutrients, lack of suitable environmental conditions, presence of inhibitory substances "e.g., lactic acid, fatty acid", presence of inhibitory organisms on the skin, age, sex

Candida species represent a good example of opportunistic organisms for, as described above, these organisms are common inhabitants of the oral, vaginal, and gastrointestinal tracts of normal humans, and in the main no ill effect results from their colonization. However, often minor and subtle defects in host defenses allow these organisms to invade and cause illness, which may be serious. The aims of this section are to give an historical review of the emergence of candidosis, and to discuss common predisposing factors and conditions which alter host defenses and predispose to *Candida* infection. In addition, it includes a review of the mechanisms by which these factors may predispose to candidosis.

A. HISTORICAL

The cases of fungal infections reported in the pre-antibiotic and steroid era were very rare. In 1945, Downing and Conant[144,145] wrote a comprehensive review of opportunistic infections and commented that systemic infections with *C. albicans* were rare. They found a total of three fatalities due to systemic moniliasis (candidosis), and two cases of mycotic endocarditis (in drug addicts) during the quarter century preceding their report. Winner and Hurley[25] reported several additional cases of systemic moniliasis and possibly four more fatalities during the pre-antibiotic and steroid period.

The first case of disseminated candidosis in an adult was reported in 1937 by Bogen and Kessel.[146] A decade thereafter the first report on *in vitro* antibacterial potentiation of fungi appeared. They reported on enhancement of growth of fungi by streptomycin and increased lethal effect of *C. albicans* in penicillin-treated chick embryos.[147,148] In 1950 Zimmerman[149] reviewed some of the factors which seemed to be responsible for the rising incidence of fungal diseases. In the early 1950s, Behrman,[150] Freeman,[151] Mangiaracine,[152] and Reiches[153] reported on the increase in oral and anogenital side effects. These authors attributed this to fungal overgrowth in patients receiving antibiotics. In particular, the broad spectrum antibiotics such as chlor- and oxytetracycline, and chloramphenicol were implicated. Also, at this time, there were several cases of severe monilial infections, causing bronchopulmonary complications, as well as reports of an increase in endocardial, spinal cord, urinary tract, and systemic fungal infections.[154-158] These fungal infections correlated with the administration of antibiotics.

In 1956, Keye and Magee[159] analyzed the clinical records, autopsy protocols, and histological and bacteriological reports relating to 15,845 autopsies from 1919 to 1955. They found a total of 88 cases of fungal infections, including *Candida*, of which 47 were regarded as secondary to treatment of the primary disease. The authors concluded that there had been no change in the incidence of primary fungal infections since 1947, whereas there had been a striking increase in secondary fungal infections. Torack[160] related the increased incidence of fungal infections to the use of antibiotics, with or without corticosteroids, the more invasive forms developing in patients taking steroids. Baker[161] and Gruhn and Sanson[162] have confirmed the increased incidence of secondary mycosis in leukemic patients. Baker[161] attributed the recent increase of mycotic fatalities in patients with leukemia to the use of anti-leukemic drugs and corticoids rather than to antibiotics. He reported that although no deaths were attributable to fungus infections among the 143 leukemic patients autopsied prior to 1953, in the 8 years thereafter, 9% of the deaths among leukemic patients were directly due to fungal infections. In a study of 103 leukamic patients, Gruhn and Sanson[162] showed that, of the total of 25 secondary fungal infections, five who died of disseminated mycosis had received multiple antibiotic therapy as well as corticosteroids.

Hutter and Collins[163] found a 15-fold increase in the number of fungal infections between 1950 and 1959 at Memorial-Sloan Kettering Cancer Center. In cases where fungal infections complicated acute leukemia, Bodey[136] reported that the frequency of fungal infections doubled in the period from 1959 to 1964, compared with the preceding 5 year period. In 1964, Schumacher et al.[164] evaluated the incidence of fungal infection in leukemia, examining the medical records of patients admitted between 1957 to 1962. Some 13.7% of patients were found to have had fungal infection. The most common fungus observed was *Candida*, which occurred in 89.3% of patients. *Candida* was a major or contributory factor in the cause of death in 24% of the patients with candidosis.[164] By then, the increased incidence of candidosis and the correlation of this increase with the various therapeutic modalities was established. Reports followed stressing this fact in different patient populations or health centres (see References 165 to 172).

Now, in an era in which there is widespread exploitation of various surgical procedures and the chemotherapies of modern medicine, systemic candidosis is a frequent complication and has long ceased to be reported. From the early seventies onwards research was directed towards understanding the mechanisms by which various predisposing factors enhance the pathogenicity of *Candida* and how to control its infection.

B. PREDISPOSING FACTORS

Infection with *Candida* follows alterations of the host defense systems (both immunological and nonimmunological), or environmental changes at particular sites in the human body which provide the correct conditions for the proliferation of this yeast. A multitude of predisposing factors have been implicated in bringing about such favorable conditions. Gentles and La Touche[47] listed nearly 40 different predisposing factors including debilitation, underlying hormonal disturbances, neoplastic and infectious diseases, chemotherapy, and others. Kostiala et al.[38] tabulated general and local predisposing factors for oral mycoses. Lopez-Martinez et al.[75] listed the opportunistic factors found in 127 cases of vaginal candidosis they examined. Odds[12] classified predisposing factors into four general headings: natural factors, dietary factors, mechanical factors, and iatrogenic medical factors. A modification of the same system was adopted by Smith.[173] In reviewing these reports, and those of others, we have compiled two tables. Table 4 lists the factors responsible for predisposing the host to candidosis in general, while Table 5 gives specific factors responsible for candidosis in the mouth, vagina, and neonates.

A point needs stressing with respect to the role of each predisposing factor in the etiology of fungal infections, in general, and candidosis in particular. The clinical setting which patients predisposed to fungal infections are exposed to, does not usually allow a clear delineation of the role of each specific factor in their susceptibility to infection. Furthermore, in a lot of situations, e.g., in cancer patients undergoing a combination therapy, these factors are intricately interwoven in such a complex manner that makes it impossible to categorize the specific role of each factor.

C. MECHANISMS POTENTIATING *CANDIDA* INFECTIONS

The mechanisms by which predisposing factors enhance the invasiveness of *Candida* is considered to be due either to a direct effect on the host, or on the microorganism, or both.

1. Effect on the Host

An understanding of the mechanisms exerted by various predisposing factors in the potentiation of infection by *Candida* spp. entails an understanding of the host factors that ordinarily protect against superficial or systemic invasion by these species. These protective mechanims may be divided into three categories: host defense mechanisms, anatomical barriers, and protection afforded by normal flora.[174]

a. Host Defense Mechanisms

The most important host defense systems are depicted in Figure 2. These include: immune system, phagocytic system, reticuloendothelial system, and the complement system.

The immune system contains two major lymphocytes cell types, the T- and B-cells. T-cells develop under the influence of the thymus gland and respond to an antigenic challenge by mounting a cell-mediated response. The B-cells, bone marrow-derived lymphocytes and plasma

TABLE 4
General Factors Predisposing to *Candida* Infection

Factors	Comments	Representative examples
Immunological factors		
Neutrophils defect	Presence of abnormally small numbers of neutrophils in the circulating blood, myeloperoxidase deficiency	Acute leukemia, chemotherapy, irradiation
T-lymphocyte, Mononuclear phagocyte	Certain diseases could lead to a defect in T-lymphocyte mononuclear phagocyte	Auto immune deficiency syndrome (AIDS), Hodgkin's disease, chemotherapy
Reticuloendothelial system (RES)	Defect of RES causes impairment in the clearance of infectious particles from the blood; this is due to congenital or surgical causes	Congenital absence or defect of the spleen, splenectomy
Chemotherapy and radiotherapy	Treatment with drugs and/or irridation that alters the composition of the endogenous microbial flora or suppresses host defenses against infection	Immunosuppressive agents, antineoplastic agents, antibiotics, corticosteroids
Interrupted integument	Membrane trauma, burns	Finger or bone marrow punctures, gastrointestinal (GI) ulcers, catheters, IV needles
	Local occlusion or maceration of tissues	Wearing dentures, dish-washing
Surgical procedures	Introduction of mechanical devices and prostheses into vessels or tissues	Heart valves replacements, tracheostomy respiratory assistance, endoscopies, renal transplant, heart operations, GI or gynecological surgery, blood transfusion
Physiological factors	Infectious, idiopathic, congenital, or other debilitating diseases and disorders	Microbial infections, endocrine dysfunctions, defect in cell-mediated immunity
	Digressions from normal physiological status	Pregnancy, infancy
Nutritional factors	Excess or deficiency of food stuff that create an environment conducive to the development of mucosal candidosis	Carbohydrate-rich diets, vitamin deficiency

TABLE 5
Specific Factors Predisposing to Neonatal, Oral, and Vaginal Candidosis

Factors predisposing the host to candidosis	Type of candidosis: Neonatal	Oral	Vaginal
Physiological factors	Prematurity, mother-placental insufficiency, exposure to infectious diseases, hyperbilirubinemia, pathogenic *E. coli* in the stool, genetic disorder, prolonged labor	Severe blood dyscarasis (leukemia, leukopenia, aplastic anemia, myeloma), malignant and chronic diseases (carcinoma, kidney diseases, chronic lung diseases), endocrine disorders (diabetes), pregnancy	Anemia, diabetes, malignancy pregnancy, sex, hormonal change
Interrupted integument	Birth trauma, catheters	Burns, oral trauma, wearing dentures	
Nutritional factors	Malnutrition, breast and bottle feeding	Malabsorption	Malnutrition
Chemotherapy and radiotherapy	Antibiotic and steroid therapy (taken by mother or infant after birth)	Broad spectrum antibiotics, corticosteroids, irradiation, or drugs with xerostomic side effects (psychopharmaceuticals), radiation therapy	Antibiotic therapy, oral contraceptive
Surgical procedures	Resuscitative procedure, intubation	Parenteral nutrition, convalescence after operations (especially transplantation and heart surgery), tooth extraction	
Other factors	Unsterile delivery, male sex, unhygienic environment, low birth weight, thumb and dummy sucking, immunological factors	Poor oral hygiene, oral leukoplakia, immunological factors	Environmental factors (humidity and warmth, wearing tight-fitting clothing), immunological factors

Data abstracted from References 6, 12, 38, 47, 64, 68, 75, 172, and 181.

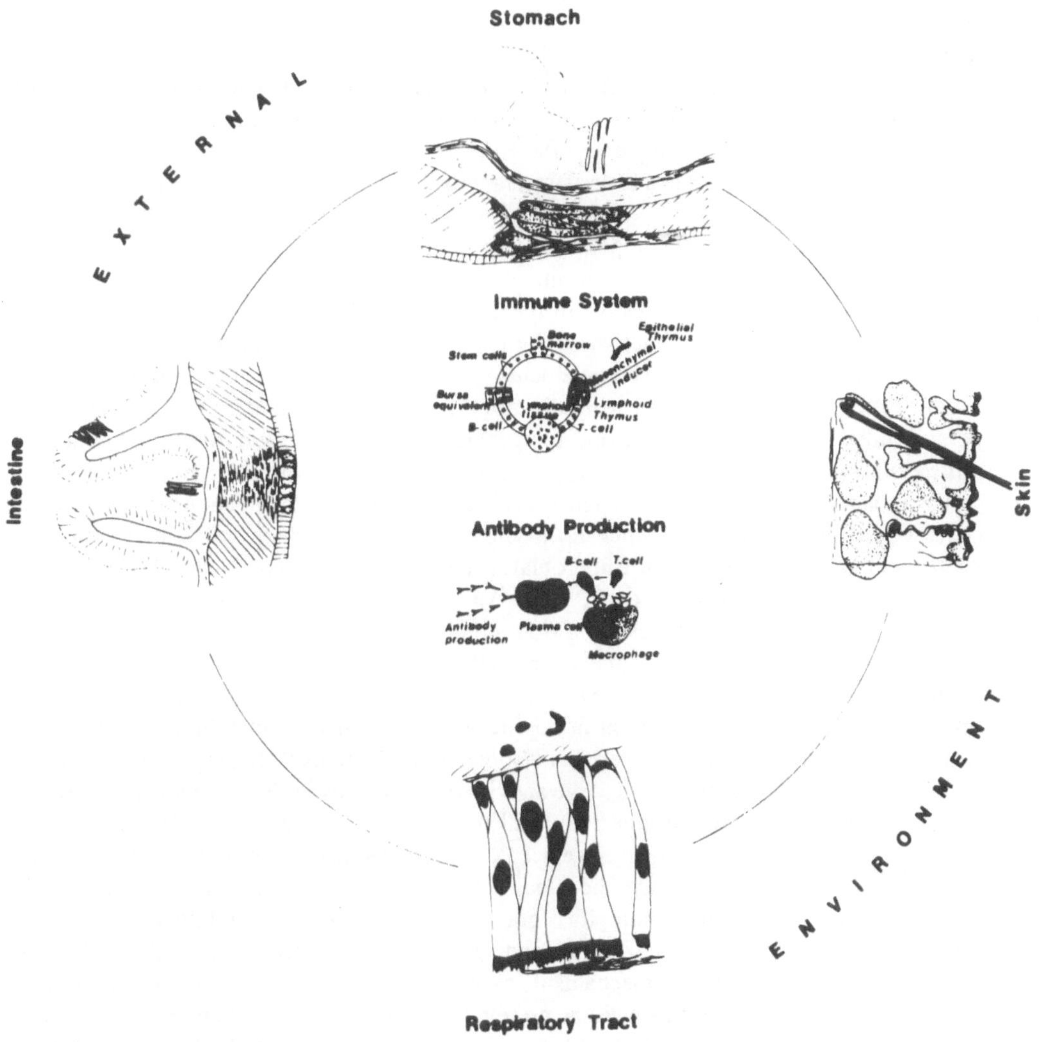

FIGURE 2. A schematic representation of the integument (outer ring) and host defense mechanisms (inside the ring).

cells, develop under the influence of the unknown human analogue of the chicken bursa of Fabricius. These cells respond to an antigenic stimulus by producing immunoglobulins. Defective cell-mediated responsiveness is associated with infections by intracellular organisms such as fungi, viruses, mycobacteria, and protozoa as opposed to extracellular pyogenic infections which are classically associated with defective antibody synthesis.[175] The reader is referred to the following reviews[175,176] for an understanding of the immune system and its functional significance.

The phagocytic system consists of cells that circulate and provide continuous surveillance against invading microbes or antigens. Phagocytic cells comprise polymorphonuclear neutrophils, eosinophils, or basophils and circulating mononuclear leukocytes. These are phagocytic cells which are released from the bone marrow and differentiate into macrophages when they arrive at the appropriate tissue sites. Adequate numbers of phagocytic cells are necessary for normal host defense. In addition, adequate phagocytic function is also required for protection against severe recurrent infections.[177]

The reticuloendothelial system (RES) describes an anatomic system which includes tissue fixed mononuclear phagocytic cells or macrophages. This system consists of pulp or reticulum cells of the spleen and lymph nodes, as well as endothelial-like cells lining sinusoids in lymph nodes, liver, spleen, and bone marrow.[178] The RES is responsible for the efficient clearance of infectious particles from the blood. This system must remove and kill, or immunologically alter, large numbers of bacteria rather rapidly since failure to control early microbial invasion may lead to sepsis and death. Consequently, the RES recognizes the particles for clearance, kills viable organisms, like bacteria, and assists in rapidly mounting additional defenses against the invader.[178]

The complement system is composed of 11 proteins, divided into nine components.[179] The first complement component is made up of a trimolecular complex C1q, C1r, and C1s. Activation of the complement results in the sequential generation of an enzymatic system resulting in the deposition of activated complement components on the antigen-antibody complex. The components of the complement system may play a key role in certain aspects of serum bactericidal activity, viral neutralization, neutrophil chemotaxis, increased vascular permeability, immune adherence, and phagocytosis. It is quite clear that these functions represent a significant part of the ability of the host to defend against microbial invasion.[179]

Disturbance of any of these host defense mechanisms could lead to infection. For an excellent coverage of the situations in which host defenses against infection are impaired and the subsequent susceptibility to specific microbial agents the reader is referred to the following three volumes: Klastersky,[180] Verhoef et al.,[181] and Allen.[182] These are an essential reading for any worker dealing with infections resulting from immunosuppression.

Attention should be focused on alterations in those host mechanisms which may predispose to infection with *Candida* (for review, see References 183, 184). Phagocytic and cellular immume factors are more important in host resistance to *Candida* than humoral immunity, which is associated more with bacterial and some viral infections.[174] In general, the clearance of *C. albicans* in blood is attributed to the activity of leukocytes.[98,183] Neutrophils, monocytes, and eosinophils are known to ingest and kill this yeast (Figure 3).[98,185,186] The interactions between these cells and *C. albicans* have been extensively studied *in vitro*.[187-190] There are two mechanisms suggested for the candidacidal activity of these cells (both of which involve proteins contained within the neutrophil's granules): the first involves an interaction between myeloperoxidase (basic protein enzyme)[177] and hydrogen peroxide (H_2O_2) and superoxide anion (Figure 4).[185,191,192] The second mechanism involves the action of certain proteins, the most studied being chymotrypsin-like cationic protein (CLCP).[193,194] This protein exhibits candidacidal activity towards *C. parapsilosis* and other *Candida* spp. *in vitro*.[195] CLCP protein is rapidly and preferentially adsorbed by *C. parapsilosis*, leading to an increase in the membrane permeability.[196]

The most common abnormality of neutrophil granules is myeloperoxidase deficiency, in which there is complete absence of leukocyte myeloperoxidase from the primary granules (see review by Klebanoff and Clark[197]). Such deficiencies could be brought about by drugs,[98,198,199] or by certain diseases such as leukemia and bone-marrow aplasia where leukopenia is known to exist. Patients suffering from acute myelogenous leukemia show mature-appearing neutrophils that are deficient in composition and candidacidal function.[198,200] Myeloperoxidase deficiency is not associated with severe clinical disease in most patients, although increased susceptibility to infection with *C. albicans* has been reported, particularly in patients who have host defense defects for other reasons (such as diabetes mellitus).[201,202] Lehrer and Cline[191] reported on a patient suffering from chronic *C. albicans* infection, whose neutrophils were lacking myeloperotidase and leukocytes were completely incapable of killing *C. albicans*. Another patient suffering from myelomatic leukemia had peripheral and bone-marrow leukocytes lacking in myeloperoxidase activity, and a neutrophil bactericidal defect was also observed.[203] Other situations associated with myeloperoxidase deficiency include bacterial infections,[204,205] refractory anemia,[206] and preleukemia.[207,208]

Defects in the capacity to generate hydrogen peroxide or superoxide anion (comprising the

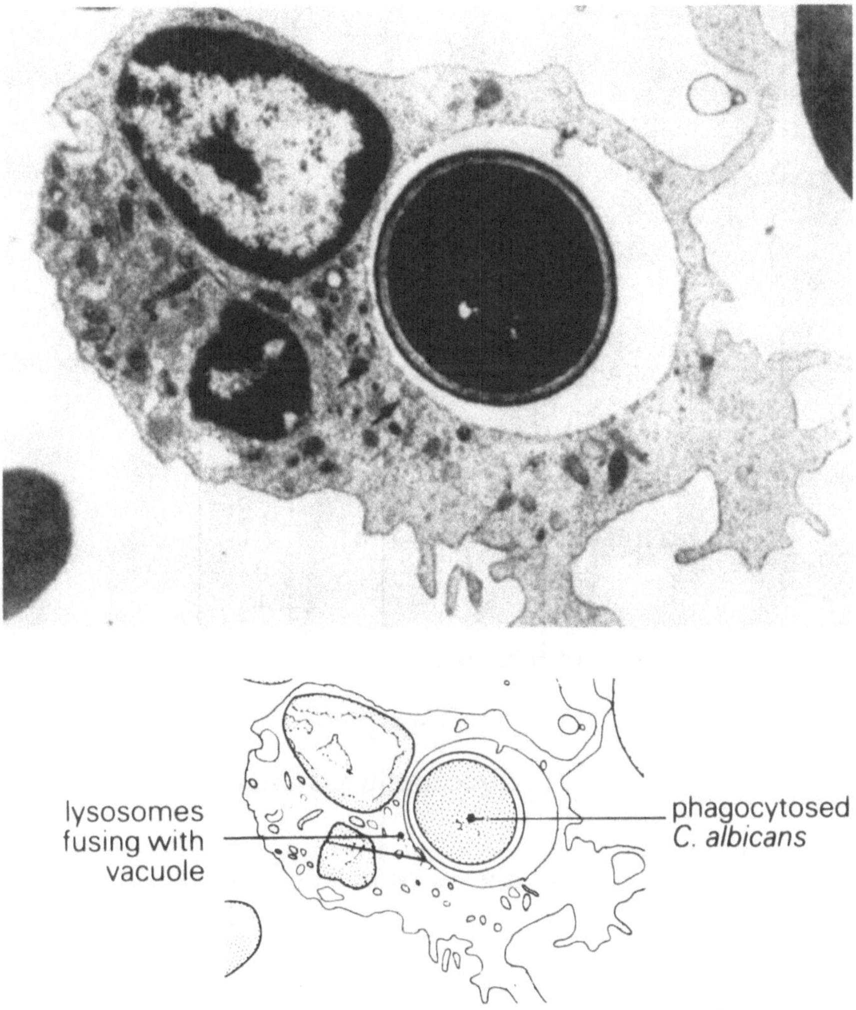

FIGURE 3. A schematic drawing showing a neutrophil containing phagocytosed *Candida albicans*. (From Roitt, I. M., Brostoff, J., and Male, D., *Immunology*, Churchill Livingstone, Edinburgh, 1985. With permission.)

other components of the reaction necessary for candidacidal activity) could also lead to failure in the killing of ingested *C. albicans*.[98,191] This disorder is often referred to as an abnormality in the intraleukocytic metabolism.[201] Several disorders related to this have been described.[197] The most extensively studied patients with this disorder are those with chronic granulomatous disease of childhood. Patients with this disease have an increased incidence of infection with microorganisms which are catalase positive, including *C. albicans*.

The most common cause of neutrophil defect is antineoplastic chemotherapy. Many treatment regimens for all types of neoplasia render patients neutropenic for such extended periods that infection regularly results.[209] Neutrophils deficient in candidacidal activity were observed in patients receiving chemotherapy for metastatic cancer[200] and renal transplant recipients receiving immunosuppressive drugs.[210] Sulfadiazine and sulfisoxazole diminish the candidacidal activity of human polymorphs by inhibiting intracellular killing mechanisms at concentrations achieved clinically.[211] These compounds interfere with myeloperoxidase-mediated reaction and with the formation of hydrogen peroxide.

A reduction in the number of neutrophils (neutropenia), besides defects in the function of

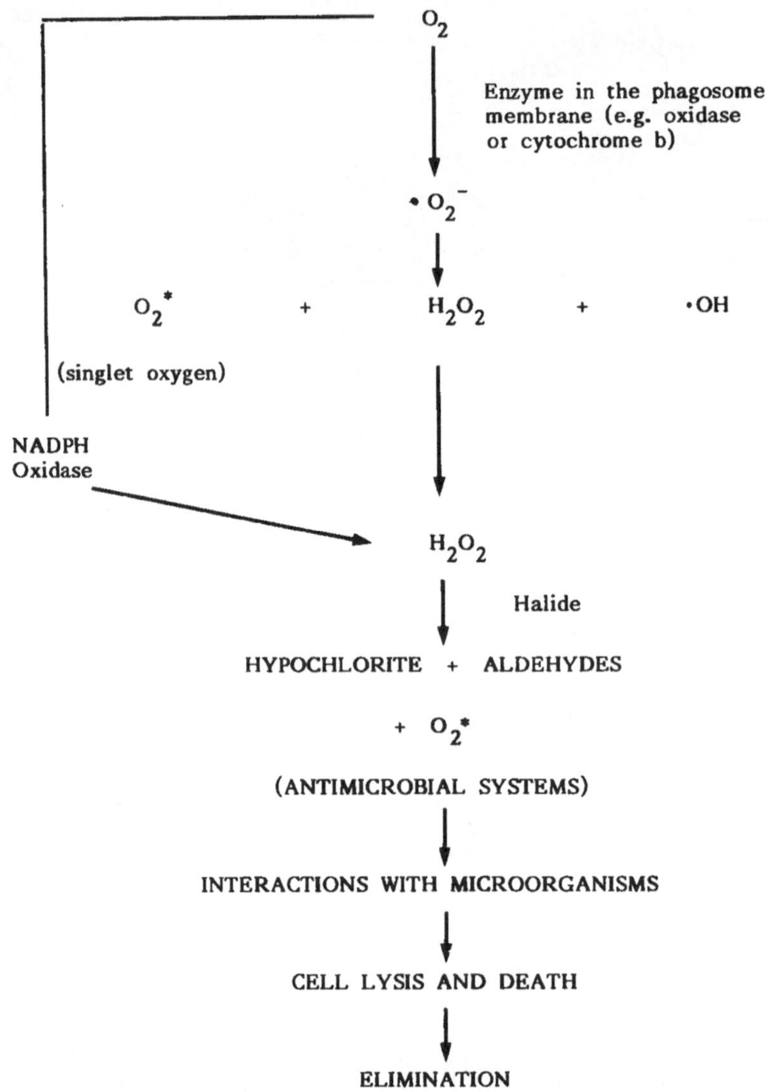

FIGURE 4. Mechanism of microbicidal activity involving myeloperoxidase, hydrogen peroxide, and superoxide anion.

these cells, is probably the most frequently encountered clinical problem of neutrophil dysfunction.[177] There is great risk of infection when the circulating number of neutrophils is less than 500/mm^3 but increased infections also occur when counts are between 500 and 1500/mm^3. The cause of neutropenia may be congenital, on the basis of an autosomal recessive disorder,[212] or due to drug therapy. Drugs may affect neutrophils differently. For example, nitrogen mustard and chlorambucil have a direct cytolytic effect on neutrophils, while azathioprine and 6-mercaptopurine affect the metabolism of these cells.[174] Neutropenia results in both cases. Neutropenia is also associated with the use of chloramphenicol, sulphonamides, methicillin, and cephaloridin.[177] Drugs such as prednisone also alter neutrophil function while cytotoxic agents, antimetabolites, and radiotherapy affect T-lymphocyte and mononuclear phagocyte function in addition to decreased neutrophil function and numbers.[209]

Defective cell-mediated responses have been associated with the occurrence of mucocutaneous candidosis. A number of patients with T-cell defects, not involving obvious thymic

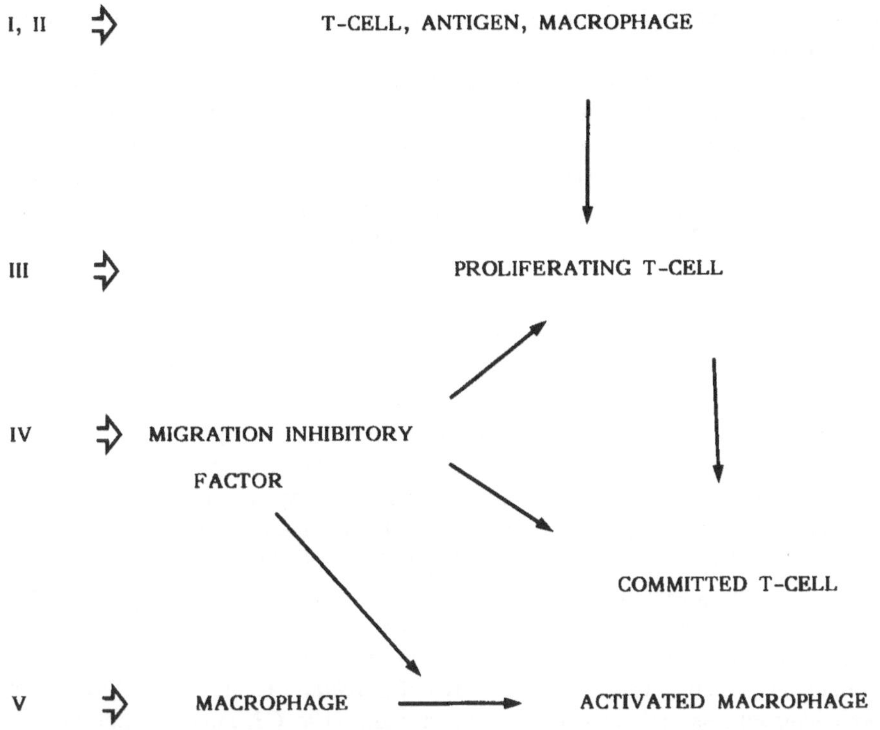

FIGURE 5. Defects of T-cell function in mucocutaneous candidosis. I, Lymphopenia; II, selective immunologic unresponsiveness to *Candida* antigen; III, absent proliferative response to *Candida* antigen; IV, absent migration inhibitory factor production in response to *Candida* antigen; and V, defective mononuclear response to chemotactic factor. (Adapted from Provost, T., in *Infection and the Compromised Host*, Allen, J.C., Ed., Williams & Wilkins, Baltimore, 1976.)

defects, have been described under mucocutaneous candidosis.[25,213-217] *C. albicans* is reported to grow luxuriantly locally in mucous membranes or on skin in the presence of a T-lymphocyte, mononuclear phagocyte defect but to seldom disseminate.[209] Burford-Mason et al.[218] suggested that separate immune responses protect against mucocutaneous and systemic candidosis and the defective immune response against mucocutaneous disease need not imply lack of protection against systemic spread. It should be pointed out here that mucocutaneous candidosis is not seen in individuals with B-cell defects in the absence of concomitant T-cell defects.[175] The subtle defects in T-cell function which have been associated with mucocutaneous candidosis are represented in Figure 5.[219] For more elaborate discussion on the defects of T-lymphocytes the reader is referred to Provost.[175]

Immunoglobulins play a minor role, if any, in the protection against candidosis, which does not usually occur in patients with specific immunoglobulin deficiences.[220] Even the suggestion that immunoglobulin G (IgG) contains effective opsonins[184] for *C. albicans* is controversial.[173] Kagaya and Fukazawa[188] reported that the addition of specific rabbit antibodies to an *in vitro* system containing *C. albicans* and neutrophils did not augment the rate of phagocytosis or killing. La Force et al.[221] showed that serum from patients with disseminated candidosis inhibited phagocytosis of *C. albicans* by human neutrophils.[221] These authors suggested that this effect was due to the presence of high-titer anticandida antibodies, which are prevalent in patients suffering from systemic or mucocutaneous candidosis.[184] Conversely, Diamond et al.[222] reported enhancement of phagocytosis of Candida by human neutrophils in normal serum due to opsonization by low levels of anti-candida IgG. Diamond and Krzesicki[223] investigated the mechanisms involved in the attachment and damage of *C. albicans* by neutrophils in the absence of serum. They concluded that *Candida* contains a molecule on its surface which is recognized by neutrophil receptors. Subsequent damage to *Candida* by neutrophils occurred primarily by oxidative mechanisms.

Another role suggested for immunoglobulin in *Candida* control is interference with adherence. Epstein et al.[224] showed that anticandidal IgA (a relatively minor immunoglobulin in serum,[224] but the predominant immunoglobulin class in external secretions)[225] antibodies in saliva can inhibit adherence of *Candida* to the oral mucosa. However, these authors concluded that the inhibition of adherence by antibodies, so clearly seen in many bacterial systems,[226-229] exists but is less striking in the *C. albicans*-human host relationship.

Although the complement system and humoral factors (serum factors) are less important in host resistance to *Candida* than are phagocytic and cellular immune systems, a discussion of what is known of their role in the control of candidosis is pertinent. Studies concerning the role of the complement system in the promotion of phagocytes *in vivo* have presented controversial evidence. Lehrer and Cline[98] showed that the candidacidal activity of blood is attributable to its leukocytes because plasma or serum do not kill *C. albicans,* despite the fact that they contain antibodies and complement components. These authors suggested that the ability of human neutrophils to kill ingested *C. albicans* does not depend on prior exposure of the organisms to heat-labile serum factors or serum factors which could be inhibited by EDTA. Morelli and Rosenberg,[230] on the other hand, showed that phagocytosis of *Candida* by mouse blood leukocytes was enhanced by a complete hemolytic complement system. They also showed that phagocytosis of this yeast was enhanced by specific antiserum. Complement-deficient mice had reduced resistance to *Candida* infections.[230] Humans with hereditary C2 deficiency were shown by Meuwissen et al.[231] to have deficient phagocytosis of *Candida*. Results obtained by Miller and Nilsson[232] stressed this and showed that C5, in addition to C3, may be important in opsonizing these organisms. Lyon et al.[233] using congenic mice, sufficient or deficient with respect to the C5 component, of complement concluded that C5 or later acting complement components are important, but not absolutely essential, to successful defense against *C. albicans*.[233] Kozel et al.[234] stated that there is little doubt that activation of the complement cascade is an important event in the pathogenesis of infection by *C. albicans* and *Cryptococcus neoformans*. They suggested that the alternative complement pathway is the primary mechanism for activation and binding of C3 fragments to *Candida albicans* and that binding of C3 to this yeast involves both ester and amide bonds.

The extent of intracellular killing of *Candida* measured with mouse peritoneal cells was not influenced by the presence of antibody,[230,235] or of a complete C system, or the C status of the animals providing the cells. However, Morelli and Rosenberg[230] did not exclude the contribution by normal antibody altogether. According to Smith,[173] although complement in fresh serum is not directly candidacidal it assists chemotaxis of leukocytes to the site of infection and enhances phagocytosis of *Candida* by non-immuno-specific mechanisms.[188,230,232] The events following ingestion by phagocytes are not, however, influenced by complement components.[98,230,236]

Other serum factors that may play a role in the control of infection by *Candida* include unbound iron,[237-240] copper,[241] transferrin,[240,242-246] lactoferrin,[245,247] protein deficiency,[248] and vitamin-A supplementation.[249] The significance of these substances in defense against *Candida* requires further investigations.

Data on the effect of macrophages on *C. albicans* is limited and contradictory. Kagaya and Fukazawa,[188] investigating the phagocytic and intracellular killing activities of normal mouse phagocytes against *C. albicans,* reported that viable *C. albicans* cells were ingested by mouse peripheral blood leukocytes and peritoneal macrophages at the same rate, with leukocytes being more effective than macrophages in killing intracellular yeast. Their results suggested that peritoneal macrophages and factors associated with the alternative pathway of complement (see Provost and Allen[250] for the alternate pathway of complement activation) and lymphokine-activated macrophages play a major role in the protection of mice against candidosis.[188] Lehrer et al.[251] and Peterson and Caldrone[252] consolidated the role of macrophages in the control of candida infection by showing that they can kill ingested *C. albicans* to the same extent as human neutrophils. Nevertheless, Stanley and Hurley[253] and Ozato and Uesaka[254] found macrophages to have only transient fungistatic activity. These contradictory results could be due to the fact that Lehrer and his co-workers used alveolar macrophages while the others used peritoneal macrophages. It is known that resident peritoneal macrophages are about half as effective as the resident alveolar cells.[184] Taschdjian et al.,[255] examining tissues from patients with systemic candidosis, found *Candida* cells with altered cell wall within tissue macrophages and sessile reticuloendothelial cells throughout the body suggesting ingestion. These contradictory reports regarding the role of macrophages in controlling candidosis and the sparsity of data necessitate further investigations before a clear role for these cells is defined.

b. Anatomical Barriers

The first line of defense that an invading microorganism meets is the anatomical or physical barrier provided by the integument (comprising the skin, together with certain specialized derivatives of the skin such as nails, hair, and several kinds of glands) and the mucous membranes. However, when the integrity of the skin is broken by, e.g., burns, eczema, or surgical incision, invasion by microorganisms may result. The protection afforded by the skin is the result of both mechanical and chemical factors.

i. Mechanical Factors

The integument is composed of two layers: the epidermis, a specialized epithelium derived from the ectoderm, which is in direct contact with the external environment and protects the more delicate underlying parts from trauma, chemical irritation, microbial invasion, and other external factors; and, beneath this, the dermis, a derivative of mesoderm, composed of vascular dense connective tissue. The dermis corresponds to the lamina propria of the mucous membrane. These two layers are firmly adherent to each other and form a barrier that varies in thickness from about 0.5 to 4 mm or more in different parts of the body.

The epidermis is composed of stratified squamous epithelium which can be divided conveniently into layers, beginning with the most exterior: the stratum corneum, or horny layer; the stratum lucidum, the pars compacta; the stratum granulosum, or granular layer; the stratum malpighii, or prickle cell layer; and the stratum germinativum, or basal cell layer. Although described in this order, the layers develop in the opposite sequence. The basal cell layer gives origin to all the others. The stratum corneum or horny layer is the outer or surface division of the epidermis; it is thickest on the palmar and plantar areas and thinnest on the eyelids, prepuce, cheeks, forehead, abdomen, and flexor surfaces of the elbows. It is composed of dead epithelium cells which have become horny or keratinized. They are flattened, dry, and non-nucleated, usually showing vacuoles. The periphery has become hardened and the prickles have persisted as dried-up spicules of keratin, which is a waterproof protein.

The dermis is composed of dense irregularly arranged connective tissue and is subdivided into two strata, the papillary layer superficially and the reticular layer beneath. The papillary layer includes the ridges and papillae that protrude into the epidermis.[256,257]

This arrangement of closely packed cells, continuous layering, and the presence of keratin would explain why intact skin provides such a formidable physical barrier to the entrance of microorganisms (Figure 6A). The intact surface of healthy epidermis is rarely, if ever, penetrated by microorganisms.

Candida is not a normal denizen of healthy skin. However, when the epithelial surface is broken (Figure 6B), as a result of trauma (as in burn patients) or increased moisture and maceration of the skin (as found in those who work by hand in water-related jobs, children who suck their thumb, as well as deformities due to rheumatoid arthritis and occlusive dressing), a subcutaneous infection often develops.[258] Protection of the host is also provided by the mucus, which bathes the mucosal surfaces. The mucus refers to the clear, viscid secretion of the mucous

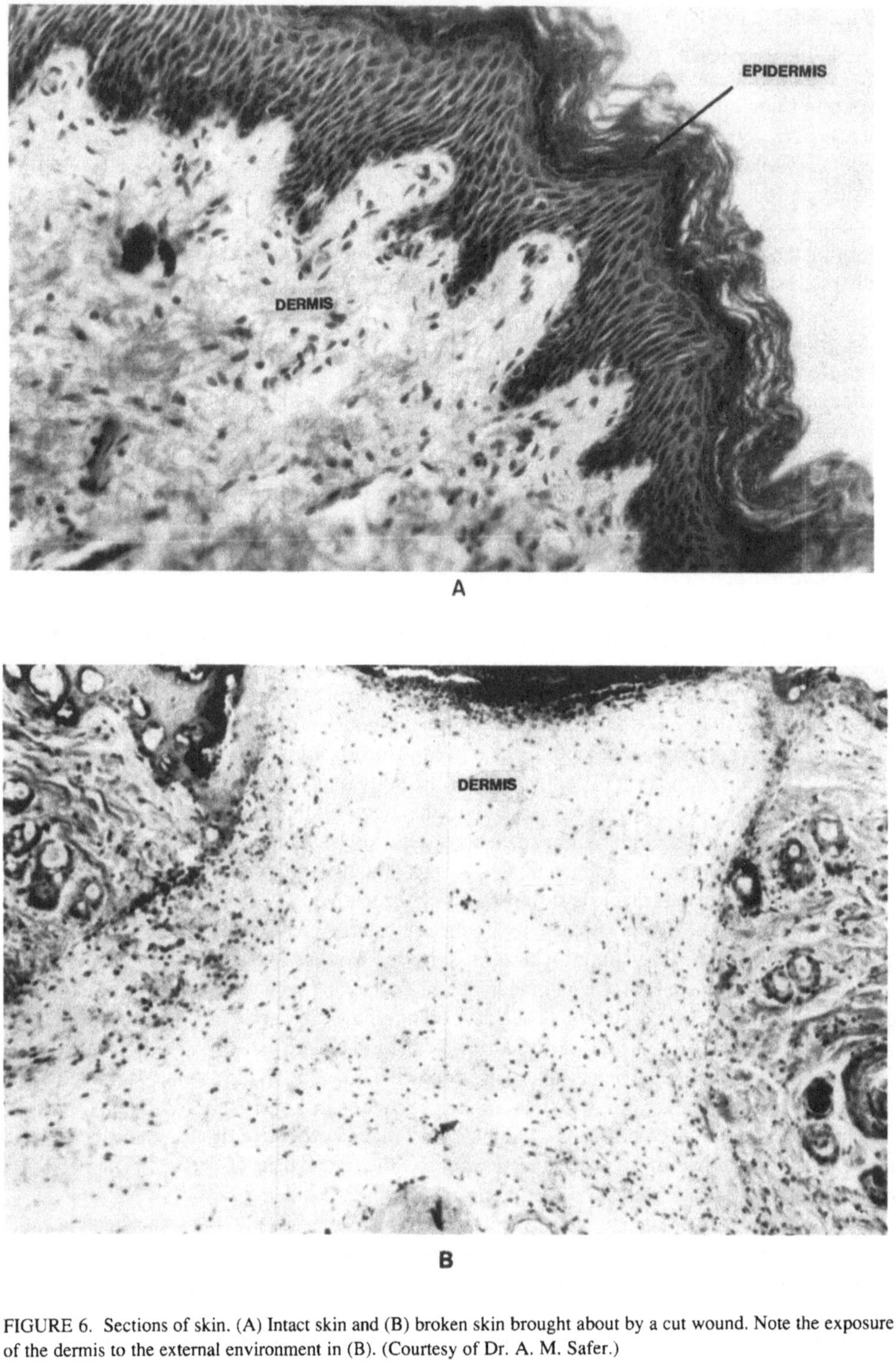

FIGURE 6. Sections of skin. (A) Intact skin and (B) broken skin brought about by a cut wound. Note the exposure of the dermis to the external environment in (B). (Courtesy of Dr. A. M. Safer.)

membranes, consisting of mucin, epithelial cells, leukocytes, and various inorganic salts suspended in water. However, the "mucus material" or "mucus gel" of the mucosal surfaces refers usually to glycoproteins and glycolipids forming this material.[259,260] For an organism to invade the epithelial mucosa it has to pass through these complex hydrated gels. The extent to which these gels prevent the entrance of microorganisms and inert colloidal particles is a matter of controversy. Some workers suggest that the mucus forms a formidable barrier to particles, but others have shown that macromolecules cross mucosal barriers with some regularity (see review by Warshaw et al.[263]).[261-263] Freter et al.[264-267] concluded that the mucus gels present a barrier to the penetration of particles in the size range of bacteria. Larger particles such as yeast cells penetrate less efficiently. This barrier appears to be imperfect, because even inert particles may penetrate to the base of the villi. A more detailed discussion of the mucous layer and its role in the adherence process is covered in Chapter five.

Other chemical factors that enhance the mucosal defense against microorganisms include tears, produced by the lacrimal apparatus; saliva, produced by salivary glands, and gastric juice.

Mucous membranes consist of an epithelial layer and an underlying connective tissue layer. They line body cavities that open to the exterior and comprise an enormous surface area. Mucosal epithelial surfaces include the entire GI and respiratory tract, salivary and lacrimal glands, biliary system, and sections of the genitourinary system. Although mucous membranes do inhibit the entrance of many microorganisms, they offer less protection than the skin.

Other factors that could be considered under the mechanical factors include desquamation, flushing action of the urine, the ciliary movement of the ciliated epithelium lining the respiratory tract, and persistaltic movement of food in the intestine. In desquamation the surface of the skin is constantly worn away, with individual cells being sloughed off as microscopic fragments.[124] The ciliated epithelium traps and sweeps away inhaled microorganisms. The movement of intestinal contents down the intestine is considered as the most important determinant of the distribution of bacteria within the intestine.[268,269]

Disorders that impair the normal persistalsis of the gut commonly lead to an increased bacterial population in the ileum.[270] The flushing action of urine helps prevent establishment of infection in the urinary tract. Urinary tract infection is associated with urinary stasis, e.g., in ureteric reflux, bladder tumor, neurological abnormalities leading to sphincter disturbance.

A number of mechanisms, both immune and nonimmune, have developed at the epithelial surfaces to combat invasion and damage by pathogenic microorganisms and toxins. Nonimmune mechanisms include the indigenous flora and the secretion of a number of biologically active substances such as lysozyme, lactoferrin, and peroxidase. The secretion of secretory IgA (sIgA) into the fluids that bathe mucous membranes represents the most important immune mechanism. An extensive review of the host defense mechanisms at mucosal surfaces is beyond the scope of this volume, and the reader is referred to an excellent review by McNabb and Tomasi.[103]

ii. Chemical Factors

Mechanical factors alone do not account for the high degree of resistance of skin and mucous membranes to microbial invasion. Certain substances also play an important role.

The skin is protected by a greasy mixture of sebum, sweat, exfoliated epidermal cells, and various additives from external sources. This surface film contains amino acids, urea, uric acid, lactic acid, ammonia, triglycerides, free fatty acids, wax alcohols, sterols, phospholipids, pentoses, complex polypeptides, and other substances.[257] These materials and the pH of this surface film act as defenses against infection, supply lubrication, work as buffering agents, and influence hydration of the corneous cells.[257]

Chemical factors, such as volatile fatty acids and secondary bile acids, present in the normal gastrointestinal environment have been reported to play a role in the protection against *Candida* infection through interfering with yeast-mucosal association[271] (see Normal Flora below). Factors that impair the intactness of skin and mucosal continuity and interfere with the mechanisms described above, in such a manner that local and general conditions become suitable, will predispose the human to microbial infection including *Candida*. However, the mechanical breakdown of the skin or mucosal discontinuity per se does not predispose the patient to *Candida* infection. It is in cases where a patient has another immune defect, such as neutropenia, that the possibility of candidosis becomes high. Accordingly, Wolfson et al.,[272] on the basis of pathophysiological events, classified candidosis with opportunistic primary skin infections. These infections follow some form of injury to the skin that has provided the opportunity for these usually nonvirulent members to invade. Then, with secondary host defenses impaired, significant local or disseminated disease may result.[272] This is best illustrated in burn patients where multiple factors contribute to the development of disseminated candidosis. These include prolonged survival of patients with severe burns, widespread use of topical and systemic antibiotics, the use of intravenous catheters, and hyperalimentation.[273,274]

Thus, one would predict that the skin would become a portal of entry for potentially serious infection under the following circumstances: when the surface layer of keratin is destroyed or bypassed by trauma or by the introduction of such foreign materials as intravascular catheters;[275] when the skin is kept moist and becomes macerated, as under occlusive dressings; and when broad-spectrum antimicrobial therapy has been administered, altering the normal flora and thus abrogating the beneficial effect of bacterial interference.[276-279] (For review consult Aly and Shinefield[280]).

Candida infections have long been associated with situations similar to the above circumstances. Thus, it is well known that *Candida* species colonize and infect around intravenous catheters, parental hyperalimenation tubing, and frequent intravenous injections[139,169,281-285] (as in the case of drug addicts[286,287]) surgical patients,[140,166,288-290] immunocompromised patients,[174,272] and burn patients.[273,274,291] These cases of candidosis are a result of interruption of the integument in patients whose other defense mechanisms are impaired.

Mucosal continuity is particularly important in preventing the entry of pathogens in patients with compromised phagocytic and lymphoid cell function. This is particularly so with *Candida* since this yeast colonizes the mouth, the gut, and the vagina in up to 80% of normal individuals.[12,292] Any discontinuity or alteration of the mucosal epithelial membrane or its environment (for a review see Ryley[68]) will allow this yeast to gain access through these surfaces and initiate an infection.[3,293-297]

Mucosal damage and irritation has been attributed to denture wearing, chemotherapy with immunosuppressive agents,[174,295] radiation treatment,[6,171,298] and, in an extreme case, chemical injury due to ingestion of concentrated sulfuric acid.[299] These agents cause changes in the mucosa of the digestive tract ranging from mild (as it is the case in denture wearers) to extremely severe injury (as with chemical ingestion)[300] and are known to predispose to candidosis.

c. Normal Flora

The role of the normal flora in the protection of the host against infection in various regions of the body, particularly in the GI tract,[301] became apparent in the 1950s with the use of broad-spectrum antibiotics such as the tetracyclines. A number of studies showed that patients receiving antibiotics had a distinct increase in the incidence of positive cultures of *Candida* from the nasopharynx, sputum, and stools.[302-306] Numerous cases exemplifying the association of *Candida* infection with antibiotics have been reported.[12,25,307-310] A general conclusion is that long-term treatment of patients with antibiotics, which reduce the numbers of certain species of the normal flora of the GI tract, leads to the disruption of the normal regulatory mechanisms. Certain opportunistic pathogens, including *Candida*, can then attain higher than normal population levels in the intestinal tract and have the ability to adhere, colonize, and disseminate. Thus, the use of antibiotics for therapy of systemic infections that have little or no effect on the normal fecal flora is desirable and should be persued.[311]

As the GI tract is considered as the main portal of entry of disseminated *Candida* infection,[3,139,169,296,297] most studies concerning the role of indigenous flora in the control of candidosis have concentrated on this organ system (few studies deal with the role played by the normal flora of skin in control of candidosis [see Selwyn[312]]). Such studies compared the resistance of germ-free and conventional animals to infection[313-316] or to the suppression of the normal flora by the use of antibiotics,[271,307,315,317,318] X-irradiation,[319] intramuscular cortisone,[320] or immunosuppresants.[321-323] Germ-free animals, which are derived originally by sterile caesarean section or sterile hatching of eggs and are reared and continuously maintained with germ-free techniques under isolator conditions, are free from all demonstrable forms of microbial life.[324] Conventional animals, in contrast, are housed in readily accessible animal rooms and harbor a complete normal flora. Germ-free animals are more readily colonized and infected by gastrointestinal pathogens than are conventional animals. These models were exploited by a number of workers to examine the effects of antibiotics on the normal microbial flora. Thus, Phillips and Balish[313] showed that the environment of the alimentary tract of certain germ-free and conventional mice is adequate to support luxuriant growth of *C. albicans*. However, a difference existed in the form in which *Candida* was isolated from these two types of animals with blastospores and hyphae being isolated from the germ-free animals while only budding yeast forms were isolated from the guts of conventional animals. A decrease in the concentration of sulfhydryl-containing compounds was observed in germ-free animals. Other differences between these two types of animals were recorded for tissue acidity and E_h values of contents from the gut. The same workers,[316] had similar results using germ-free and conventional chicks and showed germ-free chicks but not conventional ones to be infected with *C. albicans*. They concluded that the conventional intestinal flora provided protection against candidosis of the gut, a hypothesis confirmed by other workers.[169,325,326] It has also been shown that mice possessing a competitive bacterial flora are more effective than mice with an intact immune system in suppressing colonization with *C. albicans*.[327]

The ability of specific bacterial species to protect against infection by *Candida* has also been investigated. Balish and Phillips[328] established in chicks a monoflora of either *Escherichia coli* or *Streptococcus fecalis* (which are known to be constituents of the normal intestinal bacterial flora of humans) followed by oral inoculation with *C. albicans*. Examination revealed severe crop infections in chicks with a diflora containing *S. fecalis* and *C. albicans*, while chicks with diflora of *E. coli* and yeast had few microscopic crop lesions. Thus, it was concluded that *E. coli* but not *S. fecalis* provided protection against crop infection by *C. albicans*. Nishikawa et al.[314] reported that *E. coli* has the capacity to antagonize *C. albicans*. Other workers supported these findings.[315,316] Studies by Liljemark and Gibbons,[329] based on the inhibition of *C. albicans* growth *in vitro* and *in vivo*, proposed a suppressive role for *Streptococci* and mixed bacteria in saliva. Kennedy[317] showed that indigenous anaerobic bacteria, which dominate the intestinal microflora,[330,331] could exert some protective mechanism against *C. albicans* in the oral cavity. Kennedy and Volz[332] went a step further and suggested that it is the strict anaerobic bacteria and not the enteric bacilli or aerobic bacteria that are responsible for the inhibition of *C. albicans* adhesion and the resistance of gut colonization and dissemination by this organism from the GI tract. Ekenna and Sherertz[323] questioned these results and suggested that the colonization of the tract of mice by *C. albicans* can be facilitated by manipulating the aerobic, anaerobic, or both types of gut flora.

Andermont et al.[333] showed that enteric bacilli, including *E. coli*, do not suppress the growth of *C. albicans* in the alimentary tract of conventional mice. In yet another study Kennedy and Volz[271] attributed the protection against candidosis to a complex indigenous microflora. They suggested that the indigenous intestinal microflora suppresses *C. albicans* colonization and dissemination from the gut by inhibiting the *Candida*-mucosal association and reducing the yeast population levels.

Clark[315] showed that mice contaminated with *E. coli* alone or with *Bacteroides* sp., *Lactobacillus* sp., *Streptococcus lactis*, and *S. fecalis* did not simulate in germ-free mice, the conditions which in pathogen-free mice prevented long-term establishment of *C. albicans* in the GI tract. These contaminated mice allowed *C. albicans* to grow unchecked for several weeks. It is suggested that fusiform bacteria, in addition to physiological, biochemical, anatomical, or even immunological factors, may play a role in the protection against candidosis since these bacteria predominate in certain portions of the digestive tract.[106,315] In conclusion, the control of *Candida* is not due to a single bacterial species; rather it is a function of a complex indigenous microflora together with the environmental conditions prevailing in a given region.

A number of mechanisms have been proposed by which the indigenous normal flora exercises its protective action against *Candida* colonization and dissemination. The protective role of the flora may involve one or a combination of the following mechanisms. The inhibitory flora may: (1) produce a change in the physiological environment (e.g., altered pH, redox potential) which is inimical to growth of the inhibited microorganisms; (2) elaborate an antimicrobial substance; (3) deplete a vitamin, substrate, or ion which is essential for the growth or survival of the target species, or (4) interfere with the target-microorganisms-mucosal association. Each of these mechanisms has been implicated in antagonism of *Candida* by one or more constituents of the normal flora.

Possibly antibiotic-induced alterations of the local pH increase the ability of *Candida* to persist and disseminate. However, Paine[334] showed that bacterial inhibition of *C. albicans in vitro* was not influenced by pH. Furthermore, *Candida* grows well over a wide range of pH (between pH 5.5 to 9.0).[335] Sharp[336] demonstrated that the mouth pH of patients who developed positive oropharyngeal cultures during antibiotic therapy also had wide variations, and suggested that there could be no correlation between pH and Candida superinfection. A similar observation was reported by Balish and Phillips[316] who showed that the large numbers of *C. albicans* found in the crop contents of challenged germ-free chicks were not related solely to the pH of the intestinal tract. In another study they did not clarify the relationships of pH levels to protection against infection. However, they suggested that the increased pH in chicks with *E. coli* may reflect the synthesis of compounds (such as amines) by *E. coli* which, after challenge, promote hyphal growth in *C. albicans*.[328]

Similarly, changes in the redox potential (E_h), which occurred in germ-free chicks, did not account for the infectivity of *Candida*. Both pH and E_h changes were related to conditions for morphogenesis vs. inhibition of the formation of hyphal forms in germ-free animals. Thus, Balish and Phillips[316] suggested that the resistance of conventional chicks to intestinal candidosis is due to inhibition by intestinal bacteria of morphogenesis in *C. albicans*. In another study, these workers suggested that intestinal bacteria produced SH compounds in amounts sufficient to prevent hyphal formation.[313] For more details on the role of SH compounds in preventing filamentation, refer to Nickerson and co-workers.[337-340]

Evidence for the role of vitamin deficiency caused by the use of antimicrobial agents in enhancing *Candida* invasion is largely speculative. Case reports of vitamin deficiencies following prolonged administration of broad spectrum antimicrobial agents are available.[341,342] Harris[343] proposed that antimicrobial agents decrease vitamin production by intestinal bacteria, thus creating the environment for *Candida* infection. Administration of the B complex led to a reduction in the incidence of candidal lesions. However, other workers discounted the role of vitamins in the incidence of *Candida* infection. Thus, McGovern et al.[344] observed no effects of vitamins in decreasing the incidence of positive *Candida* cultures following chloramphenicol use. Lipnik et al.[303] questioned the belief that antimicrobial use led to vitamin deficiency and stated that data supporting this hypothesis are inadequate.

The normal intestinal flora seems to have some influence on the anatomy and physiology of the GI tract. This is inferred from studies using germ-free animals where a decrease in the weight of tissues in the small intestine and cecal enlargement in germ-free animals was observed.[313,315,345] Transferring germ-free animals into a conventional environment or association with bacteroides restored the cecum to normal size.[346,347] This association of bacteroides is of particular interest in view of the recent finding by Kennedy and Volz[317,332] regarding the role of anaerobic bacteria in the protection against candidosis. Clark[315] suggested that the extra capacity of the GI tract may serve as a reservoir for *Candida*.

Peristaltic flow is another factor affected by the normal flora. Dubos et al.[348,349] reported that peristalsis is slower in germ-free animals, and Abrams et al.[350] suggested that, in mice, the normal bacteria of the gut stimulates motility. The cleaning action of the movement of intestinal contents is probably the most important determinant of the distribution of bacteria within the intestine.[268,269] Abrams and Bishop[351] proposed that the normal flora contributes to the control of the population of *Salmonella typhimurium* in the small intestine primarily by stimulating peristaltic emptying. Clark[315] gave a similar explanation for the persistance of *C. albicans* in germ-free mice.

The GI environment is affected by the metabolic activities of the normal flora. These activities result in the production of certain metabolites that may interfere with *Candida*-mucosal adhesion. Such products include volatile fatty acids. These acids, together with secondary bile acids, may reduce the mucosal association of *C. albicans* by modifying *Candida* adhesin(s) or mucosal receptor(s) or both leading to inhibition of adhesion and subsequent colonization and dissemination of the yeast.[352]

Although Helstrom and Balish[327] placed less importance on the role of the intact immune system in preventing gastric candidosis, and stated that an immune deficiency may allow increased yeast colonization of the keratinized and glandular stomach epithelium,[327] Abrams and Bishop[353] reported that the normal intestinal flora increased the ability of the host to cope with infection by enhancing leukocytic mobilization. Thus, depression of the normal flora would have an indirect effect on the immune defenses of the host.

The ability of *Candida* to disseminate through the GI tract is determined by the level of this yeast in the gut,[307,323,327,354] which is controlled by the inhibitory effects of the microbial flora. However, if the gut population of *Candida* is high enough, dissemination will occur. Krause et al.[355] recovered yeast within 30 min from the blood of an adult human volunteer who drank 10 [20] *C. albicans* cells. Ekenna and Sherertz,[323] however, showed that factors other than levels of *C. albicans* in the gut or peripheral leukocyte counts, or both, may be important determinants of whether or not *C. albicans* disseminates from the gut.

Two possible mechanisms are responsible for the inhibition of *Candida* colonization and dissemination from the alimentary tract: (1) decreasing the size of *Candida* population in the gut[271] and (2) inhibiting the mucosal association of *Candida* cells by forming thick layers of bacteria in the mucous material covering the epithelium.[356] Pertinent to the first mechanism is the finding that antibiotic treatment suppresses the bacterial flora, thus permitting overgrowth of antimicrobial-resistant microorganisms including *Candida*.[1,39,304,305,357] This result was supported by the work of Kennedy and Volz[271] who showed that hamsters treated with antibiotics had high numbers of *Candida* colonizing their GI tracts with a high incidence of dissemination to visceral organs. In contrast, untreated animals with a normal flora had low numbers of *C. albicans* in their intestinal tract. Consequently, the suppression of the *Candida* population seems to be an important determinant for control of its dissemination.

The role of the indigenous normal flora in the control of *Candida* infection through interference with the adhesion process has received considerable attention in recent years. A more detailed coverage of this matter is dealt with in Chapter five. Briefly, the normal flora seems to form dense bacterial layers covering the mucosal epithelium and forming a physical barrier that inhibits *Candida* colonization and dissemination. Thus, *C. albicans* can readily associate with and pass through the gut wall of infant mice, which lack a complete bacterial flora, including the dense microbial population in the mucus gel,[354,358] initiating a systemic infection.[359,360] In contrast, adult mice, that have fully developed indigenous flora, are resistant even to colonization.[359,361] Protection against *Candida* mucosal association and subsequent dissemination has been reported in adult hamsters having an indigenous wall-associated microflora.[271] It is suggested that the mucosal association of *C. albicans* is reduced by competition for adhesion sites and by physically blocking of the larger yeast cells from penetrating the mucus gel. Scanning electron microscopy and disassociation studies[359,361] provide further evidence for this hypothesis.

The results discussed above indicate that the indigenous microflora suppress candidal population levels and contributes to the prevention of the systemic spread of *Candida*. A decision on which mechanism is responsible for this effect is a complex matter. However, it seems likely there are several inhibitory mechanisms produced by the indigenous microflora which may be acting together in suppressing the colonization and dissemination of *Candida* from the gut. The importance of a single inhibitory mechanism should not be underestimated since it may well supply that extra increment of resistance which is responsible for the protection of the host to *Candida* infection.[271]

Although we have categorized the host defense mechanisms against infection into separate categories, it is frequently difficult to define the relative contribution of each defense mechanism in the etiology of a given infection. This is due to compromised defense mechanisms resulting from the underlying disease itself, the use of multiple invasive procedures (surgery, venous and urinary catheters, artificial airways), the extensive use of antimicrobial agents, and often the use of combinations of immunosuppressive or cytotoxic agents. As a result, it is difficult or impossible to correlate occurrence of a given infection with a particular host defense defect and, therefore, with a given factor. This is best exemplified in cancer[174] or burn patients[272-274] who are simultaneously exposed to a multitude of predisposing factors with a resultant compromise in more than one defense mechanism. This confusion not only prevents a clear understanding of the functionally important impact of a given predisposing factor on defense against infection, but also impedes the progression of our understanding the particular aspects of host defense in the prevention of specific infections, in this case candidosis.

2. Effect on the Microorganism

Although the majority of studies concerning the mechanisms by which predisposing factors, particularly chemotherapeutic agents, enhance the invasiveness of fungi have concentrated on the effect on the host (with some workers suggesting that the enhanced pathogenicity is due entirely to host immune suppression),[362] some workers claim that these chemotherapeutic agents have a direct stimulating effect on the fungi. They suggested that such effects are also important in the induction of pathogenicity by the microorganism.[363,364] The main drugs studied in this respect are antibiotics, steroids, and antineoplastic agents.

The effect of antibacterial agents on *Candida* has been widely studied and the results obtained have been diverse and contradictory. Moore,[365] Pappenfort and Schnall,[366] and Huppert et al.[367,368] have suggested that some antibiotics including chlortetracycline, bacitracin, and neomycin stimulate the growth of *Candida* directly. In one series of experiments, oral chlortetracycline preparations enhanced growth whereas parenteral preparations did not, suggesting that the antibiotic per se was not a stimulating factor.[366] Other workers have not found growth potentiation with chlortetracycline or other antibiotics.[306,369] Many other researchers have carried out similar studies, and these are reviewed by Winner and Hurley.[25] They concluded that the evidence suggesting direct stimulatory effects of antibacterial agents on *C. albicans* are inconclusive and highly suspect, and see no reason for supporting this view.

As in the case of antibacterial agents, there are contradictory data on the direct effect of steroids on growth of microorganisms, with some workers showing growth stimulatory effects,[363,364,370-372] while others report growth inhibitory effects.[373-375] Yet other workers showed that steroids exhibited neither stimulatory nor inhibitory effects. Both testosterone and cortisone had no effect on *E. coli*.[376]

The *in vitro* response of *Candida* is neutral with steroids exerting no perceptible effect on this growth. Thus, oestriol, oestrone, oestradiol benzoate, ethinyl oestradiol, testosterone, and progesterone did not affect the growth of *C. albicans*.[377-380] Two reports, using dry weights as an estimate of cell growth, however, claimed that dexamethasone has a growth-enhancing effect on *C. albicans*.[363,364] We have investigated this effect ourselves and have estimated the growth by dry weight measurements as well as viable numbers.[381] Comparison of the two measurements

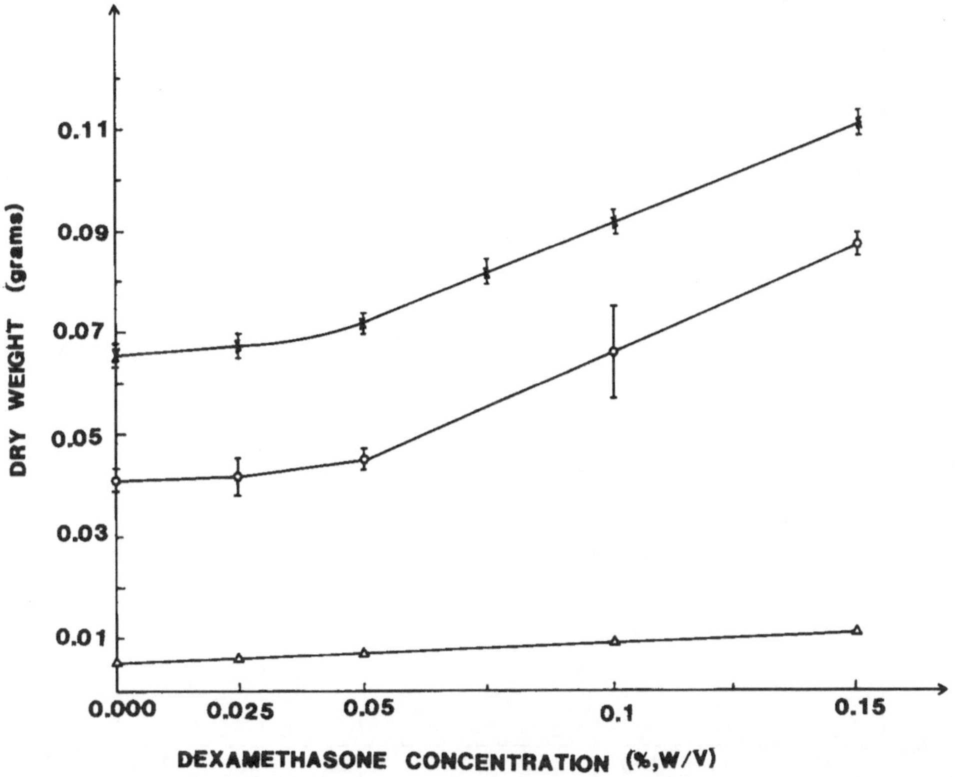

FIGURE 7. The effect of dexamethasone on the dry weight of *Candida albicans*. Cells grown in the presence of: (Δ) dexamethasone alone, (o) dexamethasone plus 0.1% glucose, and (x) dexamethasone plus 10% glucose. (From Ghannoum, M., Burns, G., and Abu Elteen, K., *Sabouraudia*, 23, 313, 1985. With permission.)

(Figures 7 and 8) and of the total nitrogen content of the cells suggested that the increases observed in the dry weight in our study, and those of others, are not the result of enhanced growth but represent residual drug which cannot be removed by the washing procedures generally employed in such studies. Thus, opportunistic infections by *C. albicans*, which are promoted by steroids, must be due entirely to effects other than growth enhancement of the organism. Further investigations of the effect of steroids on *C. albicans* were carried out in our laboratories. The results showed that dexamethasone is not metabolized by *C. albicans* to any significant extent and is largely incorporated unchanged into the cell wall and cell membrane (Table 6).[382] The presence of steroids on the outermost surface of *Candida* suggested that these compounds might promote infection, at least in part, by affecting the adherence of the organism to host tissues. This hypothesis was tested using a number of steroids known to promote candidosis.[383-388] The adherence capabilities of various *Candida* spp. to human buccal epithelial cells were affected in two different ways. Incubation with dexamethasone and triamcinolone acetonide promoted the adherence of *Candida* in general, while growth in the presence of cortisone acetate or hydrocortisone blocked the adherence.[389] Therefore, certain steroids could influence the pathogenicity of *Candida* by exerting a direct effect on its adherence ability.

Compared to antibiotics and steroids, little information is available on the antimicrobial activity of antineoplastic agents, and until recently no systematic study of this topic seems to have been undertaken.[390-393] This is surprising in view of the well-known association between neoplasm, its treatment, and *Candida* infection. Goldschmidt and Bodey,[394] in an attempt to explain the predominance of *Pseudomonas* in leukaemia patients, exposed five of the most common Gram-negative organisms encountered in cancer patients to a number of antineoplastic

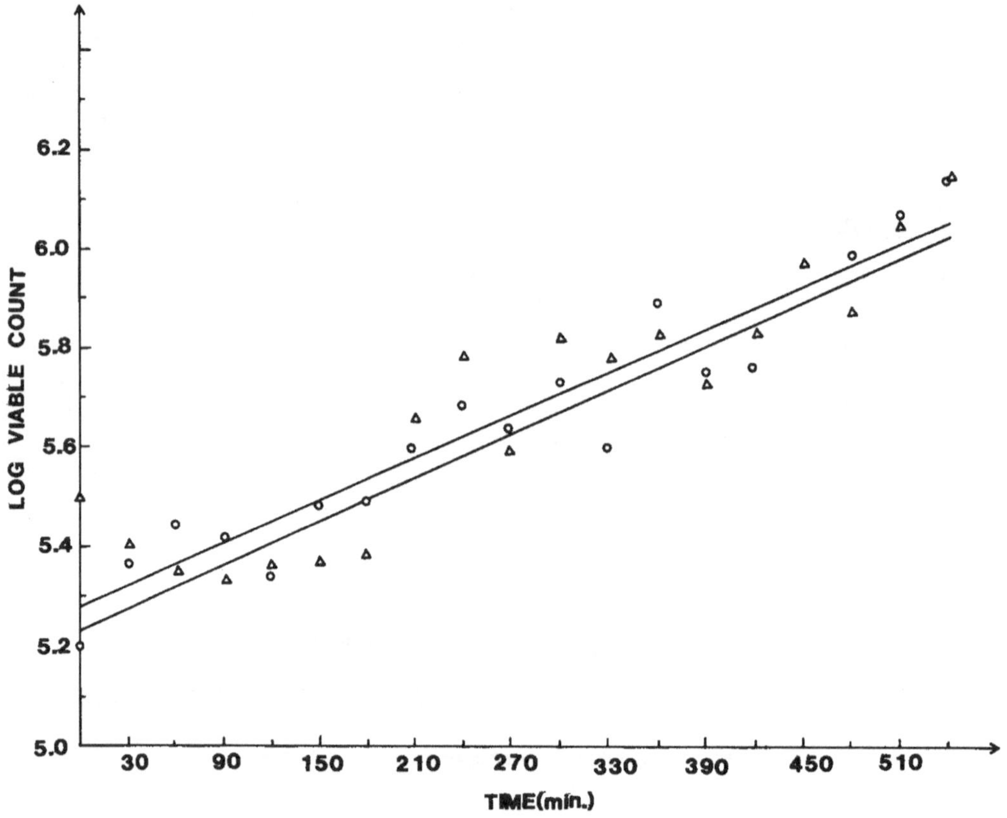

FIGURE 8. Effect of dexamethasone on the viable count of *Candida albicans*. o, 0.0% dexamethasone, Δ , 0.15% (W:V) dexamethasone. Graphs were plotted using the least squares program. (From Ghannoum, M., Burns, G., and Abu Elteen, K., *Sabouraudia*, 23, 313, 1985. With permission.)

drugs. They reported significant differences in susceptibility to these drugs with only *Pseudomonas* showing resistance to all of the test compounds. Metcalfe and Hughes[395] studied the effect of methotrexate on a number of Gram-positive, Gram-negative bacteria and *C. albicans*. They found that methotrexate inhibit Group A beta hemolytic streptococci in low concentrations (7.8 μg/ml), whereas *Pseudomonas aeruginosa*, *E. coli*, *Staphylococcus aureus*, *Serratia marcescens*, and *Klebsiella-Entrerobacter* sp. were resistant to 250 μg/ml. Combined with *in vivo* experimentation, these authors concluded that methotrexate therapy may influence the incidence of Group A beta hemolytic streptococci and the occurrence of streptococcal infections.

The direct effect of antineoplastic agents on *C. albicans* and other yeasts was investigated by a number of workers. Thus, Land et al.[396] studied the effect of several of the chemotherapeutic agents used in the treatment of cancer on the morphology of *C. albicans*. They observed that germ-tube growth of *C. albicans* 5865, a yeast highly committed to filamentation, was inhibited and forced into stasis by several drugs common to cancer therapy. Investigations carried out in our laboratory extended these studies. *In vitro* culturing of *C. albicans* and *C. glabrata* in the presence of antineoplastic agents affected both growth and morphology, to varying degrees and showed that these agents exert an influence at various cellular levels.[397,398] Methotrexate and thiotepa inhibited the growth, while vincristine sulphate and endoxan-asta showed little or no effect. Morphological studies revealed an interesting response of *C. albicans* to treatment with methotrexate. This agent stimulated the development of germ-tubes and filamentous forms from blastospores of *C. albicans* (Figure 9). This phenomenon was not reported for any of the 15 different agents studied by Land et al.[396] Similar findings were observed for *C. glabrata*.[398] These

TABLE 6
Distribution of Radioactivity in Sub-Cellular Fractions of *Candida albicans*

Fraction	Cpm[a]	Incorporated radioactivity (%)
Whole cells	3.63×10^5	100.0
Broken cells	3.60×10^5	99.2
Cell wall	2.27×10^5	62.5
Cell membrane	9.78×10^4	26.9
Mitochondria	12.60×10^3	3.5
Cytosol	15.70×10^3	4.3
Ribosomes	9.30×10^3	2.6
Recovered activity	3.62×10^5	99.8

[a]Cpm, counts per minute.

From Ghannoum, M.A., Mudher, S., and Burns, G., *Microbios*, 42, 103, 1985 With permission.

morphological alterations resulting from treatment with antineoplastic agents are of particular clinical interest for two reasons. They may explain the reason for the frequent isolation of morphologically impaired yeast isolates from cancer patients[396] and in addition, they stress the necessity of yeast speciation (a question posed by Kiehn et al.[4]) to identify the etiological agent, thereby avoiding inaccurate diagnosis. The role of chemotherapy, X-irradiation, and a combination of both in the phenomenon of adherence of yeast to buccal epithelial cells (BEC) was investigated *in vitro*.[399] Our results showed that growth of three *Candida* spp. in the presence of eleven antineoplastic agents led to reduction of adherence of these yeasts to BEC. These findings may have important clinical implications and point to the need of further study (see Chapter 6 for a full account of these investigations).

To elucidate the mechanism(s) by which antineoplastic agents enchance the pathogenicity of *C. albicans*, we carried out studies *in vivo* in which both the host (mice) and the parasite were pretreated with thiotepa or methotrexate.[400] Our results (Table 7) suggest that thiotepa enhances the pathogenicity of *C. albicans* in one of two ways: (1) by exerting an effect on the organism itself and as such making it more invasive; and (2) by exerting an effect on the host compromising its defense system, and making it more susceptible to infection by this organism. By comparison, the mechanism by which methotrexate potentiates the effect on *C. albicans* is through an effect on the host solely, and not through the organism.

Our results suggest that the enhancing effect of antineoplastic agents on candidosis represents a complex biological phenomenon in which host response, as well as stimulatory effect exerted upon the fungus, are both of significance and should be taken into consideration. This contrasts with the opinion that only the host is affected.[387] The complexity becomes apparent when it is appreciated that in addition to host and fungal factors, different chemotherapeutic agents may affect the organism in a contrasting manner. This occurs in cancer patients where immunosuppressive treatment is often continued for a long period. It is possible that the effect on the host overshadows that on the organism and may explain why most of the investigations have concentrated on the effect these agents exert on the host, and few have been directed towards studying the effect of the drugs on the infectious agents.

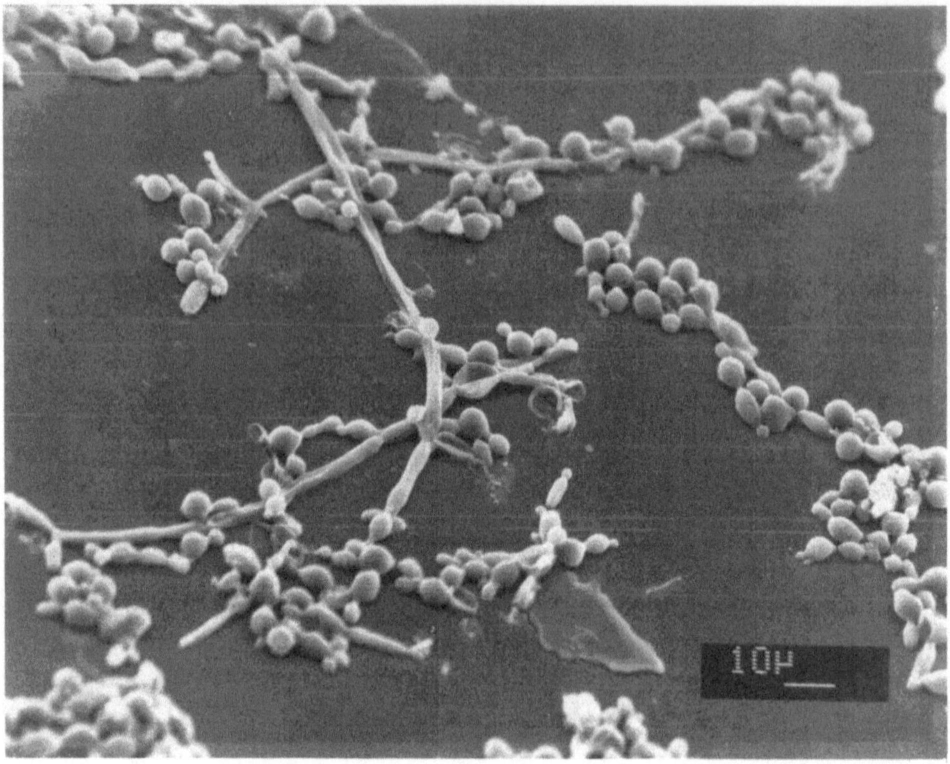

FIGURE 9. Scanning electron micrograph of *Candida albicans* grown in the presence of 500 μg/ml of methotrexate. Note the filamentation. (From Ghannoum, M. A. and Al-Khars, A., *Mykosen*, 27, 452, 1984. With permission.)

TABLE 7
Colony Forming Units (CFU) for *Candida albicans* in Mouse Tissues

Category	Time (h)	Spleen	Liver	Kidney	Heart	Lungs	Brain
Mice injected with *C. albicans* grown in the absence of drugs	3	16666	34848	1000	0	1000	0
	12	2424	25757	8575	0	2878	151
	24	0	27777	10390	0	2038	0
	48	0	4400	12077	0	916	0
	96	48	3333	30555	0	139	601
Mice injected with *C. albicans* grown in the presence of thiopeta	3	232291	147881	1409652	934	10416	0
	12	361111	1079826	559027	1033	128472	0
	24	874930	1843750	354166	28090	6944	312
	48	684375	1339583	482291	129687	158333	6250
	96	691666	729166	249652	347	333333	209
Mice injected with *C. albicans* grown in the presence of methotrexate	3	0	0	0	0	0	0
	12	0	344	69	0	34	0
	24	0	687	0	0	34	34
	48	0	1047	1527	343	0	172
	96	34	16239	375	0	69	0
Mice treated with thiotepa and injected with *C. albicans*	3	36145	379166	2114	104	16114	0
	12	83333	1156250	3468	0	197916	0
	24	0	208	312	104	0	0
	48	0	104	416	73	31	0
	96	301	1333333	161770	382	590	347
Mice treated with methotrexate and injected with *C. albicans*	3	38541	1645833	110416	12291	479166	2083
	12	3197	25729	10416	3232	2218	104
	24	3125	22916	7291	3229	656	104
	48	729	3125	6250	73	0	177
	96	34	208333	659	208	34	0

Data abstracted from Ghannoum, M.A., Khatter, M., Kayed, F., and Shridar, P., *Mykosen*, 28, 369, 1985. With permission.

REFERENCES

1. **Meyer, S.A., Ahearn, D.G., and Yarrow, D.,** Genus 4. *Candida* Berkhout, in *The Yeasts, A Taxonomic Study,* Kreger-van Rij, N.J.W., Ed., Elsevier Biochemical, Amsterdam, 1984, 585.
2. **Hurley, R., De Louvois, J., and Mulhall, A.,** Yeasts as human and animal pathogens, in *The Yeasts,* Vol. 1, 2nd ed., Rose, A.H. and Harrison, J.S., Eds., Academic Press, London, 1987, 207.
3. **Wingard, J.R., Merz, W.G., and Saral, R.,** *Candida tropicalis* a major pathogen in immunocompromised patients, *Ann. Intern. Med.,* 91, 539, 1979.
4. **Kiehn, T.E., Edwards, F.F., and Armstrong, D.,** The prevalence of yeasts in clinical specimens from cancer patients, *Am. J. Clin. Pathol.,* 73, 518, 1980.
5. **Toala, P., Schroeder, S.A., Daly, K., and Finland, M.,** *Candida* at Boston City Hospital, *Arch. Intern. Med.,* 126, 983, 1970.
6. **Ghannoun, M.A., Motawy, M.S., Mubarak, A.L., and Al-Awadhi, H.A.,** Incidence and susceptibility of yeasts isolated from cancer patients in Kuwait, *Mykosen,* 28, 619, 1985.
7. **Ghannoun, M.A., Sharif, H.F., and Al-Gharreer, H.,** Sensitivity of clinical yeast isolates in Kuwait against a number of anti-fungal agents, *Mykosen,* 27, 402, 1984.
8. **Aisner, J., Schimpff, S., Sutherland, J., Young, V., and Wiernik, P.,** *Torulopsis glabrata* infections in patients with cancer, *Am. J. Med.,* 61, 23, 1976.
9. **Dolan, T.,** A practical approach to identification of yeast-like organisms, *Am. J. Clin. Pathol.,* 55, 580, 1971.
10. **Berkowitz, I.D., Robboy, S.J., Karchmer, A.W., and Kunz, L.J.,** *Torulopsis glabrata* fungemia — a clinical pathological study, *Medicine,* 58, 430, 1979.
11. **Rippon, J.W.,** Opportunistic infections: A. yeasts: candidiasis and the pathogenic yeasts, in *Medical Mycology: The Pathogenic Fungi and the Pathogenic Actinomycetes,* Rippon, J.W., Ed., W.B. Saunders, Philadelphia, 1982, 484.
12. **Odds, F.C.,** *Candida and Candidosis,* Leicester University Press, Leicester, 1979.
13. **Hantschke, D. and Zobel, M.,** The reaction of the physiological vaginal flora to topical antimycotics, *Mykosen,* 22, 267, 1979.
14. **Strom, B.G., Beaudry, R., and Morin, F.,** Yeast overgrowth in operated stomachs, *J. Can. Assoc. Radiol.,* 29, 161, 1978.
15. **Crozier, W.J. and Coats, H.,** A case of onychomycosis due to *Candida ravautii, Aust. J. Dermatol.,* 18, 139, 1977.
16. **Nitzulescu, V. and Niculescu, M.,** Three cases of ocular candidiasis caused by *Candida lipolitica, Arch. Roum. Pathol. Exp. Microbiol.,* 35, 269, 1976.
17. **Khan, Z.U., Misra, V.C., Randhawa, H.S., and Damodaran, V.,** Pathogenicity of some ordinarily harmless yeasts for cortisone-treated mice, *Sabouraudia,* 18, 319, 1980.
18. **Barnett, J.A., Payne, R.W., and Yarrow, D.,** *Yeasts Characteristics and Identification,* Cambridge University Press, Cambridge, 1983, 29.
19. **Kreger-van Rij, N.J.W.,** Classification of yeasts, in *The Yeasts,* 2nd ed., Rose, A.H. and Harrison, J.S., Eds., Academic Press, London, 1987, 5.
20. **Yarrow, D. and Meyer, S.A.,** Proposal for amendment of the diagnosis of the genus *Candida.* Berkhout nom. cons., *Int. J. Syst. Bacteriol.,* 28, 611, 1978.
21. **Barnett, J.A., Payne, R.W., and Yarrow, D.,** *A Guide to Identifying and Classifying Yeasts,* Cambridge University Press, Cambridge, 1979.
22. **Lodder, J.,** *The Yeasts: a Taxonomic Study,* Lodder, J., Ed., Elsevier/North Holland, Amsterdam, 1970, 1.
23. **Lodder, J. and Kreger-van Rij, N.J.W.,** *The Yeasts: a Taxonomic Study,* Elsevier/North Holland, Amsterdam, 1952.
24. **Kurtzman, C.P. and Phaff, H.J.,** Molecular taxonomy, in *The Yeasts,* Vol. 1, 2nd ed., Rose, A.H. and Harrison, J.S., Eds., Academic Press, London, 1987, 63.
25. **Winner, H.I. and Hurley, R.,** Taxonomy, in *Candida Albicans,* J. & A. Churchill, London, 1964, Chap 1.
26. **Haley, L.D. and Callaway, C.S.,** *Laboratory Methods in Medical Mycology,* Center for Disease Control, Atlanta, GA, 1978.
27. **McGinnis, M.R.,** Yeast Identification, in *Laboratory Handbook of Medical Mycology,* Academic Press, New York, 1980, 337.
28. **Cooper, B.H. and Silva-Hunter, M.,** Yeasts of medical importance, in *Manual of Clinical Microbiology,* Lennette, E.H., Balows, A., Hausler, W.J., Jr., and Truant, J.P., Eds., American Society for Microbiology, Washington, D.C., 1980, 526.
29. **Chmel, H. and Louria, D.B.,** Antifungal antibiotics: mechanisms of action, resistance, susceptibility testing, and assays of activity in biological fluids, in *Antibiotics in Laboratory Medicine,* Lorian, V., Ed., Williams % Wilkins, Baltimore, MD, 1980, 170.
30. **Hopfer, R.L. and Fainstein, V.,** Antifungal agents, in *Clinical Laboratory Annual,* Homburger, H.A., Batsakis, J.G., Eds., Appleton-Century-Crofts, New York, 1984, 425.

31. **McGinnis, M.R.,** Susceptibility testing and bioassay procedures, in *Laboratory Handbook of Medical Mycology,* Academic Press, New York, 1980, 412.
32. **Shadomy, S. and Espinel-Ingroff, A.,** Susceptibility Testing with antifungal agents, in *Manual of Clinical Microbiology,* Lennette, E.H., Balows, A., Hausler, A., Jr., and Truant, J.P., Eds., American Society for Microbiology, Washington, D.C., 1980, 562.
33. **Bowman, P.I. and Ahearn, D.G.,** Evaluation of commercial systems for the identification of clinical yeast isolates, *J. Clin. Microbiol.,* 4, 49, 1976.
34. **Cooper, B.H., Johnson, J.B., and Thaxton, E.S.,** Clinical evaluation of the Uni-Yeast-Tek system for rapid presumptive identification of medically important yeasts, *J. Clin. Microbiol.,* 7, 349, 1978.
35. **Land, G.A., Harrison, B.A., Hulme, K.L., Cooper, B.H., and Byrd, J.C.,** Evaluation of a new APi 20C strip for yeasts identification against a conventional method, *J. Clin. Microbiol.,* 10, 357, 1979.
36. **Oblack, D.L., Rhodes, J.C., and Martin, W.J.,** Clinical evaluation of the Auto Microbic System Yeast Biochemical Card for rapid identification of medically important yeasts, *J. Clin. Microbiol.,* 13, 351, 1981.
37. **Shinoda, T., Kaufman, L., and Padhye, A.A.,** Comparative evaluation of the Iatron Serological *Candida* Check Kit and the APi 20C Kit for the identification of the medically important *Candida* species, *J. Clin. Microbiol.,* 13, 513, 1981.
38. **Kostiala, I., Kostiala, A., and Kahanpaa, A.,** Oral mycoses and their treatment, *Acta Odontol Scand.,* 37, 87, 1979.
39. **Arendorf, T.M. and Walker, D.M.,** Oral candidal populations in health and disease, *Br. Dent. J.,* 147, 267, 1979.
40. **Carpenter, A.,** Studies on *Candida.* I. Identification of 100 yeast-like fungi isolated from children, *Am. J. Clin. Pathol.,* 25, 98, 1955.
41. **Barlow, A.J. and Chattaway, F.W.,** Observations on the carriage of *Candida albicans* in man, *Br. J. Dermatol.,* 81, 103, 1969.
42. **Marks, M.I., Marks, S., and Brazeau, M.,** Yeast colonization in hospitalized and nonhospitalized children, *J. Pediatr.,* 87, 524, 1975.
43. **Stenderup, A. and Pederson, G.T.,** Yeasts of human origin, *Acta Pathol. Microbiol. IMB,* 54, 462, 1962.
44. **Hardie, J.,** Microbial flora of the oral cavity, in *Oral Microbiology and Infectious Disease,* 2nd ed., Schuster, G., Ed., William & Wilkins, Baltimore, 1983, 162.
45. **Vandenbussche, M. and Swinne, D.,** Yeasts oral carriage in denture wearers, *Mykosen,* 27, 431, 1984.
46. **Berdicevsky, I., Ben-Aryeh, H., Szargel, R., and Gutman, D.,** Oral Candida in children, *Oral Surg.,* 57, 37, 1984.
47. **Gentles, J.C. and La Touche, C.J.,** Yeasts as human and animal pathogens, in *The Yeasts,* Vol. 1, 1st ed., Rose, A.H. and Harrison, J.S., Eds., Academic Press, London, 1969, 107.
48. **Arendorf, T.M. and Walker, D.M.,** The prevalence and intra-oral distribution of *Candida albicans* in man, *Arch. Oral. Biol.,* 25, 1, 1980.
49. **Lehner, T.,** Oral candidosis, *Dent. Pract.,* 17, 209, 1967.
50. **Beare, J.M., Cheeseman, E.A., and MacKenzie, D.W.R.,** The association between *Candida albicans* and lesions of sebor-rhoeic eczema, *Br. J. Dermatol.,* 80, 675, 1968.
51. **Gergely, L. and Uri, J.,** The mycotic flora of the healthy mouth, *Arch. Oral. Biol.,* 3, 125, 1961.
52. **Schmitt, J.A.,** Epidemiological investigations of oral *Candida albicans, Mycopathol. Mycol. Appl.,* 43, 65, 1971.
53. **Williamson, J.J.,** Diurnal variation of *Candida albicans* counts in saliva, *Aust. Dent. J.,* 17, 54, 1972.
54. **Gergely, L. and Uri, J.,** Day-by-day variation in the mycotic flora of the mouth, *Arch. Oral. Biol.,* 11, 15, 1966.
55. **Budtz-Jörgensen, E., Stenderup, A., and Grabowski, M.,** An epidemiologic study of yeasts in elderly denture wearers, *Community Dent. Oral. Epidemiol.,* 3, 115, 1975.
56. **Cawson, R.A.,** Denture sore mouth. II. The role of *Candida, Dent. Pract.,* 16, 138, 1965.
57. **Marples, M.J. and Di Menna, M.E.,** The incidence of *Candida albicans* in Dunedin, New Zealand, *J. Pathol. Bacteriol.,* 64, 497, 1952.
58. **Shuttleworth, C.W. and Gibbs, F.J.,** The aetiological significance of *Candida albicans* in chronic angular cheilitis and its treatment with nystatin, *Br. Dent. J.,* 108, 354, 1960.
59. **Neil, D.J.,** Symposium on denture sore mouth. I. An aetiological review, *Dent. Pract. (Bristol),* 16, 135, 1965.
60. **Ritchie, G.M., Fletcher, A.M., Main, D.M.G., and Prophet, A.S.,** The etiology, exfoliative cytology, and treatment of denture stomatitis, *J. Prosthet. Dent.,* 22, 185, 1969.
61. **Van Reenen, J.F.,** Microbiologic studies on denture stomatistis, *J. Prosthet. Dent.,* 30, 493, 1973.
62. **Samaranayake, L. and MacFarlane, T.W.,** Hypothesis on the role of dietary carbohydrates in the pathogenesis of oral candidosis, *FEMS Microbiol. Lett.,* 27, 1, 1985.
63. **Abramovits, W.,** Resistant oral candidiasis in an infant due to pacifier contamination, *Clin. Pediatr.,* 20, 393, 1981.
64. **Manning, D.J., Coughlin, R.P., and Poskitt, E.M.,** *Candida* in mouth or on dummy, *Arch. Dis. Child.,* 60, 381, 1985.

65. **Castellani, A.,** A case of vaginitis probably due to *Monilia pinoyi* cast. 1910, *J. Trop. Med. Hyg.*, 19, 89, 1916.
66. **Hesseltine, H.C.,** Biologic and clinical import of vulvo-vaginal mycoses, *Am. J. Obstet. Gynecol.*, 34, 855, 1937.
67. **Hesseltine, H.C.,** Vulval and vaginal mycosis and trichomoniasis, *Am. J. Obstet. Gynecol.*, 40, 641, 1940.
68. **Ryley, J.F.,** Pathogenicity of *Candida albicans* with particular reference to the vagina, *J. Med. Vet. Mycol.*, 24, 5, 1986.
69. **Harris, L.J., Pritiken, H.G., Laski, B., Eisen, A., Steiner, J.W., and Schack, L.,** The effect of nystatin (mycostatin) on neonatal candidiasis (thrush): a method of eradicating thrush from hospital nurseries, *Can. Med. Assoc. J.*, 79, 891, 1958.
70. **Hurley, R. and Morris, E.D.,** The pathogenicity of *Candida* species in the human vagina, *J. Obstet. Gyn. Br. Comm.*, 71, 692, 1964.
71. **Gardner, H.L., Dampeer, T.K., and Dukes, C.D.,** The prevalence of vaginitis. A study in incidence. *Am. J. Obstet. Gynecol.*, 73, 1080, 1957.
72. **Raphael, M. and Levy, B.,** The value of vaginal cultures in general practice, *J. R. Coll. Gen. Pract.*, 27, 349, 1977.
73. **Oriel, J.D., Partridge, B.M., Denny, M.J., and Coleman, J.C.,** Genital yeast infections, *Br. Med. J.*, 4, 761, 1972.
74. **Rohatiner, J.J. and Grimble, A.,** Genital Candidiasis and oral contraceptives, *J. Obstet. Gyn. Br. Comm.*, 77, 1013, 1970.
75. **Lopez-Martinez, R., Ruiz-Sanchez, D., and Vertiz-Chavez,** Vaginal candidosis, *Mycopathologia*, 85, 167, 1984.
76. **Ohm, M.J. and Galask, R.P.,** Bacterial flora of the cervix from 100 prehysterectomy patients, *Am. J. Obstet. Gynecol.*, 122, 683, 1975.
77. **Neary, M.P., Allen, J., Okubadejo, O.A., and Payne, D.J.H.,** Preoperative vaginal bacteria and post operative infections in gynaecological patients, *Lancet*, 2, 1291, 1973.
78. **De Louvois, J., Hurley, R., and Stanley, V.C.,** Microbial flora of the lower genital tract during pregnancy: relationship to morbidity, *J. Clin. Pathol.*, 28, 731, 1975.
79. **Slotnick, I.J., Stelluto, M., and Prystowski, H.,** Microbiology of the female genital tract. III. Comparative investigation of the cervical flora of parturients receiving either rectal or vaginal examination, *Am. J. Obstet. Gynecol.*, 85, 519, 1963.
80. **Morris, C.A. and Morris, D.F.,** Normal vaginal microbiology of women of childbearing age in relation to the use of oral contraceptives and vaginal tampons, *J. Clin. Pathol.*, 20, 636, 1967.
81. **Corbishley, C.M.,** Microbial flora of the vagina and cervix, *J. Clin. Pathol.*, 30, 745, 1977.
82. **Goldacre, M.J., Watt, B., Loudon, N., Milne, L.J.R., Loudon, J.D.O., and Vessey, M.P.,** Vaginal microbial flora in normal young women, *Br. Med. J.*, 1, 1450, 1979.
83. **Leegaard, M.,** The incidence of *Candida albicans* in the vagina of "healthy young women": how often do they have symptoms? Possible etiological factors, *Acta Obstet. Gyn. Scand.*, 63, 85, 1984.
84. **Hurley, R., Stanley, V.C., Leask, B.G.S., and de Louvois, J.,** Microflora of the vagina during pregnancy, in *The Microbial Flora of Man*, Skinner, F.A. and Carr, J.G., Eds., Academic Press, London, 1974, 155.
85. **Anyon, C.P., Desmond, F.B., and Eastcott, D.F.,** A study of candida in one thousand and seven women, *N.Z. Med. J.*, 73, 9, 1971.
86. **Hilton, A.L. and Warnock, D.W.,** Vaginal candidiasis and the role of the digestive tract as a source of infection, *Br. J. Obstet. Gynecol.*, 82, 929, 1975.
87. **Wied, G.L., Davis, M.E., Frank, R., Segal, P.B., Meier, P., and Rosenthal, E.,** Statistical evaluation of the effect of hormonal contraceptives on the cytologic smear pattern, *Obstet. Gynecol.*, 27, 327, 1966.
88. **Davis, B.A.,** Vaginal moniliasis in private practice, *Obstet. Gynecol.*, 34, 40, 1969.
89. **Hurley, R., Leask, B.G.S., Faktor, J.A., and de Fonseka, C.P.,** Incidence and distribution of yeast species and of *Trichomonas vaginalis* in the vagina of pregnant women. *J. Obstet. Gyn. Br. Comm.*, 80, 252, 1973.
90. **Mizuno, S.,** Vulvovaginal candidiasis: survey of vulvovaginal candidiasis in Japan, in *Tokyo: Research Committee of Candidiasis*, Education Ministry of Japan, 1961, 19.
91. **Miles, M.R., Olsen, L., and Rogers, A.,** Recurrent vaginal candidiasis, importance of an intestinal reservoir, *JAMA*, 238, 1836, 1977.
92. **Cruickshank, R. and Sharman, A.,** The biology of the vagina in the human subject. I. Glycogen in the vaginal epithelium and its relation to ovarian activity, *J. Obstet. Gynaecol. Br. Emp.*, 41, 190, 1934.
93. **Cruickshank, R. and Sharman, A.,** The biology of the vagina in the human subject II. The vaginal epithelium, *J. Obstet. Gynaecol. Br. Emp.*, 41, 208, 1934.
94. **Benham, R.W.,** Species of *Candida* most frequently isolated from man: methods and criteria for their identification, *J. Chron. Dis.*, 5, 460, 1957.
95. **Wade, J.C. and Schimpff, C.,** Epidemiology and prevention of *Candida* infections, in *Candidiasis*, Bodey, G. and Fainstein, V., Eds., Raven Press, New York, 1985, 111.

96. **Roberts, S.O., Hay, R.J., and MacKenzie, D.W.R.,** The superficial mycoses: Candidosis (candidiasis), in *A Clinician's Guide to Fungal Disease,* Roberts, S.O., Hay, R.J., and MacKenzie, D.W.R., Eds., Marcel Dekker, New York, 1984, chap. 2.
97. **Carmo-Sousa, L.,** Distribution of yeasts in nature, in *The Yeasts,* Vol. 1, 1st ed., Rose, A.H. and Harrison, J.S., Eds., Academic Press, London, 1969, 79.
98. **Lehrer, R.I. and Cline, M.J.,** Interaction of *Candida albicans* with human leukocytes and serum, *J. Bacteriol.,* 98, 996, 1969.
99. **Cohen, R., Roth, F.J., Delgado, E., Ahearn, D.G., and Kalser, M.H.,** Fungal flora of the normal human small and large intestine, *N. Engl. J. Med.,* 280, 638, 1969.
100. **Macy, J.M., Yu, I., Caldwell, C., and Hungate, R.E.,** Reliable sampling method for analysis of the ecology of the human alimentary tract, *Appl. Environ. Microbiol.,* 35, 113, 1978.
101. **Drasar, B.S. and Hill, M.J.,** *Human Intestinal Flora,* Academic Press, New York, 1974.
102. **Bolivar, R. and Bodey, G.P.,** Candidiasis of the gastrointestinal tract, in *Candidiasis,* Bodey, G.P. and Fainstein, V., Eds., Raven, New York, 1985, 118.
103. **McNabb, P.C. and Tomasi, T.B.,** Host defense mechanisms at mucosal surfaces, *Annu. Rev. Microbiol.,* 35, 477, 1981.
104. **Savage, D.C. and Dubos, R.J.,** Localization of indigenous yeast in the murine stomach, *J. Bacteriol.,* 94, 1811, 1967.
105. **Dubos, R., Schaedler, R.W., Costello, R., and Hoet, P.,** Indigenous, normal and autochtonous flora of the gastrointestinal tract, *J. Exp. Med.,* 122, 67, 1965.
106. **Savage, D.C., Dubos, R., and Schaedler, R.W.,** The gastrointestinal epithelium and its autochtonous bacterial flora, *J. Exp. Med.,* 127, 67, 1968.
107. **Brownlee, A. and Moss, W.,** The influence of diet on lactobacilli in the stomach of the rat, *J. Pathol. Bacteriol.,* 82, 513, 1961.
108. **Rosen, T.,** Cutaneous Candidiasis, in *Candidiasis,* Bodey, G.P. and Fainstein, V., Eds., Raven, New York, 1985, 227.
109. **Drouhet, E.,** Quelques aspects mycologiques, pathogeniques et therapeutiques des candidoses, *Bull. Soc. Fr. Dermatol. Syphiligr.,* 67, 646, 1960.
110. **Benham, R.V. and Hopkins, A. McH.,** Yeast-like fungi found on the skin and in the intestine of normal subjects. Survey of 100 persons, *Arch. Dermatol. Syphilol.,* 28, 532, 1933.
111. **Downing, J.G., Nye, R.N., and Cousins, S.M.,** Investigation of the fungous flora of apparently normal skins, *Arch. Dermatol. Syphilol.,* 35, 1087, 1937.
112. **MacKinnon, J.E.,** *Zimologia Medica,* El Siglo Illustrado, Montevideo, 1946.
113. **Di Menna, M.E.,** Non-pathogenic yeasts of the human skin and alimentary tract: a comparative survey, *J. Pathol. Bacteriol.,* 68, 89, 1954.
114. **Croft, C.C. and Black, L.A.,** Yeast-like fungi isolated from normal skins, *J. Lab. Clin. Med.,* 23, 1259, 1938.
115. **Marwin, R.M.,** Relative incidence of *Candida albicans* on the skins of persons with and without skin diseases, *J. Invest. Dermatol.,* 12, 229, 1949.
116. **Granits, J.,** Pilzkulturen aus Reihenuntersuchungen von gesunder epidermis, *Dermatol. Wochenschr.,* 130, 1352, 1954.
117. **Clayton, Y.M. and Noble, W.C.,** Observations on the epidemiology of *Candida albicans, J. Clin. Pathol.,* 19, 76, 1966.
118. **Aly, R., Britz, M.B., and Maibach, H.I.,** Quantitative microbiology of human vulva, *Br. J. Dermatol.,* 101, 445, 1979.
119. **Ludlam, G.B. and Henderson, J.L.,** Neonatal thrush in a maternity hospital, *Lancet,* 1, 64, 1942.
120. **Marples, M.,** The normal microbial flora of the skin, in *The Normal Microbial Flora of Man,* Skinner, F.A. and Carr, J.G., Eds., Academic Press, London, 1974, 7.
121. **Marples, M. and Somerville, P.,** Oral and cutaneous distribution of *Candida albicans* and other yeasts in Rarotonga, Cook Islands, *Trans. R. Soc. Trop. Med.,* 62, 256, 1968.
122. **Marples, M.,** Some observations on the ecology of *Candida albicans,* a potential mammmalian pathogens, *Proc. N. Z. Ecol. Soc.,* 13, 29, 1966.
123. **Kanan, W.M.,** Cement dermatitis and atmospheric parameters in Kuwait, *Br. J. Dermatol.,* 86, 155. 1972.
124. **Noble, W.C. and Somerville, D.A.,** The fungal flora, in *Microbiology of Human Skin,* Noble, W.C., Ed., W.B. Saunders , London, 1974, 206.
125. **Aly, R., Maibach, H.I., Shinefield, H.R., and Strauss, W.G.,** Survival of pathogenic microorganisms on human skin, *J. Invest. Dermatol.,* 58, 205, 1972.
126. **Noble, W.C. and Willie, J.A.,** Carriage of inhibitor-producing organisms on human skin, *J. Med. Microbiol.,* 13, 329, 1980.
127. **Aly, R., Maibach, H.I., Rahman, R., Shinefield, H.R., and Mandel, A.D.,** Correlation of human *in vivo* and *in vitro* cutaneous antimicrobial factors, *J. Infect. Dis.,* 131, 579, 1975.

128. **Bodey, G.P. and Luna, M.**, Skin lesions associated with disseminated candidiasis, *JAMA*, 229, 1466, 1974.
129. **Barnetson, R. and Milne, L.J.**, Skin sampling for *Candida* with adhesive tape, *Br. J. Dermatol.*, 88, 487, 1973.
130. **Connell, G.H. and Skinner, C.E.**, The external surface of the human body as a habitat for non-fermenting, non-pigmented yeasts, *J. Bacteriol.*, 66, 627, 1953.
131. **MacKenzie, D.W.R.**, Yeasts from human sources, *Sabouraudia*, 1, 8, 1961.
132. **Seebacher, C., Hübner, U., and Blaschke-Hellmessen, R.**, Vergleichende Untersuchungen zum Vorkommen von Sprosspilzen auf gesunder und Krankhaft veränderter Haut, *Mykosen*, 14, 371, 1971.
133. **Marples, M.J. and Somerville, D.A.**, 1966 cited by Marples, M.J., The resident coccal flora of human skin, *J. Invest. Dermatol.*, 52, 397, 1969.
134. **Somerville, D.A.**, The normal flora of the skin in different age groups, *Br. J. Dermatol.*, 81, 248, 1969.
.135. **Kirck, J.A. and Remington, J.S.**, Opportunistic invasive fungal infections in patients with leukemia and lymphoma, *Clin. Haematol.*, 5, 249, 1976.
136. **Bodey, G.P.**, Fungal infections complicating acute leukemia, *J. Chron. Dis.*, 19, 667, 1966.
137. **Young, R.C., Bennett, J.E., Geelhoed, G.W., and Levine, A.S.**, Fungemia with compromised host resistance. A study of 70 cases, *Ann. Intern. Med.*, 80, 605, 1974.
138. **McLeod, R. and Remington, J.S.**, Postoperative fungal endocarditis, in *Infections of Prosthetic Heart Valves and Vascular Grafts*, Duma, R.J., Ed., University Park Press, Baltimore, 1977, 163.
139. **Louria, D.B., Stiff, D.P., and Bennett, B.**, Disseminated moniliasis in the adult, *Medicine (Baltimore)*, 41, 307, 1962.
140. **Bernhardt, H.E., Orlando, J.C., Benfield, J.R., Hirose, F.M., and Foos, R.Y.**, Disseminated candidiasis in surgical patients, *Surg. Gynaecol. Obstet.*, 134, 819, 1972.
141. **Rifkind, D., Marchioro, T.L., Schneck, S.A., and Hill, R.B., Jr.**, Systemic fungal infections complicating renal transplantation and immunosuppressive therapy. Clinical, microbiologic, neurologic and pathologic features, *Am. J. Med.*, 43, 28, 1967.
142. **Hughes, J.M. and Remington, J.S.**, Systemic candidiasis, a diagnostic challenge, *Calif. Med.*, 116, 8, 1972.
143. **Edwards, J.E., Lehrer, R.I., Stiehm, E.R., Fischer, T.J., and Young, L.S.**, Severe candidal infections: clinical perspective, immune defense mechanisms, and current concepts of therapy, *Ann. Intern. Med.*, 89, 91, 1978.
144. **Downing, J.G. and Conant, N.F.**, Medical progress: mycotic infections, *N. Engl. J. Med.*, 233, 153, 1945.
145. **Downing, J.G. and Conant, N.F.**, Medical progress: mycotic infections, *N. Engl. J. Med.*, 233, 181, 1945.
146. **Bogen, E. and Kessel, J.**, Monilial meningitis, *Arch. Pathol.*, 23, 909, 1937.
147. **Campbell, C.C. and Saslaw, S.**, Enhancement of growth of certain fungi by streptomycin, *Proc. Soc. Exp. Biol. Med.*, 70, 562, 1949.
148. **Foley, G.E. and Winter, W.D. Jr.**, Increased mortality following penicillin therapy of chick embryos infected with *Candida albicans var. stellatoidea*, *J. Infect. Dis.*, 85, 268, 1949.
149. **Zimmerman, L.E.**, *Candida* and *Aspergillus* endocarditis: with comments on the role of antibiotics in dissemination of fungus disease, *Arch. Pathol.*, 50, 591, 1950.
150. **Behrman, H.T.**, Pruritus vulvae due to aureomycin, *JAMA*, 144, 995, 1950.
151. **Freeman, H.E.**, Pigmented hairy tongue accompanying antibiotherapy, *Arch. Dermatol.*, 65, 99, 1952.
152. **Mangiaracine, A.**, Oral moniliasis following antibiotic therapy — a warning, *N. Engl. J. Med.*, 244, 655, 1951.
153. **Reiches, A.J.**, Antibiotic sensitivity and moniliasis, *Arch. Dermatol. Syphilol.*, 64, 604, 1951.
154. **Wolff, F.W.**, Moniliasis pneumonia following aureomycin therapy, *Lancet*, 1, 1236, 1952.
155. **Kunstadter, R.H., MacLean, H., and Greengard, J.**, Mycotic endocarditis due to *Candida albicans*, *JAMA*, 149, 829, 1952.
156. **Wybel, R.E.**, Mycosis of cervical spinal cord following intrathecal penicillin therapy; report of a case simulating cord tumor, *Arch. Pathol.*, 53, 167, 1952.
157. **Taylor, H. and Rundle, J.A.**, Acute moniliasis of the urinary tract, *Lancet*, 1, 1236, 1952.
158. **Gausewitz, P.L., Jones, F.S., and Worley, G., Jr.**, Fatal generalized moniliasis report of a case, *Am. J. Clin. Pathol.*, 21, 41, 1951.
159. **Keye, J.D., Jr. and Magee, W.E.**, Fungal diseases in a general hospital: a study of 188 patients, *Am. J. Clin. Pathol.*, 26, 1235, 1956.
160. **Torack, R.M.**, Fungus infections associated with antibiotic and steroid therapy, *Am. J. Med.*, 22, 872, 1957.
161. **Baker, R.D.**, Leuukopenia and therapy in leukemia as factors predisposing to fatal mycoses, *Am. J. Clin. Pathol.*, 37, 358, 1962.
162. **Gruhn, J.G. and Sanson, J.**, Mycotic infections in leukemic patients at autopsy, *Cancer*, 16, 61, 1963.
163. **Hutter, R.V.P. and Collins, H.S.**, The occurrence of opportunistic fungus infection in a cancer hospital, *Lab. Invest.*, 11, 1035, 1962.
164. **Schumacher, H.R., Ginns, D.A., and Warren, W.J.**, Fungus infection complicating leukemia, *Am. J. Med. Sci.*, 247, 313, 1964.
165. **Wolfson, J.S., Sober, A.J., and Rubin, R.H.**, Dermatologic manifestations of infections in immunocompromised patients, *Medicine*, 64, 115, 1985.

166. **Solomkin, J.S., Flohr, A., and Simmons, R.L.,** *Candida* infections in surgical patients, *Ann. Surg.*, 195, 177, 1982.
167. **Meunier-Carpentier, F., Kiehn, T.E., and Armstrong, D.,** Fungemia in the immunocompromised host: changing patterns, antigenemia, high mortality, *Am. J. Med.*, 71, 363, 1981.
168. **Cho, S.Y. and Choi, H.Y.,** Opportunistic fungal infection among cancer patients, *Am. J. Clin. Pathol.*, 72, 617, 1979.
169. **Myerowitz, R.L., Pazin, G.J., and Allen, C.M.,** Disseminated candidiasis: change in incidence, underlying diseases, and pathology, *Am. J. Clin. Pathol.*, 68, 29, 1977.
170. **Mazumdar, P.K. and Marks, M.I.,** *Candida albicans* infections in hospitalized children: a survey of predisposing factors, *Clin. Pediatr.*, 14, 123, 1975.
171. **Chen, T.Y. and Webster, J.H.,** Oral monilia study on patients with head and neck cancer during radiotherapy, *Cancer*, 39, 246, 1974.
172. **Klein, J.D., Yamauchi, T., and Horlick, S.P.,** Neonatal candidiasis, meningitis, and arthritis: observations and a review of literature, *J. Pediatr.*, 81, 31, 1972.
173. **Smith, C.B.,** Candidiasis: pathogenesis, host resistance, and predisposing factors, in *Candidiasis*, Bodey, G.P. and Fainstein, V., Eds., Raven, New York, 1985, 53.
174. **Tattersall, M.H.,** Agressive cancer treatment and its role in predisposing to infection, *Eur. J. Cancer*, 11, 9, 1975.
175. **Provost, T.,** Infections associated with T-cell dysfunction, in *Infection and The Compromised Host*, Allen, J.C., Ed., Williams & Wilkins, Baltimore, 1976, 49.
176. **Sharma, S.D. and Remmington, J.S.,** The role of cell-mediated immunity in resistance to infection in the immunocompromised host, in *Proc. 1st Int. Symp. Infections in the Immunocompromised Host—Pathogenesis, Prevention and Therapy*, Verhoef, J., Peterson, P.K., and Quie, P.G., Eds., Elsevier/North-Holland Biomedical Press, Amsterdam, 1980, 59.
177. **Quie, P.G.,** Infection associated with phagocytic cell deficiences, in *Infection and The Compromised Host*, Allen, J.C., Ed., Williams & Wilkins, Baltimore, 1976, 81.
178. **Drachman, R.H.,** Splenic and reticulo-endothelial function and infection, in *Infection and The Compromised Host*, Allen, J.C., Ed., Williams & Wilkins, Baltimore, 1976, 103.
179. **Ruddy, S., Gigli, I., and Austen, K.F.,** The complement system of man (Three Parts), *N. Engl. J. Med.*, 287, 489, 495, 545, 642, 695, 1972.
180. **Klastersky, J.,** *Infections in Cancer Chemotherapy*, Symp. held at the Inst. Jules Bordet, Brussels, Belgium, Pergamon, Oxford, 1975.
181. **Verhoef, J., Peterson, P.K., and Quie, P.G.,** *Proc. 1st Int. Symp. Infections in the Immunocompromised Host — Pathogensis, Prevention and Therapy*, Elsevier/North-Holland Biomedical Press, Amsterdam, 1980.
182. **Allen, J.C.,** *Infection and The Compromised Host*, Williams & Wilkins, Baltimore, 1976.
183. **Rogers, T. and Balish, E.,** Immunity to *Candida albicans*, *Microbial Rev.*, 44, 660, 1980.
184. **Edwards, J.E., Lehrer, R., Stiehm, E.R., Fischer, T.J., and Young, L.,** Severe candidal infections: clinical perspective, immune defense mechanisms, and current concepts of therapy, *Ann. Intern. Med.*, 89, 91, 1978.
185. **Lehrer, R.I.,** The fungicidal mechanisms of human monocytes. I. Evidence for myeloperoxidase-linked and myeloperoxidase-independent candidacidal mechanisms, *J. Clin. Invest.*, 55, 338, 1975.
186. **Lehrer, R.I.,** Measurement of candidacidal activity of specific leukocyte types in mixed cell populations. II. Normal and chronic granulomatous disease eosinophils, *Infect. Immun.*, 3, 800, 1971.
187. **Louria, D.B. and Brayton, R.G.,** Behavior of *Candida* cells within leukocytes. *Proc. Soc. Exp. Biol. Med.*, 115, 93, 1964.
188. **Kagaya, K. and Fukazawa, Y.,** Murine defense mechanism against *Candida albicans* infection. II. Opsonisation, phagocytosis and intracellular killing of *C. albicans*, *Microbiol. Immun.*, 25, 807, 1981.
189. **Richardson, M.D. and Smith, H.,** Resistance of virulent and attenuated strains of *Candida albicans* to intracellular killing by human and mouse phagocytes, *J. Infect. Dis.*, 144, 565, 1981.
190. **Yamamura, M. and Valdimarsson, H.,** Participation of C3 in intracellular killing of *Candida albicans*, *Scand. J. Immunol.*, 6, 591, 1977.
191. **Lehrer, R.I. and Cline, M.J.,** Leukocyte myeloperoxidase deficiency and disseminated candidiasis: the role of myeloperoxidase in resistance to *Candida* infection, *J. Clin. Invest.*, 48, 1478, 1969.
192. **Lehrer, R.I.,** Antifungal effects of peroxidase systems, *J. Bacteriol.*, 99, 361, 1969.
193. **Odeberg, H. and Olsson, I.,** Antibacterial activity of cationic proteins from human granulocytes, *J. Clin. Invest.*, 56, 1118, 1975.
194. **Odeberg, H. and Olsson, J.,** Mechanisms for microbicidal activity of cationic proteins of human granulocytes, *Infect. Immun.*, 14, 1269, 1976.
195. **Lehrer, R.I., Ladra, K.M., and Hake, R.B.,** Nonoxidative fungicidal mechanisms of mammalian granulocytes: demonstration of components with candidacidal activity in human, rabbit and guinea pig leukocytes, *Infect. Immun.*, 11, 1226, 1975.

196. **Drazin, R.E. and Lehrer, R.I.,** Fungicidal properties of a chymotrypsin-like cationic protein from human neutrophils: adsorption to *Candida parapsilosis, Infect. Immun.,* 17, 382, 1977.
197. **Klebanoff, S.J. and Clark, R.A.,** *The Neutrophil: Function and Clinical Disorders,* Elsevier/North-Holland, New York, 1978.
198. **Kernbaum, S. and Bastin, R.,** Corticoides et. fonctions phagocytaires des polynucleaires neutrophiles humains, *Nouv. Presse Med.,* 3, 2386, 1974.
199. **Xanthou, M., Valassi-Adam, E., Kintzonidou, E., and Matsaniotis, N.,** Phagocytosis and killing ability of *Candida albicans* by blood leucocytes of healthy term and preterm babies, *Arch. Dis. Child.,* 501, 72, 1975.
200. **Lehrer, R.I. and Cline, M.J.,** Leukocyte candidacidal activity and resistance to systemic candidiasis in patients with cancer, *Cancer,* 27, 1211, 1971.
201. **Gallin, J.,** Disorders of phagocyte function, in Proc. *1st Int. Symp. Infections in the Immunocompromised Host — Pathogenesis, Prevention and Therapy,* Verhoef, J., Peterson, P.K., and Quie, P., Eds., Elsevier/North-Holland Biomedical Press, Amsterdam, 1980, 37.
202. **Raith, L., Csato, M., and Dobozy, A.,** Decreased *Candida albicans* killing activity of granulocytes from patients with diabetes mellitus, *Mykosen,* 26, 557, 1983.
203. **Davis, A.T., Brunning, R.D., and Quie, P.G.,** Polymorphonuclear leukocyte myeloperoxidase deficiency in a patient with myelomonocytic leukemia, *N. Engl. J. Med.,* 285, 789, 1971.
204. **Graham, G.S.,** The neutrophilic granules of the circulating blood in health and disease — a preliminary report, *N.Y. State J. Med.,* 20, 46, 1920.
205. **Cocchi, P., Mori, S., and Ravina, A.,** Myeloperoxidase-deficient leucocytes in streptococcal infections, *Helv. Paediat. Acta,* 28, 79, 1973.
206. **Lehrer, R.I., Goldberg, L.S., Apple, M.A., and Rosenthal, N.P.,** Refractory megaloblastic anemia with myeloperoxidase-deficient neutrophils, *Ann. Intern. Med.,* 76, 447, 1972.
207. **Breton-Gorius, J., Houssay, D., and Dreyfus, B.,** Partial myeloperoxidase deficiency in a case of preleukaemia. I. Studies of fine structure and peroxidase synthesis of promyelocytes, *Br. J. Haematol.,* 30, 273, 1975.
208. **Breton-Gorius, J., Houssay, D., Vilde, J.L., and Dreyfus, B.,** Partial myeloperoxidase deficiency in a case of pre-leukaemia. II. Defects of degranulation and abnormal bactericidal activity of blood neutrophils, *Br. J. Haematol.,* 30, 279, 1975.
209. **Armstrong, D.,** Infections in patients with neoplastic disease, in *Proc. 1st Int. Symp. Infections in the Immuno-compromised Host — Pathogenesis, Prevention and Therapy,* Verhoef, J., Peterson, P.K., and Quie, P.G., Eds., Elsevier/ North-Holland Biomedical Press, 1980, 129.
210. **Lehrer, R.I.,** The fungicidal activity of human leukocytes, in *Phagocytic Mechanisms in Health and Disease,* Williams, R.C., Jr. and Fudenberg, H.H., Eds., Intercontinental Medical, New York, 1972, 151.
211. **Lehrer, R.I.,** Inhibition by sulfonamides of the candidacidal activity of human neutrophils, *J. Clin. Invest.,* 50, 2498, 1971.
212. **Kostman, R.,** Infantile genetic agranulocytosis, *Acta Paediat. Scand.,* 45 (Suppl), 105, 7, 1956.
213. **Hitzig, W.H.,** The Swiss type of agammaglobulinemia, in *Immunologic Deficiency Diseases in Man,* Bergsma, D. and Good, R.A., Eds., National Foundation-March of Dimes, *Birth Defects,* 4, 82, 1968.
214. **DiGeorge, A.M.,** Congenital absence of the thymus and its immunologic consequences: concurrence with congenital hypoparathyroidism, in *Immunologic Deficiency Diseases in Man,* Bergsma, D. and Good, R.A., Eds., National Foundation-March of Dimes, *Birth Defects,* 4, 116, 1968.
215. **Nezelof, C.,** Thymic dysplasia with normal immunoglobulins and immunologic deficiency: pure alymphocytosis, in *Immunologic Deficiency Diseases in Man,* Bergsma, D. and Good, R.A., Eds., National Foundation-March of Dimes, *Birth Defects,* 4, 104, 1968.
216. **Schoch, E.P.,** Thymic conversion of *Candida albicans* from commensalism to pathogenism, *Arch. Dermatol.,* 103, 311, 1971.
217. **Montes, L.F., Carter, R.E., Moreland, N., and Ceballos, R.,** Generalized cutaneous candidiasis associated with diffuse myopathy and thymoma, *Am. J. Med.,* 204, 351, 1968.
218. **Burford-Mason, A.P., Mathews, R.C., and Williams, J.R.B.,** Transient abrogation of immunosuppression in a patient with chronic mucocutaneous candidiasis following vaccination with *Candida albicans, J. Infect.,* 14, 147, 1987.
219. **Provost, T., Garrettson, L., Zeschke, R.H., Rose, N., and Tomasi, T.B., Jr.,** Combined immune deficiency, autoantibody formation and mucocutaneous candidiasis, *Clin. Immunol. Immunopathol.,* 1, 429, 1973.
220. **Hermans, P.E. and Wilson, W.R.,** Immunoglobulin in deficiency-pathogenesis and types of infections, in *Proc. 1st Int. Symp. Infections in The Immunocompromised Host — Pathogenesis, Prevention and Therapy,* Verhoef, J., Peterson, P.K., and Quie, P., Eds., Elsevier/North-Holland Biomedical Press, Amsterdam, 1980, 5.
221. **LaForce, F.M., Mills, D.M., Iverson, K., Cousins, R., and Everett, E.D.,** Inhibition of leukocyte candidacidal activity by serum from patients with disseminated candidiasis, *J. Lab. Clin. Med.,* 86, 657, 1975.

222. **Diamond, R.D., Krzesicki, R., and Wellington, J.,** Damage to pseudohyphal forms of *Candida albicans* in the absence of serum *in vitro, J. Clin. Invest.*, 61, 349, 1978.

223. **Diamond, R.D. and Krzesicki, R.,** Mechanisms of attachment of neutrophils to *Candida albicans* pseudohyphae in the absence of serum, and subsequent damage to pseudohyphae by microbicidal processes of neutrophils *in vitro, J. Clin. Invest.*, 61, 360, 1978.

224. **Epstein, J.B., Kimura, L.H., Menard, T.W., Truelove, E.L., and Pearsall, N.N.,** Effect of specific antibodies on the interaction between the fungus *Candida albicans* and human oral mucosa, *Arch. Oral Biol.*, 27, 469, 1982.

225. **Waldman, R.H. and Ganguly, R.,** Role of immune mechanisms on secretory surfaces in prevention of infections, in *Infection and The Compromised Host*, Allen, J.C., Ed., Williams & Wilkins, Baltimore, 1976, 29.

226. **Eden, C.S.,** Attachment of *Escherichia coli* to human urinary tract epithelial cells, an *in vitro* test system applied in the study of urinary tract infections, *Scand. J. Infect. Dis.*, Suppl. 15, 5, 1978.

227. **Fubara, E.S. and Freter, R.,** Protection against enteric bacterial infection by secretory IgA antibodies, *J. Immunol.*, 111, 395, 1973.

228. **Jones, G.W. and Rutter, J.M.,** Role of the K88 antigen in the pathogenesis of neonatal diarrhea caused by *Escherichia coli* in piglets, *Infect. Immun.*, 6, 918, 1972.

229. **Williams, R.C. and Gibbons, R.J.,** Inhibition of bacterial adherence by secretory immunoglobulin A: a mechanism of antigen disposal, *Science*, 177, 697, 1972.

230. **Morelli, R. and Rosenberg, L.T.,** The role of complement in the phagocytosis of *Candida albicans* by mouse peripheral blood leukocytes, *J. Immunol.*, 107, 476, 1971.

231. **Meuwissen, H.J., Rhee, M.S., Rynes, R.I., and Pikering, R.J.,** Phagocytosis, chemoluminescence, and intracellular killing of fungi by phagocytes from subjects with deficiency of the second component of complement, *Int. Arch. Allerg. Appl. Immunol.*, 68, 22, 1982.

232. **Miller, M.E. and Nilsson, U.R.,** A familial deficiency of the phagocytosis-enhancing activity of serum related to dysfunction of the fifth component of complement (C5), *N. Engl. J. Med.*, 282, 354, 1970.

233. **Lyon, F.L., Hector, R.F., and Domer, J.E.,** Innate and acquired immune responses against *Candida albicans* in Congenic B10.D2 mice with deficiency of the C5 complement component, *J. Med. Vet. Mycol.*, 24, 359, 1986.

234. **Kozel, T.R., Brown, R.R., and Pfrommer, G.S.,** Activation and binding of C3 by *Candida albicans, Infect. Immun.*, 55, 1890, 1987.

235. **Sterzl, J.,** The opsonic activity of complement in sera without antibody, *Folia Microbiol.*, 8 , 240, 1963.

236. **Kernbaum, S.,** Pouvois phagocytaire et fongicide envers *Candida albicans* des polynucleaires neutrophiles humainsen presence de serum depourvu de C3 et C4, *Ann. Microbiol. (Inst. Pasteur)*, 126, 75, 1975.

237. **Abe, F., Tateyma, M., Shibuya, H., Azumi, N., and Ommura, Y.,** Experimental candidiasis in iron overload, *Mycopathologia*, 89, 59, 1985.

238. **Fletcher, J., Mather, J., Lewis, M., and Whiting, G.,** Mouth lesions in iron-deficient anemia: relationship to *Candida albicans* in saliva and to impairment of lymphocyte transformation, *J. Infect. Dis.*, 131, 44, 1975.

239. **Elin, R.J. and Wolff, S.M.,** Effect of pH and iron concentration on growth of *Candida albicans* in human serum, *J. Infect. Dis.*, 127, 705, 1973.

240. **Caroline, L., Rosner, R., and Kozinn, P.,** Elevated serum iron, low unbound transferrin and candidiasis in acute leukemia, *Blood*, 34, 441, 1969.

241. **Vaughn, V. and Weinberg, E.,** *Candida albicans* dimorphism and virulence: role of copper, *Mycopathologia*, 64, 39, 1978.

242. **Schade, A.L. and Caroline, L.,** An iron-binding component in human blood plasma, *Science*, 104, 340, 1946.

243. **Roth, F.J., Jr., Boyd, C.C., Sagami, S., and Blank, H.,** An evaluation of the fungistatic activity of serum, *J. Invest. Dermatol.* 32, 549, 1959.

244. **Esterly, N.B., Brammer, S.R., and Crounse, R.G.,** *In vitro* inhibition of candidal growth by human serum, *J. Invest. Dermatol.*, 49, 246, 1967.

245. **Masson, P.L.,** *La Lactoferrine. Proteine Des Secretiones Externes et des Leucocytes Neutrophiles*, Editions Arsica S.A., Bruxelles, Belgium, 1970.

246. **Leffell, M.S. and Spitznagel, J.K.,** Fate of human lactoferrin and myeloperoxide in phagocytizing human neutrophils: effects of immunoglobulin G subclasses and immune complexes coated on latex beads, *Infect. Immun.*, 12, 813, 1975.

247. **Odeberg, H., Oloffson, T., and Olsson, I.,** Primary and secondary granule contents and bactericidal capability of neutrophils in acute leukaemia, *Blood Cells*, 2, 543, 1976.

248. **Jakab, G.J., Warr, G.A., and Astry, C.L.,** Alterations of pulmonary defense mechanisms by protein depletion diet, *Infect. Immun.*, 34, 610, 1981.

249. **Cohen, B.E. and Elin, R.J.,** Enhanced resistance to certain infections in vitamin A treated mice, *Plast. Reconstr. Surg.*, 54, 192, 1974.

250. **Provost, T. and Allen, J.C.,** Susceptibility to infection related to deficiencies in the complement system, in *Infection and the Compromised Host,* Allen, J.C., Ed., Williams & Wilkins, Baltimore, 1976, 133.
251. **Lehrer, R.I., Ferrari, L.G., and Patterson-Delafield, J.,** The fungicidal activity of rabbit macrophages (abstract), *Clin. Res.,* 25, 490A, 1977.
252. **Peterson, E.M. and Calderone, R.A.,** Growth inhibition of *Candida albicans* by rabbit alveolar macrophages, *Infect. Immun.,* 15, 910, 1977.
253. **Stanley, V.C. and Hurley, R.,** The growth of Candida species in cultures of mouse peritoneal macrophages, *J. Pathol.,* 97, 357, 1969.
254. **Ozato, K. and Uesaka, I.,** The role of macrophages in *Candida albicans* infection *in vitro, Jpn. J. Microbiol.,* 18, 29, 1974.
255. **Taschdjian, C.L., Toni, E.F., Hsu, K.C., Seelig, M.S., Cuesta, M.B., and Kozinn, P.J.,** Immunofluorescence studies of *Candida* in human reticuloendothelial phagocytes: implications for immunogenesis and pathogenesis of systemic candidiasis, *Am. J. Clin. Pathol.,* 56, 50, 1971.
256. **Leeson, C.R. and Leeson, T.S.,** *Histology,* 3rd ed., W.B. Saunders, Philadelphia, 1976, 299.
257. **Domonkos, A.N.,** *Andrews' Diseases of the Skin — Clinical Dermatology,* 6th ed., W.B. Saunders, Philadelphia, 1971, 1.
258. **Warin, R.P.,** Fungal infections of the skin, *Br. Med. J.,* 2, 1307, 1966.
259. **Freter, R.,** Mechanisms of association of bacteria with mucosal surfaces, in *Ciba Found. Symp. 80, Adhesion Microorganism Pathogenicity,* Pitman Medical, Marshfield, MA, 1981, 36.
260. **Reid, L. and Clamp, J.R.,** The biochemical and histochemical nomenclature of mucus, *Br. Med. Bull.,* 34, 5, 1978.
261. **Edwards, P.A.W.,** Is mucus a selective barrier to macromolecules? *Br. Med. Bull.,* 34, 55, 1978.
262. **Schrank, G.D. and Verwey, W.F.,** Distribution of cholera organisms in experimental *Vibrio cholerae* infections: proposed mechanisms of pathogenesis and antibacterial immunity, *Infect. Immun.,* 13, 195, 1976.
263. **Warshaw, A.L., Bellini, C.A., and Walker, W.A.,** The intestinal mucosal barrier to intact antigenic protein: difference between colon and small intestine, *Am. J. Surg.,* 133, 55, 1977.
264. **Freter, R., Allweiss, B., O'Brien, P.C.M., and Halstead, S.A.,** The role of chemotaxis in the virulence of cholera vibrios, in *Proc. 13th Joint U.S.-Japan Conf. Cholera,* (DHEW Publ. No. NIH 78-1590), Government Printing Office, Washington, D.C., 1978, 152.
265. **Freter, R., O'Brien, P.C.M., and Halstead, S.A.,** Adhesion and chemotaxis as determinants of bacterial association with mucosal surfaces, *Adv. Exp. Med.,* 107, 429, 1978.
266. **Freter, R., O'Brien, P.C.M., and Macsai, M.S.,** Effect of chemotaxis in the interaction of cholera vibrios with intestinal mucosa, *J. Clin. Nutr.,* 32, 128, 1979.
267. **Freter, R., O'Brien, P.C.M., and Macsai, M.,** Correlation between *in vitro* chemotactic attraction of vibrios into the deep intervillous spaces of the intestinal mucosa, in *Proc. 14th Joint U.S.-Japan Conf. Cholera: Symp. Cholera,* Takeya, K. and Zinnaka, Y., Eds., (Japanese Cholera Panel), U.S.-Japan Cooperative Medical Science Program, 1979, 94.
268. **Donaldson, R.M.,** *Handbook of Physiology (Section 6),* Vol. 5, American Physiological Society, Washington, D.C., 1968, 2807.
269. **Gorbach, S.L., Nahas, L., Plaut, A.G., Weinstein, L., Patterson, J.F., and Levitan, R.,** Studies of intestinal microflora. Population control in small bowel, *Gut,* 8, 530, 1967.
270. **Salen, G., Goldstein, F., and Wirts, C.W.,** Malabsorption in intestinal scleroderma. Relation to bacterial flora and treatment with antibiotics, *Ann. Intern. Med.,* 64, 834, 1966.
271. **Kennedy, M.J. and Volz, P.A.,** Ecology of *Candida albicans* gut colonization: inhibition of *Candida* adhesion, colonization, and dissemination from the gastrointestinal tract by bacterial antagonism, *Infect. Immun.,* 49, 654, 1985.
272. **Wolfson, J.S., Sober, A.J., and Rubin, R.H.,** Dermatologic manifestations of infections in immunocompromised patients, *Medicine,* 64, 115, 1985.
273. **Law, E.J., Kim, O., Stieritz, D., and MacMillan, B.G.,** Experience with systemic candidiasis in the burned patient, *J. Trauma,* 12, 543, 1972.
274. **MacMillan, B.G., Law, E.J., and Holder, I.A.,** Experience with *Candida* infections in the burn patient, *Arch. Surg.,* 104, 509, 1972.
275. **Collins, R.N., Braun, P.A., Zinner, S.H., and Kass, E.H.,** Risk of local and systemic infection with polyethylene intravenous catheters, *N. Engl. J. Med.,* 279, 340, 1968.
276. **Carlisle, J.R., Millet, R.E., Cho, C.T., and Vats, T.S.,** Primary cutaneous aspergillosis in a leukemia child, *Arch. Dermatol.,* 114, 78, 1978.
277. **Dennis, J.E., Rhodes, K.H., Cooney, R.D., and Roberts, G.D.,** Nosocomal *Rhizopus* infection (zygomycosis) in children, *J. Pediatr.,* 96, 824, 1980.
278. **Gartenberg, G., Bottone, E., Keusch, G., and Weitzman, I.,** Hospital-acquired mucormycosis *(Rhizopus rhizopodiformis)* of skin and subcutaneous tissue, *N. Engl. J. Med.,* 299, 1115, 1978.

279. **Keys, T., Haldorson, A., Rhodes, K., Roberts, G., and Fifer, E.,** Nosocomial outbreak of *Rhizopus* infections associated with elastoplast wound dressings, *Minn. Morbid Mortal Week. Rep.*, 27, 33, 1978.
280. **Aly, R. and Shinefield, H.R.,** *Bacterial Interference*, CRC Press, Boca Raton, FL, 1982.
281. **Montgomerie, J.Z. and Edwards, J.E. Jr.,** Association of infection due to *Candida albicans* with intravenous hyperalimentation, *J. Infect. Dis.*, 137, 197, 1978.
282. **Curry, C.R. and Quie, P.G.,** Fungal septicemia in patients receiving parenteral hyperalimentation, *N. Engl. J. Med.* 285, 1221, 1971.
283. **Dewilde, T., DeVroey, C., Bosmans, J., Hubens, A., and Vanbreuseghem, R.,** Prevalence of *Candida albicans* in patients receiving total parenteral nutrition, *Sabouraudia*, 20, 169, 1982.
284. **Abel, R., Jr. and Baxter, T.,** *Candida albicans* endophthalmitis due to catheter tip sepsis, *Dent. Med. J.*, 48, 121, 1976.
285. **Maki, D.G. and Martin, W.T.,** Nationwide epidemic of speticemia caused by contaminated infusion products. IV. Growth of microbial pathogens in fluids for intravenous infusion, *J. Infect. Dis.*, 131, 267, 1975.
286. **Dupont, B. and Drouhet, E.,** Cutaneous, ocular, and osteoarticular candidiasis in heroin addicts: new clinical and therapeutic aspects in 38 patients, *J. Infect. Dis.*, 152, 577, 1985.
287. **Podzamczer, D. and Gudiol, F.,** Systemic candidiasis in heroin abusers, *J. Infect. Dis.*, 153, 1182, 1986.
288. **Shapira, Y., Drucker, M., Russell, A., and Yarom, R.,** Candida endocarditis and encephalitis in an infant, *Clin. Pediatr.*, 13. 542, 1974.
289. **Katzenstein, A.A. and Maksem, J.,** *Candida* infection of gastric ulcers: histology, incidence, and clinical significance, *Am. J. Clin. Pathol.*, 71, 137, 1979.
290. **Binder, R.J. and Nelson, J.A.,** *Candida albicans* ulcer within hiatus hernia sac presenting as an ulcerated mass, *Gastroenterology*, 68, 587, 1975.
291. **Spebar, M.J. and Pruitt, B.A.,** Candidiasis in the burned patient, *J. Trauma*, 21, 237, 1981.
292. **Gorbach, S.L., Nahas, L., Lerner, P.I., and Weinstein, L.,** Studies of intestinal microflora, *Gastroenterology*, 53, 845, 1967.
293. **Sharon, A., Berdicevsky, I., Ben-Arych, H., and Gutman, D.,** The effect of chlorhexidine mouth rinses on oral candida in a group of leukemia patients, *Oral Surg.*, 44, 201, 1977.
294. **Allen, J.C.,** Infection complicating neoplastic disease and cytotoxic therapy, in *Infection and The Compromised Host*, Allen, J.C., Ed., Williams & Wilkins, Baltimore, 1976, 151.
295. **Newman, K., Schimpff, S.C., and Wade, J.C.,** Antibiotic prophylaxis of infection for patients with granulocytopenia, in *Proc. 1st Int. Symp. Infection in the Immunocompromised Host—Pathogenesis, Prevention and Therapy*, Verhoef, J., Peterson, P.K., and Quie, P.G., Elsevier/North-Holland Biomedical Press, Amsterdam, 1980, 187.
296. **Bayer, A.S., Edwards, J.E., Seidel, J.S., and Guze, L.B.,** Candida meningitis. Report of seven cases and review of the English literature, *Medicine*, 55, 477, 1976.
297. **De Vita, V.T., Utz, J.P., Williams, T., and Carbone, P.P.,** Candida meningitis, *Arch. Intern. Med.*, 117, 527, 1966.
298. **Silverman, S., Jr., Luangjarmekron, L., and Greenspan, D.,** Occurrence of oral candida in irradiated head and neck cancer patients, *J. Oral Med.*, 39, 194, 1984.
299. **Citron, B.P., Pincus, I.J., Geokas, M.C., and Haverback, B.J.,** Chemical trauma of esophagus and stomach, *Surg. Clin. N. Am.*, 48, 1303, 1968.
300. **Khuroo, M.S., Naik, S.R., Sehgal, S.C., Naik, S., Kaushik, S.P., Katariya, R.N., Datta, D.V., and Mehta, S.K.,** *Candida* infection of the upper gastrointestinal tract superadded upon chemical injury with acids, *Am. J. Gastroenterol.*, 72, 276, 1979.
301. **Tannock, G.W.,** Control of gastrointestinal pathogens by normal flora, in *Proc. Third Int. Symp. Microbiol Ecology*, Klug, M.J. and Reddy, C.A., Eds., *American Society for Microbiology*, Washington, D.C., 1984, 374.
302. **Brown, R.L.,** The gastrointestinal tract and stool following aureomycin therapy, *Antibiot. Chemother.*, 2, 5, 1952.
303. **Lipnik, M.J., Kligman, A.M., and Strauss, R.,** Antibiotics and fungus infections, *J. Invest. Dermatol.*, 18, 247, 1952.
304. **Marshall, H.C., Palmer, W.L., and Kirsner, J.B.,** Effects of chemotherapeutic agents on fecal bacteria in patients with chronic ulcerative colitis, *JAMA*, 144, 900, 1950.
305. **Meads, M., Rowe, W.P., and Haslam, N.M.,** Alterations in the bacterial flora of the throat during oral therapy with aureomycin, *AMA Arch. Intern. Med.*, 87, 533, 1951.
306. **Woods, J.W., Manning, I.H., and Patterson, C.N.,** Monilial infections complicating the therapeutic use of antibiotics, *JAMA*, 145, 207, 1951.
307. **Kennedy, M.J. and Volz, P.A.,** Dissemination of yeasts after gastrointestinal inoculation in antibiotic-treated mice, *Sabouraudia*, 21, 27, 1983.
308. **Giunchi, G.,** Candidosis following antibiotic therapy, *Sci. Med. Ital.* (English ed.), 6, 580, 1958.
309. **Seelig, M.S.,** Mechanisms by which antibiotics increase the incidence and severity of candidiasis and alter the immunological defenses, *Bacteriol. Rev.*, 30, 442, 1966.

310. **Seelig, M.S.,** The rationale for preventing antibacterial-induced fungal overgrowth, *Med. Times*, 96, 689, 1968.
311. **Sutter, V.L. and Finegold, S.M.,** The effect of antimicrobial agents on human faecal flora: studies with Cephalexin, Cyclacillin and Clindamycin, in *The Normal Microbial Flora of Man*, Skinner, F.A. and Carr, J.G., Eds., Academic Press, London, 1974, 229.
312. **Selwyn, S.,** Natural antibiosis among skin bacteria as a primary defence against infection, *Br. J. Dermatol.*, 93, 487, 1975.
313. **Phillips, A.W. and Balish, E.,** Growth and invasiveness of *Candida albicans* in the germ-free and conventional mouse after oral challenge, *Appl. Microbiol.*, 14, 737, 1966.
314. **Nishikawa, T., Hatano, H., Ohnishi, N., Sasaki, S., and Nomur, T.,** Establishment of *Candida albicans* in the alimentary tract of the germ-free mice and antagonism with *Escherichia coli* after oral inoculation, *Jpn. J. Microbiol.*, 13, 263, 1969.
315. **Clark, J.D.,** Influence of antibiotics or certain intestinal bacteria on orally administered *Candida albicans* in germ-free and conventional mice, *Infect. Immun.*, 4, 731, 1971.
316. **Balish, E. and Phillips, A.W.,** Growth, morphogenesis, and virulence of *Candida albicans* after oral inoculation in the germ-free and conventional chick, *J. Bacteriol.*, 91, 1736, 1966.
317. **Kennedy, M.J.,** Inhibition of *Candida albicans* by the anaerobic oral flora of mice *in vitro*, *Sabouraudia*, 19, 205, 1981.
318. **Mulligan, M.E., Citron, D.M., McNamara, B.T., and Finegold, S.M.,** Impact of cefoperazone therapy on fecal flora, *Antimicrob. Agents Chemother.*, 22, 226, 1982.
319. **Turner, J.R., Butler, T.F., Johnson, M.E., and Gordee, R.S.,** Colonization of the intestinal tract of conventional mice with *Candida albicans* and treatment with anti-fungal agents, *Antimicrob. Agents Chemother.*, 9, 787, 1976.
320. **De Maria, A., Buckley, H., and Von Lichtenberg, F.,** Gastrointestinal candidiasis in rats treated with antibiotics, cortisone, and azathioprine, *Infect. Immun.*, 13, 1761, 1976.
321. **Ecknauer, R. and Lohrs, U.,** The effect of a single dose of cyclophosphamide on the jejunum of specified pathogen-free and germ-free rats, *Digestion*, 14, 269, 1976.
322. **Guentzel, M.N. and Herrera, C.,** Effects of compromising agents on candidosis in mice with persistent infections initiated in infancy, *Infect. Immun.*, 35, 222, 1982.
323. **Ekenna, O. and Sherertz, R.J.,** Factors affecting colonization and dissemination of *Candida albicans* from the gastrointestinal tract of mice, *Infect. Immun.*, 55, 1558, 1987.
324. **Pleasants, J.R.,** Gnobiotics, in *Handbook of Laboratory Animal Science*, Vol. I, Melby, E.C., Jr. and Altman, N.H., Eds., CRC Press, Cleveland, OH, 1974, 119.
325. **Stone, H.H., Geheber, C.E., Kolb, L.D., and Kitchens, W.R.,** Alimentary tract colonization by *Candida albicans*, *J. Surg. Res.*, 14, 273, 1973.
326. **Stone, H.H., Kolb, L.D., Currie, C.H., Geheber, C.E., and Cuzzell, J.F.,** *Candida* sepsis: pathogenesis and principles of treatment, *Ann. Surg.*, 179, 697, 1974.
327. **Helstrom, P.B. and Balish, E.,** Effect of oral tetracycline, the microbial flora, and the athymic state on gastrointestinal colonization and infection of BALB/c mice with *Candida albicans*, *Infect. Immun.*, 23, 764, 1979.
328. **Balish, E. and Phillips, A.W.,** Growth and virulence of *Candida albicans* after oral inoculation in the chick with a monoflora of either *Escherichia coli* or *Streptococcus faecalis*, *J. Bacteriol.*, 91, 1744, 1966.
329. **Liljemark, W.F. and Gibbons, R.J.,** Suppression of *Candida albicans* by human oral *Streptococci* in gnotobiotic mice, *Infect. Immun.*, 8, 846, 1973.
330. **Freter, R., Brickner, H., Botney, M., Cleven, D., and Aranki, A.,** Mechanisms that control bacterial populations in continuous-flow culture models of mouse large intestinal flora, *Infect. Immun.*, 39, 676, 1983.
331. **Syed, S.A., Abrams, G.D., and Freter, R.,** Efficiency of various intestinal bacteria in assuming normal functions of enteric flora after association with germ-free mice, *Infect. Immun.*, 2, 376, 1970.
332. **Kennedy, M.J. and Volz, P.A.,** Effect of various antibiotics on gastrointestinal colonization and dissemination by *Candida albicans*, *Sabouraudia*, 23, 265, 1985.
333. **Andermont, A., Raiband, P., and Tancrede, C.,** Effect of erythromycin on microbial antagonism: a study in gnotobiotic mice associated with a human fecal flora, *J. Infect. Dis.*, 148, 579, 1983.
334. **Paine, T.F.,** The inhibitory actions of bacteria on *Candida* growth, *Antibiot. Chemother.*, 8, 273, 1958.
335. **Karnaky, K.J.,** Hydrogen ion concentration (pH) of *Monilia Albicans* infection and treatment, *South. Med. J.*, 39, 731, 1946.
336. **Sharp, J.L.,** The growth of *Candida albicans* during antibiotic therapy, *Lancet*, 266, 390, 1954.
337. **Jillson, O.F. and Nickerson, W.J.,** Mutual antagonism between pathogenic fungi. Inhibition of dimorphism in *Candida albicans*, *Mycologia*, 40, 369, 1948.
338. **Nickerson, W.J. and Van Rij, N.J.W.,** The effect of sulfhydryl compounds, penicillin, and cobalt on the cell division mechanism of yeasts, *Biochim. Biophys. Acta*, 3, 461, 1949.
339. **Nickerson, W.J. and Mankowski, Z.,** Role of nutrition in the maintenance of yeast shape in *Candida*, *Am. J. Bot.*, 40, 584, 1953.

340. **Nickerson, W.J. and Chung, C.,** Genetic block in cellular division mechanism of a morphological mutant of a yeast, *Am. J. Bot.,* 41, 114, 1954.
341. **Gewin, H.M. and Friou, G.J.,** Manifestations of vitamin deficiency during aureomycin and chloramphenicol therapy of endocarditis due to *Staphylococcus aureus, Yale J. Biol. Med.,* 23, 332, 1951.
342. **Smith, D.T.,** The disturbance of the normal bacterial ecology by the administration of antibiotics with the development of new clinical syndromes, *Ann. Intern. Med.,* 37, 1135, 1952.
343. **Harris, H.J.,** Aureomycin and chloramphenicol in brucellosis, *JAMA,* 142, 161, 1950.
344. **McGovern, J.J., Parrott, R.H., Emmons, C.W., Ross, S., Burke, F.G., and Rice, E.C.,** The effect of aureomycin and chloramphenical on the fungal and bacterial flora of children, *N. Engl. J. Med.,* 248, 397, 1953.
345. **Luckey, T.D.,** *Germ-free Life and Gnotobiology,* Academic Press, New York, 1963
346. **Schaedler, R.W., Dubos, R., and Costello, R.,** Association of germ-free mice with bacteria isolated from normal mice, *J. Exp. Med.,* 122, 77, 1965
347. **Wagner, M.,** Serologic aspects of germ-free life, *Ann. N. Y. Acad. Sci.,* 78, 261, 1959.
348. **Dubos, R.J., Savage, D.C. and Schaedler, R.W.,** The indigenous flora of the gastrointestinal tract, *Dis. Colon Rectum,* 10, 23, 1967.
349. **Dubos, R.J., Schaedler, R.W., and Costello, R.,** Composition, alterations, and effects of the intestinal flora, *Fed. Proc.,* 22, 1322, 1963.
350. **Abrams, G.D., Bauer, H., and Sprinz, H.,** Influence of the normal flora on mucosal morphology and cellular renewal in the ileum. A comparison of germ-free and conventional mice, *Lab. Invest.,* 122, 355, 1963.
351. **Abrams, G.D. and Bishop, J.E.,** Effect of the normal microbial flora on the resistance of the small intestine to infection, *J. Bacteriol.,* 92, 1604, 1966.
352. **Freter, R. and Jones, G.W.,** Models for studying the role of bacterial attachment in virulence and pathogenesis, *Rev. Infect. Dis.,* 5, 5647, 1983.
353. **Abrams, G.D. and Bishop, J.E.,** Normal flora and leukocyte mobilization, *Arch. Pathol.,* 79, 213, 1965.
354. **Schaedler, R.W., Dubos, R., and Costello, R.,** The development of the bacterial flora in the gastrointestinal tract of mice, *J. Exp. Med.,* 122, 59, 1965.
355. **Krause, W., Matheis, H., and Wulf, K.,** Fungemia and funguria after oral administration of *Candida albicans, Lancet,* 1, 598, 1969.
356. **Dubos, R., Schaedler, R.W., Costello, R., and Hoet, P.,** Indigenous, normal and antochothonous flora of the gastrointestinal tract, *J. Exp. Med.,* 122, 67, 1965.
357. **Dearing, W.H. and Heilman, F.R.,** The effect of antibacterial agents on the intestinal flora of patients. The use of aureomycin, chloromycetin, dihydrostreptomycin, sulfasuxidine and sulfathalidine, *Gastroenterology,* 16, 12, 1950.
358. **Davis, C.P., McAllister, J.S., and Savage, D.C.,** Microbial colonization of the intestinal epithelium in suckling mice, *Infect. Immun.,* 7, 666, 1973.
359. **Field, L.H., Pope, L.M., Cole, G.T., Guentzel, M.N., and Berry, L.J.,** Persistence and spread of *Candida albicans* after intragastric inoculation of infant mice, *Infect. Immun.,* 31, 783. 1981.
360. **Freter, R.,** Bacterial association with the mucus gel system of the gut, in *Microbiology-1982,* Schlessinger, D., Ed., *American Society for Microbiology,* Washington, D.C., 1982, 278.
361. **Pope, L., Cole, G., Guentzel, M., and Berry, L.,** Systemic and gastrointestinal candidiasis of infant mice after intragastric challenge, *Infect. Immun.,* 25, 702, 1979.
362. **Folb, P.I. and Trounce, J.R.,** Immunological aspects of *Candida* infection complicating steroid and immunosuppressive drug therapy, *Lancet,* 2, 1112, 1970.
363. **Hasany, S.M., Basu, P.K., and Kazdan, J.J.,** Production of corneal ulcer by opportunistic and saprophytic fungi. I. The effect of pretreatment of fungi with steroid, *Can. J. Ophthalmol.,* 8, 119, 1973.
364. **Khosla, P.K., Chawla, K.S., Prakash, P., and Mahajan, V.M.,** The effect of dexamethasone and oxytetracyclin on the growth of *Candida albicans, Mykosen,* 21, 342, 1978.
365. **Moore, M.,** *In vivo* and *in vitro* effect of aureomycin hydrochloride on *Syringosporia (Monila, Candida albicans), J. Lab. Clin. Med.,* 37, 703, 1951.
366. **Pappenfort, R.B. and Schnall, E.S.,** Moniliasis in patients treated with aureomycin. Clinical and laboratory evidence that aureomycin stimulates the growth of *Candida albicans, AMA Arch. Intern. Med.,* 88, 729, 1951.
367. **Huppert, M., MacPherson, D.A., and Cazin, J.,** Pathogenesis of *Candida albicans* infection following antibiotic therapy. I. The effect of antibiotics on the growth of *Candida albicans, J. Bacteriol.,* 65, 171, 1953.
368. **Huppert, M. and Cazin, J.,** Pathogenesis of *Candida albicans* infection following antibiotic therapy. II. Further studies of the effect of antibiotics on the *in vitro* growth of *Candida albicans, J. Bacteriol.,* 70, 435, 1955.
369. **Robinson, H.R., Jr.,** Moniliasis complicating antibiotic therapy, *AMA Arch. Dermatol. Syphilol.,* 70, 640, 1954.
370. **Langcake, P., Cornford, C.A., and Pryce, R.J.,** Sterols in potato leaves and their effect on growth and sporulation of phytophthora-infestans, *Trans. Br. Mycol. Soc.,* 63, 573, 1974.
371. **Prickett, P.S. and Massengale, O.N.,** Mycobacterium, ergosterol content of certain species and effect of ergosterol on their growth, *J. Infect. Dis.,* 49, 297, 1931.

372. **Buetow, D.E. and Levedahl, B.H.**, The relationships between steroid molecular structure and cell growth, *Arch. Biochem. Biophys.*, 86, 34, 1960.
373. **Jefferson, W.E., Jr. and Sisco, G.**, The influence of steroids on the growth of *Aspergillus niger* and *Torula utilis*, *J.Gen. Physiol.*, 44, 1029, 1961.
374. **Hechter, O. and Lester, G.**, Cell permeability and hormone action, *Recent Prog. Horm. Res.*, 16, 139, 1960.
375. **Lester, G., Stone, D., and Hechter, O.**, The effect of deoxycorticosterone and other steroids on *Neurospora crassa*, *Arch. Biochem.*, 75, 196, 1958.
376. **Hennes, A.R., Muchmore, H.G., McClure, H.G., and Hammersten, J.F.**, *In vitro* inhibition of growth of *M. tuberculosis* by certain 11-oxygenated steroids, *Proc. Soc. Exp. Biol. Med.*, 101, 145, 1959.
377. **Kaben, U. and Neumann, G.**, Do sex steroids, amniotic fluid and lochial secretions favor the growth of *Candida albicans* and *Torulopsis glabrata? Wiss. Z. Univ. Rostock Math. Naturwiss. Reihe*, 19, 707, 1970.
378. **Moursi, H. and Refai, M.**, The effect of hormones on some fungi *in vitro*, *Mykosen*, 11, 747, 1968.
379. **Neumann, G. and Kaben, U.**, Einfluss β von Östriol, Östradiolbenzoat, Progesteron, Lochialsekret und Fruchtwasser auf das Wachstun von Hefepilzen *in vitro*, *Zentrdalbl. Gynaekol.*, 83, 1147, 1971.
380. **Yotis, W. and Stanke, R.**, Bacteriostatic action of progesterone on staphylococci and other microorganisms, *J. Bacteriol.*, 92, 1285, 1966.
381. **Ghannoum, M., Burns, G., and Abu-Elteen, K.**, Growth of *Candida albicans* in dexamethasone-supplemental media, *Sabouraudia*, 23, 313, 1985.
382. **Ghannoum, M.A., Mudher, S., and Burns, G.**, Incorporation of dexamethasone by *Candida albicans*, *Microbios*, 42, 103, 1985.
383. **Kass, E.H., Kendrick, M., and Finland, M.**, Effect of certain corticosteroids and of growth hormones of nucleoproteins of lymph nodes, *Ann. N.Y. Acad. Sci.*, 56, 737, 1953.
384. **Pingleton, W., Bone, R., Kerby, G., and Ruth, W.**, Orpharyngeal candidiasis in patients treated with triamcinolone acetonide aerosol, *J. Allerg. Clin. Immunol.*, 60, 254, 1977.
385. **Raab, W.**, Effects of local corticosteroids in skin infections, *Dermatologica*, 152 (Suppl. 1), 67, 1976.
386. **Seligmann, E.**, Virulence enhancement of *Candida albicans* by antibiotics and cortisone, *Proc. Soc. Exp. Biol. Med.*, 83, 778, 1953.
387. **Stern, K. and Davidsohn, I.**, Effect of estrogen and cortisone on immune hemo-antibiotics in mice of inbred strains, *J. Immunol.*, 74, 479, 1955.
388. **Winter, W. and Foley, G.**, Enhancement of *Candida* infection: differential distribution of renal lesions in mice treated with aureomycin, *J. Infect. Dis.*, 98, 150, 1956.
389. **Ghannoum, M.A. and Abu-Elteen, K.**, Effect of growth of *Candida* spp. in the presence of various glucocorticoids on the adherence to human buccal epithelial cells, *Mycopathologia*, 98, 171, 1987.
390. **Michel, J., Jacobs, J.Y., and Sacks, T.**, Bactericidal effect of combinations of antimicrobial drugs and antineoplastic antibiotics against gram-negative bacilli, *Antimicrob. Agents Chemother.*, 16, 761, 1979.
391. **Jacobs, J.Y., Michel, J., and Sacks, T.**, Bactericidal effect of combinations of antimicrobial drugs and antineoplastic antibiotics against *Staphylococcus aureus*, *Antimicrob. Agents Chemother.*, 15, 580, 1979.
392. **Moody, M.R., Morris, M.J., Young, V.M., Moye L.A., III, Schimpff, S.C., and Wiernik, P.H.**, Effect of two cancer chemotherapeutic agents on the antibacterial activity of three antimicrobial agents, *Antimicrob. Agents Chemother.*, 14, 737, 1978.
393. **Hamilton-Miller, J.M.T.**, Antimicrobial activity of 21 antineoplastic agents, *Br. J. Cancer*, 49, 367, 1984.
394. **Goldschmidt, M.C. and Bodey, G.**, Effect of chemotherapeutic agents upon microorganisms isolated from cancer patients, *Antimicrob. Agents Chemother.*, 1, 348, 1972.
395. **Metcalfe, D. and Hughes, W.**, Effect of methotrexate on group A Beta hemolytic *Streptococci* and Streptococcal infection, *Cancer*, 30, 588, 1972.
396. **Land, G.A., Hulme, K.L., and Chaffin, W.L.**, The effect of selected antineoplastic agents on the morphology of *Candida albicans* 5865, *Can. J. Microbiol.*, 26, 812, 1980.
397. **Ghannoum, M.A. and Al-Khars, A.**, Effect of antineoplastic agents on the growth and ultrastructure of *Candida albicans*, *Mykosen*, 27, 452, 1984.
398. **Ghannoum, M.A.**, Effect of antineoplastic agents on growth, morphology and metabolism of *Torulopsis glabrata*, *Mycopathologia*, 95, 175, 1985.
399. **Ghannoum, M.A., Abu-Elteen, K.H., and Motawy, M.S.**, The role of cancer chemotherapy and X-irradiation on the adherence of *Candida* spp. to human buccal epithelial cells, Cong. Int. Soc. Human Animal Mycology (ISHAM), Barcelona, 1988.
400. **Ghannoum, M.A., Khattar, M., Kayed, F., and Shridar, P.**, The influence of antineoplastic agents on experimental candidosis in mice, *Mykosen*, 28, 369, 1985.

Chapter 3

PATHOGENICITY DETERMINANTS IN *CANDIDA*

TABLE OF CONTENTS

I. INTRODUCTION

The association between the host and the parasite is often classified into three categories: beneficial parasitism, where both benefit and which encompasses relationships such as mutualism and symbiosis; parasitism injurious to the host (referred to sometimes as antibiosis),[1] where one partner benefits at the expense of the other; and commensalism, where only one organism derives benefit, living in close proximity to or on the surface of the other organism without causing any damage. The margins of these categories are not sharply defined and aspects of each may exist to some degree in the others. An apparent commensal association often proves, on closer examination, to be actually injurious or symbiotic. This complex relationship becomes more intricate when the balance controlling the host-parasite equilibrium is upset.

In Chapter 2 the host-parasite relationship which exists between *Candida* and man, and the factors which disturb the "detente" existing between them, with resultant injurious parasitism, have been described. It is the purpose of this chapter to review, in detail, the pathogenicity determinants and pathogenicity pattern of *Candida*.

II. PATHOGENICITY DETERMINANTS

Pathogenicity is defined as the capacity of microorganisms to cause disease. Whether disease occurs or is prevented depends on the outcome of interactions between the microbe and the host. This in itself is decided by multifactorial determinants produced by the pathogen, and the susceptibility of the host.[2] The concept that pathogenicity is a microbial property alone is inadequate as a basis for attempting to understand the process of infection.[3] This is particularly true for *Candida* infection where even the most virulent *Candida* species, *C. albicans,* occurs ordinarily as a commensal in the human host, and is able to invade and damage tissues only when host defenses are impaired.[4] In this review we shall deal only with the pathogenicity determinants of *Candida*.

A. SPECIES AND STRAINS

In the early 20th century, only *C. albicans* of the genus *Candida* was believed to be pathogenic.[5,6] However, more careful investigations performed in the 50s and 60s showed that pathogenicity extends to other species in this genus.[7-13]

Based on clinical experience and various experimental models, ten species of the genus *Candida* are considered pathogenic (see also Chapter 2).[4,7-9,14-22] Figure 1 shows scanning electron micrographs of six different pathogenic *Candida* species which predominate in clinical samples in Kuwait. Other "nonpathogenic" *Candida* species are likely to be associated with pathological conditions in the future. This will become prevalent with the increasing use of pharmaceuticals and/or surgical practices which render individuals immunologically incompetent for purposes of organ transplantation and for therapy of autoimmune disease.[23] Hurley and co-workers[18,24] demonstrated the adaptation of saprophytic yeasts to colonization of hosts. Similar work carried out by Khan et al.[25] emphasized the pathogenic potential of ordinarily harmless fungi to immunosuppressed hosts.

Although some workers could not detect virulence differences between *C. albicans* strains, others reported strain variations in the virulence of this yeast for experimental animals.[9,26-29] Since pathogens must go through a number of processes (i.e., enter the host, multiply in host tissues, resist or not stimulated host defenses, and damage the host) to accomplish disease,[2] and given that each process is complex, several determinants are usually involved in the overall effect. Lack of one member of the full complement may result in considerable attenuation of the strain concerned.[30] Taking this into consideration, investigators tried to relate the differences in virulence between strains of *C. albicans* to differences in pathogenicity determinants, namely, adherence, proteinase, phospholipase, and germ-tube production.[31]

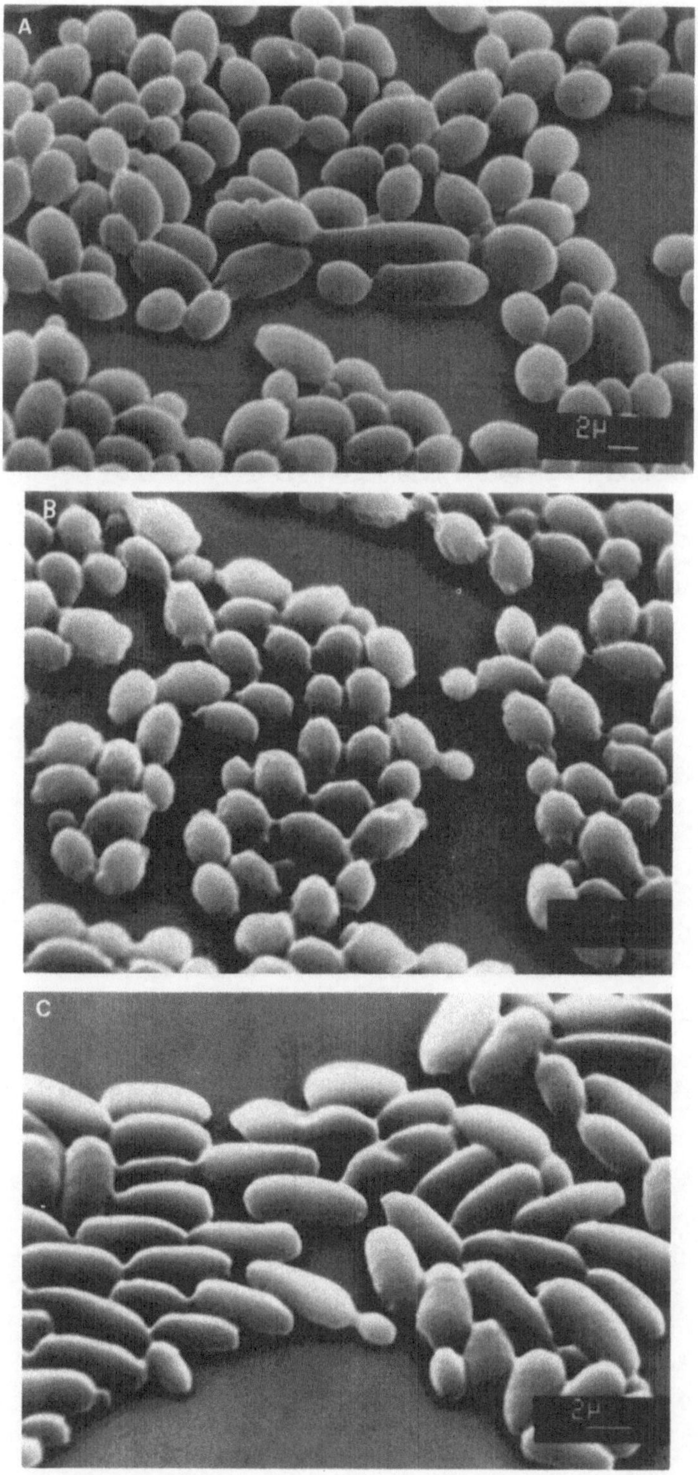

FIGURE 1. Scanning electron micrographs of blastosperes of six different pathogenic *Candida* species. (A) *C. albicans*, (B) *C. tropicalis*, (C) *C. krusei*, (D) *C. pseudotropicalis*, (E) *C. parapsilosis*, (F) *C. lusitania*. All micrographs are of the same magnification.

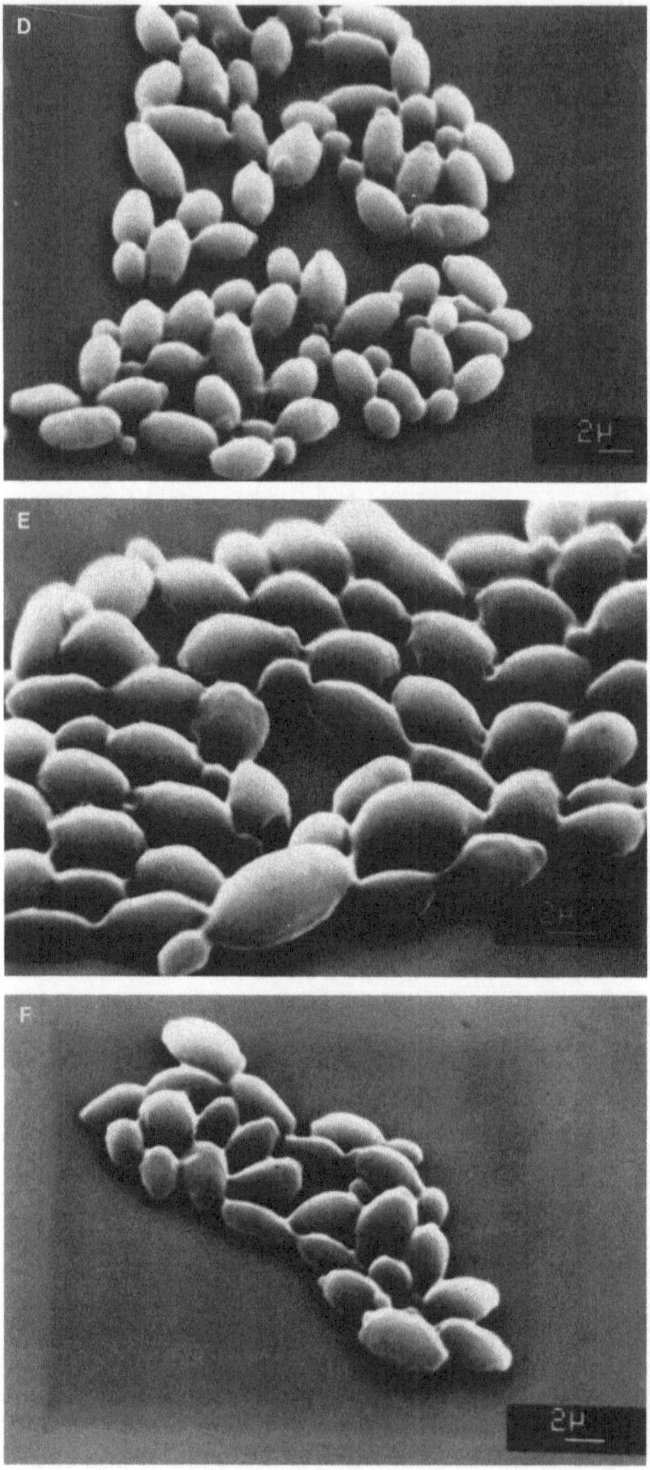

FIGURE 1D, 1E, and 1F

McCourtie and Douglas[32] compared the adherence to acrylic and human buccal epithelial cells of seven strains of *C. albicans* isolated from active infections (I strains) and two strains

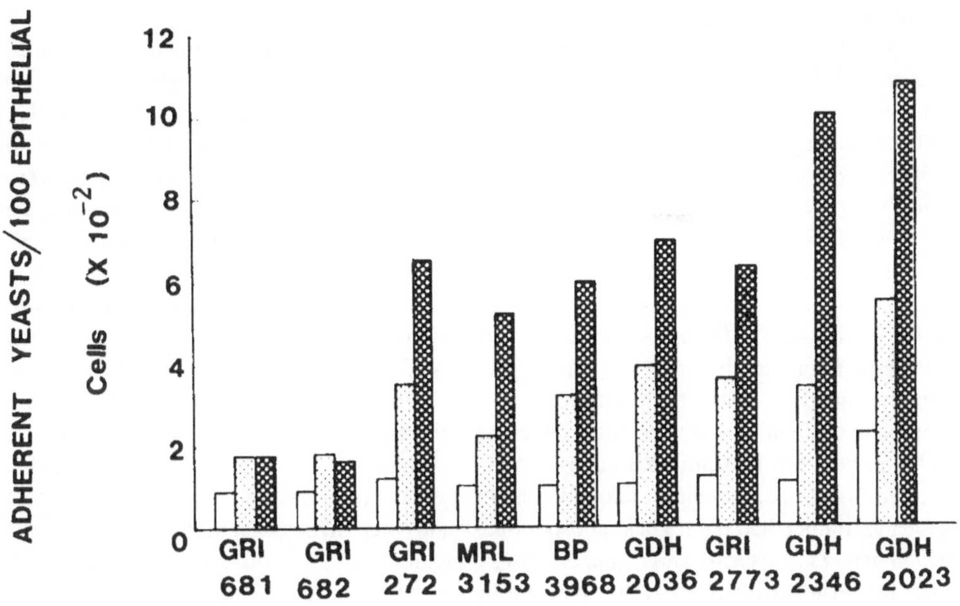

FIGURE 2. Adherence of nine strains of *C. albicans* to buccal epithelial cells after growth in medium containing 50 m*M* glucose, 500 m*M* sucrose, or 500 m*M* galactose. Standard errors range from 2.3 to 9.3% of mean value (From McCourtie, J. and Douglas, L.J., *Infect. Immun.*, 45, 6, 1984. With permission.)

obtained from asymptomatic carriers (C strains). The adherence of these strains was compared following growth in the presence of low or high concentrations of sugars (Figure 2). No difference in the adherence of these strains was detected when growth in the presence of low concentrations was carried out. In contrast, these strains showed significant variation in adherence when growth was at high concentrations of sugars (Figure 2). The interpretation given was that I strains are able to modify their surface composition in response to growth in the presence of high sugar concentrations. Such modification can enhance both their ability to adhere to surfaces and their virulence. C strains lack this capability, or possess it to a lower degree, which is reflected in lower adheribility and virulence. Examining the variability of the adherence of *C. albicans* strains to human buccal epithelial cells, Kearns et al.[33] detected slight differences in adherence between four laboratory strains of this yeast which differed in virulence to mice, and between fresh commensals and thrush isolates. These strains were used after only two subcultures in the laboratory, to avoid the possibility that changes in surface properties related to adherence may have occurred, thus affecting the results. However, such differences did not consistently show that the thrush isolates had higher adherence values than commensal strains. They concluded that to try and explain virulence differences between strains on the basis of variations in adherence alone is not possible and stressed the fact that other properties of the pathogenic strains, or of the host from which they were isolated may be responsible for the pathogenicity.

Proteinase activity related to virulence of various strains of *C. albicans* is another factor which was investigated. Since the majority of strains of *C. albicans* excrete an inducible proteinase, a number of workers thought it essential to evaluate this in relation to virulence.[34-38] Thus, MacDonald and Odds,[39] using a proteinase-secreting strain of *C. albicans* and a proteinase-deficient mutant, suggested that extracellular proteinase is an important factor in the degree of virulence. Similarly, Kwon-Chung et al.[40] studied the relationship between extracellular proteinase and the virulence for mice in *C. albicans* using a set of three isolates: proteinase-producing parent (C9), proteinase-deficient mutant (C9M1) derived from C9, and a spontaneous revertant (C9M1M). Isolate C9 produced a high level of proteinase *in vitro* and caused 100% mortality within 21 d. Isolate C9M1 produced no detectable enzymes, and all mice survived until day 22, with 30% mortality occuring by day 30. The proteinase activity of isolate C9M1M was about 44% that of the wild type and it caused fatal infection in 90% of the mice infected by day 30. These findings show that production of extracellular proteinase is a vital determinant in pathogenicity of various strains. Schreiber et al.[41] contradict these findings. Although they were able to detect proteinase in 97% of *C. albicans* isolates studied, they were unable to correlate the amount of proteolytic activity and the degree of invasiveness of these strains. It is our belief that these differences are due to the fact that most of these studies on the pathogenesis of *Candida* infections have focused on one factor alone, and related it to virulence, and/or that only one, two, or a few strains of *C. albicans* (most often without strain differentiation) were used in such studies. Exceptions are the studies by Barrett-Bee et al.[42] and Ghannoum and Abu-Elteen.[43] The former measured the amount of phospholipase activity in different isolates of *C. albicans* and attempted a correlation between their adherence abilities and pathogenicity. They found a correlation between phospholipase activity, pathogenicity, and adherence to epithelial cells. Thus, *C. albicans* isolates which adhered most strongly to buccal epithelial cells and were most pathogenic in mice had the highest phospholipase activities. In contrast, nonpathogenic yeast *(Saccharomyces cerevisiae* and *C. parapsilosis),* including *C. albicans* isolates which did not adhere and did not kill mice, had lower phospholipase activities. Some 53 isolates of *C. albicans,* representing seven strain types (differentiation based on resistance to sodium selenite, boric acid, cetrimide, sodium periodate, and silver nitrate),[44,45] were tested in our laboratory for their proteinase production and this parameter was correlated with adherence to buccal epithelial cells in 23 isolates and lethality to mice for 14 isolates. Our results showed a variation in proteinase production and adherence both among isolates of the same strain type and different strain types. All isolates tested, irrespective of strain type or source, excreted an inducible proteinase and showed a tendency towards adherence, albeit to different extent (Table 1). This suggests that these factors are equally important for both pathogenicity and commensalism, an observation made by Kearns et al.[33] However. *C. albicans* isolates which adhered most strongly to buccal epithelial cells had the highest relative proteinase activities and showed higher tissue colonization.

Recently, Borg and Ruchel[259] traced an acid proteinase from *Candida* spp. in the initial stages of the pathogenesis of candidosis. Using immunoscanning electron microscopy these authors detected proteinase antigens on the surface of adhering blastoconidia and invading filamentous cells of *C. albicans* serotype A. *C. tropicalis* isolates showed the same distribution of proteinase antigens. In contrast, filamentous cells of serotype B, as well as, *C. parapsilosis* carried little or no proteinase antigen. In the same study, these authors found a correlation between the ability to produce proteinase, adherence, and tissue invasion.[259]

Factors, such as the host immune status, debilitated state, therapy, and other environmental conditions which promote adhesion as well as proteinase and phospholipase production are important in the mechanism of pathogenicity.[48-52] Thus, although *C. albicans* occurs in both patients and healthy individuals, the propensity for the yeast to remain saprophyte or become infective, depends on a dynamic balance between the various parameters involved.

To what extent the ability to form germ-tubes by some strains is important in relation to their ability to invade tissue is a controversial point. Thus, a clear relationship between the ability of four strains of *C. albicans* to form germ-tubes and their virulence to mice, was reported by Richardson.[46] Sobel et al.,[47] studying the role of germ-tube formation in the pathogenesis of candidal vaginitis, concluded that although the capacity of *C. albicans* to produce hyphae appears to be an important factor, it is not absolutely essential to be able to cause infection. These workers based their conclusion on studies using a variant strain of *C. albicans* which completely failed to germinate *in vivo*. However, successful colonization of the rat vagina with this strain needed a higher inoculum and the infection tended to be mild. Judging the degree of virulence

TABLE 1
Correlation of Growth, Proteinase Production, and Adherence in Various Strains of *C. albicans*

Strain type	Isolate number	Relative dry weight (%)[a,b]	Relative proteinase (%)[a,b]	No. of yeast cells adhering to 100 BEC ± SE for duplicate	Relative adherence (%)[a]
- -C- -	1	100	100	430 ± 28	100
	2	197	81	NT	NT
	3	93	80	NT	NT
	4	169	73	NT	NT
	5	146	69	300 ± 25	70
	6	115	65	NT	NT
	7	93	58	NT	NT
	8	91	57	270 ± 23	63
	9	73	53	NT	NT
	10	105	52	NT	NT
	11	101	52	NT	NT
	12	112	45	200 ± 20	47
	13	82	44	NT	NT
	14	136	42	NT	NT
	15	103	42	NT	NT
	16	117	42	NT	NT
	17	148	41	NT	NT
	18	101	40	220 ± 21	51
	19	96	38	150 ± 18	35
	20	106	36	NT	NT
	21	90	33	NT	NT
	22	103	32	NT	NT
	23	96	29	NT	NT
	24	129	16	120 ± 16	28
-BC- -	25	111	56	330 ± 23	77
	26	103	56	NT	NT
	27	116	53	NT	NT
	28	107	52	NT	NT
	29	112	51	NT	NT
	30	102	50	NT	NT
	31	122	47	NT	NT
	32	106	42	220 ± 19	51
	33	102	39	NT	NT
	34	93	39	NT	NT
	35	17	35	NT	NT
	36	81	30	NT	NT
	37	107	22	165 ± 18	38
	38	111	19	NT	NT
	39	102	13	140 ± 17	33
- - - - -	40	58	70	370 ± 26	86
	41	100	63	285 ± 25	66
	42	115	49	240 ± 20	56
	43	45	44	NT	NT
	44	88	43	140 ± 16	33
	45	92	43	NT	NT
ABC- -	46	67	55	260 ± 25	60
	47	107	48	190 ± 21	44
	48	72	11	160 ± 17	35
-BCD-	49	81	50	290 ± 21	67
	50	98	33	200 ± 17	47
-B- - -	51	82	53	120 ± 14	28

TABLE 1 (continued)
Correlation of Growth, Proteinase Production, and Adherence in Various Strains of *C. albicans*

Strain type	Isolate number	Relative dry weight (%)[a,b]	Relative proteinase (%)[a,b]	No. of yeast cells adhering to 100 BEC ± SE for duplicate	Relative adherence (%)[a]
	52	22	24	100 ± 11	23
A-C- -	53	82	20	125 ± 14	29

Note: NT, Not tested.

[a] Percentage relative to isolate no. 1, strain type - -C- - =100%.
[b] Values are the average of three separate experiments.

From Ghannoum, M.A., and Abu Elteen, K., *J. Med. Vet. Mycol.*, 24, 407, 1986. With permission.

of strains solely on the basis of their ability to form germ-tubes is not significant, unless a major difference exists, i.e., one strain possesses this ability while the other lacks it completely. Minor differences will not suffice. (See Reference 261.)

B. ADHERENCE

Adherence is a phenomenon shared by both pathogenic and commensal microorganism, which allows them to lodge themselves in tissue surfaces in an effort to establish themselves at such sites.[53] Consequently, commensal microorganisms associate with epithelial surfaces in all areas of the body exposed to the external environment and live in harmony with their host, in equilibrium with the host's resistance mechanism. Some pathogens adhere to epithelial cells only transiently as an initial phase in a step process in which they invade the cells.[54]

Evidence that the adherence of microorganisms to mucosal surfaces is a prerequisite for the pathogenesis of serious microbial infections came from work with bacteria. Work by Gibbons [55-57] and associates on the selective nature of attachment of bacteria to various surfaces and niches of oral and dental tissues gave the first indication with regard to the importance of bacterial adherence to pathogenesis. The role of adherence as a pathogenicity determinant has received much attention, with the publication of a considerable number of research papers, reviews, and books. The reader is referred to several volumes which deal comprehensively with various aspects of microbial adhesion to surfaces.[58-66] Evidence supporting the role of adherence in the infection process is derived from three categories of investigation: (1) studies related to bacterial adherence *in vitro* vs. infectivity *in vivo*.[63] These compare the ability of bacterial strains capable of adhesion *in vitro* to their adherence *in vivo*,[67,68] (2) studies related to sites of infection vs. adherence of bacteria to target tissue cells. These compare bacteria known to colonize a specific organ or tissue site *in vivo* with its ability to do so *in vitro*,[69-72] (3) studies related to the capacity of epithelial cells from infection-prone hosts to bind bacterial pathogens. These compare the binding ability of epithelial cells obtained from hosts susceptible to infection with epithelial cells obtained from hosts not prone to infection.[73-79] The results of these studies strongly suggest that epithelial cell adherence is a vital step in the pathogenesis of bacterial infection.

Compared to the enormous amount of information available on bacterial adhesion, reports on the adherence of fungi to mucosal surfaces are few, with most studies involving the opportunistic fungus *C. albicans*. Indications suggesting the involvement of adherence in the colonization process of *C. albicans* came first from work carried out by Liljemark and Gibbons.[80] These workers observed the attachment of *C. albicans* to rat tongue and cheek cells of conventional and germ-free animals and found that this yeast adheres in lower numbers to

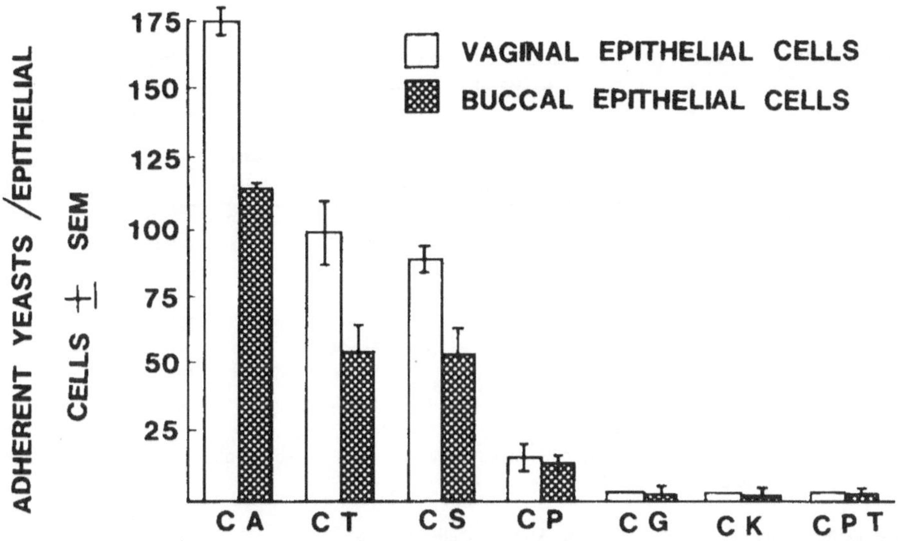

FIGURE 3. Quantitative adherence of the various *Candida* species to vaginal and buccal epithelial cells. (CA) *C. albicans;* (CT) *C. tropicalis;* (CS) *C. stellatoidea;* (CP) *C. parapsilosis;* (CG) *C. guilliermondii;* (CK) *C. krusei;* (CpT) *C. pseudotropicalis;* SEM, standard error of the mean. (From King, R.D., Lee, J.C., and Morris, A.L., *Infect. Immun.*, 27, 667, 1980. With permission.)

epithelial cells from conventional rats than to those from germ-free rats. This suggested that indigenous oral microorganisms interfere with attachment and colonization by *Candida*. In addition, these authors showed that buccal surfaces of germ-free mice were colonized after oral challenge with *C. albicans*. From these results the authors suggested the participation of adherence in the colonization process. Evidence that adherence may be the initial step in mucocutaneous candidosis came from the work of King et al.[81] They examined the adherence capabilities of seven *Candida* species *in vitro* and showed that *Candida* species readily adhered to vaginal and buccal epithelial cells, with *C. albicans* adhering to a significantly greater degree than other species tested (Figure 3). *C. tropicalis* and *C. stellatoidea* showed significant adherence to mucosal cells, with *C. parapsilosis* adhering to a slight degree only. *C. guilliermondii*, *C. krusei*, and *C. pseudotropicalis* failed to interact with collected epithelial cells.

This type of comparative investigation among various *Candida* species, regarding the relationship of adherence to pathogenicity, was carried out by others with results agreeing generally with the previous findings, that the pathogenic species *C. albicans*, *C. tropicalis*, and *C. stellatoidea* adhered more strongly than the nonpathogenic species *C. parapsilosis*, *C. guilliermondii*, *C. krusei*, and *C. pseudotropicalis*.[31,82-84] This difference in adherence was significant irrespective of whether the adherence assay was performed in saline, or phosphate buffer, regardless of the time of incubation,[82] and irrespective of the type of cells used (buccal, vaginal, or skin corneocytes).[82-84]

Considering the relative pathogenicity of various *Candida* species outlined in Chapter 2, the fact that *C. tropicalis* and *C. stellatoidea* occasionally colonize, or infect the host, or both,[85] and that the nonadherent species are rarely involved in the candidal disease process, it is reasonable to assume that adherence plays a determinant role in the pathogenesis of this disease. However, it is important to remember that adherence alone is rarely, if ever, responsible for disease initiation and that factors other than attachment must be considered in the pathogenesis of candidal infection.

Although both yeast and mycelial forms of *C. albicans* are found in infected tissue,[4,86] the hyphal form has been associated with active symptomatic infection by *C. albicans* whereas

TABLE 2
Adherence of Nongerminated and Germinated Forms of *C. albicans* to Buccal Epithelial Cells

Time (min)	Attached *C. albicans* cells per 100 epithelial cells ± SE (% of total attached fungi) Nongerminated	Germinated	% Germinated fungi
0	267 ± 24 (75)	87 ± 11 (25)	6 ± 1
30	376 ± 25 (65)	206 ± 18 (35)	27 ± 2
60	435 ± 28 (56)	344 ± 21 (44)	56 ± 2
90	165 ± 16 (30)	387 ± 28 (70)	79 ± 2
120	188 ± 17 (28)	482 ± 29 (72)	93 ± 3

Note: Fungal cell concentration used in the adherence assay was 1×10^7cells/ml.
SE = Standard Error.

From Ghannoum, M. and Abu Elteen, K., Unpublished data, 1986.

saprophytic *C. albicans* is almost invariably of the yeast form. Several investigators have suggested that the germinated (mycelial) form is more virulent.[87-89] Bearing this in mind, the relation between germination and adherence has been considered by a number of investigators and the difference in adherability to epithelial cells between yeasts and hyphal forms of *C. albicans* has been claimed by several authors. Kimura and Pearsall[90] found that *C. albicans* adhered to human buccal epithelial cells better under conditions conducive to germ-tube formation than when germ-tubes were not formed. In a subsequent study, these workers gave evidence for a selectively greater potential for adherence of *C. albicans* hyphae.[91] Incubating *C. albicans* under conditions favourable for germination increased its adherence to buccal epithelial cells. On the other hand, partial inhibition of germination led to lower adherence. In the same study, competition assays with germinated and nongerminated yeast showed the occurrence of a selective attachment favoring the former. Work carried out in our laboratory confirms these findings (Table 2).[92] Evidence supporting the preferential adherence of germinated yeasts came also from work by Sandin et al.,[93] who demonstrated that germinated *C. albicans* had significantly greater adherence to buccal cells than blastospores, with germ-tubes being 50 times more adherent than yeast cells. In addition, germinated yeasts were more susceptible to adherence inhibition by concanavalin A than nongerminated *C. albicans*. Sobel et al.[47] found that in prolonged adherence tests, a variant strain of *C. albicans* adhered less well to vaginal epithelia than its hypha-forming wild type. Anderson and Odds,[94] evaluating the difference in the adherence of different morphological forms of *C. albicans* in terms of fungal biomass and surface areas, rather than cell numbers attached per epithelial cell, observed that hyphae attached significantly better than germ-tubes, with yeast forms adhering the least (Table 3).

It may not be absolutely essential for germination to be necessary before invasion can take place, since both blastospores and germ-tubes were shown to be capable of invading superficial epithelial cells.[95] Experiments *in vivo* by Sobel et al.[47] showed that an agerminative variant of *C. albicans,* which completely failed to germinate, was able to successfully colonize the rat vagina. Thus, the ability to germinate may be an asset to the enhancement of the adherence phenomenon, but its absence will not prevent the invasion process.

C. DIMORPHISM

C. albicans is a dimorphic yeast which is capable of growing in culture as an ellipsoidal bud (referred to usually as a blastospore or yeast form), as a pseudohypha of elongated yeast-like

TABLE 3
Attachment of *C. albicans* 73/079 and 73/055 to Vaginal Epithelia from Pregnant and Nonpregnant Females

Basis of estimation	*C. albican* growth (form/age)	Epithelia from nonpregnant		Epithelia from pregnant	
		C. a. 079[a]	*C. a.* 055[b]	*C. a.* 079	*C. a.* 0.55
ATP concentration per epithelium (pM)	yeasts/stationary	0.35 ± 0.13	0.33 ± 0.13	0.35 ± 0.09	0.38 ± 0.10
	yeasts/2 h	0.41 ± 0.08	0.38 ± 0.08	NT[c]	NT[c]
	yeasts/6 h	0.45 ± 0.14	0.43 ± 0.09	0.42 ± 0.20	0.45 ± 0.08
	germ-tubes/2 h	0.43 ± 0.15	0.41 ± 0.11	0.40 ± 0.08	0.40 ± 0.10
	hyphae/6 h	0.96 ± 0.32	0.96 ± 0.48	1.44 ± 0.56	1.36 ± 0.56
Fungal surface area per epithelium (m^2)	yeasts/stationary	320 ± 120	300 ± 120	320 ± 80	340 ± 90
	yeasts/2 h	370 ± 70	340 ± 70	NT	NT
	yeasts/6 h	400 ± 130	390 ± 80	380 ± 180	400 ± 70
	germ-tubes/2 h	650 ± 230	600 ± 120	620 ± 170	600 ± 150
	hyphae/6 h	880 ± 290	880 ± 440	1310 ± 510	1240 ± 510

Note: Data are expressed in terms of ATP concentration (equivalent to viable biomass) per epithelial cell and fungal surface area per epithelial cell on the basis of ATP per cell and surface area per cell.

[a] *C. albicans* isolate 73/079.
[b] *C. albicans* isolate 73/055.
[c] NT = not tested.

From Anderson, M.L. and Odds, F.C., *Mykosen*, 28, 531, 1985. With permission.

cells, or as a true septate hypha (referred to as filamentous form) (Figure 4). Mycelium production is the only reproducible property known to belong exclusively to *C. albicans* among *Candida* spp.[4] Both the true hypha and pseudohypha may regenerate the yeast phase by producing clusters of yeast-like blastoconidia.[96] The growth phenotype depends on environmental conditions and the growth history of the cells.[4,97-106]

A large volume of literature deals with these forms, the differences between them, and the effects of different environmental factors (temperature, pH, growth media) on *Candida* growth and dimorphism. For review see Odds,[4] Soll,[106] and the review by Soll[107] dealing with the molecular basis of phenotypic conversion between the two growth forms.

Budding cells are round to ellipsoidal and are encased in a cell wall (from 100 to 300 nm thick), and have a diameter between 2 to 5 μm.[108-110] The process of budding involves the formation of an evagination, as an outpocketing of cytoplasmic membrane and cell wall, which enlarges in volume to a size just smaller than that of the parent cell. During this process, a septum contiguous with the wall develops centripetally at the junction of the parent and daughter cell units. This is followed by cell separation. Blastospores contain various organelles, including: nucleus, mitochondria, endoplasmic reticulum, ribosomes, lipid granules, and vacuoles.[110-113]

The pseudohyphal morphology of *C. albicans* is an intermediate between the ellipsoidal blastospore and elongated hypha. Although a pseudohypha resembles a true hypha microscopically, its mode of formation is different.[4] An individual pseudohypha is typically 2 to 5 μm in diameter and 26 to 30 μm in length.[114]

True hyphal cells of *C. albicans* are compartmentalized tubes or filaments with diameters measuring typically 0.8 to 1.3 μm, which is about one third the diameter of a comparable blastospore.[107,110,114] Enclosed within each hyphal compartment are typical organelles similar to those found in blastospores. Hyphae grow apically and can arise as branches of existing hyphae or by germination of spores.[115] Formation of hyphae from blastospores is initiated by the

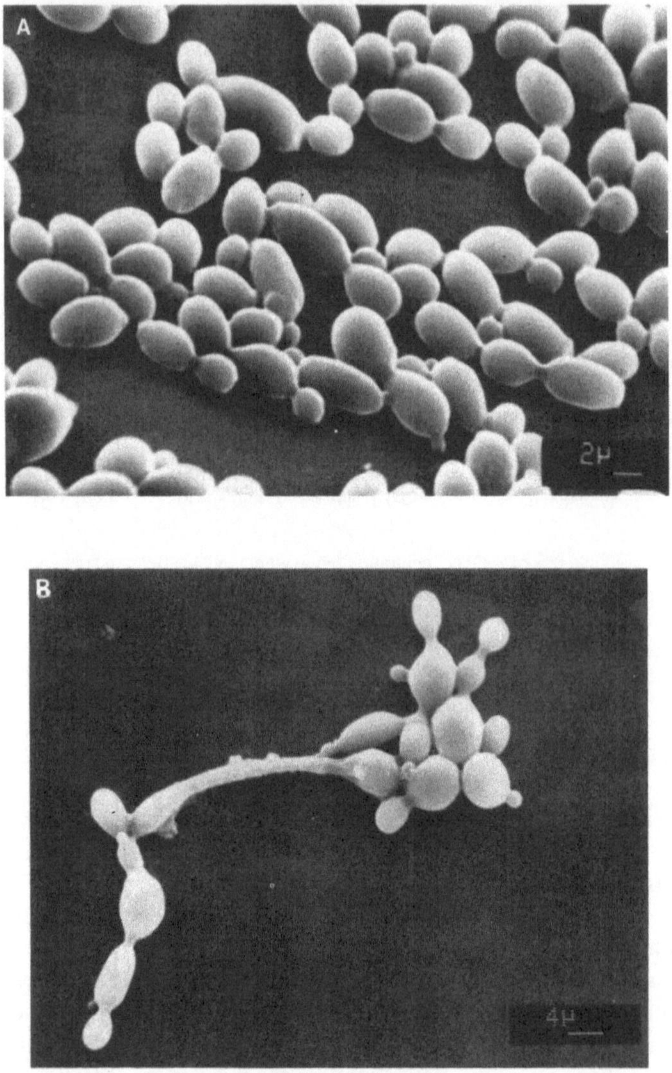

FIGURE 4. Scanning electron micrographs of *C. albicans*. (A) Blastospores; (B) Pseudohypha, and (C) Mycelia.

emergence of germ-tubes which grow continuously by extension. Significant differences exist between the surface layers of blastospores, pseudohyphae, and hyphae.[4,109,110,116-118]

In addition to the three forms mentioned so far, *C. albicans* can exist as thick-walled ellipsoidal cells referred to as chlamydospores.[119,120] These cells may represent a stationary phase or even a terminal phenotype of blastospores.[121-123] Electron microscopic observations showed that chlamydospores lack ribosomes and mitochondria, possessing only lipoid inclusions, occasional membrane-bound structures, and a thick cell wall.[112,122,123]

Szaniszlo[124] subdivided dimorphic fungi into three groups according to their morphology *in vivo:* (1) fungi that exhibit predominantly yeast development *in vivo* and have extensive sporulating hyphal phases under saprophytic conditions. Species included in this group are *Ajellomyes (Blastomyces) dermatitidis, Emmonsiella capsulata (Histoplasma capsulatum), Paracoccidioides brasiliensis,* and *Sporothrix schenckii.* (2) Fungi that have a predominantly yeast-like saprophytic phase, can be induced *in vitro* to carry out distinct yeast-hyphal transitions, and tend to exhibit both yeasts and hyphae *in vivo*. Species included in this group

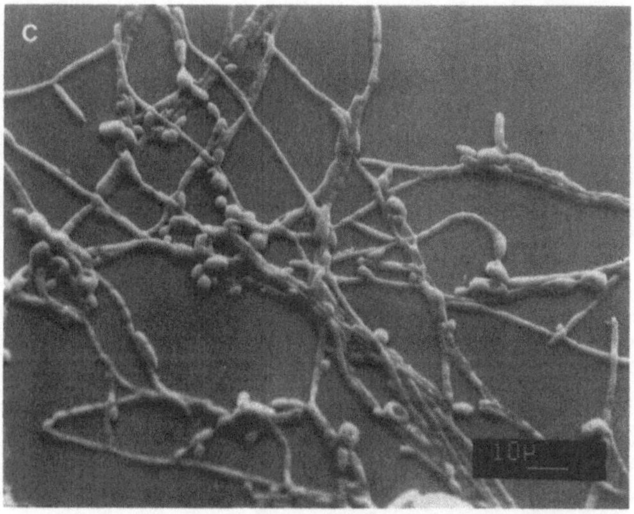

FIGURE 4C

are *C. albicans, Exophiala werneckii,* and *Wangiella dermatitidis.* (3) Fungi that have well-developed saprophytic hyphal phases, but do not develop yeast phases *in vivo* or *in vitro.* Instead the *in vivo* morphology consists of mostly isotropically enlarged forms. Species included in this group are *Phialophora verrucosa, Chrysosporium parvum var. parvum, C. parvum var crescens,* and *Coccidioides immitis.*

The ability of *Candida albicans* to exist in the two forms *in vivo* has received much attention, particularly since a number of workers associated the presence of one form or the other with pathogenicity.[28,86,87,125,126] Thus, a large body of literature claims a causal relationship between mycelial formation and infection.[4,28,29,89,127] This is based on the assumption that hyphae penetrate tissues more readily than yeast cells, are more difficult to ingest, and form emboli.[4] Also filamenteous forms of *C. albicans* are often observed in the scrapings from lesions. However, there is evidence for increased virulence of the yeast form of *C. albicans* over the filamentous form.[128-133] Some authors could not show any difference in the infectivity between mycelium and yeast forms.[14,134,135] This controversial issue was reviewed thoroughly by Odds,[4] who concluded that an association between mycelium production in *C. albicans* and enhanced virulence of this species does exist. However, this association is clearly relative and by no means absolute.

Evidence presented by Shepherd,[136] using morphological mutants of *C. albicans,* indicate that both yeast and mycelial forms of *C. albicans* are capable of invading soft tissues, causing systemic infection which results in the death of mice. In reviewing the literature *in toto* we tend to support his conclusion. This is based on a number of observations: (1) *C. albicans* yeast cells were reported by Klotz et al.[137] to be able to burrow into the endothelial surfaces by a mechanism that was independent of germ-tube formation, (2) the ability of the yeast form to resist intracellular killing by phagocytes more efficiently than pseudohyphae,[138-140] and (3) the ability of phagocytosized yeast cells to form germ-tubes capable of thrusting out of the phagocytes.[20,139,141]

Taking these observations collectively it is possible to suggest that, at early stages of infection yeast forms of *C. albicans* burrow into the epithelial surfaces (Figure 5). There they resist the killing action of the polymorphonuclear phagocytes, considered as virulence determinants probably operating at early stages of infection, with a possible selection of yeast cells resistant to intracellular killing.[139] This is followed shortly by the formation of germ-tubes which have the ability to mechanically burst through the phagocytic cell, escaping and spreading to oth tissues. Once germ-tubes have been formed, they are more susceptible to the phagocytic killing action of other polymorphonuclear phagocytes. It is possible, under these circumstances, that

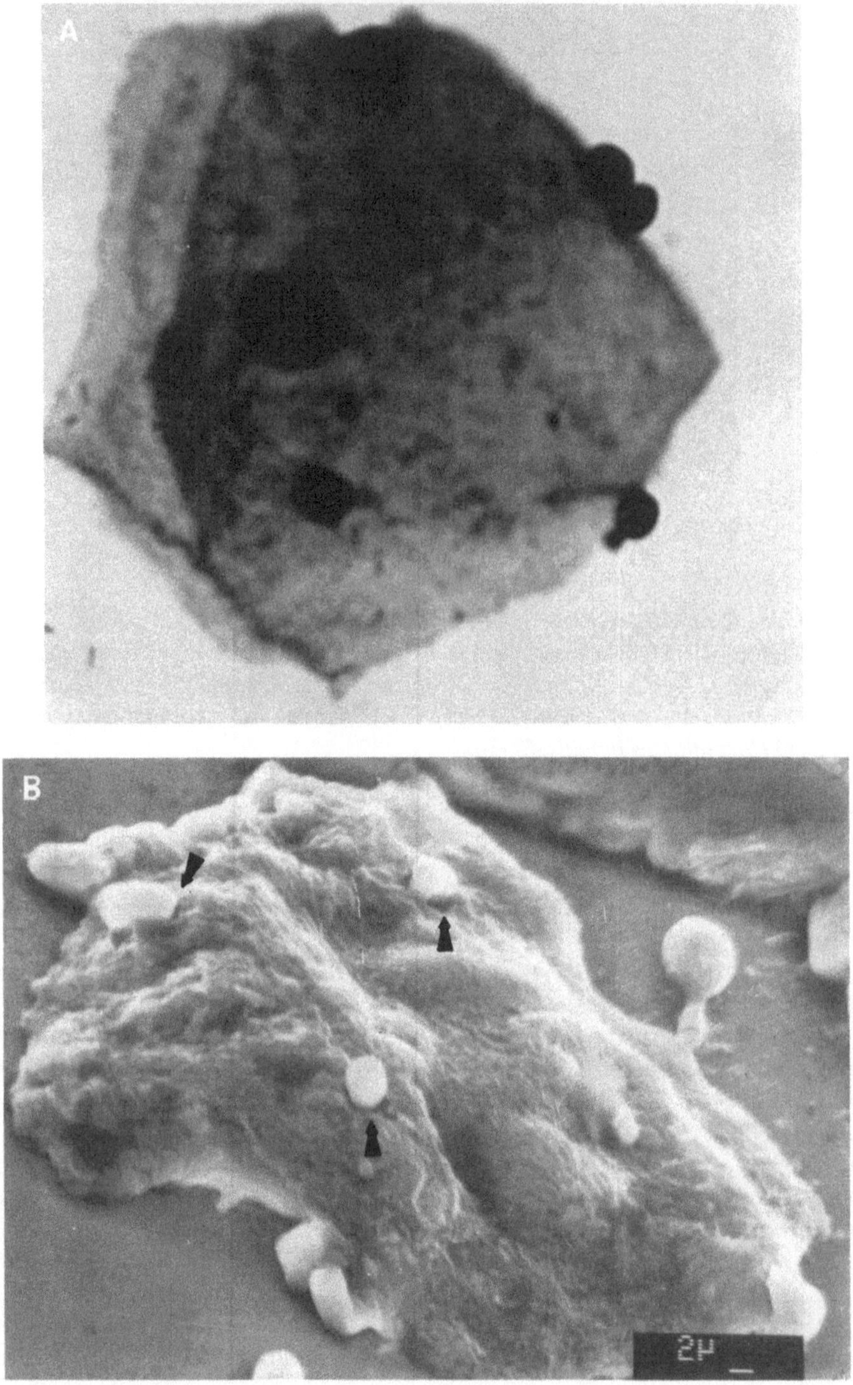

FIGURE 5. (A) A light micrograph and (B) a scanning electron micrograph of *C. albicans* cells adhering to buccal epithelial cells. Note that the yeast cells are embedded in shallow depressions of the epithelial cell.

C. albicans converts again to produce the more resistant blastospores. Thus, the ability of this yeast to form blastospores is possibly a mechanism evolved by *C. albicans* to combat phagocytes. Accordingly, both forms function together and are important in the pathogenicity of this yeast. Recovery from infection is brought about by the eventual inability of *C. albicans* to withstand host defenses after these defenses have been strengthened by both antibody and cell-mediated immunity.[142,143]

D. TOXINS

Fungal toxins (mycotoxins) are natural products produced by fungi which evoke toxic response when introduced in low concentrations to higher vertebrates and other animals by a natural route.[144] The role of bacterial and fungal toxins as pathogenicity determinants in the establishment or acceleration of diseases in humans is well recogonized for both bacterial and fungal toxins.[145,146]

The presence of endotoxin-like substances in *C. albicans* was first proposed by Henrici.[147] This was followed by two studies by Salvin,[148,149] who reported on hemolysin from the yeast-like phases of *C. albicans* and endotoxin production. There followed a number of publications on high molecular weight toxins from *C. albicans*.[146,150-166]

Toxins produced by *C. albicans* can be divided into two groups: (1) High molecular weight toxins and (2) low molecular weight toxins.

1. High Molecular Weight Toxins

Included in this category are glycoprotein toxins and canditoxin.

a. Glycoprotein Toxins

Toxic substances containing carbohydrates, namely mannose and glucose, and protein as toxic components were reported by Cutler et al.[165] These workers tested intact cells of *C. albicans* as well as fractions thereof for toxicity using four systems ordinarily employed with bacterial endotoxins. They found that both fungal cell walls and intact cells were pyrogenic in rabbits and lethal to actinomycin-D-treated mice. Only cell walls were lethal to chicken embryos and neither coagulated a hemolysate of horseshoe crabs. In contrast, bacterial endotoxin was strongly active in all four systems used and at much lower concentrations. They proposed that the carbohydrate moiety causes toxicity in mice, and that the relative amount of protein determines pyrogenicity. Yamamoto and Iwata[167] isolated three toxic glycoproteins from cells of a virulent strain of *C. albicans*. Chemical characterization showed the sugar moiety to be *D*-mannose. Full chemical analysis was carried out on only one of these three substances (designated Fr II-2) because of its high yield. The ultracentrifugal profile of this fraction showed a single peak with an $S_{20,w}$ value of 3.18 and its molecular weight was approximately 50,000 Da. Elemental analysis yielded the following values: N(21%), C(39.5%), and H(6.2%). The protein to sugar ratio was 13:86 by weight. Toxicity testing showed this fraction to have a potent and acute toxicity for mice.[168] It is suggested that both protein and sugar moieties of the glycoprotein molecule would be necessary for the toxic activity, with mannan playing a major role and protein acting as a helper.[168] The toxic activity of glycoproteins was also reported in other investigations (see References 155, 169 to 173). In addition to their role in toxicity, *C. albicans* glycoproteins, particularly mannoproteins, have gained prominence as adhesins involved in colonizing body surfaces.[174-176] Although these compounds may themselves be nontoxic, they are referred to as aggressins.[145] Another property attributed to cell wall glycoprotein from hyphae and pseudohyphae of *C. albicans* is the ability to inhibit attachment of neutrophils to live hypha, and to impair neutrophil functions beyond attachment.[177,178] The possible mechanism suggested for this adherence inhibition is competitive binding of these glycoproteins to receptors on the neutrophil surface.[177]

b. Canditoxin

The isolation and characterization of this high molecular weight toxin from a virulent strain of *C. albicans* is the result of intensive investigation carried out by Yamamoto and Iwata.[167] Canditoxin is an acidic protein having an ultracentrifugal pattern of a single peak with an $S_{20,w}$

value of 4.6 and a molecular weight of approximately 75,000 Da. It is heat labile, being inactivated completely by heating for 30 min at 50°C.[179-181] This protein is made up of four subunits linked in sequence: two α-subunits (carboxypeptidase), one β-subunit (phosphomonoesterase), and one γ-subunit (which has not been identified).[182,183] All four subunits are essential for toxicity. Canditoxin is lethal to mice by intravenous injection (LD_{50} value equal 0.3 μg/g body weight). In addition, it shows versatile activities, viz. cytotoxic (toward cultured mammalian cells), pharmacological, immunological, enzymatic, and infection-enhancing activities.[146]

2. Low Molecular Weight Toxins (Tremorgenic Toxins)

The isolation of six different low molecular weight toxins from canditoxin-producing *C. albicans* strain was reported by Iwata.[146] These compounds were closely related and shared shock-evoking and/or lethal activities. Only one of these compounds (designated as substance E) was characterized as having a melting point of 133 to 135°C with the following elemental analysis: C(64.23%), H(5.06%), and O(30.71%), on a weight basis. Its molecular weight was 164 and chemical formula:  $-CH_2-CO-COOH$. Substance E was lethal to mice by intravenous injection (LD_{50} value equal 5 μg/g body weight).[146]

Evidence supporting the role of toxins as pathogenicity determinants has accumulated from both animal experimentation and clinical observations. Iwata[168] examined the possible role of high and low molecular weight fungal toxins in the etiopathology of experimental infections following various criteria necessary to establish toxin effect such as (1) the infection enhancing activity of the isolated toxins; (2) the infection-suppressing activity of antisera or immunoglobulin from anti-toxin area; (3) demonstration of *in vivo* production of toxin during the course of infection with a fungal strain producing it, and (4) influence of toxins on host defense mechanisms involving humoral and/or cellular immunity.[168] The results of this investigation confirmed the role of glycoproteins and canditoxins in pathogenicity. Clinical evidence was derived from patients with systemic candidosis who occasionally manifest a clinical syndrome indistinguishable from Gram-negative sepsis.[184] Such cases offer the best evidence in humans that *Candida* toxins can play an important role in the pathogenesis of the disease.[184]

E. ENZYMES

To aid invasion of host tissues, microbial cells possess constitutive and inducible hydrolytic enzymes which destroy or derange constituents of membranes leading to membrane disfunction and/or physical disruption. Since membranes are made up from lipids and proteins, it is obvious that these biochemicals constitute the target of enzyme attack. A number of membrane-damaging enzymes are known to be secreted by bacteria.[145] A variety of such enzymes produced by pathogenic fungi, almost exclusively by *C. albicans*, have been reported.[34,179-181,183,185-195] Enzymes secreted by *C. albicans* and thought to be concerned with pathogenesis are categorized into two main types: (1) proteinases which hydrolyze peptide bonds, and (2) phospholipases which hydrolyze phosphoglycerides and lysophospholipases which hydrolyse lysophosphoglycerides. The relationship of proteinase and phospholipase activity to virulence of various strains and species of *Candida* has already been discussed (see Section II.A). In this section we will deal with the different types of enzymes reported for *C. albicans* with a description of their chemical nature and relation to pathogenicity.

1. Proteinases

C. albicans and some other *Candida* species secrete proteinases when they are grown in media containing proteins as sole nitrogen sources.[38,196,197] The ability of *C. albicans* to produce extracellular proteinases, was discovered by Staib and his group,[34,191] who were the first to purify such enzymes. Isolation and characterization of proteinases from species of *Candida (albicans, tropicalis,* and *pseudotropicals)* have been carried out.[4] Table 4 summarizes some of the physical and chemical properties of such enzymes.

Evidence that secretory proteolytic activity is an important determinant of pathogenicity is derived from the following avenues (for review see Reference 198): (1) the enzyme is secreted only by the most pathogenic of the *Candida* species associated with human diseases. Thus, *C. albicans* isolates are strongly proteolytic, followed by *C. tropicalis* isolates which are usually proteolytic to a lesser extent. Moderate proteolytic activity is detected with strains of *C. parapsilosis*.[35,199,200] In contrast, the other *Candida* species of clinical relevance like *C. glabrata, C. krusei, C. pseudotropicalis,* and *C. guilliermondii* do not appear to produce extracellular proteinases.[199,200] This mirrors the order of virulence for man as suggested by Ahearn,[201] (2) *C. albicans* strains with significantly higher proteolytic activity showed more lethality and/or colonization for mice,[37,39,40,43,191,202] and (3) the ability of *C. albicans* to secrete proteinase *in vivo*. This was indicated by the observation of a halo around yeasts in infected animal tissue stained with fluorescent antibody to purified proteinase.[196] The presence of *Candida* secretory proteinase in a case of human acronecrosis was also detected by immunofluorescence.[203]

Germaine et al.,[204,205] basing their conclusions on the denaturation of *Candida* proteinases at pH values higher than pH 6, suggested that proteinases are not significant contributors to the pathogenicity of *Candida* in the oral cavity. Other evidence, however, proposed by Samaranayake et al.,[206] who performed experiments similar to those of Germaine and Tellefson,[204] showed that conditions which are conducive to the production, activity, and stability of *C. albicans* proteases, exist in the human oral cavity.[206] In addition, it is known that the salivary pH becomes more acidic in the oral cavity of certain patients, e.g., cancer patients, and in carious lesions.[57,207] Under these conditions there is an increase in the magnitude of yeast growth, as well as more suitable conditions for enzyme activity.[207,208] Yet another important support for the role of proteinases in pathogenicity is the finding that neutral proteinases are secreted by *C. albicans* which allow proteolysis to proceed at pH values near neutrality.[209]

The overall impression gained so far is that proteinases contribute to pathogenicity of *Candida*.[260] However, other factors (related to the yeast and/or host) must be considered simultaneously. Thus, the ability to produce proteinase by a given strain is not a guarantee for virulence. Even though *C. parapsilosis* is known to produce proteinases *in vitro*, it lacks the ability to induce their production under conditions of infection, which may account for the low virulence of most isolates of this species.[210]

2. Phospholipases and Lysophospholipases

The presence in *C. albicans* of phospholipase activity was first detected by Costa et al.[211,212] and Werner[213] by growing this yeast on media containing egg yolk and lecithin. Attempts to identify the types of phospholipases produced suggest the presence of phospholipases A, B, and C.[192,214-217] No evidence was found for the presence of phospholipase D.[216] In addition, the presence of lysophospholipase and lysophospholipase-transacylase was reported.[216,217] Evidence for the presence of these enzymes was based on the extraction and separation of products of phospholipase activity such as glycerylphosphocholine, phosphocholine, and lysolecithin,[215] or by the use of various pure substrates for analysis of these products.[217] Figure 6 shows the sites of action of phospholipases and lysophospholipases produced by *C. albicans* and their breakdown products.

Reports of phospholipase production by different *Candida* species claim that this activity is confined to *C. albicans*. Thus, Samaranayake et al.[218] using a plate assay method, described originally by Price et al.,[219] screened the phospholipase activity of 41 *Candida* isolates (comprising four species). Their results showed that 79% of the *C. albicans* isolates were phospholipase producers, whereas none of the *C. tropicalis, C. glabrata,* or *C. parapsilosis* isolates produced the enzyme as assayed for. This is the only paper available to us which compares the phospholipase activity between *Candida* species. Therefore, further studies, using isolates from different patient groups and centers, are necessary to substantiate this conclusion. In addition, possibly the use of more sensitive methods may be required.

TABLE 4
Characteristics of Proteinases Isolated From *Candida* spp. That May Contribute to Pathogenicity

Isolated from	Group	M_r	Isoelectric point	Denaturation pH	Optimum pH	Type of Activity	Temp. of inactivation	Isolated and characterized by
C. albicans and *C. tropicalis*	Carboxyl proteinase	45,000	4.45	>8.4	2.5—3.9	Proteolytic	45°C	Rüchel[224]
C. albicans	Carboxyl proteinase	42,000	4.0	<2.5 & >6.0	4.0	Keratinolytic	>50°C	Negi et al.[225,226]
C. albicans	Neutral proteinase	—	—	—	6—7.5	Proteolytic	—	Logan[209]
C. albicans	—	46,000	4.2	>6.0	3.5—4.0	Collagen-olytic	70°C	Kaminishi et al.[227]
C. parapsilosis	Aspartic proteinase	33,000	5.3	>7.0	4.3	Proteolytic	—	Rüchel et al.[210]

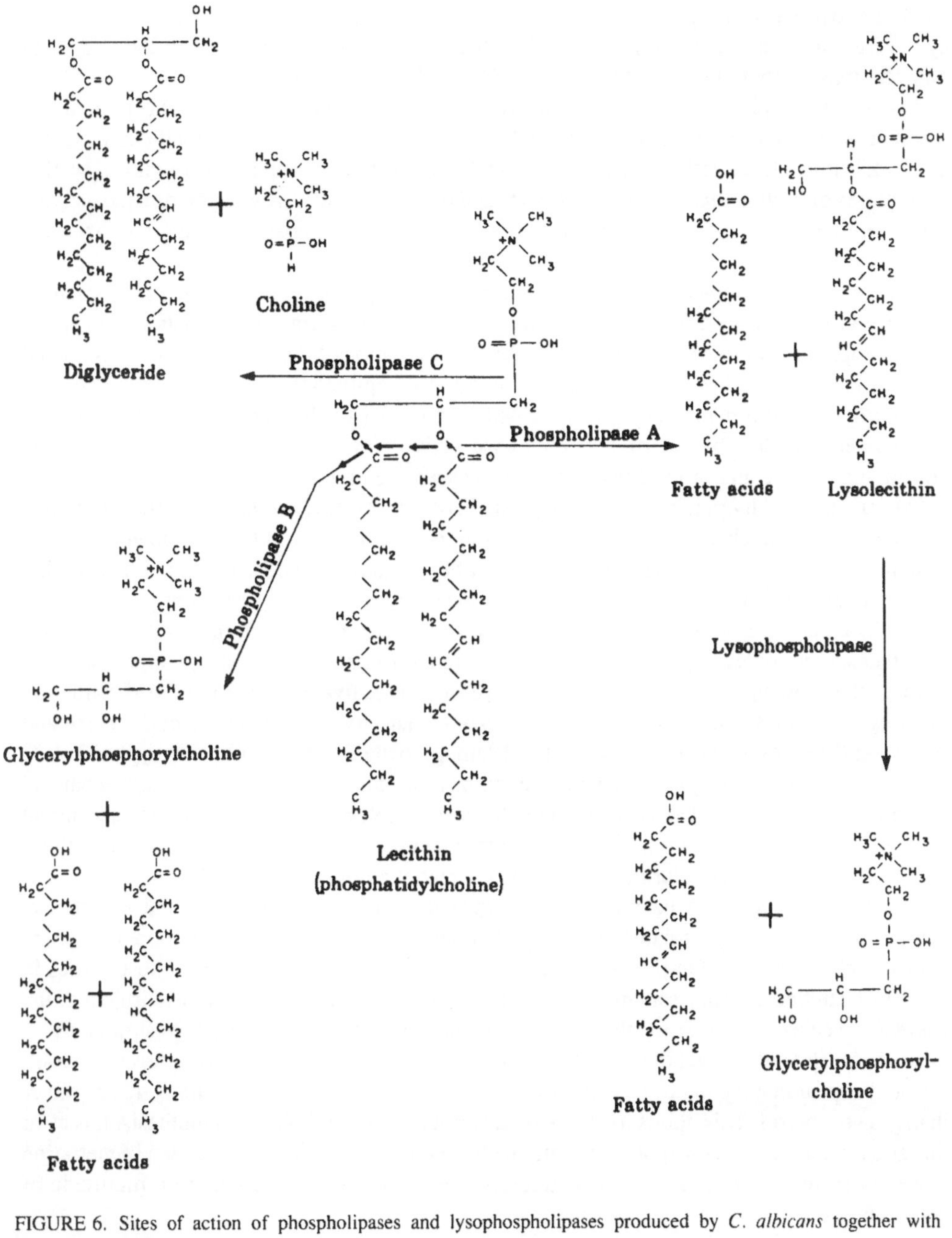

FIGURE 6. Sites of action of phospholipases and lysophospholipases produced by *C. albicans* together with breakdown products.

Variation of phospholipase activity among *C. albicans* isolates has been reported. The three studies available to us compare various *C. albicans* strains using the method of Price et al.[219] or a modification of it. Price et al. found a large variation in phospholipase activity between different isolates of *C. albicans*. However, the Pz value (the ratio of colony diameter to diameter of a dense white zone of precipitation which forms around colonies of phospholipase-positive isolates) is constant for any one isolate regardless of the site from which it is recovered in the patient. *C. albicans* isolates positive for phospholipase represent between 30 to 55%, depending on whether they are blood, wound, or urine isolates. Modifying the technique of Price et al.,[219] Samaranayake and co-workers[218,220] increased the percentage of *C. albicans* isolates with positive phospholipase activity to 79% in one study, and 94% in another. Attempts to exploit the ability to produce phospholipases by *C. albicans* as a criterion for the biotyping, as suggested by several workers, were hampered by the observation of high variations in the degree of phospholipase activity upon repeated testing of *C. albicans* isolates.[220]

Localization of phospholipases in *C. albicans* and *C. albicans* infected chick chorio-allantoic membranes (CAM) has been carried out by Pugh and Cawson.[216,221] They initially investigated the cytochemical localization of phospholipase A and lysophospholipase in growing as well as old cultures of *C. albicans*.[216] These authors found that in a rapidly growing culture the enzyme activity is associated with bud formation, while in older cultures phospholipase A appears at the periphery of the cell and is secreted into the medium. Lysophospholipase was present at the same cell sites as phospholipase A. Demonstrating phospholipases at the site of bud formation suggests a role for these enzymes in cell growth and possibly germ-tube formation, and points to the complexity of these processes. The second investigation of Pugh and Cawson[221] is related more to the role of phospholipases in the invasion of epithelial tissues, and, as such, is a pathogenicity determinant. Using chick CAM as a model for human lesions, these authors investigated the distribution of phospholipase activity in the inoculated blastospores, in developing hyphae and in the fungi which penetrated the epithelial cells.[17,222,223] Infection of CAM was initiated by placing stationary phase blastospores of *C. albicans* on the membrane. Observations of the phospholipase localization in these cells showed that the enzyme activity of individual blastospores varies greatly in quantity, with a few cells showing a heavy stain deposit at the cell periphery and some showing localized activity. Observations of the cells, 6 h after inoculation showed many blastospores with intense enzyme activity while other yeast cells had started to develop into hyphae with a slight concentration of phospholipase activity towards the growing tip. A small number of yeast cells actively invade the CAM while the majority of the blastospores produce a high enzyme activity over the entire cell surface and subsequently autolyse with resultant localized damage to the CAM. Usually only hyphae invade the membrane successfully where enzyme secretion is limited to the growing points. Mechanical support of the mature hyphae is afforded by the host cytoplasm which is exposed to a minimal damage at this stage. Phospholipase activity in the older portions of these hyphae is intracellular and is concerned with cellular metabolism and eventual autolysis (Figure 7).[221]

It is clear from the above that phospholipases produced by *C. albicans* are of dual importance. On the one hand, they are related to the growth control in the yeast and the remodeling of the parasite cell membrane. On the other hand, they are related to the invasion of host tissues in lesions of candidosis. This is a unique property, limited to *C. albicans,* the most virulent of the *Candida* species. Considering that other *Candida* species undergo budding similar to *C. albicans* (in spite of the fact that, so far, they have been shown to be devoid of phospholipase), it would appear that the major contribution of these enzymes is related to pathogenicity rather than growth control. This points to the need for further work to verify this hypothesis. It is clear that an association between hyphal formation, phospholipase production, and tissue penetration is necessary for tissue invasion. This suggests that virulence is governed by a multitude of determinants.

F. CELL SURFACE

The majority of host-parasite interactions involve surfaces; either those of microbes, those of host cells, or those of both. The outcome of these interactions may have an effect on microbial pathogenicity. Smith[2] divided the microbial surfaces into five different components, depending on their contribution to pathogenicity:

1. Components which contribute to entry to the host
2. Components which contribute to multiplication *in vivo*
3. Components which contribute to interference with host defences
4. Components which contribute to host and tissue specificity
5. Components which contribute to host damage

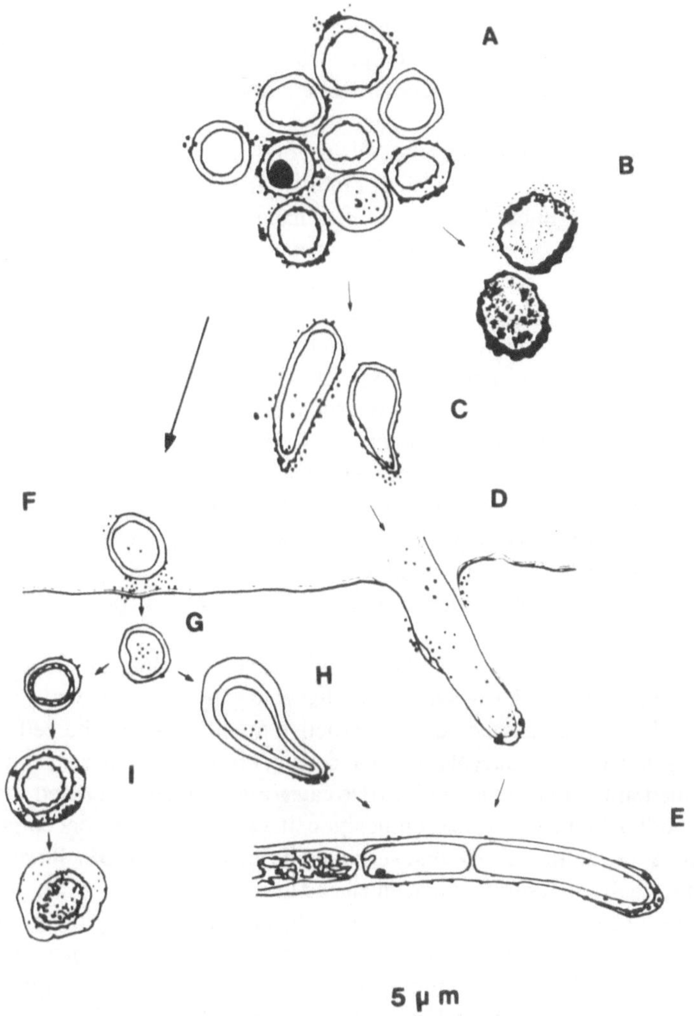

FIGURE 7. Schematic representation of the changes in distribution of phospholipase activity in *C. albicans* infecting the CAM. (A) Blastospores from an 18 h culture used to infect the chick CAM. The enzyme activity of individual cells varies greatly in quantity. A few cells show a heavy stain deposit at the cell periphery and some show localized activity. (B) At 6 h after inoculation of chick membrane many blastospores show intense phospholipase activity. The heavy stain deposit obscures the cell outline in many cells. (C) Some blastospores develop into hyphae. There is slight concentration of enzyme activity towards the growing tip. (D) A hypha penetrating the chick epithelium. The enzyme activity is concentrated in the apical region. There is a space between the two tissues except at the apex; this is formed by the breakdown of chick cytoplasm during the advance of the hyphal tip. (E) A hypha established in the CAM extending by apical growth. There is phospholipase activity at the hyphal apex but only slight activity along the lateral walls of the fungus which are closely opposed to the host cytoplasm. Cross wall form in long hyphae and intracellular enzyme activity is found in the older autolysing segments. (F) Blastospore lying on the CAM and secreting phospholipase. (G) Blastospore closely enveloped by chick cytoplasm and secreting enzyme. (H) Blastospore embedded in CAM forming a germ-tube. Activity is greatest at the growing point. A vacuole is formed around the blastospore. (I) Blastospore within the CAM secrete phospholipase and break down the surrounding host cytoplasm to form vacuoles. This is followed by autolysis of the blastospores. (From Pugh, D. and Cawson, R.A., *Sabouraudia*, 15, 29, 1977. With permission.)

It is obvious from Smith's review[2] that information on the role of fungal surface components in general, and *C. albicans* components in particular is largely ambiguous. However, with the rising interest in *Candida* adherence to both host tissues and inanimate materials, studies regarding the relationship between cell surface composition and virulence of *C. albicans* gained impetus. The chemical nature and composition of the *Candida* cell surfaces is discussed in Chapter 4. Also, the specific chemical components involved in adherence are covered in the same chapter. The discussion in this section will confine itself to the relationship of *Candida* surface layers to pathogenicity.

Toxic components originating from the cell wall of *C. albicans* have been reported by Cutler et al.[165] *Candida* cell walls were pyrogenic in rabbits, lethal to actinomycin-D-treated mice, and to chicken embryos. Wright et al.[228] showed that mannan derived from the cell walls of yeast differentially affected several functions of human neutrophils. These included inhibition of the respiratory burst and release of myeloperoxidase stimulated by phagocytosis of serum-opsonized zymosan *in vitro,* with a resultant effect on the host defense mechanisms.

A mucus sheath secreted by the cells of *C. albicans* is considered to be the outermost layer covering the cells. This layer contains polysaccharides and secreted enzymes, as well as, in mature cultures, dead cells and cell debris. It is thought to play a role in the formation of plaque and the penetration of yeast cells into infected host tissues.[229]

Work on the adherence of yeast led to the identification of cell wall components of *C. albicans* which may contribute to mucosal surface infection and penetration. Maisch and Calderone,[230] working on the role of surface mannan on adherence of *C. albicans* to fibrin-platelet clots, suggested a role for this wall constituent in the adherence process. Sandin et al.[93] demonstrated that mannose-containing moieties on the surface of *C. albicans* and buccal cells could mediate *in vitro* adherence.

A more direct relationship between the cell surface composition, adherence ,and virulence of *C. albicans* came from work carried out simultaneously by Tronchin et al.[231] and McCourtie and Douglas.[32] The former gave evidence for active modification of the cell wall coat of *C. albicans* during adherence to buccal epithelial cells in response to external stimuli (Figure 8). The latter divided strains of *C. albicans* into two categories. The first included strains which had the ability to modify their cell surface composition in response to high concentrations of sugars in the medium. These authors found that such a modification promoted adherence of the yeast to acrylic and epithelial cell surfaces and enhanced its virulence for mice. The second category included strains which either completely lacked the ability to modify their outer surfaces or possessed it to a much lower degree. Strains in this category had a lower pathogenic potential.[32]

Alterations in the cell surface features of *C. albicans,* including mode of budding, bud scar morphology, surface topography, and intercellular matrix were reported by Kulkarni et al.[232] to vary with the carbon source. Changes in the cell surface hydrophobicity of *C. albicans* and *C. glabrata* were shown to be affected by growth conditions including temperature, growth phase, and growth medium.[233] The role of this general cell surface property in adherence and subsequent virulence has been demonstrated in bacteria.[233-238] To what extent the alterations of cell surfaces of *C. albicans* observed *in vitro* are reflected *in vivo* requires investigation. In this respect, it is clear that the surface changes induced in *C. albicans* by high concentrations of different sugars may be important *in vivo*.[32] Sucrose, glucose, and maltose are common dietary sugars,[239] while galactose may be produced in the mouth as a result of lactose degradation by oral bacteria.[240] Thus development and persistence of oral candidosis in persons on carbohydrate-rich diets may be related to these changes.[241,242]

The role of *Candida* surface components in multiplication *in vivo* is not clear, although Elin and Wolff[243] reported that this yeast may possess siderochromes which are believed to be involved in the microbial uptake of iron. They demonstrated that the mortality rate in experimentally infected mice was increased by simultaneous injections of *C. albicans* and ferrous ammonium citrate.

The role of surface components of *C. albicans* as a pathogenicity determinant is clear, although many investigations have so far been directed towards the relationship to adherence. Other aspects, as outlined by Smith,[2] should be investigated in order that a more definite role for these constituents may be defined. The alterations in the cell surface observed *in vitro* and their relationship to pathogenicity should be extended to *in vivo* studies.

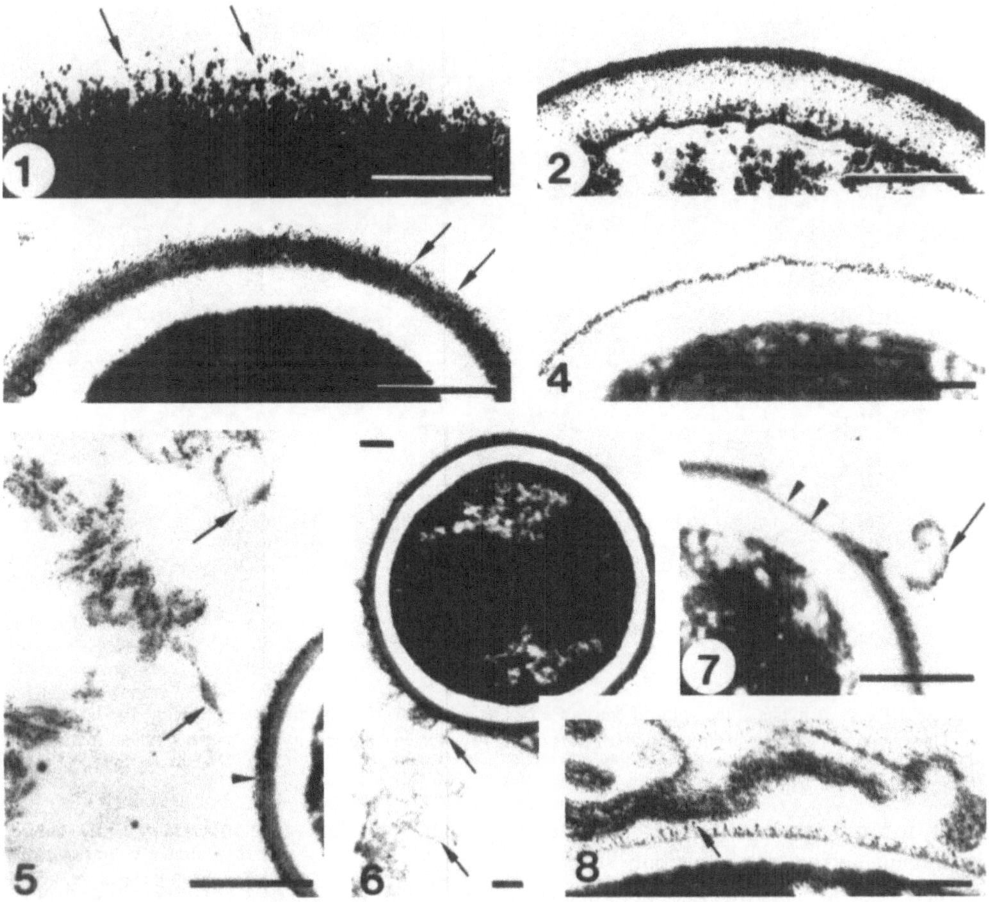

FIGURE 8. Ultrastructural aspects of the (arrows) cell wall coat of blastospores adhering to (1 to 3) epithelial cells after 30 min of incubation compared to (2 and 4, 1 and 2) blastospores harvested from Sabouraud medium after culture for 24 h. Modification of the (5 and 7) cell wall coat during adherence assay. In some cases (arrows) detachment of numerous isolated fragments only labeled on their external face, left behind a few or (arrowheads) unreactive cell wall surface. Detachment of (arrows) an outer sublayer of (6) cell wall coat components. (8) Yeast-epithelial cell interactions. Attachment of blastospores via (arrow) external granules lodged into the bilayer membrane of a buccal cell; bars represent 0.5 μm. (From Tronchin, G., Poulain, D., and Vernes, A., *Arch. Microbiol.*, 139, 221, 1984. With permission.)

III. PATHOGENICITY PATTERN

Microorganisms and their products are associated with a wide spectrum of pathogenetic patterns. Briody[244] divided these patterns into six different classes including: (1) intoxication; (2) adherence and growth; (3) adherence, growth, and toxin production; (4) adherence, penetration, and growth; (5) lymphatic spread; and (6) systemic distribution. The most common patterns are adherence, growth, and toxin production; and adherence, penetration, and growth. These two determine the course of more than two thirds of all infectious diseases in man.

The pathogenetic pattern for *Candida* infection is a combination of patterns (3) and (4). It involves adherence to and multiplication on mucosal surfaces, with filamentation and germ-tube formation in the case of *C. albicans*.[80,245-248] This is followed by production of enzymes (phospholipases and proteinases), tissue damage (Figure 9), penetration, and provocation of an inflammatory response in the underlying tissues.[133,249,250] This may culminate, occasionally, in systemic colonization, depending on the immunological state of the host and the microbial

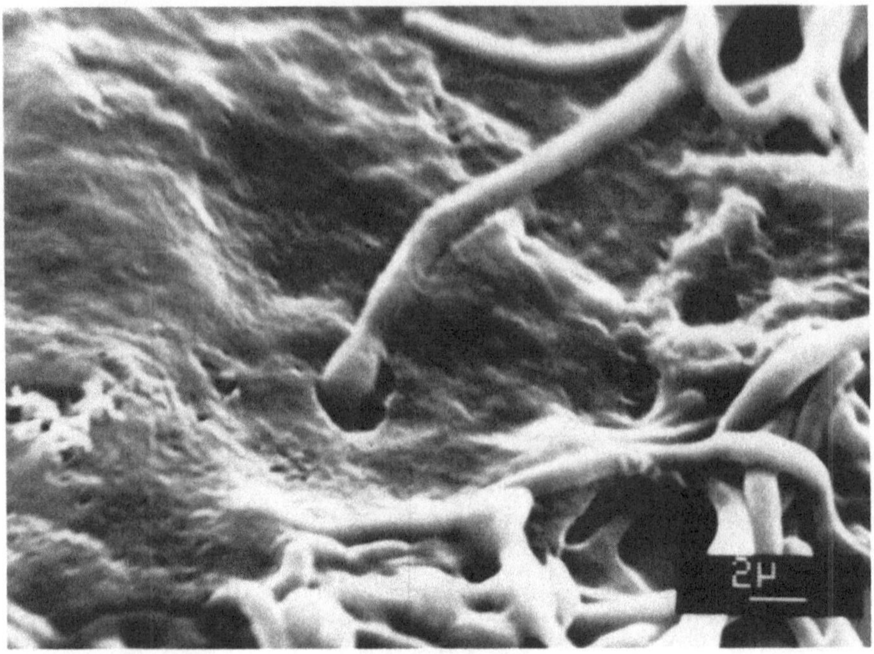

FIGURE 9. A hypha at the point of entry of a buccal epithelial cell. The host cell membrane has been broken down by the advancing hypha, which is probably due to enzyme activity known to be highest at the tip of the hypha.

ability to proliferate and alter its immediate environment, thus damaging host tissues, extending its domain, and causing infection.[251]

Samaranayake and MacFarlane,[52] in an effort to explain the role of diet-rich carbohydrates in the pathogenesis of oral candidosis, proposed a unifying hypothesis of the events which lead to mucosal colonization and invasion by *Candida* (Figure 10). This hypothesis was adopted by Tomsikova et al.[83] to describe the sequential events occurring in oral as well as vaginal candidosis. These include the enhancement of growth, multiplication and adherence of *Candida* spp. to epithelial surfaces, and the production of excessive quantities of short chain carboxylic acids (mainly pyruvates and acetates) as byproducts of sugar metabolism. The resultant acidic milieu could influence the disease process in several ways:

1. Direct irritation of the mucosal surface by the acidic metabolites which leads to inflammatory responses
2. Activation of the acidic proteinases of *Candida*, which lead to direct action on the mucosal surfaces and cleaving of the secretory IgA (this plays a major role in preventing *Candida* adhesion to epithelial cells)[252]
3. Activation of the phospholipases of *C. albicans*, which are capable of destroying the host cell membranes[216]
4. Encouragement of growth of aciduric flora (e.g., lactobacilli), and inhibition of commensal, which prefer neutral environmental conditions (these commensals play an important role in the prevention of yeast adhesion to epithelial cells)

In addition, this acidic environment is reported to enhance the adhesion of *Candida* to epithelial and acrylic surfaces.[51,253]

Once *Candida* has established itself in the mucosal surfaces and the process of epithelial cell

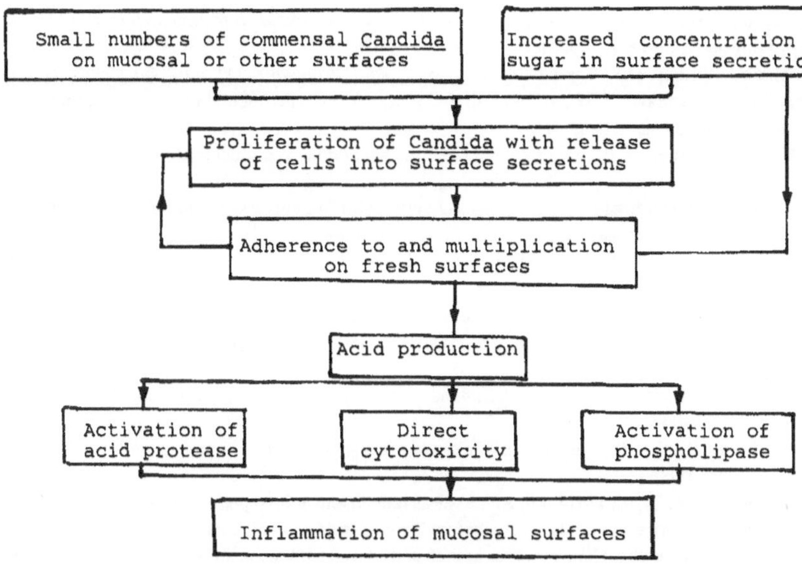

FIGURE 10. Possible mechanisms by which high concentrations of carbohydrates could influence the pathogenesis of inflammatory lesions observed in candidosis. (From Samaranayake, L.P. and MacFarlane, T.W., *FEMS Microbiol. Lett.*, 27, 1, 1985. With permission.)

layer penetration is complete, the yeast, like other microorganisms, encounters the basement membrane. This membrane acts as a filter and can, to some extent, hold up infection, but its functional integrity is soon broken by inflammation or epithelial cell damage.[254] This brings *Candida* face to face with the following host defense systems: (1) the tissue fluids, (2) the lymphatic system leading to the lymph nodes, and (3) phagocytic cells. When the fungus has overcome the obstacle of phagocytosis, systemic mycosis may develop.[259] Events occurring subsequently are beyond the scope of this review. The reader is referred to References 4, 133, 138, 140, 245, 254 to 258.

REFERENCES

1. **Smith, T.**, *Parasitism and Princeton*, Princeton University Press, Princeton, NJ, 1934.
2. **Smith, H.**, Microbial surfaces in relation to pathogenicity, *Bacteriol. Rev.*, 41, 475, 1977.
3. **Smith, H.**, The little-known determinants of microbial pathogenicity, *Symp. Soc. Gen. Microbiol.*, 22, 1, 1972.
4. **Odds, F.C.**, *Candida and Candidosis*, Leicester University Press, Leicester, 1979.
5. **Redaelli, P.**, Experimental moniliasis, *J. Trop. Med. Hyg.*, 27, 211, 1924.
6. **Benham, R.W.**, Certain monilias parasitic on man: their identification by morphology and agglutination, *J. Infect. Dis.*, 49, 183, 1931.
7. **Hasenclever, H.F.**, Comparative pathogenicity of *Candida albicans* for mice and rabbits, *J. Bacteriol.*, 78, 105, 1959.
8. **Hasenclever, H.F. and Mitchell, W.O.**, Pathogenicity of *Candida albicans* and *Candida tropicalis*, *Sabouraudia*, 1, 16, 1961.
9. **Hasenclever, H.F. and Mitchell, W.O.**, Antigenic studies of *Candida*. III. Comparative pathogenicity of *Candida albicans* group A, group B and *Candida stellatoidea*, *J. Bacteriol.*, 82, 578, 1961.
10. **Hurley, R.**, The Relationship between Host and Parasite in Systemic Moniliasis: a Clinical and Experimental Study, M.D. thesis, University of London, 1962.
11. **Hurley, R.**, The pathogenicity of *Candida stellatoidea*, *J. Pathol. Bacteriol.*, 90, 351, 1965.

12. **Hurley, R.**, Pathogenicity of the genus *Candida*, in *Symposium on Candida Infections*, Winner, H.I. and Hurley, R., Eds., Livingstone, London, 1966, 13.
13. **Hurley, R. and Winner, H.I.**, The pathogenicity of *Candida tropicalis*, *J. Pathol. Bacteriol.*, 84, 33, 1962.
14. **Dastidar, S.G., Purandare, N.M., and Desai, S.C.**, Studies on experimental cutaneous candidiasis, *Indian J. Exp. Biol.*, 9, 240, 1971.
15. **Ray, T.L. and Wuepper, K.D.**, Experimental cutaneous candidiasis in rodents, *J. Invest. Dermatol.*, 66, 29, 1976.
16. **Howlett, J.A.**, The infection of rat tongue mucosa *in vitro* with five species of *Candida*, *J. Med. Microbiol.*, 9, 309, 1976.
17. **Patridge, B.M., Athar, M.A., and Winner, H.I.**, Chick embryo inoculation as a pathogenicity test for *Candida* species, *J. Clin. Pathol.*, 24, 645, 1971.
18. **Hurley, R. and Stanley, V.C.**, Cytopathic effects of pathogenic and nonpathogenic species of *Candida* on cultured mouse epithelial cells: relation to the growth rate and morphology of the fungi, *J. Med. Microbiol.*, 2, 63, 1969.
19. **Stanley, V.C. and Hurley, R.**, Growth of *Candida* species in cultures of mouse epithelial cells, *J. Pathol. Bacteriol.*, 94, 301, 1967.
20. **Stanley, V.C. and Hurley, R.**, The growth of *Candida* species in cultures of mouse peritoneal macrophages, *J. Pathol.*, 97, 357, 1969.
21. **Hurley, R., deLouvois, J., and Mulhall, A.** Yeasts as human and animal pathogens, in *The Yeast*, Vol. 1, 2nd ed., Rose, A.H. and Harrison, J.S., Eds., Academic Press, London, 1987, Chap. 7.
22. **Gentles, J.C. and LaTouche, C.J.**, Yeasts as human and animal pathogens, in *The Yeast*, Vol. 1, 1st ed., Rose, A.H. and Harrison, J.S., Eds., Academic Press, London, 1969, Chap. 4.
23. **Seabury, J.H. and Buechner, H.A.**, Management of rare pulmonary mycoses, in *Management of Fungus Diseases of the Lungs*, Buechner, H.A., Ed., Charles C Thomas, Springfield, Illinois, 1971, 212.
24. **Hurley, R. and deLouvois, J.**, Ecological aspects of yeast-like fungi of medical importance — pathogenic potential in the genus *Candida*, *Zbl. Bakt. Hyg. I. Abt.*, Suppl. 8, 67, 1980.
25. **Khan, Z.U., Misra, V.C., Randhawa, H.S., and Damodaran, V.N.**, Pathogenicity of some ordinarily harmless yeasts for cortisone-treated mice, *Sabouraudia*, 18, 319, 1980.
26. **Albano, M.M. and Schmitt, J.A.**, Pathogenicity in mice of strains of *Candida albicans* (Robin) Berk. isolated from burn patients, *Mycopathol. Mycol. Appl.*, 49, 283, 1973.
27. **Al-Doory, Y.**, Studies on the pathogenicity of *Candida albicans* isolated from baboon, *Can. J. Microbiol.*, 14, 443, 1968.
28. **Louria, D.B., Brayton, R.G., and Finkel, G.**, Studies on the pathogenesis of experimental *Candida albicans* infections in mice, *Sabouraudia*, 2, 271, 1963.
29. **Saltarelli, C.G., Gentile, K.A., and Mancuso, S.C.**, Lethality of *Candida* strains as influenced by the host, *Can. J. Microbiol.*, 21, 648, 1975.
30. **Smith, H. and Pearce, J.H.**, Microbial pathogenicity in man and animals, in *22nd Symp. Soc. Gen. Microbiol.*, Cambridge University Press, Cambridge, 1972.
31. **Macura, A.B., Pawlik, B., and Wita, B.**, *Candida* adherence to mucosal epithelial cells with regard to its pathogenicity, *Zbl. Bakt. Hyg. I. Abt. A*, 254, 561, 1983.
32. **McCourtie, J. and Douglas, L.J.**, Relationship between cell surface composition, adherence, and virulence of *Candida albicans*, *Infect. Immun.*, 45, 6, 1984.
33. **Kearns, M.J., Davies, P., and Smith, H.**, Variability of the adherence of *Candida albicans* strains to human buccal epithelial cells: inconsistency of differences between strains related to virulence, *Sabouraudia*, 21, 93, 1983.
34. **Staib, F.**, Serum-proteins as nitrogen source for yeast-like fungi, *Sabouraudia*, 4, 187, 1965.
35. **Budtz-Jorgensen, E.**, Proteolytic activity of *Candida* spp. as related to the pathogenesis of denture stomatitis, *Sabouraudia*, 12, 266, 1971.
36. **Chattaway, F.W., Odds, F.C., and Barlow, A.J.E.**, An examination of the production of hydrolytic enzymes and toxins by pathogenic strains of *Candida albicans*, *J. Gen. Microbiol.*, 67, 225, 1971.
37. **Staib, F., Grosse, G., Blisse, A., and Senska, M.**, Proteolysing *Candida albicans* strains *in vitro* and in animal experiments, in *Yeasts and Yeast-like Microorganisms in Medical Science*, Iwata, K., Ed., University of Tokyo Press, Tokyo, 1972, 201.
38. **Rüchel, R., Tegeler, R., and Trost, M.**, A comparison of secretory proteinase from different strains of *Candida albicans*, *Sabouraudia*, 20, 233, 1982.
39. **MacDonald, F. and Odds, F.C.**, Virulence for mice of a proteinase-secreting strain of *Candida albicans* and a proteinase-deficient mutant, *J. Gen. Microbiol.*, 129, 431, 1983.
40. **Kwon-Chung, K.J., Lehman, D., Good, C., and Magee, P.T.**, Genetic evidence for role of extracellular proteinase in virulence of *Candida albicans*, *Infect. Immun.*, 49, 571, 1985.
41. **Schreiber, B., Lyman, C.A., Gurevich, J., and Needham, C.A.**, Proteolytic activity of *Candida albicans* and other yeasts, *Diagn. Microbiol. Infect. Dis.*, 3, 1, 1985.

42. **Barrett-Bee, K., Hayes, Y., Wilson, R.G., and Ryley, J.F.,** A comparison of phospholipase activity, cellular adherence and pathogenicity of yeasts, *J. Gen. Microbiol.*, 131, 1217, 1985.
43. **Ghannoum, M. and Abu Elteen, K.,** Correlative relationship between proteinase production, adherence and pathogenicity of various strains of *Candida albicans*, *J. Med. Vet. Mycol.*, 24, 407, 1986.
44. **Warnock, D.W., Speller, D., Day, J., and Farrell, A.,** Resistogram method for differentiation of strains of *Candida albicans*, *J. Appl. Bacteriol.*, 46, 571, 1979.
45. **Ghannoum, M.A., Motawy, M.S., Al-Mubarek, A.L., and Al-Awadhi, H.A.,** *Candida albicans* strain differentiation in cancer patients undergoing therapy, *Mykosen*, 28, 388, 1985.
46. **Richardson, M.D. and Smith, H.,** Production of germ tubes by virulent and attenuated strains of *Candida albicans*, *J. Infect. Dis.*, 144, 565, 1981.
47. **Sobel, J.D., Muller, G., and Buckey, H.R.,** Critical role of germ tube formation in the pathogenesis of *Candida* vaginitis, *Infect. Immun.*, 44, 576, 1984.
48. **Ghannoum, M., Khattar, M., Kayed, F., and Shridar, P.,** The influence of antineoplastic agents on experimental candidosis in mice, *Mykosen*, 28, 369, 1985.
49. **Makrides, H. and MacFarlane, T.,** An investigation of the factors involved in increased adherence of *Candida albicans* to epithelial cells mediated by *E. coli*, *Microbios*, 38, 177, 1983.
50. **Persi, M., Burnham, J., and Duhring, J.,** Effects of carbon dioxide and pH on adhesion of *Candida albicans* to vaginal epithelial cells, *Infect. Immun.*, 50, 82, 1985.
51. **Samaranayake, L. and MacFarlane, T.,** Factors affecting the *in vitro* adherence of the fungal oral pathogen *Candida albicans* to epithelial cells of human origin, *Arch. Oral Biol.*, 27, 869, 1982.
52. **Samaranayake, L. and MacFarlane, T.,** Hypothesis: on the role of dietary carbohydrates in the pathogenesis of oral candidosis, *FEMS Lett.*, 27, 1, 1985.
53. **Arbuthnott, J.P. and Smyth, C.J.,** Bacterial adhesion in host/pathogen interactions in animals, in *Adhesion of Micro-Organisms to Surfaces*, Ellwood, D.C., Melling, J., and Rutter, P., Eds., Academic Press, London, 1979, 165.
54. **Savage, D.C.,** Effects on host animals of bacteria adhering to epithelial surfaces, in *Bacterial Adhesion, Mechanisms and Physiological Significance*, Savage, D.C. and Fletcher, M., Eds., Plenum, New York, 1985, 437.
55. **Gibbons, R.J. and van Houte, J.,** Bacterial adherence and the formation of dental plaques, in *Bacterial Adherence, Receptors and Recognition*, Series B, Vol. 6, Beachey, E.H., Ed., Chapman and Hall, London, 1980, 60.
56. **Gibbons, R.J. and van Houte, J.,** Formation of dental plaques, *J. Periodont.*, 44, 347, 1973.
57. **Gibbons, R.J. and van Houte, J.,** Bacterial adherence in oral microbial ecology, *Annu. Rev. Microbiol.*, 29, 19, 1975.
58. **Stiles, H.M., Loesche, W.J., and O'Brien, T.C.,** *Microbial Aspects of Dental Cares*, Vol. 1, 2, and 3, Information Retrieval, London, 1976.
59. **Ellwood, D.C., Melling, J., and Rutter, P.,** *Adhesion of Microorganisms to Surfaces*, Published for the Society for General Microbiology, Academic Press, London, 1979.
60. **Beachey, E.H.,** *Bacterial Adherence, Receptors and Recognition*, Series B, Vol. 6, Chapman and Hall, London, 1980.
61. **Bitton, G. and Marshall, K.C.,** *Adsorption of Microorganisms to Surfaces*, John Wiley & Sons, New York, 1980.
62. **Berkeley, R.C.W., Lynch, J.M., Melling, J., Rutter, P.R., and Vincent, B.,** *Microbial Adhesion to Surfaces*, Ellis Horwood, Chichester, U.K., 1980.
63. **Beachey, E.H., Eisenstein, B.I., and Ofek, I.,** *Bacterial Adherence in Infectious Diseases*, Upjohn Company, Kalamazoo, Michigan, 1982.
64. **Elliott, K.,** *Adhesion and Microorganism Pathogenicity*, Ciba Foundation Symposium 80, Pitman Medical, London, 1981.
65. **Marshall, K.C.,** *Microbial Adhesion and Aggregation*, Springer-Verlag, Berlin, 1984.
66. **Savage, D.C. and Fletcher, M.,** *Bacterial Adhesion, Mechanisms and Physiological Significance*, Plenum , New York, 1985.
67. **Satterwhite, T.K., DuPont, H.L., Evans, D.G., and Evans, D.J., Jr.,** Role of *Escherichia coli*, colonization factor antigen in acute diarrhoea, *Lancet*, 2, 181, 1978.
68. **Middeldorp, J.M. and Witholt, B.,** K88-mediated binding of *Escherichia coli* outer membrane fragments to porcine intestinal epithelial cell brush borders, *Infect. Immun.*, 31, 42, 1981.
69. **Alkan, M., Ofek, I., and Beachey, E.H.,** Adherence of pharyngeal and skin strains of group A streptococci to human skin and oral epithelial cells, *Infect. Immun.*, 18, 555, 1977.
70. **Candy, D.C.A., Chadwick, J., Leung, T., Phillips, A., Harris, J.T., and Marshall, W.C.,** Adhesion of Enterobacteriaceae to buccal epithelial cells, *Lancet*, 2, 1157, 1978.
71. **Gould, K., Ramirez-Ronda, C.H., Holmes, R.K., and Sanford, J.P.,** Adherence of bacteria to heart valves *in vitro*, *J. Clin. Invest.*, 56, 1364, 1975.

72. **Ofek, I., Bar-Shavit, Z., Goldman, R., Mirelman, D., and Sharon, N.,** Studies on the interaction between *Escherichia coli* and mannose receptors on mammalian cell surfaces, *Isr. J. Med. Sci.*, 13, 952, 1977.
73. **Aly, R., Litz, C., Shinefield, H., Maibach, H.I., and Strauss, W.,** Binding of *Staphylococcus aureus* to human epithelial cells: role of teichoic acid, *Clin. Res.*, 27, 40A, 1979.
74. **Kallenius, G. and Winberg, J.,** Bacterial adherence to periurethral epithelial cells in girls prone to urinary-tract infections, *Lancet*, 2, 540, 1978.
75. **Kallenius, G., Mollby, R., and Winberg, J.,** *In vitro* adhesion of uropathogenic *Escherichia coli*, to human periurethral cells, *Infect. Immun.*, 28, 972, 1980.
76. **Sanford, B.A., Shelokov, A., and Ramsey, M.A.,** Bacterial adherence to virus-infected cells: a cell culture model of bacterial superinfection, *J. Infect. Dis.*, 137, 176, 1978.
77. **Sanford, B.A., Smith, N., Shelokov, A., and Ramsey, M.A.,** Adherence of group B streptococci and human erythrocytes to influenza A virus-infected MDCK cells (40424), *Proc. Soc. Exp. Biol. Med.*, 160, 226, 1979.
78. **Sanford, B.A., Smith, N., Shelokov, A., and Ramsey, M.A.,** Adherence of influenza A viruses to group B streptococci, *J. Infect. Dis.*, 141, 496, 1980.
79. **Sellwood, R., Gibbons, R.A., Jones, G.W., and Rutter, J.M.,** Adhesion of enteropathogenic *Escherichia coli* to pig intestinal brush borders: the existence of two pig phenotypes, *J. Med. Microbiol.*, 8, 405, 1975.
80. **Liljemark, W.F. and Gibbons, R.J.,** Suppression of *Candida albicans* by human oral streptococci in gnotobiotic mice, *Infect. Immun.*, 8, 846, 1973.
81. **King, R.D., Lee, J.C., and Morris, A.L.,** Adherence of *Candida albicans* and other *Candida* species to mucosal epithelial cells, *Infect. Immun.*, 27, 667, 1980.
82. **Macura, A.B.,** Adherence of *Candida* to mucosal epithelial cells, *Acta Microbiol. Pol.*, 34, 55, 1985.
83. **Tomsikova, A., Kotal, L., and Novackova, D.,** The role of adherence of *Candadida* species in the pathogenesis of vaginal and oral candidiasis, *Mykosen*, 30, 74, 1986.
84. **Ray, T.L., Digre, K.B., and Payne, C.D.,** Adherence of *Candida* species to human epidermal corneocytes and buccal mucosal cells: correlation with cutaneous pathogenicity, *J. Invest. Dermatol.*, 83, 37, 1984.
85. **Goldstein, E. and Hoeprich, P.D.,** Candidosis, in Infectious Diseases, Hoeprich, P.D., Ed., Harper and Row, Hagerstown, MD, 1977, 372.
86. **MacKenzie, D.W.,** Morphogenesis of *Candida albicans in vivo, Sabouraudia*, 3, 225, 1964.
87. **Hill, D.W. and Gerbhardt, L.P.,** Morphological transformation of *Candida albicans* in tissues of mice, *Proc. Soc. Exp. Biol. Med.*, 92, 640, 1956.
88. **Kozinn, P.J., Taschdjian, C.L., and Burchall, J.J.,** Transmission of ^{32}P labelled *Candida albicans* to newborn mice at birth, *Am. J. Dis. Child.*, 99, 31, 1960.
89. **Young, G.,** The process of invasion and the persistence of *Candida albicans* injected intraperitoneally into mice, *J. Infect. Dis.*, 102, 114, 1964.
90. **Kimura, L.H. and Pearsall, N.N.,** Adherence of *Candida albicans* to human buccal epithelial cells, *Infect. Immun.*, 21, 64, 1978.
91. **Kimura, L.H. and Pearsall, N.N.,** Relationship between germination of *Candida albicans* and increased adherence to human buccal epithelial cells, *Infect. Immun.*, 28, 464, 1980.
92. **Ghannoum, M.A. and Abu-Elteen, K.,** Unpublished data, 1986.
93. **Sandin, R.L., Rogers, A.L., Patterson, R.J., and Beneke, E.S.,** Evidence for mannose-mediated adherence of *Candida albicans* to human buccal cells *in vitro*, *Infect. Immun.*, 35, 79, 1982.
94. **Anderson, M.L. and Odds, F.C.,** Adherence of *Candida albicans* to vaginal epithelia: significance of morphological form and effect of ketoconazole, *Mykosen*, 28, 531, 1985.
95. **Garcia-Tamayo, J., Castillo, G., and Martinez, A.J.,** Human genital candidiasis: histochemistry, scanning and transmission electron microscopy, *Acta Cytol.*, 26, 7, 1982.
96. **Szaniszlo, P.J., Jacobs, C.W., and Geis, P.A.,** Dimorphism: morphological and biochemical aspects, in *Fungi Pathogenic for Humans and Animals, Part A, Biology*, Howard, D.H., Ed., Marcel Dekker, New York, 1983, 323.
97. **Evans, E.G., Odds, F.C., Richardson, M.D., and Holland, R.T.,** Effect of growth medium on filament production in *Candida albicans, Sabouraudia*, 12, 112, 1974.
98. **Evans, E.G., Odds, F.C., and Holland, K.T.,** Optimum conditions for initiation of filamentation in *Candida albicans, Can. J. Microbiol.*, 21, 338, 1975.
99. **Chaffin, W.L. and Sogin, S.L.,** Germ-tube formation from zonal rotor fractions of *Candida albicans, J. Bacteriol.*, 126, 771, 1976.
100. **Shepherd, M.G. and Sullivan, P.A.,** The production and growth characteristics of yeast and mycelial forms of *Candida albicans* in continuous culture, *J. Gen. Microbiol.*, 93, 361, 1976.
101. **Soll, D.R. and Bedell, G.W.,** Bud formation and the inducibility of pseudo-mycelium outgrowth during release from stationary phase in *Candida albicans, J. Gen. Microbiol.*, 108, 173, 1978.
102. **Mitchell, L. and Soll, D.R.,** Commitment to germ-tube or bud formation during release from stationary phase in *Candida albicans, Exp. Cell Res.*, 120, 167, 1979.
103. **Bell, G. and Chaffin, W.L.,** Nutrient-limited yeast growth in *Candida albicans:* effects on yeast mycelial transition, *Can. J. Microbiol.*, 26, 102, 1980.

104. **Buffo, J., Herman, M., and Soll, D.R.,** A characterization of pH-regulated dimorphism in *Candida albicans*, Mycopathologia, 85, 21, 1984.
105. **Soll, D.R. and Herman, M.,** Growth and inducibility of mycelium formation in *Candida albicans:* a single cell analysis using a perfusion chamber, *J. Gen. Microbiol.*, 129, 2809, 1983.
106. **Soll, D.R.,** The role of zinc in *Candida* dimorphism, in *Current Topics in Medical Mycology*, Vol. 1, McGinnis, M.R., Ed., Springer-Verlag, New York, 1985.
107. **Soll, D.R.,** *Candida albicans*, in *Fungal Dimorphism, with Emphasis on Fungi Pathogenic for Humans*, Szaniszlo, P.J., Ed., Plenum, New York, 1985, chap. 7.
108. **Poulain, D., Tronchin, G., Dubremetz, J.F., and Biguet, J.,** Ultrastructure of the cell wall of *Candida albicans* blastospores: Study of its constitutive layers by use of a cytochemical technique revealing polysaccharides, *Ann. Microbiol.*, 129, 141, 1978.
109. **Yamaguchi, H.,** Effect of biotin insufficiency on composition and ultrastructure of cell wall of *Candida albicans* in relation to its mycelial morphology, *J. Gen. Appl. Microbiol.*, 20, 217, 1974.
110. **Yamaguchi, H., Kanda, Y., and Osmi, M.,** Dimorphism in *Candida albicans*. II. Comparison of fine structure of yeast-like and filamentous phase growth, *J. Gen. Appl. Microbiol.*, 20, 101, 1974.
111. **Shannon, J.L. and Rothman, A.H.,** Transverse septum formation in budding cells of the yeast-like fungus *Candida albicans*, *J. Bacteriol.*, 106, 1026, 1971.
112. **Miller, S.E., Spurlock, B.O., and Michaels, G.E.,** Electron microscopy of young *Candida albicans* chlamydospores, *J. Bacteriol.*, 119, 992, 1974.
113. **Persi, M.A. and Burnham, J.C.,** Use of tunnic acid as a fixative-mordant to improve the ultrastructural appearance of *Candida albicans* blastospores, *Sabouraudia*, 19, 1, 1981.
114. **Silva-Hunter, M.,** *Yeasts*, in *Manual of Clinical Microbiology*, Blair, J., Lennette, E.H., and Truant, J.P., Eds., Williams & Wilkins, Baltimore, 1970, 352.
115. **Braun, P.C. and Calderone, R.A.,** Chitin synthesis in *Candida albicans:* comparison of yeast and hyphal forms, *J. Bacteriol.*, 135, 1472, 1978.
116. **Mardon, D., Balish, E., and Phillips, A.W.,** Control of dimorphism in a biochemical variant of *Candida albicans*, *J. Bacteriol.*, 100, 701, 1969.
117. **Chattaway, F.W. and Holmes, M.R.,** Cell wall composition of the mycelial and blastospore forms of *Candida albicans*, *J. Gen. Microbiol.*, 51, 367, 1968.
118. **Yamaguchi, H.,** Mycelial development and chemical alteration of *Candida albicans* from biotin insufficiency, *Sabouraudia*, 12, 320, 1974.
119. **Winner, H.I. and Hurley, R.,** *Candida albicans*, J. & A. Churchill, London, 1964.
120. **Jansons, V.K. and Nickerson, W.J.,** Induction, morphogenesis, and germination of the chlamydospore of *Candida albicans*, *J. Bacteriol.*, 104, 910, 1970.
121. **Nickerson, W.J. and Mankowski, Z.,** Role of nutrition in the maintenance of the yeast-shape in *Candida*, *Am. J. Bot.*, 40, 584, 1953.
122. **Miller, S.E. and Finnerty, W.R.,** Age-related physiological studies comparing *Candida albicans* chlamydospores to yeast, *Can. J. Microbiol.*, 25, 765, 1979.
123. **Shannon, J.L.,** Scanning and transmission electron microscopy of *Candida albicans* chlamydospores, *J. Gen. Microbiol.*, 125, 199, 1981.
124. **Szaniszlo, P.J.,** *Fungal Dimorphism with Emphasis on Fungi Pathogenic for Humans*, Plenum, New York, 1985, chap. 1.
125. **Winblad, B.,** Experimental renal candidiasis in mice and guinea pigs, *Acta Pathol. Microbiol. Scand. A*, 83, 406, 1975.
126. **Parker, J.C., McCloskey, J.J., and Knauer, K.A.,** Pathobiologic features of human candidiasis. A common deep mycosis of the brain, heart and kidney in the altered host, *Am. J. Clin. Pathol.*, 65, 991, 1976.
127. **Gresham, G.A. and Whittle, C.H.,** Studies on the invasive, mycelial form of *Candida albicans*, *Sabouraudia*, 1, 30, 1961.
128. **MacKinnon, J.E.,** Dissociation in *Candida albicans*, *J. Infect. Dis.*, 66, 59, 1940.
129. **Evans, Z.A. and Mardon, D.N.,** Organ localization in mice challenged with a typical *Candida albicans* strain and pseudohyphal variant, *Proc. Soc. Exp. Biol. Med.*, 155, 234, 1977.
130. **Simonetti, N. and Strippoli, V.,** Pathogenicity of the Y form as compared to M form in experimentally induced *Candida albicans* infections, *Mycopathol. Mycol. Appl.*, 51, 19, 1973.
131. **Winsten, S. and Murray, T.J.,** Virulence enhancement of a filamentous strain of *Candida albicans* after growth on media containing cysteine, *J. Bacteriol.*, 71, 738, 1956.
132. **Mardon, D.N., Gunn, K.L., and Robinett, E.,** Variation in the lethal response in mice to yeast-like and pseudohyphal forms of *Candida*, *Can. J. Microbiol.*, 21, 1681, 1975.
133. **Evans, Z.A.,** Tissue responses to the blastospores and hyphae of *Candida albicans* in the mouse, *J. Med. Microbiol.*, 14, 307, 1980.
134. **O'Grady, F., Pennington, J.H., and Stansfield, A.G.,** Delayed hypersensitivity in mouse-thigh candidosis, *Br. J. Exp. Pathol.*, 48, 196, 1967.

135. **Russell, C. and Jones, J.H.**, Effects of oral inoculation of *Candida albicans* in tetracycline-treated rats, *J. Med. Microbiol.*, 6, 275, 1973.
136. **Shepherd, M.G.**, Pathogenicity of morphological and auxotrophic mutants of *Candida albicans* in experimental infections, *Infect. Immun.*, 50, 541, 1985.
137. **Klotz, S.A., Drutz, D.J., Harrison, J.L., and Huppert, M.**, Adherence and penetration of vascular endothelium by *Candida* yeasts, *Infect. Immun.*, 42, 374, 1983.
138. **Diamond, R.D., Krzesicki, R., and Jao, W.**, Damage to pseudohyphal forms of *Candida albicans* by neutrophils in the absence of serum *in vitro, J. Clin. Invest.*, 61, 349, 1978.
139. **Richardson, M.D. and Smith, H.**, Resistance of virulent and attenuated strains of *Candida albicans* to intracellular killing by human and mouse phagocytes, *J. Infect. Dis.*, 144, 557, 1981.
140. **Cockayne, A. and Odds, F.C.**, Interactions of *Candida albicans* yeast cells, germ tubes and hyphae with human polymorphonuclear leucocytes *in vitro, J. Gen. Microbiol.*, 130, 465, 1984.
141. **Louria, D.B. and Brayton, R.G.**, Behaviour of *Candida* cells within leukocytes, *Proc. Soc. Exp. Biol. Med.*, 115, 93, 1964.
142. **Yamamura, M. and Valdimarsson, H.**, Participation of C_3 in intracellular killing of *Candida albicans, Scand. J. Immunol.*, 6, 591, 1977.
143. **MacKenzie, D.W.R.**, Immune responses to fungal infections, in *Immunological Aspects of Infectious Diseases*, Dick, G., Ed., MTP Press, Lancaster, England, 1979, 21.
144. **Bennett, J.W.**, Mycotoxins, mycotoxicoses, mycotoxicology and mycopathologia, *Mycopathology*, 100, 3, 1987.
145. **Stephen, J. and Pietrowski, R.A.**, *Bacterial Toxins, Aspects of Microbiology 2*, American Society for Microbiology, Washington, D.C., 1986.
146. **Iwata, K.**, Fungal toxins and their role in etiopathology of fungal infections, in *Recent Advances in Medical and Veterinary Mycology*, Iwata, K., Ed., University of Tokyo, Tokyo, 1975, 15.
147. **Henrici, A.T.**, An endotoxin from *Aspergillus fumigatus, J. Immunol.*, 36, 319, 1939.
148. **Salvin, S.B.**, Hemolysin from the yeast-like phase of some pathogenic fungi, *Proc. Soc. Exp. Biol. Med.*, 76, 852, 1951.
149. **Salvin, S.B.**, Endotoxin in pathogenic fungi, *J. Immunol.*, 69, 88, 1952.
150. **Braude, A.I., McConnell, J., and Douglas, H.**, Fever from pathogenic fungi, *J. Clin. Invest.*, 39, 1266, 1960.
151. **Dobias, B.**, Moniliasis in pediatrics, *J. Dis. Child.*, 94, 234, 1957.
152. **Mankowski, Z.T.**, The experimental pathogenicity of various species of *Candida* in Swiss mice, *Trans. N. Y. Acad. Sci.*, 19, 545, 1957.
153. **Roth, F.J. and Murphy, W.H., Jr.**, Lethality of cell-free extract of *Candida albicans, Proc. Soc. Exp. Biol. Med.*, 94, 530, 1957.
154. **Meister, H., Heymer, B., Schaffer, H., and Haferkamp, O.**, Role of *Candida albicans* in granulomatous tissue reactions. I. *In vitro* degradation of *C. albicans* and immunospecificity of split products, *J. Infect. Dis.*, 135, 224, 1977.
155. **Mankowski, Z.T.**, Production of glycoprotein by *Candida albicans* in a synthetic medium and its influence on the growth of newborn mice, *Mycopathol. Mycol. Appl.*, 36, 113, 1968.
156. **Hasenclever, H.F. and Mitchell, W.O.**, Endotoxin induced tolerance to toxic manifestation of *Candida albicans, J. Bacteriol.*, 85, 1088, 1963.
157. **Holder, I.A. and Nathan, P.**, Effect in mice of injection of variable *Candida albicans* and a cell-free sonic extract on circulating platelets, *Infect. Immun.*, 7, 468, 1973.
158. **Isenberg, H.D., Allerhand, J., and Berkman, J.I.**, An endotoxin-like fraction extracted from the cells of *Candida albicans, Nature (London)*, 197, 516, 1963.
159. **Iwata, K., Uchida, K., and Endo, H.**, 'Canditoxin', a new toxic substance isolated from a strain of *Candida albicans*. I. Relationship between strains and virulence and conditions for toxin substance production, *Med. Biol.*, 74, 345, 1967.
160. **Iwata, K. and Uchida, K.**, 'Canditoxin', a new toxic substance isolated from a strain of *Candida albicans*. II. Isolation and purification of the toxic substance, *Med. Biol.*, 74, 351, 1967.
161. **Iwata, K. and Uchida, K.**, 'Canditoxin', a new toxic substance isolated from a strain of *Candida albicans*. III. Purification of the toxic substance and physico-chemical properties of the purified toxin, *Med. Biol.*, 75, 192, 1967.
162. **Sikl, D., Masler, L., Bauer, S., Sandula, J., Tomsikova, A., and Zevazal, V.**, Allergen aktivitat der polysaccharid-protein komplexe von *Candida, Z. Immun. Allerg. Klin. Immun.*, 138, 207, 1969.
163. **Svec, R.**, On the mechanism of action of glycoprotein from *Candida albicans, J. Hyg. Epid. Microbiol. Immunol.*, 18, 373, 1974.
164. **Nosal, R., Novotny, J., and Sikl, D.**, The effect of glycoprotein from *Candida albicans* on isolated rat mast cells, *Toxicon*, 12, 103, 1974.
165. **Cutler, J.E., Friedman, L., and Milner, K.C.**, Biological and chemical characterization of toxic substances from *Candida albicans, Infect. Immun.*, 6, 616, 1972.

166. **Iwata, K. and Yamamoto, Y.**, Glycoprotein toxins produced by *Candida albicans*, in *Proc. 4th Int. Conf. Mycoses*, Brasilia, 1977.
167. **Yamamoto, Y. and Iwata, K.**, A newly isolated toxin from a strain of *Candida albicans*, *J. Jpn. Bacteriol.*, 31, 178, 1976.
168. **Iwata, K.**, Fungal toxins as a parasitic factor responsible for the establishment of fungal infections, *Mycopathology*, 65, 141, 1978.
169. **Mankowski, Z.T.**, The pathological activity of metabolic products of *Candida albicans* on newborn mice, *Mycopathol. Mycol. Appl.*, 17, 165, 1962.
170. **Mankowski, Z.T.**, Seizure induction by pentylenetetrazol in rodents receiving *Candida albicans* glycoprotein orally, *Mycopathol. Mycol. Appl.*, 45, 131, 1971.
171. **Masler, L., Sikl, D., Bauer, S., and Sandula, J.**, Extracellular polysaccharide-protein complexes produced by selected strains of *Candida albicans*, *Folia Microbiol.*, 11, 373, 1966.
172. **Sandula, J. and Merkel, M.**, Chemical structure of *Candida albicans*, *Bull. Acad. Pol. Sci.*, 13, 463, 1965.
173. **Zaikina, N.A. and Elinov, N.P.**, Influence of fungal plasmocoagulase on some microbial infections, *Mycopathol. Mycol. Appl.*, 39, 187, 1969.
174. **Al-Bassam, T., Poulain, D., Giummelly, P., Lematre, J., and Bonaly, R.**, Chemical and antigenic alterations of *Candida albicans* cell walls related to the action of amphotericin B sub-inhibitory doses, *J. Antimicrob. Chemother.*, 15, 265, 1985.
175. **McCourtie, J. and Douglas, L.J.**, Extracellular polymer of *Candida albicans*: isolation, analysis and role in adhesion, *J. Gen. Microbiol.*, 131, 495, 1985.
176. **Elorza, V., Murgui, A., and Sentandreu, R.**, Dimorphism in *Candida albicans*: contribution of mannoproteins to the architecture of yeast and mycelial cell walls, *J. Gen. Microbiol.*, 131, 2209, 1985.
177. **Davies, R.R. and Denning, T.J.V.**, *Candida albicans* and the fungicidal activity of the blood, *Sabouraudia*, 10, 301, 1972.
178. **Porter, P.S. and Lyle, J.S.**, Yeast vulvovaginitis due to oral contraceptives, *Arch. Dermatol. Syphilol.*, 93, 402, 1966.
179. **Iwata, K. and Uchida, K.**, 'Canditoxin', a new toxic substance isolated from a strain of *Candida albicans*. V. Enzymatic activity with special reference to phosphatase, *Med. Biol.*, 77, 159, 1968.
180. **Iwata, K. and Uchida, K.**, Denaturation of canditoxin and its relation to phosphatase activation, *Med. Biol.*, 83, 283, 1971.
181. **Uchida, K. and Iwata, K.**, Enzymatic properties of the extracellular alkaline phosphatase, *Med. Biol.*, 77, 171, 1968.
182. **Uchida, K. and Iwata, K.**, The role of protease on canditoxin denaturation, *Med. Biol.*, 83, 287, 1971.
183. **Uchida, K. and Iwata, K.**, Dissociation of canditoxin protein into subunits, *Med. Biol.*, 83, 291, 1971.
184. **Louria, D.B.**, *Candida* infection in experimental animals, in *Candidiasis*, Bodey, G.P. and Fainstein, V., Ed., Raven, New York, 1985, 29.
185. **Kim, Y.P., Adachi, K., and Chow, D.**, Leucine aminopeptidase in *Candida albicans*, *J. Invest. Dermatol.*, 38, 115, 1962.
186. **Kurup, P.V.**, Histochemical demonstration of alkaline phosphatase in *Candida*, *Curr. Sci.*, 32, 408, 1963.
187. **Zaikina, N.A.**, The chemical nature of coagulase from yeast-like fungi, *Vop. Med. Khim.*, 13, 130, 1967.
188. **Zaikina, N.A. and Elinov, N.P.**, Fungal plasmocoagulase, *Mycopathol. Mycol. Appl.*, 35, 10, 1966.
189. **Zaikina, N.A. and Elinov, N.P.**, Mode of action of plasma coagulase obtained from some microbes, *Z. Microbiol. Epid. Immunobiol.*, 43, 58, 1966.
190. **Zaikina, N.A. and Elinov, N.P.**, Influence of fungal plasmo-coagulase on some microbial infections, *Mycopathol. Mycol. Appl.*, 39, 187, 1969.
191. **Remold, H., Fasold, H., and Staib, F.**, Purification and characterization of a proteolytic enzyme from *Candida albicans*, *Biochim. Biophys. Acta*, 167, 399, 1968.
192. **Costa, A.L., Misefari, A., and Amato, A.**, On the enzymatic activity of certain fungi. VII. Phosphatase activity on media containing sheep's blood of pathogenic strains of *Candida albicans*, *Atti Soc. Peloritana Sci. Fis. Mat. Nat.*, VIV, 93, 1968.
193. **Uchida, K. and Iwata, K.**, 'Canditoxin', a new toxic substance isolated from a strain of *Candida albicans*. VI. Kinds of the intracellular phosphatases and their fluctuation during the growth of the toxin-producing organism, *Med. Biol.*, 77, 165, 1968.
194. **Chattaway, F.W. and Odds, F.C.**, An examination of the production of hydrolytic enzymes and toxins by pathogenic strains of *Candida albicans*, *J. Gen. Microbiol.*, 67, 255, 1971.
195. **Odds, F.C. and Hierholzer, J.C.**, Purification and properties of a glycoprotein acid phosphatase from *Candida albicans*, *J. Bacteriol.*, 114, 257, 1973.
196. **MacDonald, F. and Odds, F.C.**, Inducible proteinase of *Candida albicans* in diagnostic serology and in the pathogenesis of systemic candidosis, *J. Med. Microbiol.*, 13, 423, 1980.
197. **Rüchel, R.**, Properties of a purified proteinase and activation of blood coagulation *in vitro*, *Zbl. Bakt. Hyg. I. Abt. A*, 255, 368, 1983.

198. **Odds, F.C.,** *Candida albicans* proteinase as a virulence factor in the pathogenesis of *Candida* infections, *Zbl. Bakt. Hyg. A,* 260, 539, 1985.
199. **MacDonald, F.,** Secretion of inducible proteinase by pathogenic species, *Sabouraudia,* 22, 79, 1984.
200. **Rüchel, R., Uhlemann, K., and Böning, B.,** Secretion of acid proteinases by different species of the genus *Candida, Zbl. Bakt. Hyg. I. Abt. A,* 255, 537, 1983.
201. **Ahearn, D.G.,** Medically important yeasts, *Annu. Rev. Microbiol.,* 32, 59, 1978.
202. **Staib, F.,** Proteolysis and pathogenicity of *Candida albicans* strains, *Mycopathol. Mycol. Appl.,* 37, 345, 1969.
203. **Rüchel, R.,** On the role of proteinases from *Candida albicans* in the pathogenesis of acronecrosis, *Zbl. Bakt. Mikrobiol. Hyg. A,* 255, 524, 1983.
204. **Germaine, G.R. and Tellefson, L.M.,** Effect of pH and human saliva on protease production by *Candida albicans, Infect. Immun.,* 31, 323, 1981.
205. **Germaine, G.R., Tellefson, L.M., and Johnson, G.L.,** Proteolytic activity of *Candida albicans:* action on human salivary proteins, *Infect. Immun.,* 22, 861, 1978.
206. **Samaranayake, L.P., Hughes, A., and MacFarlane, T.W.,** The proteolytic potential of *Candida albicans* in human saliva supplemented with glucose, *J. Med. Microbiol.,* 17, 13, 1984.
207. **Shipman, B.,** Clinical evaluation of oral *Candida* in cancer chemotherapy patients, *J. Prosthet. Dent.,* 41, 63, 1979.
208. **Young, G., Krasner, R.I., and Yudkofsky, P.L.,** Interaction of oral strains of *Candida albicans* and lactobacilli, *J. Bacteriol.,* 72, 525, 1956.
209. **Logan, D.A.,** Neutral proteinase activity of *Candida albicans, Exp. Mycol.,* 10, 157, 1986.
210. **Rüchel, R., Böning, B., and Borg, M.,** Characterization of a secretory proteinase of *Candida parapsilosis* and evidence for the absence of the enzyme during infection *in vitro, Infect. Immun.,* 53, 411, 1986.
211. **Costa, A.L., Misefari, A., and Amato, A.,** Enzymatic activities of mycetes. I. Enzymatic activity of *Candida albicans* in egg yolk-containing media, *Atti XIV Cong. Naz. Microbiol.,* Messina Taormina.
212. **Costa, A.L., Misefari, A., and Amato, A.,** Enzymatic activities of mycetes. II. Phosphatidase activity in *Candida albicans* growing in media containing lecithin, *Atti XIV Cong. Naz. Microbiol.,* Messina Taormina.
213. **Werner, H.,** Untersuchungen uber die Lipase-Aktivitat bei Hefen und Hefeahnlichen Pilzen, *Zbt. Bakt. Hyg. I. Abt. A,* 200, 113, 1966.
214. **Price, M.F.,** M.I. Biol., Thesis, Institute of Biology, London, 1972.
215. **Price, M.F. and Cawson, R.A.,** Phospholipase activity in *Candida albicans, Sabouraudia,* 15, 179, 1977.
216. **Pugh, D. and Cawson, R.A.,** The cytochemical localization of phospholipase A and lysophospholipase in *Candida albicans, Sabouraudia,* 13, 110, 1975.
217. **Banno, Y., Yamada, T., and Nozawa, Y.,** Secreted phospholipases of the dimorphic fungus, *Candida albicans;* separation of three enzymes and some biological properties, *J. Med. Vet. Mycol.,* 23, 47, 1985.
218. **Samaranayake, L.P., Raeside, J.M., and MacFarlane, T.W.,** Factors affecting the phospholipase activity of *Candida* species *in vitro, Sabouraudia,* 22, 201, 1984.
219. **Price, M.F., Wilkinson, I.D., and Gentry, L.O.,** Plate method of phospholipase and lysophospholipase in *Candida albicans, Sabouraudia,* 20, 7, 1982.
220. **Williamson, M.I., Samaranayake, L.P., and MacFarlane, T.W.,** Phospholipase activity as a criterion for biotyping *Candida albicans, J. Med. Vet. Mycol.,* 24, 415, 1986.
221. **Pugh, D. and Cawson, R.A.,** The cytochemical localization of phospholipase in *Candida albicans* infecting the chick chorio-allantoic membrane, *Sabouraudia,* 15, 29, 1977.
222. **Wain, W.H., Price, M.F., and Cawson, R.A.,** The fate of *Candida albicans* inoculated on the chick chorio-allantoic membrane, *J. Dent. Res.,* 53, 1066, 1974.
223. **Wain, W.H., Price, M.F., and Cawson, R.A.,** Factors affecting plaque formation by *Candida albicans* infecting chick chorio-allantoic membrane, *Sabouraudia,* 14, 149, 1975.
224. **Rüchel, R.,** Properties of a purified proteinase from the yeast *Candida albicans, Biochim. Biophys. Acta,* 659, 99, 1981.
225. **Negi, M., Tsuboi, R., Matsui, T., and Ogawa, H.,** Isolation and characterization of proteinase from *Candida albicans:* substrate specificity, *J. Invest. Dermatol.,* 83, 32, 1984.
226. **Hattori, M., Yoshiura, K., Negi, M., and Ogawa, H.,** Keratinolytic proteinase produced by *Candida albicans, Sabouraudia,* 22, 175, 1984.
227. **Kaminishi, H., Hagihara, Y., Hayashi, S., and Cho, T.,** Isolation and characteristics of collagenolytic enzyme produced by *Candida albicans, Infect. Immun.,* 53, 312, 1986.
228. **Wright, C.D., Herron, M.J., Gray, G.R., Holmes, B., and Nelson, R.D.,** Influence of yeast mannan on human neutrophil functions: inhibition of release of myeloperoxidase related to carbohydrate-binding property of the enzyme, *Infect. Immun.,* 32, 731, 1981.
229. **Pugh, D. and Cawson, R.A.,** The surface layer of *Candida albicans, Microbios,* 23, 19, 1978.
230. **Maisch, P.A. and Calderone, R.A.,** Role of surface mannan in the adherence of *Candida albicans* to fibrin-platelet clots formed *in vitro, Infect. Immun.,* 32, 92, 1981.

231. **Tronchin, G., Poulain, D., and Vernes, A.**, Cytochemical and ultrastructure studies of *Candida albicans*. III. Evidence for modifications of the cell wall coat during adherence to human buccal epithelial cells, *Arch. Microbiol.*, 139, 221, 1984.
232. **Kulkarni, R.K., Hollingsworth, P.J., and Voltz, P.A.**, Variation in cell surface features of *Candida albicans* with respect to carbon sources, *Sabouraudia*, 18, 255, 1980.
233. **Miörner, H., Johansson, G., and Kronvall, G.**, Lipoteichoic acid is the major cell wall component responsible for surface hydrophobicity of group A Streptococci, *Infect. Immun.*, 39, 336, 1983.
234. **Magnusson, K.E., Stendahl, O., Tagesson, C., Edebo, L., and Johansson, G.**, The tendency of smooth and rough *Salmonella typhimurium* bacteria and lipopolysaccharide to hydrophobic and ionic interaction as studied in aqueous polymer two-phase systems, *Acta Pathol. Microbiol. Scand. Sect. B*, 85, 212, 1977.
235. **Perers, L., Andoker, L., Edebo, L., Stendahl, O., and Tagesson, C.**, Association of some enterobacteria with the intestinal mucosa of mouse in relation to their partition in aqueous polymer two-phase systems, *Acta Pathol. Microbiol. Scand. Sect. B*, 85, 308, 1977.
236. **Smyth, C.J., Jonsson, P., Olsson, E., Söderlind, O., Rosengren, J., Hjerten, S., and Wadström, T.**, Differences in hydrophobic surface characteristics of porcine enteropathogenic *Escherichia coli* with or without K88 antigen as revealed by hydrophobic interaction chromatography, *Infect. Immun.*, 22, 462, 1978.
237. **Van Oss, C.J.**, Phagocytosis as a surface phenomenon, *Annu. Rev. Microbiol.*, 32, 19, 1978.
238. **Wadström, T., Smyth, C.J., Faris, A., Jonsson, P., and Freer, J.**, Hydrophobic adsorptive and hemagglutinating properties of enterotoxigenic *Escherichia coli* with different colonization factor antigens and adherence factor, in *Neonatal Diarrhoea, VIDO Symp.*, Stuart Bundle Printing Press, Saskatoon, Canada, 1978, 29.
239. **Page, L. and Friend, B.**, Level of use of sugars in the United States, in *Sugars in Nutrition*, Sipple, H.L. and McNutt, K.W., Eds., Academic Press, New York, 1974, 94.
240. **Chauncey, H.H., Lionetti, F., Winer, R.A., and Lisanti, V.F.**, Enzymes of human saliva. I. The determination, distribution and origin of whole saliva enzymes, *J. Dent. Res.*, 33, 321, 1954.
241. **Ritchie, G.M., Fletcher, A.M., Main, D.M.G., and Prophet, A.S.**, The etiology, exfoliative cytology and treatment of denture stomatitis, *J. Prosthet. Dent.*, 22, 185, 1969.
242. **Shuttleworth, C.W. and Gibbs, F.J.**, The aetiological significance of *Candida albicans* in chronic angular cheilitis and its treatment with nystatin, *Br. Dent. J.*, 108, 354, 1960.
243. **Elin, R.J. and Wolff, S.M.**, The role of iron in nonspecific resistance to infection induced by endotoxin, *J. Immunol.*, 112, 737, 1974.
244. **Briody, B.A.**, Factors that determine pathogenicity, in *Microbiology and Infectious Disease*, Briody, B.A. and Gillis, R.E., Eds., McGraw-Hill, New York, 1974, Chap. 5.
245. **Smith, C.B.**, Candidiasis: pathogenesis, host resistance and predisposing factors, in *Candidiasis*, Bodey, G.P. and Fainstein, V., Eds., Raven, New York, 1985, 53.
246. **Howlett, J.A. and Squire, C.A.**, *Candida albicans* ultrastructure: colonization and invasion of oral epithelium, *Infect. Immun.*, 29, 252, 1980.
247. **Marrie, T.J. and Costerton, J.W.**, The ultrastructure of *Candida albicans* infections, *Can. J. Microbiol.*, 27, 1156, 1981.
248. **Farrell, S.M., Hawkins, D.F., and Ryder, T.A.**, Scanning electron microscope study of *Candida albicans* invasion of cultured human cervical epithelial cells, *Sabouraudia*, 21, 251, 1983.
249. **Montes, L.F. and Wilborn, W.H.**, Ultrastructural features of host-parasite relationship in oral candidiasis, *J. Bacteriol.*, 96, 1349, 1968.
250. **Cawson, R.A. and Rajasingham, K.C.**, Ultrastructural features of the invasive phase of *Candida albicans*, *Br. J. Dermatol.*, 87, 435, 1972.
251. **Garber, E.D.**, The host as growth medium. *Ann. N. Y. Acad. Sci.*, 88, 1187, 1960.
252. **Epstein, J.B., Pearsall, N.N., and Truelove, E.L.**, Oral candidiasis. Effect of antifungal therapy upon clinical signs and symptoms, salivary antibody, and mucosal adherence of *Candida albicans*, *Oral Surg.*, 51, 32, 1981.
253. **Samarayanake, L.P.**, D.D.S. thesis, University of Glasgow, Scotland, 1981.
254. **Mims, C.A.**, *The Pathogenesis of Infectious Disease*, 2 ed. Academic Press, London, 1982, Chap. 3.
255. **Ishikawa, T., Yu, M.C., and Arbersman, C.E.**, Electron microscopic demonstration of phagocytosis of *Candida albicans* by human eosinophilic leukocytes, *J. Allerg. Clin. Immunol.*, 50, 183, 1972.
256. **Iannini, P.B., Arai, G.D., and LaForce, F.M.**, Vascular clearance of blastospore and pseudomycelial phase *Candida albicans*, *Sabouraudia*, 15, 201, 1977.
257. **Trnovec, T., Sikl, D., Zemanek, M., Faberova, V., Bezek, S., Gajdosik, A., and Koprda, V.**, The distribution in mice of intravenously administered labelled *Candida albicans*, *Sabouraudia*, 16, 299, 1978.
258. **Stone, H.H., Kolb, L.D., Currie, C.A., Geheber, C.E., and Cuzzell, J.Z.**, *Candida* sepsis: pathogenesis and principles of treatment, *Ann. Surg.*, 179, 697, 1974.
259. **Borg, M. and Ruchel, R.**, Expression of extracellular acid proteinase by proteolytic *Candida* spp. during experimental infection of oral mucosa, *Infect. Immun.*, 56, 626, 1988.
260. **Cassone, A., De Bernardis, F., Mondello, F., Ceddia, T., and Agatensi, L.**, Evidence for a correlation between proteinase secretion and vulvovaginal candidosis, *J. Infect. Dis.*, 156, 777, 1987.

REFERENCE ADDED IN PROOF

261. **Kennedy, M. J., Rogers, A. L., Hanselman, L. R., Soll, D. R., and Yancey, R. J.,** Variation in adhesion and cell surface hydrophobicity in *Candida albicans* white and opaque phenotypes, *Mycopathol.*, 102, 149, 1988.

Chapter 4

THE ADHERENCE PROCESS

TABLE OF CONTENTS

I. INTRODUCTION

Infection is the process whereby the parasite enters into a relationship with the host. This process has two essential steps which include: (1) entrance of the parasite into the host, and (2) establishment and multiplication of the parasite within the host. If these steps are accomplished they will culminate in the production of progressive lesions (see Chapter 1). Adherence of pathogenic microorganisms to host tissues is believed to be the first step in the colonization and subsequent dissemination of the parasite.[1-4] In other words, microbial adhesion could be considered a prerequisite to the genesis of infection.[1,2] Thus, elucidation of mechanisms that mediate adherence of microbes to host tissues is paramount and may ultimately lead to the development of new approaches for the modification, control, or prevention of infectious diseases. Interest in the study of adherence of microorganisms to various surfaces has increased tremendously in recent years. Most early work on adherence was confined to bacteria, consequently quite a number of reviews on bacterial adherence are now available.[5-14] Work on the adherence of fungi and yeast has followed and the last decade in particular has witnessed an upsurge in research regarding the adherence of fungi, especially the yeast-like genus *Candida* and particularly *Candida albicans*.[15] This is only natural considering that this organism is a prime associate of certain diseases and disorders and has been implicated as one of the causes of mortality, particularly in immunocompromised patients (see Chapter 2). Actually, superficial candidosis now ranks among the most common of all infectious diseases, with oral and vaginal infections as its most prevalent forms.[15,16]

In this chapter we (1) review the physicochemical aspects of the adherence process of *Candida*, including the specific chemicals involved; (2) discuss the current understanding of the mechanisms of association of *Candida* with surfaces, and (3) present the test systems used to quantitate adherence of *Candida*.

II. PHYSICOCHEMICAL ASPECTS

A. INTERFACE

The initial interaction between a microbial cell and a surface, be it living or inanimate, depends largely upon the nature of this surface. This in turn depends upon the way that molecules of the surface interact with molecules of the suspending medium. This intercourse defines the characteristics of the interface. An interface is created whenever two immiscible phases come into contact. It may be created at a variety of junctions including solid/liquid, liquid/gas, and gas/solid. A discussion of the properties of interfaces is beyond the scope of this review, and the reader is referred to the chapters by Fletcher and other workers discussing some characteristics of both biological and nonbiological interfaces that may affect microbial adhesion.[17-20] Nonetheless, we briefly consider two situations which are of relevance to *Candida* adhesion to both inanimate solids, e.g., acrylic strips, and host tissues.

Once placed in an aqueous environment, a solid acquires surface charge. The nature of this charge is dependent upon the composition of the aqueous system and varies considerably with changing pH and the presence of hydrated metal ions.[21-24] Subsequently, the charged surface attracts oppositely charged ions present in the aqueous environment. Running counter to this attractive force is the thermal motion force of these opposing ions. The result of these two forces acting in opposition to each other is the formation of a diffuse electric double layer in the vicinity of the charged surface. In this layer, the concentration of the oppositely charged ions is higher than elsewhere in the aqueous system.[20] In general, this leaves solid surfaces suspended in aqueous environments with an overall negative charge.[17] Other surface properties such as interfacial free energy, wettability, hydrophobic forces, etc, important to *Candida* adhesion will be reviewed in Chapter 5.

Due to the biological properties characterizing mammalian cells, their behavior in aqueous environment is much more complex than that existing in inanimate objects. This complexity stems from the cytoplasmic membrane surrounding the cell, and obviously the boundary involved in adhesion. This is a phospholipid bilayer in which cholesterol and various kinds of protein molecules are embedded.[25] In addition, glycoproteins and glycolipids are found on the outer surface of the membrane. Membranes are dynamic structures where lipids and proteins undergo redistribution in the plane of the membrane, with a multitude of factors influencing this molecular composition and organization. Thus, biological cells are far from being "ideal" particles and an understanding of the different forces involved in their adhesion is confined, of necessity, to the semiquantitative one.[20]

The surface charge of mammalian cells is determined largely by the glycoproteins and glycolipids contributing towards the surface polysaccharide layer (glycocalyx layer). Carboxyl, phosphate, and sulphate groups provide a negative charge, while amine groups provide a positive charge. Although the overall net charge is determined by the relative presence of these groups, carboxyl groups are quantitatively more abundant than others. Consequently, mammalian cells are largely negatively charged. Like inanimate materials, an electrical double layer usually forms at aqueous interface with negatively charged animal cells.

B. CHEMICAL CONSIDERATIONS

Candida cells adhere through their walls, whereas epithelial cells do so through their plasma membranes. Therefore, the study of the chemical composition making up the surface envelopes of these two cell types is an essential prerequisite in understanding the molecular bases of the process. Adherence of *Candida* to epithelial cells is also known to be mediated by specific chemical substances associated with the walls and membranes. The chemistry of such substances is also to be considered here.

1. *Candida* Cell Wall Components

Candida is a dimorphic fungus capable of growing both as ellipsoidal buds or filamentous hyphae (see Chapter 3).[16,26] Cell walls of both forms can be prepared for chemical analysis relatively easily.[27-29] For a review regarding cell wall preparation and criteria of purity the reader is referred to Reiss[29] and Arnold.[30] Several reviews have appeared on the chemistry of yeast cell walls (see References 31 to 33) with particular emphasis on *Candida* (References 16, 26). Readers interested in cell wall chemistry of yeasts and other fungi can also refer to other papers.[29,34-39] Furthermore, there are numerous studies on the chemical characterization and ultrastructure of blastospores and mycelium of *C. albicans*, and, to a lesser extent, other *Candida* species.[40-60]

a. Glucans

Candida cell wall consists predominantly of glucans and mannans; lesser amounts of chitin, proteins, and lipids are also present.[36,61-63,68] Glucans and mannans are frequently combined with proteins in cell wall extracts as glycoproteins, mannoproteins, and glycomannoproteins.[36,65,67] Such substances are believed to originate in the cell walls of *Candida* since similar molecules are often detected in culture metabolites.[69,70]

Glucans in fungal cells are normally polymers of D-glucose, linked by β-glycosidic bonds. Only a small number of fungi have in their cell walls D-glucans with α-glycosidic linkage. In most cases glucans of fungal walls contain mixed bonds, and in yeast walls these bonds are predominantly of the types β-1,3 and β-1,6.[71,72] Cell wall glucans have been the subject of several articles.[31,73-76] Regarding the cell wall glucans in *Candida* they contain β-1,6 linked glycopyranose residues with β-1,3 branch points.[31,77-83] Shepherd et al.[84] proposed that in *C. albicans* there are three types of glucan: a mixed β-1,3/β-1,6 glucan complexed to chitin adjacent to the plasma membrane, and two highly branched glucans with either β-1,3 or β-1,6 linkages as the predominant structure.

Glucan appears to be the most abundant polymer in the cell wall of *C. albicans*,[77,84-87] and forms a partially fibrillar network at the most internal layers of the wall.[86,88] Osumi et al.[89] provided evidence showing that *de novo* synthesized β–glucan chains might initially form granular particles and then be transformed into thin fibrils with single to multiple strands oriented parallel to each other so that they develop into fibrillar structures with increasing width and length and, ultimately, into wide band-like structures.

Little work has been done on the molecular mechanism(s) of glucan biosynthesis in *Candida* and other fungi. Earlier workers have not always succeeded in demonstrating the *in vitro* biosynthesis of glucans from various nucleoside diphosphate glucose precursors in a cell-free system.[90] However, a few other investigators could isolate from mechanically disintegrated *Phytophthora cinnamomi* mycelia (a plant fungal pathogen in which glucans make some 90% of the hyphal wall) a fraction which catalyzed the incorporation of radioactive glucose from UDP-[^{14}C] glucose into a glucan containing both β-1,3- and β-1,6-linkages.[91] Interestingly, the highest activity of glucan biosynthesis was found in cell-wall-rich fractions. Later, it was revealed that the wall-free membrane fraction from this fungus could catalyze active biosynthesis of β-glucan microfibrils from UDP-glucose.[92] The biosynthesis of glucans containing β-1,3 and β-1,6 bonds from UDP-[^{14}C] glucose by cellular homogenates of *Cochliobolus miyabeanus* was also demonstrated.[93,94] Similarly, the incorporation of glucose from UDP-[^{14}C] glucose into β-1,3-glucan was demonstrated in whole cells of *Saccharomyces cerevisiae*, made permeable by a mixture of toluene-ethanol.[95] The enzyme system, which could also catalyze the biosynthesis of β-glucans from GDP-[^{14}C] glucose, was prepared from cells of *S. cerevisiae*.[96] Cell-free preparations from this yeast have also been shown to synthesize glucan from UDP-glucose.[96-98] No activity was reported in the wall fraction. In contrast, the incorporation activity of glucose into β-glucan from UDP-glucose was associated in *Neurospora crassa* with the cell wall.[99] Similar results have been obtained for the localized incorporation of glucan into the wall of a number of yeasts including *S. cerevisiae* and *Schizosaccharomyces pombe*.[100-102] In *Saprolegnia monoica* the biosynthetic activity of β-glucans from UDP-glucose was associated with the cell wall and microsomal fractions.[103,104] For full biosynthetic activity of β-glucan from UDP-glucose in *Saccharomyces cerevisiae*, glycerol, bovine serum albumin, ATP, and GTP were found to be needed (for review see Sentandreu et al.[32]). It would appear that most of the enzyme activity was associated with the plasma membrane, particularly with its internal side.[105]

López-Romero and Ruiz-Herrera[97] found glucan synthetase activities in the membrane- and wall-rich fractions of *S. cerevisiae*, with the membrane fraction having a higher total activity but a lower specific activity than the wall fraction. It is apparent that the cellular location of glucan synthetase in yeasts and other fungi is still obscure. This may be due to technical difficulties.[32,38,97,106-108] (See Parodi[109] for review on glucan biosynthesis in yeasts).

Membrane fractions obtained from *S. cerevisiae* protoplasts showed a β,1-3 glucan synthetase activity.[110] This enzyme was isolated and characterized by Shematek et al.[110] Evidence presented by these workers suggests that the activity of this enzyme is subject to a sophisticated regulatory mechanism(s).[106]

b. Mannoproteins

The second component in cell walls of *Candida* is the mannoproteins.[32,73] Polysaccharide-protein complexes, in which polysaccharides are covalently linked to proteins, are important constituents of cell walls in yeasts.[31,37,73,111] Mannoproteins are known to occur in many other biological systems, including mammals.[112] For this reason, yeast mannoproteins have been suggested as appropriate models to investigate problems related to the structure, biosynthesis, and function of these compounds.[38]

Mannoproteins have been localized in cell walls of *Candida albicans* using specific antibodies, lectin binding, and extraction with solvents, scanning and transmission electron microscopy and digestion with enzymes. The results of these studies indicate that such compounds occur predominantly in the external surface of the cell wall and next to the plasma membrane.[113] The results of recent studies on protoplasts of *C. albicans* indicate that walls are formed in two steps; the first involving a chitin microfibrillar skeleton formation and the second, the addition of glucan-mannoprotein complexes.[114,115] In a subsequent study, work from the same laboratory suggested that chitin fibrils represent the fundamental component of the inner layers at early stages of wall regeneration, while mannoproteins are predominant in the outer layers of the wall.[116]

The chemical composition of mannoproteins and their biosynthesis in various yeasts have been described.[32,38,109] Two types are commonly differentiated; the nonenzymatic mannoproteins directly involved in the architecture of the wall, and the enzymatic mannoproteins mainly involved in the degradation of molecules to facilitate their uptake.

i. Structural Mannoproteins

The best studied structure is mannan of *S. cerevisiae*.[117-120] The role of these compounds in yeasts is to hold together various components of the yeast cell wall.[121] Evidence to support this functional role of mannoproteins comes from two observations: (1) changes in a single sugar residue affect the overall conformation of a macromolecule, and (2) the carbohydrate moiety of glycoconjugates changes during development, differentiation, and transformation.[115,122] Chemical analyses of *S. cerevisiae* mannans revealed that there are two types of bonds among the sugars and polypeptides depending on the molecular weight of the polysaccharide moiety.[120,123-130] Mannose polymers with up to only five monomers are linked to the peptides with O-glycosidic bonds on the (OH) groups of threonine and serine.[124,131,132] High molecular weight mannose polymers, that are also highly branched, are linked to the peptide by a N-glycosidic bond between N-acetylchitobiose and asparagine.[124,133] Two regions can be differentiated in the polysaccharide moiety with respect to their relative location to the peptide part (Figure 1), the inner core (containing 12 to 17 mannose and two N-acetylglucosamine units joined to an asparagine residue of the protein), and the outer chain (formed by 100 to 150 mannose).[117-119,133] Mannose moieties are also linked to some of the branches in the outer chain through phosphodiester bonds.[134]

Particular attention was paid to the mannans in cell walls of *Candida* (mannan is the generic name for mannose-containing glycoproteins which, in yeast cell wall, are present as amorphous material) with different workers concentrating on various aspects.[135-138] Mannans derived from the cell walls of *C. albicans* contain predominantly α-1,6 and α-1,2 linked mannopyranose residues. They commonly contain less phosphorus than the mannans of *S. cerevisiae*.[77,139-145] A very few α-1,3 bonds have also been recorded between the mannose units.[142,144] There is a polysaccharide with α-1,6 mannan backbone with side chains containing up to six hexose monomers, linked by α-1,2 and a few α-1,3 bonds. Variations between mannans of serotypes A and B of *C. albicans* and of mannans between *C. albicans* and *C. stellatoidea* are related to the extent of branching,[32] the side chain length, and the frequency of α-1,3 bonds in the side chains.

There appears to be a common similarity in the glucans and mannans from cell walls of various *Candida* species.[141,142] Recently, Koga et. al. carried out a comparative study of three cell wall mannans of *C. albicans* serotypes A and B and *C. parapsilosis*.[146] They reported yeast mannans as having an α-1,6- linked backbone with some units being substituted at O-2 with oligosacchrides joined by α-1,2 and, to a lesser extent, by α-1,3 glycosidic bonds. Branching points in the side chains of *C. albicans* mannans were found in substantial proportions for the first time.[146] Another study compared the wall mannoproteins of the yeast and mycelial cells of *C. albicans*.[147] Zymolyase-released mannoproteins contained 92% carbohydrate (86% mannose and 6% glucose) and 7% protein. These values closely resemble those found in extracellular polymers obtained from the supernatant of *C. albicans* by McCourtie and Douglas[148] and Abu Elteen et al.[149] A great majority (86 to 88%) of the carbohydrate was N-glycosidically linked to

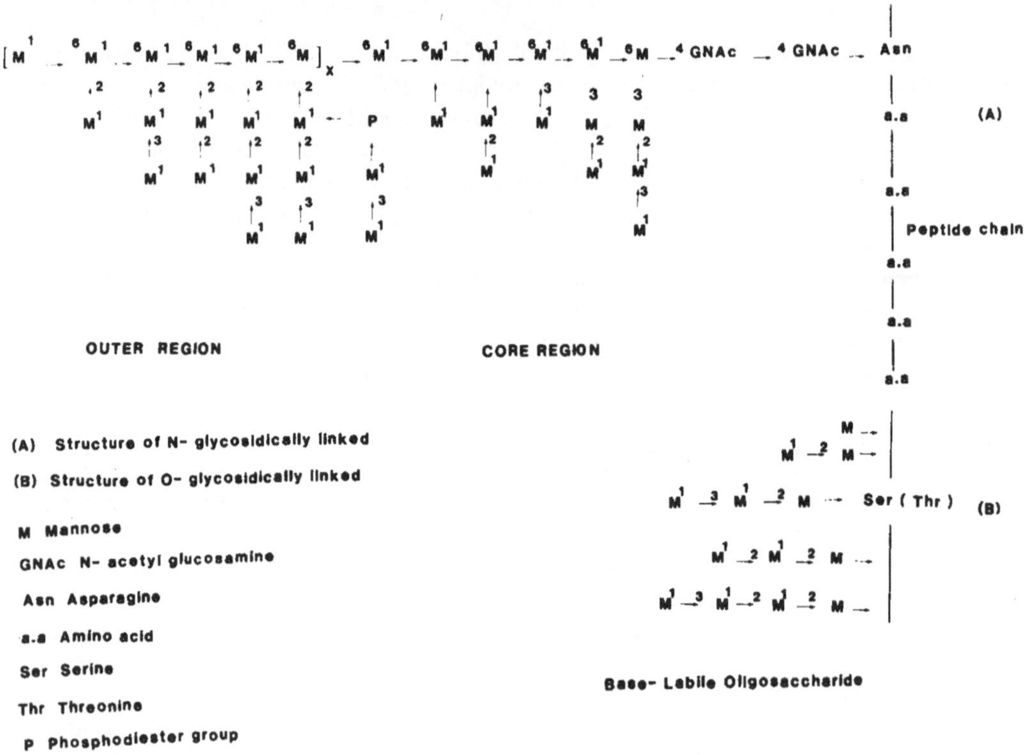

FIGURE 1. (A) Structure of *N*-glycosidically linked and (B) *O*-glycosidically linked mannan in *Saccharomyces cerevisiae*.

the protein, with the remaining linked O-glycosidically. The yeast mannoproteins had a molecular mass from 20 to 500 kDa with an average of 100 kDa (125 to 3000 mannose residues; average 600 residues) while the mycelial mannans had a molecular mass from 20 to 400 kDa, average 50 kDa (125 to 2500 mannose residues, average 300 residues).[147,150]

ii. Enzymatic Mannoproteins

The periplasmic space of the yeast cell usually contains other mannoproteins which seem to play important catalytic roles. Examples of such compounds are shown in Table 1. Evidence for the presence of invertase was gained from investigations showing that there was a loss in invertase activity of *S. cerevisiae* during its transformation into protoplasts.[151] Both invertase and acid phosphatase from cell wall of *S. cerevisiae* contain mannose (40 to 50%) and glucosamine (3 to 5%).[152-157]

Much information has now accumulated on the mechanism(s) of biosynthesis of mannoproteins in yeasts (for review see References 32, 109). It is now known that the biosynthesis of these compounds involves the production of a protein part which becomes linked to the polysaccharide moiety. Factors inhibiting protein biosynthesis inhibit the production of mannoproteins.[158-160] Mannose becomes available to the biosynthesis as GDP-Man derivatives; yet the ultimate link to the protein moiety may involve complex reactions. Individual monosaccharides may be added sequentially to the growing chain that is covalently linked to the protein. Alternatively, the polysaccharide is synthesized first and then linked to the peptide. Polyprenols may act as sugar carriers to the peptide moiety during the synthesis of asparagine-linked polysaccharides in mannoproteins. Polyprenols are also involved in the biosynthesis of O-glycosidic bonds with serine and threonine in the peptide.[161,162] Lipids have also been reported to be involved in glycosylation of proteins.[163-165] In yeast the lipid intermediate is a polyprene called dolichol. Evidence indicates that the initial reactions in the biosynthesis of N-glycosidically linked

TABLE 1
Extracytoplasmic Enzymes in *Saccharomyces cerevisiae*

Enzyme
Invertase
Acid phosphatase
Asparaginase
α-Galactosidase
Chitinase
Several exo-β-glucanases
Several endo-β-glucanases

From Sentandreu, R., Herrero, E., and Elorza, M. V., in *Microbial Cell Wall Synthesis and Autolysis.*, Nombela, C., Ed., Elsevier Science, Amsterdam, 1984, 51. With permission.

polysaccharides in *S. cerevisiae* involve the synthesis of dolichylpyrophosphate-N-acetylchitobiose.[166-170] Figure 2 summarizes the possible pathways involved in the biosynthesis of mannoproteins in *S. cerevisiae*.

c. *Chitin*

Chitin, a (1,4) linked polymer of 2-acetamido-2-deoxy-β-D-glucose, is a major structural component of the exoskeleton of invertebrates and the cell wall of fungi including *Candida*.[61,115,171-174] In yeasts, chitin forms a minor component and is restricted to the bud scars.[73,74,90,109,175] Chitin chains are usually found as a lattice of highly crystalline microfibrils within an amorphous polysaccharide or protein matrix.

Recent studies indicate that the formation of new walls in fungal (*Aspergillus*, *Saccharomyces*, and *Candida*) protoplasts is initiated by deposition of a chitin skeleton which allows incorporation of other wall constituents.[122,176,177] Sentandreu et al.[150] used the simile of chitin being the foundation of a building, responsible for the maintenance of the whole structure. These authors hypothesize that the first step in wall formation is to build ground for the cell wall, i.e., form a chitinous network on the cell surface. Glucans and high molecular weight mannoproteins are then deposited giving rise to complexes which might produce definite frameworks resulting in cell wall.[150] There is evidence to indicate that chitin in fungal walls is covalently linked to glucans.[115,178] Speculation has arisen that in *C. albicans*,[115] as in *Schizophyllum commune*,[179,180] there is a progressive crosslinking between S-glucan and chitin during cell extension, leading to a change in the solubility characteristics of the glucan. However, the formation of the initial chitin network by protoplasts of *C. albicans* is independent of the other wall polymers; the mannoprotein incorporation occurs only in a coordinated way with the synthesis of I-glucan.[115] Gow and Gooday[181] showed that the chitin extracted from *C. albicans* was α-chitin (i.e., a polysaccharide chain folded in an antiparallel conformation) and is of the short fibril type. Chattaway et al.[61] compared the cell wall of *C. albicans* yeast and mycelial form and reported that mycelial walls contained three times more chitin than those of yeast cells. This finding was subsequently suppported by the observation that homogenates from *C. albicans* mycelial cells contained twofold more chitin synthetase activity than those obtained from yeast cells.[173]

This enzyme was present in a zymogenic state in both morphological forms.[173] The biosynthesis of chitin appears to occur in a similar pattern in various biological systems (for review see Reference 32). It occurs through the following reaction:

$$\text{UDP-N-acetylglucosamine} + [\beta(1\text{-}4)\text{-N-acetylglucosamine}]_n \rightarrow [\beta(1\text{-}4)\text{-N-acetylglucosamine}]_{n+1} + \text{UDP}$$

This reaction is catalyzed by chitin synthetase. The enzyme has been isolated both from the

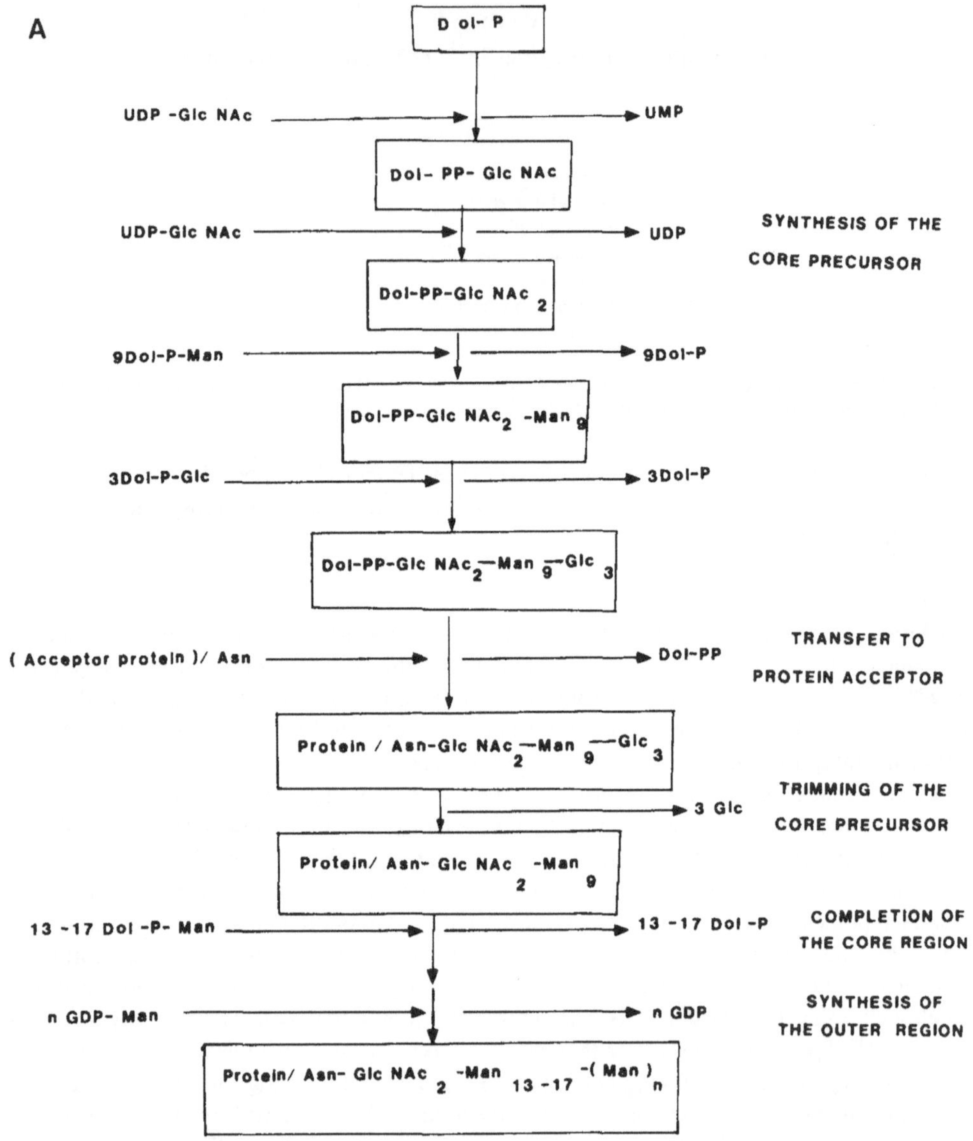

FIGURE 2. Schematic representation of the pathways leading to the synthesis of the saccharide moieties of *Saccharomyces cerevisiae* mannoprotein. The following abbreviations are employed: Dol, dolichol; P, phosphate; UDP, uridine diphosphate; Glc NAc, N-acetylglucosamine; UMP, uridine monophosphate; GDP, guanosine diphosphate; Man, mannose; Glc, glucose; Prot, protein; Asn, asparagine; Ser, serine; and Thr, threonine.

mycelia as well as from the yeast forms of dimorphic fungi, including *C. albicans*. It would appear that the only difference between chitin synthetase from mycelial and yeast forms is the stability to temperature. At 28°C the enzyme from the mycelial forms is deactivated rapidly, whereas the enzyme obtained from the yeast forms retains its activity at that temperature for hours.[107]

Studies on the subcellular distribution of chitin synthetase showed contradictory results, with some workers suggesting that it is present in the cell wall fractions, with others suggesting the membrane fraction as its location.[99,182-185] This difference in results may be due to severe disruption of cells, leading to membrane contamination. Having circumvented this problem, it was shown that chitin synthetase was disposed towards the inner cytoplasmic side of the plasma

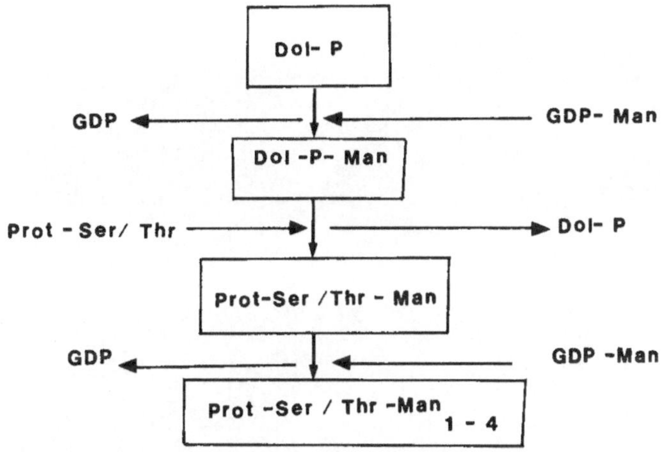

FIGURE 2B

membrane of *Saccharomyces cerevisiae* and *C. albicans.*[173,185] Alternatively, it has been reported that the enzyme is associated with small cellular particles called "chitosomes" with a diameter of 40 to 70 nm.[186-188] Chitosomes were isolated by high-speed centrifugation of a suspension of mechanically disrupted yeast cells of *Mucor rouxii.*[189] Similar structures have also been found in *S. cervisiae,* as well as other fungal cells.[190] Interestingly, the polyene antibiotics amphotericin B and nystatin, which are known to interact with sterols, inhibit the chitosome-mediated biosynthesis of chitin.[191,192] This suggests that chitosomes may possess some kind of membrane, and may be involved in the regulation of chitin biosynthesis by transferring the enzyme from the endoplasmic reticulum to the site of synthesis at the cell surface[193] (For more detail on chitin biosynthesis specifically in *C. albicans* the reader may refer to References 109,194 to 200).

d. Lipids

Lipids represent the last constituent of walls of *Candida*. Only a limited number of papers deal with this subject. Nevertheless, the occurrence of lipids in *Candida* walls has been reported for a relatively long time.[36,111] According to these reports, the lipid content of the wall is low, about 1% of the dry weight. It is tentatively assumed by many authors that lipids isolated from yeast cell wall preparations are merely contaminants from other subcellular fractions, such as the plasma membrane.[201] However, reports have appeared on the lipids in cell walls of *S. cerevisiae* and *C. albicans.*[27,56,202,203] Recently, we have attempted to isolate a membrane-free preparation of yeast cell walls (Figure 3).[27] The purity of this preparation was confirmed by the fact that its lipids revealed a fatty acid pattern basically different from that of lipids from a plasma membrane-rich fraction of the same cells (Table 2). This cell wall preparation contained 1.6% total lipids on dry weight basis. These lipids consisted of almost equal proportions of apolar and polar compounds. The apolar lipid classes were triacylglycerols (12.8% of the total lipids), fatty acids (5.6%), sterols (7.7%) and steryl esters (19.5%). The major polar lipid classes were diacylglycerophosphocholines (27.4%), diacylglycerophosphoethanolamines (18.7%), diacylglycerophosphoinositols (6.9%), ceramide monohexosides (3.3%) and ceramide dihexosides (4.1%).[27] Our work, and that of Suomalainen and Nurminen,[203] presents an alternative viewpoint to the view held that cell walls of most yeasts are devoid of lipid.

The ultrastructure of the cell wall of *Candida* has been the subject of several interesting studies.[43,45,57-60,204] Interactions between various components making up the cell wall determine the way in which these component molecules are located in the microbial cell and confer rigidity to this structure. Both covalent and non-covalent interactions have been reported to be present in the cell wall of yeast.[205] Covalent bondings exist between both glucan and chitin and glucan

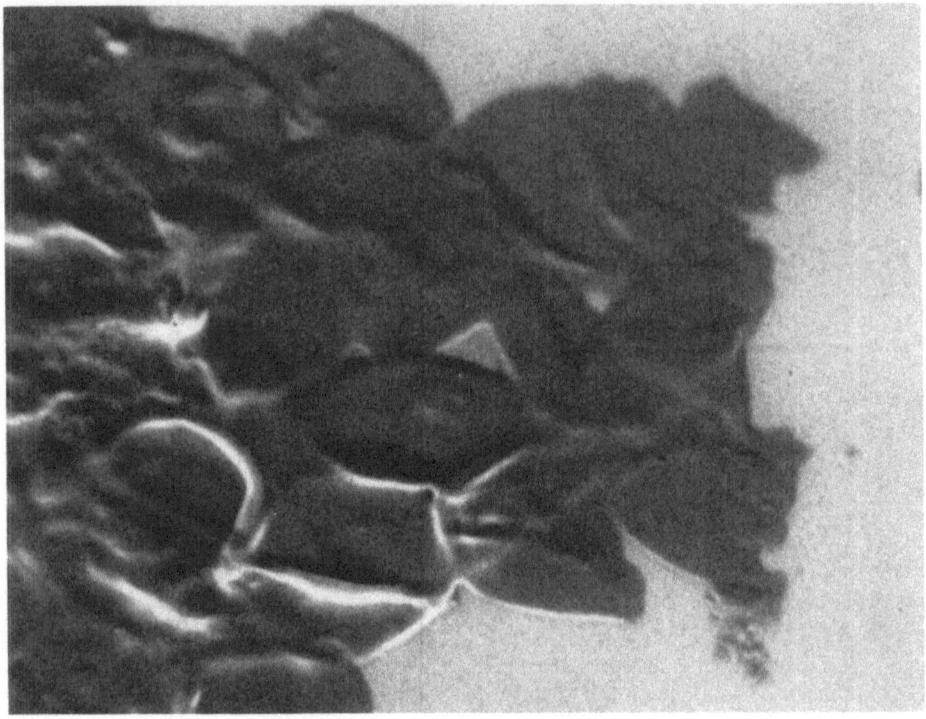

FIGURE 3. A scanning electron micrograph of isolated cell walls of *Candida albicans*. (From Ghannoum, M. A., Abu Elteen, K., and Radwan, S. S., *Mykosen*, 30, 371, 1987. With permission.)

TABLE 2
Fatty Acid Patterns of Total Lipids from Cell Walls and Cell Membranes of *Candida albicans*

Fatty acid[a]	Cell walls[b]	Cell-membrane-rich preparation[b]
14:0	27.8	5.4
16:0	29.9	26.8
16:1	11.0	Tr
16:2	tr.	Tr
16:3	5.7	15.4
18:0	12.1	24.7
18:1	11.4	3.5
18:2	ND	4.1
18:3	ND	1.5

Note: Tr, Trace; ND, not detected.

[a] Chain length: number of double bonds.
[b] Data are expressed in Wt%.

From Ghannoum, M. A., Abu Elteen, K., and Radwan, S. S., *Mykosen*, 30, 371, 1987. With permission.

and mannan.[205,206] Disulphide links, -S-S-, also play a role in holding together different mannoprotein residues which lie on or near the cell surface.[207,208] Non-covalent interactions in *S. cerevisiae* and *C. albicans* include hydrogen bonds, ionic, and cationic interactions.[205-208] Sentandreu et al.[205] speculated that differences in the cell wall of the two morphological forms of *C. albicans*, i.e., yeast and mycelium, could partially be accounted for by differences in the nature of interactions existing in their cell wall mannoproteins.

FIGURE 4. The scheme of the cell wall of *Candida albicans* yeast phase cells proposed by Poulain et al. (From Poulain, D., Tronchin, G., Dubremetz, J. F., and Biguet, J., *Ann. Microbiol. (Inst. Past.)*, 129A, 141, 1978. With permission.)

Since the cell wall of *Candida* is the site of attachment, a description of its constitutive layers is particularly relevant to this volume. The following discussion is based mainly on the paper by Poulain et al.[43] Using the polysaccharide detection technique, described by Thiery,[209] these authors suggested eight constitutive layers in the cell wall of the blastospore of *C. albicans* (Figures 4 and 5). Starting from the external layer inwards the layers are

- Layer 1. Composed of thin filaments arranged at right angles to the cell surface. Dispersed among this layer are small electron-dense granules. This layer is about 40 nm thick.
- Layer 2. Continuous, well outlined; measures 10 nm in thickness.
- Layer 3. Characterized by its low electron density; measures 40 nm in thickness.
- Layer 4. This is the innermost of these surface layers. Although it is of the same thickness as layer 3, unlike the former it is electron dense.
- Layer 5. Not well delineated towards the inside and is visible only in oblique sections. It is approximately 50 nm thick.
- Layer 6. This is unevenly distributed around the cell and is not detected in every cell of a given population. It measures about 25 nm.
- Layer 7. Shows no electron-dense deposition with the periodic acid-thio-carbohydrazide silver proteinate reaction. Consequently, this layer is not clearly contrasted from the outer layer except when layer 6 is present. Layer 7 measures about 80 nm in thickness.
- Layer 8. This is the innermost layer and lies immediately against the plasma membrane and along its entire surface. It is of irregular thickness (10 to 25 nm) but well defined.

Thus, the total thickness of all the eight layers, i.e., the whole cell wall, varies from 200 to 300 nm.

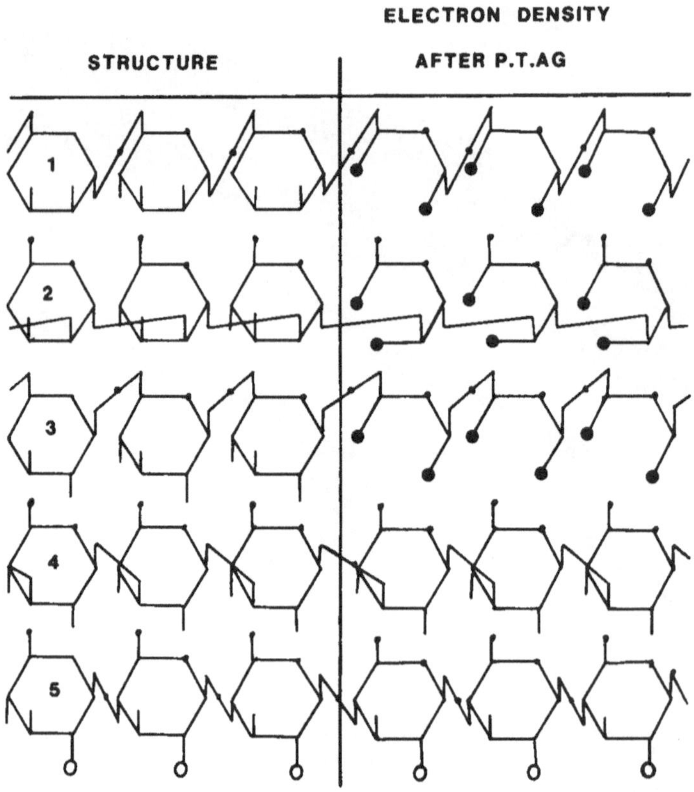

FIGURE 5. Structure and reactivity, with the Thiery method (PATAg) of polysaccharides known to be present in the cell wall of *C. albicans* blastospores. Schematic representation: straight simple lines, C-C bindings; straight lines including a little black point, O bindings; straight vertical lines, OH; straight vertical lines ending with a little black point, CH_2OH; white spheres, $NHCH_3$; black spheres, silver precipitates; 1, α-1-6-mannan; 2, α-1-2-mannan; 3, β-1-6-glucan; 4, β-1-3-glucan; 5, chitin. (From Poulain, D., Tronchin, G., Dubremetz, J. F., and Biguet, J., *Ann. Microbiol. (Inst. Past.)*, 129A, 141, 1978. With permission.)

Based on immunoelectonmicroscopic observations, Takamiya et al.[45] proposed a modification to the described scheme of the cell wall of *C. albicans* blastospores. These authors suggested the addition of channels for mannan transportation (Figure 6).

Cytochemical staining showed that each layer contains a dominant polysacchaside (Figure 4). Thus, the fifth and seventh layers were assumed to contain chitin β-1,3 and β-1,3 glucans while the outermost four layers and sometimes the sixth one appeared to contain mannans and β-1,6 glucans. Furthermore, the first and third layers appear to be rich in mannans. The innermost layer (layer 8) of the *Candida* wall is believed to contain high amounts of polysaccharides of low polymerization and in all cases non-chitinous in nature.[43]

Important differences have been recorded between the walls of the two morphological forms of *C. albicans*. It has also been demonstrated that there are temporal and spatial differences in cell wall expansion during bud and mycelial formation.[210,211] Thus, during the first two thirds of bud growth a small apical zone accounted for about 70% of surface expansion, after which expansion ceased. During mycelial growth, about 90% of expansion resulted from an apical growth zone which did not shut down. The walls of mycelial forms were found to contain three times more chitin than walls of blastospores, but only about one third as much protein.[61] Furthermore, there were differences in the amino acid composition of proteins as well as in carbohydrates from walls of both forms.[26] Two antigens were recently indentified in the extract

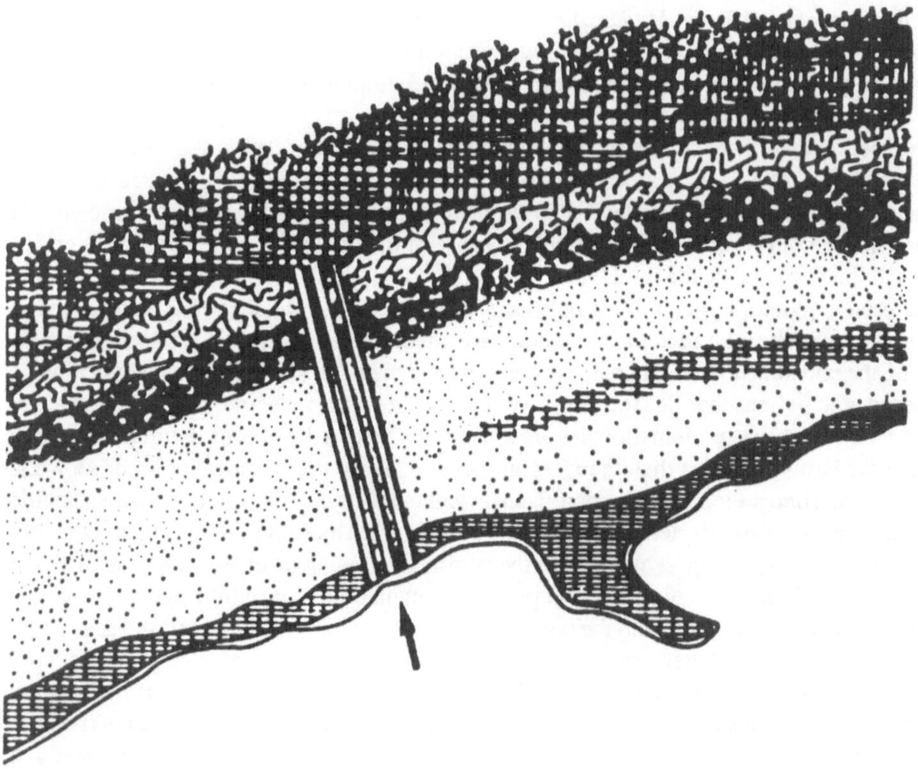

FIGURE 6. Scheme of the *Candida albicans* cell wall architecture modified from Poulain et al.[43] with suggested channels of (arrow) mannan transportation. 1, mannans, glucans; 2, mannans, glucans; 3, mannans, glucans; 4, mannans, glucans; 5, glucans, chitins; 6, mannans? proteins; 7, chitins, glucans; 8, chitins, proteins, polysaccharides. (From Takamiya, H., Vogt, A., Batsford, S., Kuttin, E. S., and Muller, J., *Mykosen*, 28, 17, 1984. With permission.)

from mycelial forms, but not in that from the yeast forms.[212] The first antigen would seem to be a 19 kDa protein, whereas the second was a polysaccharide-rich, high molecular weight antigen, connected through di-N-acetylchitobiose groups to proteins.

2. Plasma Membrane

a. Plasma Membrane of Eukaryotes

i. General Considerations

There are many excellent reviews on the plasma membrane in eukaryotic cells.[213-225] The plasma membrane is the actual biological barrier between the cytoplasm and the external environment. Only a number of very small molecules may passively penetrate this membrane. Otherwise, the exchange of the majority of substances is made possible through continuous molecular changes within the plasma membrane.[226]

In many cases, active transport across the membrane is mediated by carrier molecules localized in the membrane itself. Ingestion (and excretion) of some other substances is affected by separation of membrane vesicles from the plasma membrane (endocytosis and exocytosis). The plasma membrane is a dynamic structure,[217] whose constituent substances are continuously added to and lost. Chemically the eukaryotic membrane, irrespective of its origin, consists of lipids and proteins. The proteins serve as enzymes, or biological catalysts, and provide the membrane with its distinctive functional properties, while the lipids provide the gross structural properties of the membrane. The major lipid classes are phospholipids and sterols. Prokaryotic membranes differ from eukaryotic membranes in lacking sterols.[226,227]

ii. Architecture of the Plasma Membrane

Our knowledge of membrane chemistry and architecture has expanded in recent years. Unfortunately, most earlier studies were carried out on membranes of mammalian erythrocytes whose properties are now known to be more or less unique.

The most commonly accepted model of membrane structure today is the so-called fluid-mosaic model (Figure 7).[225,228] According to this model, the plasma membrane consists of a bimolecular lipid layer whose surface is interrupted by protein globules. The extrinsic protein globules are inserted peripherally at the polar surface of the lipid layer, whereas the integral or intrinsic globules may span the lipid bilayer entirely.[225] The peripheral parts of the extrinsic and integral protein globules often radiate polysaccharide chains and should hence be regarded as glycoproteins. Such polysaccharide moieties are involved in several physiological events including the phenomenon of adherence. Some lipids at the outer surface of the bilayer may be complexed with polysaccharides forming glycolipids. The latter molecules are also involved in adherence. The extrinsic protein globules are loosely attached to the plasma membrane and can thus be more easily removed than the integral globules. From the chemical point of view the peripherally inserted proteins are characterized by higher levels of amino acids with hydrophilic side chains than proteins inserted within the lipid bilayer. Such hydrophilic side chains certainly allow for interaction of these proteins with the surrounding aqueous phase as well as with the polar ends of the lipid molecules. Integral proteins contain both hydrophilic and hydrophobic regions.[226] The former interact with the polar ends of the lipid molecules at both surfaces of the membrane, whereas the latter are predominantly buried in the nonpolar ends of the lipid molecules, probably forming hydrophobic bonds with them. It is assumed that hydrophilic moieties (polysaccharides and amino acids) at both ends of the integral protein globules prevent reorientation of the protein in the direction of the acyl core of the lipid bilayer. The outer and inner sides of the plasma membrane vary both quantitatively and qualitatively in their extrinsic and integral proteins. In other words, proteins are asymmetrically distributed on both surfaces of the lipid bilayer, a phenomenon commonly called membrane asymmetry. Furthermore, membrane proteins may move about laterally by diffusion through the bilayer. However, not all protein globules do so. There is evidence indicating that certain integral globules are "arrested" within the lipid bilayer by a protein network present directly under the inner surface of the membrane. Some enzymatic activity is due to membrane proteins. Plasma membrane enzymes of various cells include phosphatases, peptidase, cellobiase, sterol esterases, and lipases. Workers in this field differentiate between ectoenzymes whose activities are associated with the outer surface of the membrane, and the endoenzymes whose activities are associated with the inner surface.

As previously mentioned, phospholipids are the major lipid class present in the plasma membrane of eukaryotes (and prokaryotes).[229,230] In addition, sterols and steryl esters are usually present.[214,217] There are also reports indicating that other neutral lipids, such as triacylglycerols, may be present in small amounts. The major phospholipids present in the membrane vary according to the organism as well as to the cell type. In most cases diacylglycerophosphocholines, diacylglycerophosphoethanolamines, diacylglycerophosphoserines, diacylglycerophosphoinositols, and sphingomyelins are major constituents of the plasma membrane (Figure 8).[219] Membrane lipids move about through the bilayer.[214] Their mobility is even more active than that of the proteins. One lipid molecule may travel laterally several microns within one second. Transverse mobility is rare. Apparently the movement of membrane protein and lipid constituents is largely controlled by the membrane's fluidity.[231,232] The latter in turn depends on the degree of unsaturation of the acyl moieties of phospholipids in the bilayer.[214]

The constituent fatty acids of membrane phospholipids are normally a mixture of saturated and unsaturated, with up to 5 double bonds or more per molecule. The fluidity of the membrane is directly proportional to the number of double bonds in the acyl moieties. In lipid layers containing predominantly saturated acyl moieties, the latter become "packed" next to one

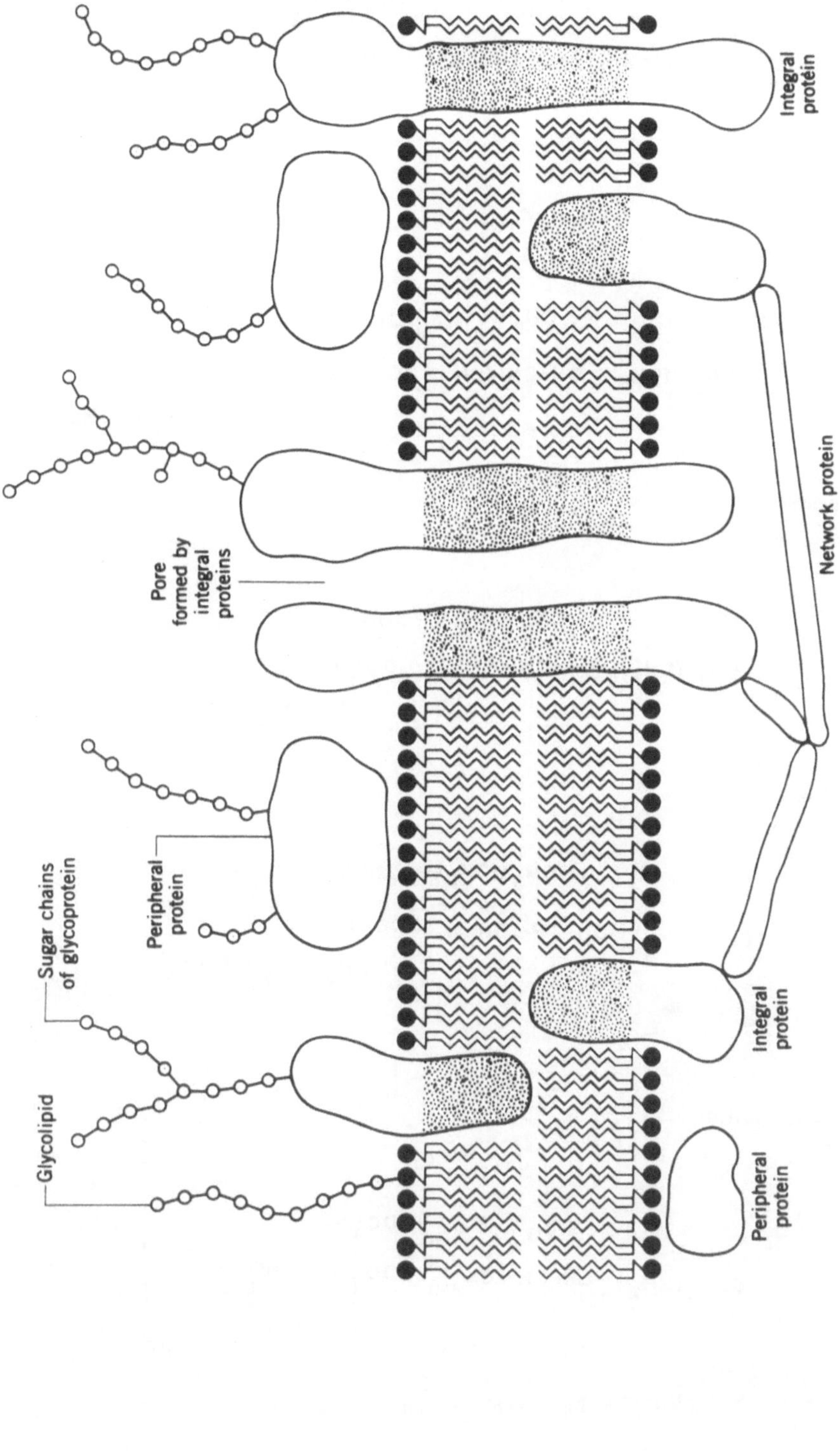

FIGURE 7. The fluid-mosaic model of membrane structure.

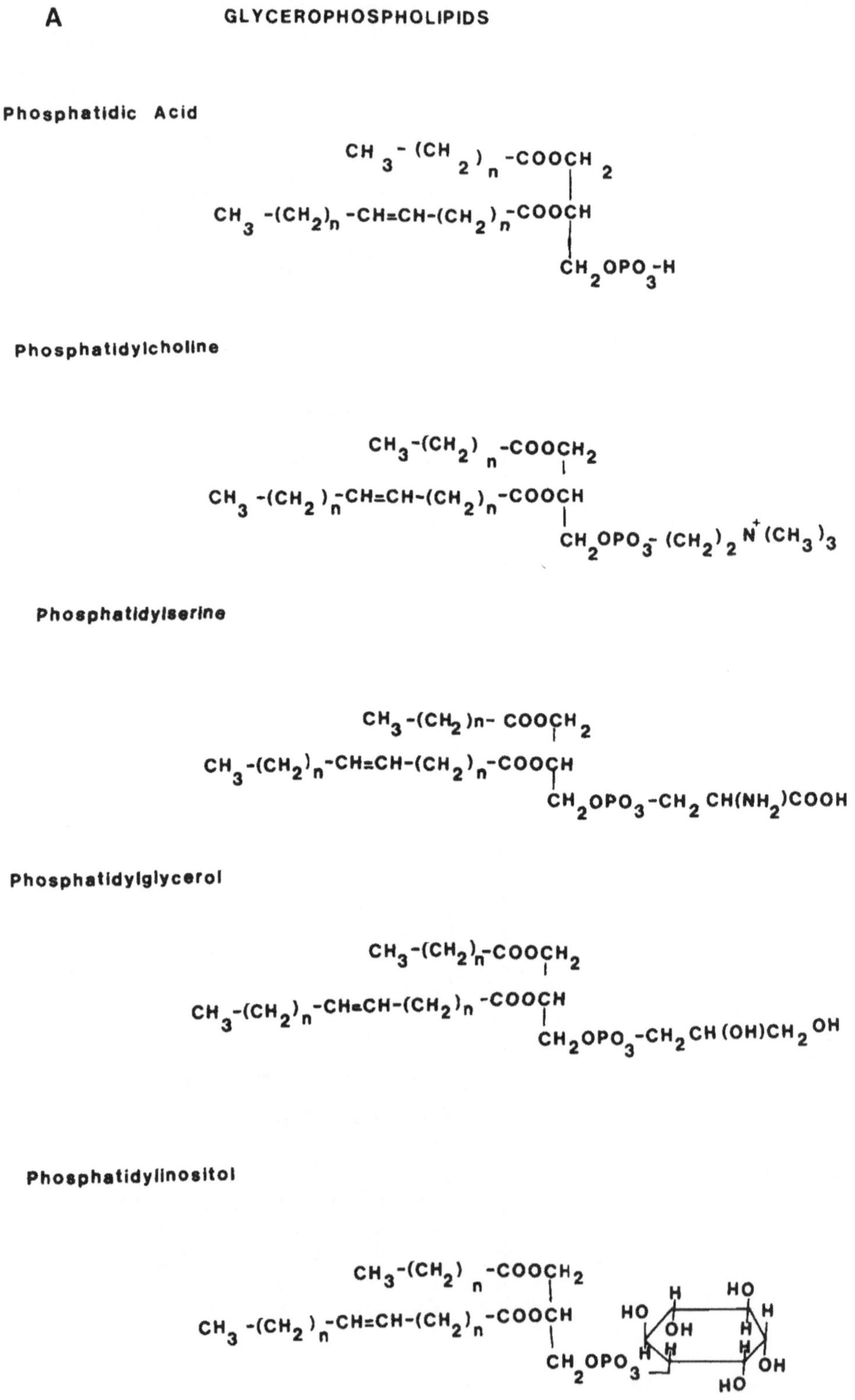

FIGURE 8. Lipids in biological membranes.

SPHINGOLIPIDS

Ceramide

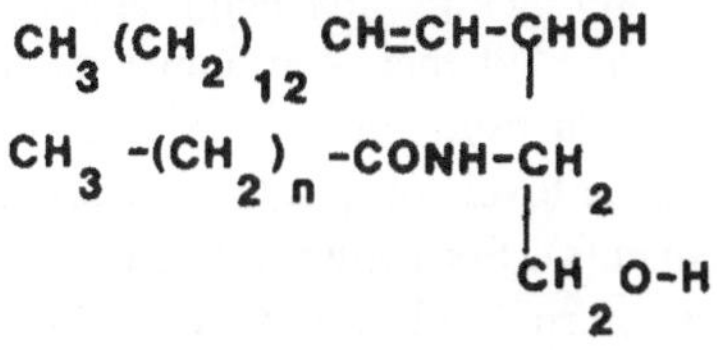

Sphingomyelin

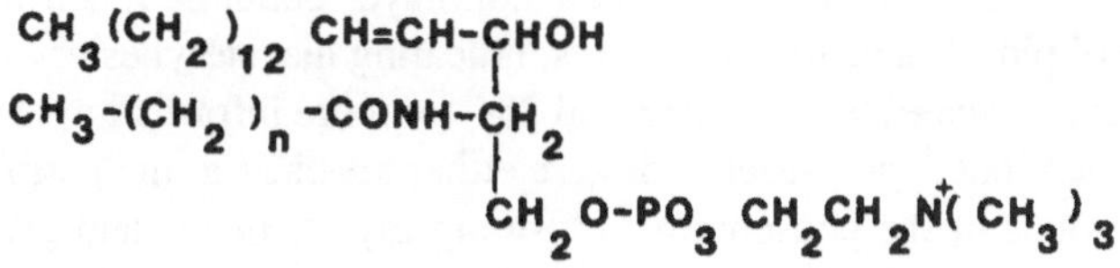

Glycosphingolipid

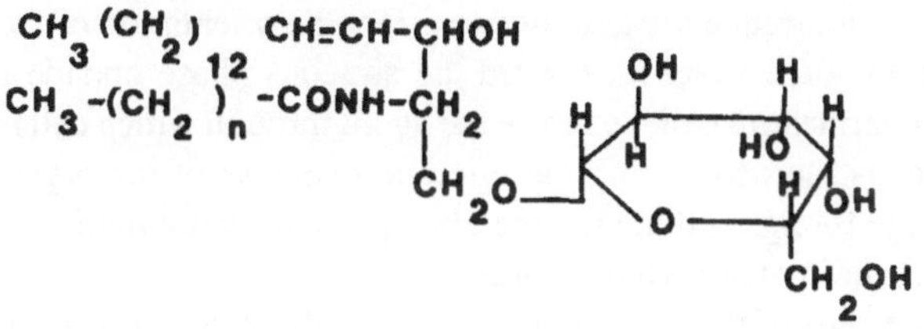

STEROLS

Cholesterol

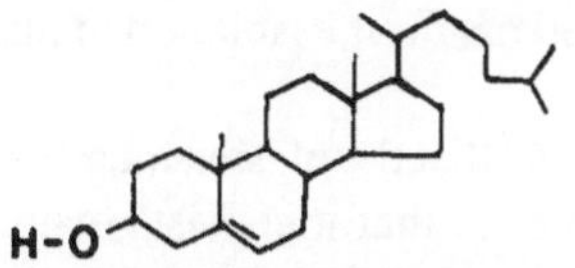

FIGURE 8B.

another in an ordered crystalline pattern resulting in a relatively rigid unit.[223] Alternatively, in lipid layers containing higher levels of unsaturated acyl moieties, the packing of adjacent acyl residues is less "perfect" because the double bonds produce bends in the carbon chain. Such bends lead to structural deformations and the result is a more fluid membrane; the degree of fluidity obviously dependent upon the degree of unsaturation.[223] It is through controlling the content of unsaturated fatty acids in membrane phospholipids that psychrophilic and thermophilic organisms can adjust the fluidity of their plasma membranes to the level suitable for active transport. Various analyses have indicated that membranes of thermophiles are richer in saturated fatty acids than membranes of psychrophiles. Similarly, incubating cells at lower temperatures usually leads to increased levels of unsaturated acyl lipids, which again can be considered as a regulatory mechanism(s). For further information on the organization of the animal cell membrane the reader is referred to References 218 to 221.

The plasma membrane also shows some degree of asymmetry, as far as the lipid composition is concerned. The type of such asymmetry varies from one membrane to another. In the erythrocyte membrane it has been observed that the outer monolayer is rich in diacylglycerophosphocholines whereas the inner monolayer contains much higher levels of the aminophospholipids. There are also reports indicating that steryl esters are predominant in the outer monolayer. (See References 233 and 234 for more information on membrane lipids.)

As mentioned, polysaccharides are either attached to the peripheral protein globules or to polar groups of the phospholipids forming glycoproteins and glycolipids, respectively.[235,236] Each polysaccharide chain is relatively short and may branch. The constituent monosaccharides in membrane glycoproteins and glycolipids include D-galactose, D-mannose, L-fucose, N-acetyl neuraminic acid, N-acetyl-D-glucosamine, and N-acetyl-D-galactosamine. The functions of these oligosaccharides are not completely understood. In view of their hydrophilic nature they are frequently assumed to affect the orientation of the protein and lipid molecules, to which they are attached, toward the aqueous phase outside the cell.[237-239] The membrane oligosaccharides are believed to be the agent through which cells of an organism can recognize foreign cells, thus triggering the immune response of the organism. Recognition is possible because the foreign cell membrane glycoproteins contain different carbohydrate markers than those present in their own membrane.

Among cells of the same tissue it is generally assumed that the surface polysaccharides act as an adhesive, linking these cells together. Oligosaccharides of the glycosphingolipids are believed to play a role in cell-to-cell recognition. Furthermore, they are probably involved in the transmembrane transport and in providing the antigenic properties of the membrane.[319]

b. *Candida* Plasma Membrane

There are a number of reports on the yeast plasma membrane, and that of *C. albicans* in particular.[240-245] A method has been described for the isolation of plasma membranes from protoplasts of *C. albicans*.[244] The protoplasts were first osmotically lysed and then fractionated by using a combination of differential and discontinuous sucrose density-gradient flotation centrifugation. The purity of the plasma membrane preparation was determined by radioactive dansylation and iodination of membranes of intact protoplasts followed by localization of radioactivity upon fractionation. Using this method a claim was made for less than 4% contamination by other membranous components.[244] This is an improvement when compared to techniques followed by other workers for the isolation of plasma membrane from yeasts such as *S. cerevisiae*.[246-248]

Chemical analyses have confirmed that the plasma membrane of *C. albicans* resembles in its composition that of other eukaryotes in that it consists mainly of proteins and lipids.[243,244] In addition, small amounts of carbohydrates and nucleic acids are also present (Table 3). There are interesting differences in the chemical composition of plasma membranes from yeast and mycelial forms of this fungus.

TABLE 3
Chemical Composition of Purified Plasma Membranes of *Candida albicans*

	Composition (%)		
	Yeast form		Mycelial form
Composition	Exponential phase	Stationary phase	Exponential phase
Protein	52 ± 2	43 ± 3	45 ± 3
Lipid	43 ± 3	31 ± 2.5	31 ± 3
Carbohydrate	9 ± 0.5	26 ± 3	25 ± 4
Nucleic acid	0.3 ± 0.1	0.2 ± 0.2	0.5 ± 0.1
Phospholipids			
Phosphatidyl-ethanolamine	70 ± 4	34 ± 2	50 ± 4
Phosphatidylserine (+ phosphatidylcholine)	11 ± 2	11 ± 3	—
Phosphatidylcholine	4 ± 1	44 ± 3.5	50 ± 4
Sphingolipid	15 ± 4	10 ± 1	—
Neutral Lipids			
Sterol esters	40 ± 4	47 ± 4	28 ± 1
Triglycerides	24 ± 3	29 ± 2.5	36 ± 2
Free fatty acids	17 ± 3	12 ± 1	27 ± 4
Free sterol	19 ± 2	13 ± 1	9 ± 1

From Kerridge, D., Koh, T. Y., Marriott, M. S., and Gale, E. F., *Microbial and Plant Protoplasts*, Peberdy, J. F., Rose, A. H., Rogers, H. J., and Cocking, E. C., Eds., Academic Press, London, 1976, 23. With permission.

The constituent amino acids in proteins from plasma membranes of yeast and mycelial forms were studied and compared.[244] The results in Table 4 show only slight differences in the amino acid composition. There were no striking differences in the amino sugar and neutral sugar contents of membranes between the two morphological forms (Table 5).[244] On the other hand, major differences in the composition of lipids and the constituent fatty acids of lipids from membranes of yeast and mycelial forms have been found (Table 3).[243] Membranes from yeast forms were much richer in sterols, steryl esters, and diacylglycerophosphocholines than those from mycelial forms. Moreover, diacylglycerophosphoserines and sphingolipids were present in membranes from the yeast forms but not from mycelial forms. The predominant sterol in plasma membranes of *C. albicans* was ergosterol, irrespective of the morphological form (Tables 6 and 7). Analysis of lipids from whole cells confirmed that yeast forms were richer in sterols and steryl esters than the mycelial forms (Table 8).[249] However, lipids from mycelial forms contained much larger amounts of sterols bound as steryl glycosides, predominantly cholesteryl mannoside than the yeast forms. Table 9 shows the identities and relative proportions of sterols from the two different forms. Acyl lipids from membranes of yeast forms contained higher proportions of linoleic acid than those from membranes of mycelial forms.[244] Alternatively, apolar and polar lipid fractions from whole mycelial forms were found to contain higher levels of linoleic and linolenic acids but lower levels of oleic acids than the corresponding fractions from whole yeast forms (Table 10).[249] These differences in the composition of the plasma membrane could be of significance as far as the adherability is concerned.

TABLE 4
Amino Acid Composition of Yeast — and Mycelial — Form Plasma Membranes of *Candida albicans*

Amino acid	Yeast form (mol %)	Mycelial form (mol %)
Alanine	8.4 ± 0.1	8.8 ± 0.2
Arginine	3.8 ± 0.1	3.8 ± 0.3
Aspartic acid	11.3 ± 0.2	10.6 ± 0.1
Glutamic acid	9.4 ± 0.5	10.0 ± 0.0
Glycine	7.5 ± 0.1	7.6 ± 0.0
Histidine	2.5 ± 0.1	1.8 ± 0.4
Isoleucine	4.6 ± 0.1	5.1 ± 0.1
Leucine	9.4 ± 0.0	9.2 ± 0.0
Lysine	9.3 ± 0.6	10.9 ± 0.1
Methionine	0.6 ± 0.3	0.3 ± 0.0
Phenylalanine	4.3 ± 0.0	4.3 ± 0.2
Proline	5.9 ± 0.6	5.5 ± 1.0
Serine	8.4 ± 0.1	8.2 ± 0.1
Threonine	6.4 ± 0.0	6.3 ± 0.1
Tryptophan	0.7 ± 0.2	0.4 ± 0.0
Tyrosine	2.8 ± 0.0	3.3 ± 0.1
Valine	4.7 ± 0.0	5.1 ± 0.1

From Marriott, M. S., *J. Gen. Microbiol.*, 86, 115, 1975. With permission.

TABLE 5
Amino and Neutral Sugar Composition of Yeast — and Mycelial — Form Plasma Membranes of *Candida albicans*

	Component or relative retention time[a]	Yeast form (%)	Mycelial form (%)
Amino sugars	0.867	35.0 ± 2.0	36.0 ± 1.0
	Glucosamine	65.0 ± 2.0	64.0 ± 1.0
Neutral sugars	0.765	6.0 ± 1.0	8.0 ± 4.0
	0.777	20.0 ± 2.0	10.0 ± 0.0
	Mannose	18.0 ± 0.0	20.0 ± 0.0
	Glucose	55.0 ± 0.0	63.0 ± 0.0

[a] Relative to inositol = 1.00.

From Marriot, M. S., *J. Gen. Microbiol.*, 86, 115, 1975. With permission.

III. SPECIFIC CHEMICALS INVOLVED IN ADHERENCE

The initial contact between microorganisms and epithelial tissues is most probably a random and instantaneous process which depends on general rather than specific surface characteristics.[8,250] This type of adhesion is reversible (since the microbial cells can be dislodged by the shearing effects of water jets)[251] and serves to stabilize the microbial cell at that site and prepares for the more permanent association between the microorganism and the epithelial cell. This permanent interaction, known as irreversible adhesion (since the microbial cells are firmly adhering to a surface and could not be dislodged by washing)[251] is a specific interaction mediated by (macro)- molecules on the microbial surface (referred to as adhesins) that combine with complementary structures on the host cell surfaces (referred to as receptors).[252]

TABLE 6
Sterol Composition of the Sterol Fraction from Yeast — and Mycelial — Form Plasma Membranes *Candida albicans*

Component	Yeast form (%)	Mycelial form (%)
Unidentified	8.0 ± 3.0	4.0 ± 0.0
Unidentified	6.0 ± 2.0	2.0 ± 0.0
Unidentified	12.0 ± 3.0	5.0 ± 2.0
Zymosterol	32.0 ± 5.0	36.0 ± 3.0
Ergosterol	10.0 ± 2.0	10.0 ± 1.0
(24,28-Dehydroergosterol)[a]	10.0 ± 2.0	12.0 ± 1.0
(3-β Hydroxy-24-methyl-cholesta-5, 7-diene)	7.0 ± 2.0	6.0 ± 0.0
(3-β Hydroxy-24-methyl-cholesta-7-ene)	ND	6.0 ± 0.0
(Fecosterol)	4.0 ± 1.0	ND
(Episterol)	8.0 ± 3.0	19.0 ± 2.0

[a] Identified by U.V. spectrum.

From Marriott, M. S., *J. Gen. Microbiol.*, 86, 115, 1975. With permission.

TABLE 7
Free Sterol Composition of Yeast — and Mycelial — Form Plasma Membranes of *Candida albicans*

Component	Yeast form (%)	Mycelial form (%)
Squalene	2.0 ± 0.0	2.0 ± 0.1
Breakdown products of ergosterol	{7.0 ± 1.0 5.0 ± 1.0	{11.0 ± 3.0 6.0 ± 2.0
Calciferol	12.0 ± 4.0	Trace
Zymosterol	ND	16.0 ± 2.0
Ergosterol	50.0 ± 5.0	42.0 ± 6.0
(24,28-Dehydroergosterol)[a]	7.0 ± 0.0	11.0 ± 3.0
(3-β-Hydroxy-24-methyl-cholesta-5, 7-diene)	7.0 ± 2.0	9.0 ± 2.0
(4,4′-Dimethyl zymosterol)	6.0 ± 2.0	ND

[a] Identified by U.V. spectrum.

From Marriott, M. S., *J. Gen. Microbiol.*, 86, 115, 1975. With permission.

A. ADHESINS

Ultrastructural evidence indicates that specific interaction between *Candida* and epithelial cells is mediated by a floccular-fibrillar adhesin layer present on the outer surface of the yeast. This layer has been repeatedly defined by electron microscopy.[253,256] Different workers have referred to this surface by different names including: fibrillar,[257,258] floccular,[259,260] fibrillar-floccular,[261] or fibrogranular cell wall coat.[260] Although Kennedy[262] suggests that these represent two different morphological classes of adhesins (floccular and fibrillar), we believe they are one and the same layer. Two points support our conclusion: first the degree of development of this

Table 8
Comparison of Lipids from Yeast and Mycelial Forms of *Candida albicans* ATCC10231

Lipids	12 h cultures Yeast forms	12 h cultures Mycelial forms	96 h cultures Yeast forms	96 h cultures Mycelial forms
Apolar compounds				
Hydrocarbons	Tr	Tr	Tr	Tr
Steryl esters	Tr	Tr	8.5	25.3
Alkyl esters	Tr	Tr	3.8	8.8
Triacylglycerols	Tr	Tr	6.4	16.5
Fatty acids	Tr	Tr	10.4	11.2
Sterols	14.8	Tr	9.3	Tr
Polar compounds				
Esterified steryl glycosides	ND	ND	Tr	6.5
Monogalactosyldiacylglycerols	Tr	Tr	Tr	2.7
Steryl glycosides	Tr	16.2	Tr	9.1
Ceramide monohexosides	10.3	Tr	8.8	Tr
Phosphatidylethanolamines	11.4	12.7	9.6	Tr
Phosphatidylglycerols	9.2	Tr	7.3	Tr
Phosphatidylcholines	25.6	28.8	15.0	4.1
Digalactosyldiacylglycerols	Tr	10.2	Tr	6.2
Phosphatidylserines	8.8	9.9	6.9	ND
Phosphatidylinositols	10.9	11.4	8.9	ND
Phosphatidic acids	9.0	7.3	4.8	ND

Note: Values are expressed as a percentage of the total lipids. Tr, traces; ND, not detected.

From Ghannoum, M. A., Janini, G., Khamis, L., and Radwan, S. S., *J. Gen. Microbiol.*, 132, 2367, 1986. With permission.

TABLE 9
Identities and Relative Proportions (%) of Sterols in Total Lipids[a] From Yeast and Mycelial Forms of Candida Albicans ATCC 10231

Sterol	Yeast forms	Mycelial forms
Degradation product of ergosterol	4.5	ND
Cholesterol	ND	37.8
Zymosterol	17.0	13.7
Ergosterol	14.6	5.6
24,28-Dehydroergosterol	42.7	18.8
3-β-Hydroxy-24-methyl-cholesta-5, 7-diene	11.7	22.2
4,4′-Dimethylzymosterol	9.5	2.0

Note: Cultures were 12 h old. ND, not detected.

[a] Subjected to acid hydrolysis.

From Ghannoum, M. A., Janini, G., Khamis, L., and Radwan, S. S., *J. Gen. Microbiol.*, 132, 2367, 1986. With permission.

TABLE 10
Fatty Acid Composition of Apolar and Polar Lipid Fractions from Yeast and Mycelial Forms of *Candida albicans* ATCC 10231

	12 h cultures				96 h cultures			
	Yeast forms		Mycelial forms		Yeast forms		Mycelial forms	
Fatty acid	Apolar lipids (%)	Polar lipids (%)	Apolar lipids (%)	Polar lipids (%)	Apolar lipids (%)	Polar lipids (%)	Apolar lipids (%)	Polar lipids (%)
14:0	4.3	4.2	4.2	3.6	6.6	7.0	6.6	6.5
14:1	1.8	6.2	Tr	Tr	5.1	2.0	5.7	8.2
16:0	11.8	13.1	14.5	15.3	28.5	25.4	28.4	20.7
16:1	11.4	9.6	12.2	13.2	10.4	10.0	14.6	Tr
16:2	Tr	Tr	Tr	Tr	8.7	3.1	6.0	9.9
18:0	14.0	10.3	16.7	15.2	14.5	7.4	10.4	12.1
18:1	37.6	41.5	26.7	26.7	14.7	22.4	15.9	8.2
18:2	12.0	10.7	16.3	13.5	5.5	18.8	2.7	Tr
18:3	7.5	4.4	9.4	12.5	5.7	3.9	9.3	33.7

Note: Apolar lipid fractions contained free fatty acids, triacylglycerols, and steryl esters; polar lipid fractions contained phospholipids and glycolipids. Tr, traces.

From Ghannoum, M. A., Janini, G., Khamis, L., and Radwan, S. S., *J. Gen. Microbiol.*, 132, 2367, 1986. With permission.

layer was found to be dependent on the nutrients made available to *Candida* (Figure 9), especially on galactose or sucrose concentrations.[43] Second, cytochemical evidence shows that both types of adhesins suggested by Kennedy stain with ruthenium red,[255] which has an affinity for anionic polymers such as polysaccharides.[263]

Montes and Wilborn[253] were the first to observe the presence of this floccular material on the outer surface of *C. albicans* cell wall. Description of this layer in literature varied with some workers describing it as thin filamentous structures (2 to 10 nm in diameter) arranged perpendicularly to the cell surface and evenly distributed around the entire cell,[12,257,258] while others reporting it as a thick (approximately 100 to 400 nm) fuzzy, amorphous cell wall coat.[259,260] This adhesin layer is reported, in some studies, as being unevenly distributed on the cell surface or localized only at an adhesive site,[255,256,259,261] while in other studies it is reported to have an ordered alignment around the cell wall.[260]

There is experimental evidence to indicate that the degree of development of this outermost layer is related to the adherability, cells with thicker layers being more adherent.[256,258,261,264,265] It has been suggested that this layer may represent appendages functioning as bacterial fimbriae do during adherence.[12] It has been mentioned that *C. albicans* produces fimbriae.[266] Interesting modifications of this outermost layer during *C. albicans* adherence to buccal epithelial cells have been observed.[260] The attachment of the yeast to epithelial cells is reported to involve spatial rearrangement of the yeast cell wall surface. The surface becomes enriched with concanavalin A (Con A) binding sites and attachment to the epithelial cells is mediated by fibrillar structures or polysaccharide granules distributed on the cell wall coat. Tokunaga et al.[267] have shown that the adhesion of *Candida* cells to human buccal epithelial cells corresponds with increased density of the fibrillar structure of the outermost *Candida* cell wall layer (Figure 10). In another study adherence was associated with the condensation or disposal of this layer, facilitating contact between epithelial cells and the deeper layer of the fungal wall.[256] In several scanning

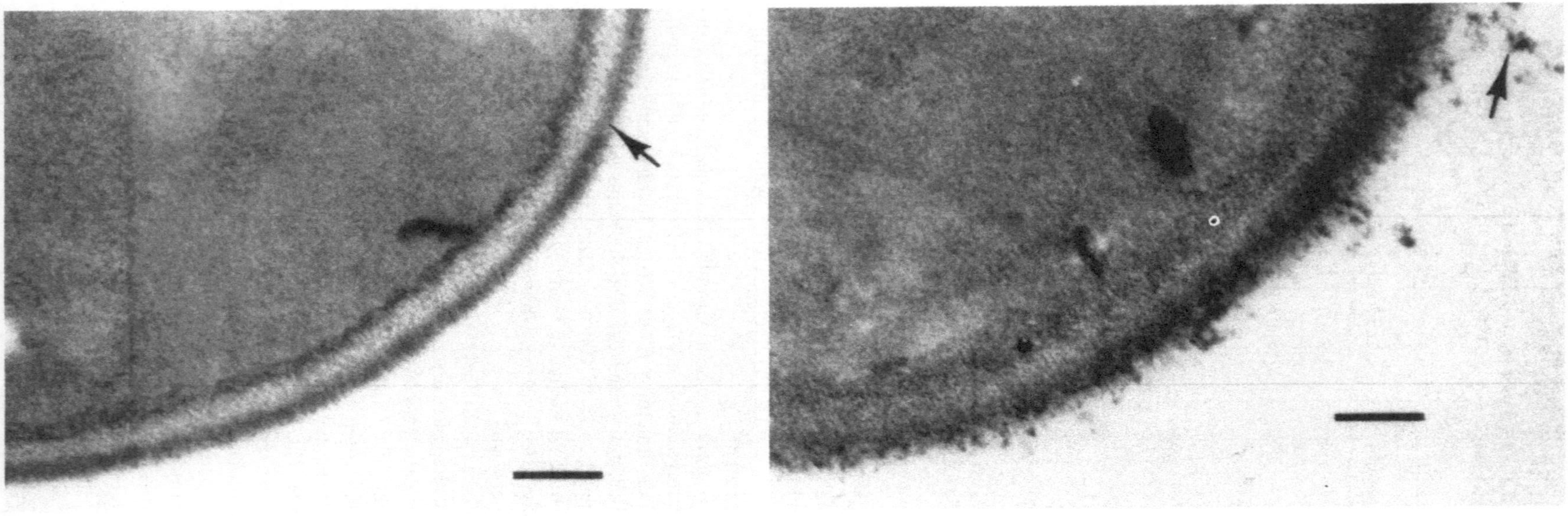

FIGURE 9. Electron micrographs of thin sections of *C. albicans* AK785. Cells harvested from (A) Tryptic soy broth +4% glucose showing (arrow) an outer electron dense layer. (B) Yeast nitrogen base +500 m*M* galactose showing (arrow) an outer floccular layer. (C) Tryptic soy broth +4% glucose showing (arrow) an outer floccular layer. (D) Trypticase soy broth showing a thin cell wall with no "extra" cell wall layers or obvious adhesive appendages. Note the differences in cell wall ultrastructure using Tryptic soy broth from different manufacturers or different environmental parameters for cell growth. Bar = 200 nm. (From Kennedy, M. J. and Sandin, R. L., *J. Med. Vet. Mycol.*, 26, 79, 1988. With permission.)

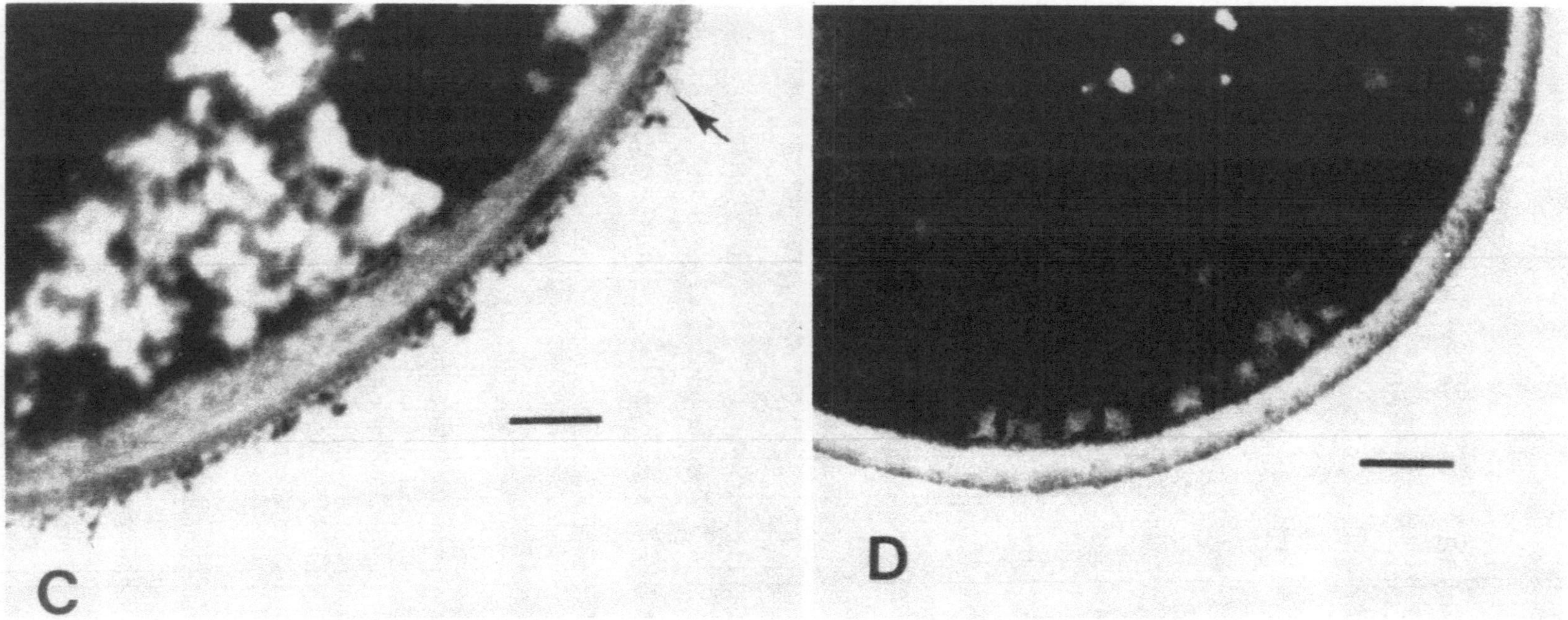

FIGURE 9C and D

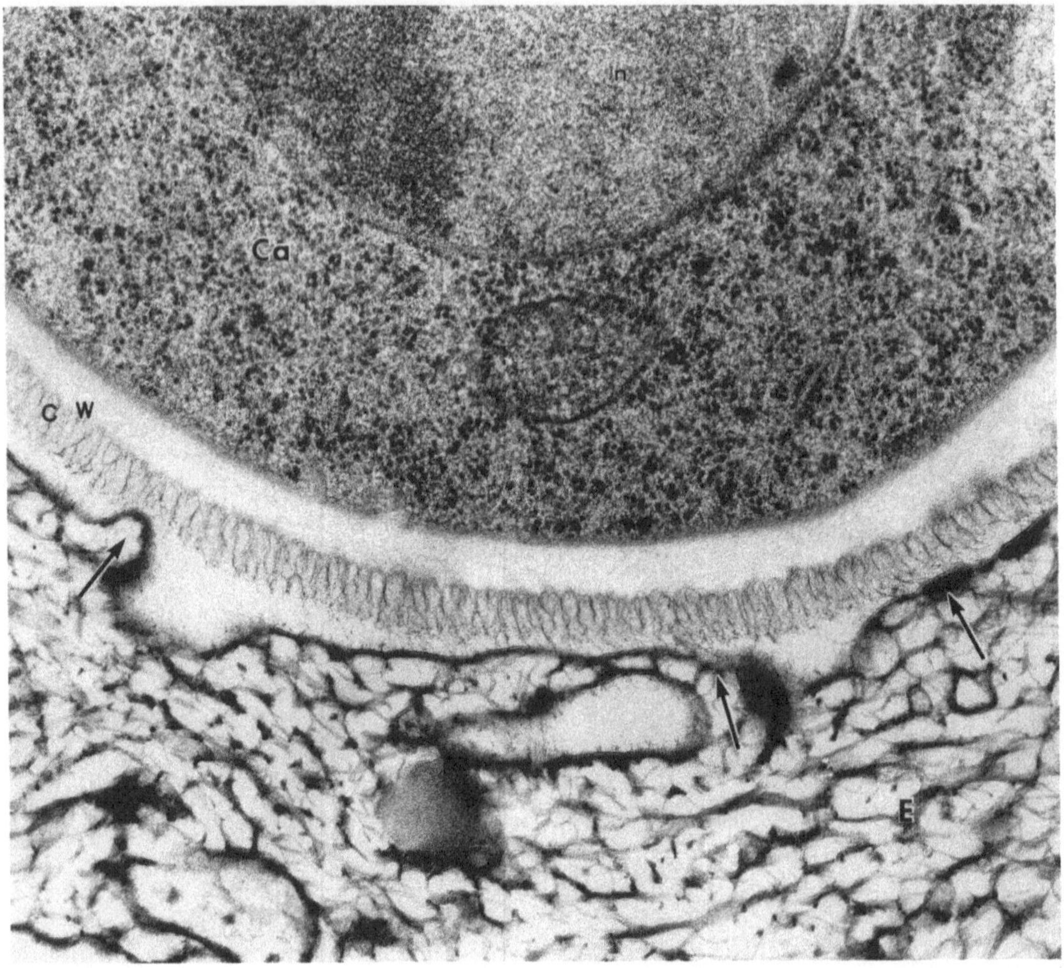

FIGURE 10. FSF-TEM micrograph of an ultrathin section of a *Candida* cell adhering to buccal epithelial cell. The region of adhesion is clearly localized between the ends of the fibrils of the yeast cell wall and the projections of the membrane interdigitations of the epithelial cells. Magnification was × 100,800. (Courtesy of Tokunaga, M., Niimi, M., and Koike, H., 10th Cong.-Int. Soc. Human Anim. Mycol., 1988.)

electron microscopy studies it was demonstrated that fibrils mediate *Candida* attachment to epithelial cells *in vitro*.[60,268,269] It is quite clear that this fibrillar-floccular layer may indeed mediate *Candida* adhesion (Figure 11). (See Reference 471.)

The chemical nature of *Candida* adhesin(s) is unknown, with some reports proposing that *C. albicans* produce more than one adhesin.[259,270-272] Based on different experimental approaches a number of compounds have been suggested as possible *Candida* adhesins (Table 11).

1. Sugars and Glycoproteins

Sugars and glycoproteins, particularly mannoproteins, received major attention by research workers as possible *Candida* adhesins. This is understandable since they make the outermost layer of the yeast cell wall. Also, such chemical substances are recognized as being involved in adherence of bacterial cells to human epithelial cells.[10,13,273,274] Various research groups adopted different approaches in revealing the involvment of sugars and glycoproteins in *Candida* adherence to host tissues. Some authors measured the adherence of cells grown in media containing various sugars. Others tested the effect of individual sugars and sugar derivatives

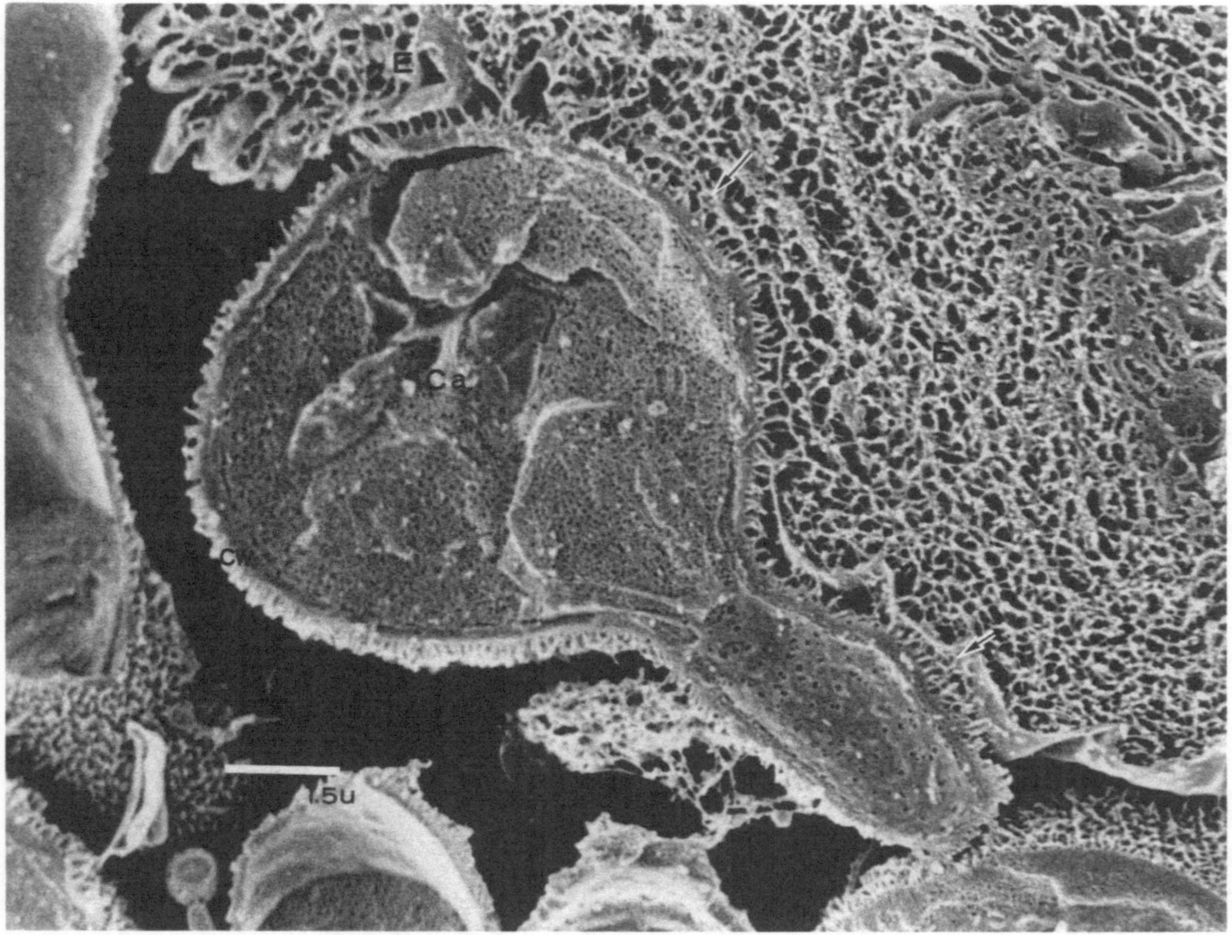

FIGURE 11. Cryo-Scanning electron micrograph showing *Candida albicans* germ-tube attached to buccal epithelial cell. Intimate contact is mediated by the outer cell wall fibrillar layer of the yeast. (Courtesy of Tokunaga, M., Niimi, M., and Koike, H., 10th Cog. Int. Soc. Human Anim. Mycol., 1988. With permission.)

TABLE 11
Possible Cell Surface Components Suggested to Serve as *Candida* Adhesins

Possible adhesin moieties	Examples of inhibitors	Ref.
Chitin	Glucosamine	280
	Chitin-soluble extract	280
	N-acetylglucosamine	280
	Mannosamine	280
Mannan/mannoprotein	Mannose	294
	α-mannosidase	294
	Tunicamycin	278
	α-methylmannosidase	284
	Concanavalin A	294
	D-methylmannopyranoside	294
Glucan	Glucose	
	Glucan	
Protein	Papain	284
	Pronase	284
	Pepsin	284
	Trypsin	284, 285
	Chymotrypsin	284, 285
Lipid	Sterols	282

From Kennedy, M. J., in *Current Topics in Medical Mycology,* Vol. 2., McGinnis, M. R., Ed., Springer-Verlag, New York, 1988, 73. With permission.

directly included in the assay medium on the adherence process. Another approach in determining adhesins is by treatment with specific lectins. These are proteins or glycoproteins with multiple sugar-binding sites. The sugar(s) bound by a particular lectin is usually quite specific; binding of a particular lectin can generally be inhibited by including specific mono- or oligosaccharide in the reaction mixture.[276] A fourth approach involves studying the effects of substances known to interfere with sugar and glycoprotein metabolism on adherence. Having identified the possible adhesins involved, workers pursue methods for their characterization. Table 12 is a summary of these methods.

McCourtie and Douglas[261] showed that the adherence of yeast to acrylic surfaces *in vitro* was increased after growth of the yeast in a medium containing high levels of different sugars, particularly galactose. Douglas et al.[264] confirmed this result on *C. albicans* adherence to buccal epithelial cells. Samaranayake and MacFarlane[277] showed that various strains of *C. albicans* preincubated in a medium containing glucose, sucrose, galactose, xylitol, or maltose exhibited enhanced adherence to HeLa epithelial monolayers and to buccal epithelial cell. The most effective sugar in this respect was maltose and the least effective glucose, although different patterns were obtained with other strains.[277] The same authors concluded that the medium sugars modify the adhesive properties of *Candida* cells. In these studies it was noted that adhesion was directly proportional to the sugar concentration. Moreover, enhanced adherence usually correlated with increased resistance to sphaeroplast formation and was associated with a surface, fibrillar mannoprotein layer on the cell surface.[148,278] McCourtie and Douglas[279] found that *C. albicans* grown on 500 m*M* galactose adhered to acrylic surfaces at a maximal linear rate throughout an incubation period of 1 h, whereas nonlinear adhesion rates were observed with cells grown on 500 m*M* sucrose, 50 m*M* glucose, or 50 m*M* galactose. It was reported from the latter cultures that populations of yeasts were isolated showing increased adhesion and increased resistance to sphaeroplast formation.[279] The authors concluded that there was an unequal distribution of adhesins among cells in such cultures.[279]

The results of many recent studies, although contradictory, revealed that the inclusion of

TABLE 12
Methods for the Identification and Characterization of *C. albicans* Adhesion Mechnisms and Their Role in Human Disease

Step	Purpose	Methodology
1.	Identify and describe adhesion mechanisms (and if there is more than one) that can be tested in simple *in vitro* systems.	Histological and ultramicroscopic examination of infected specimens followed by testing in simple *in vitro* adhesion assays (e.g., exfoliated epithelical cell assays).
2.	Determine whether adhesion in simple *in vitro* assays is of the same nature as that of more complex models (that may more closely represent the situation *in vivo.)*	Proceed to more complex *in vitro* models (e.g., intestinal slices, tissue explants), and determine if the same inhibitors block adhesion.
3.	Attempt to describe the adhesion at the molecular level.	Alter physiochemical parameters of the adhesion assay, and isolate the adhesin and/or tissue receptor.
4.	Test whether inhibitors of adhesion that act in simple models also inhibit *in vivo* adhesion, and whether they affect the progress of colonization and/or experimental disease.	Proceed to animal model assay.
5.	Determine whether the adhesins discovered in step 3 are synthesized *in vivo,* and whether the natural host produces similar receptors.	Test whether mutants lacking a given cell surface component are avirulent or are unable to colonize in animal models. Determine whether revertants regain virulence or colonizing ability.
6.	Determine what factors modify adhesion or association *in vivo.*	Use a number of different animal models, or manipulate animals, and examine biochemical and physiochemical differences of the surrounding environment where differences are noted (e.g., pH, eH, VFA, etc).
7.	Determine where in the body *Candida* cells are multiplying, dying, or both, what the *population dynamics of association* disassociation are, and what the growth rate of the organism is *in vivo.*	Do quantitative counts on, for example lumen contents and mucosal surface. Perform *in vitro* tests under parameters similar to that *in vivo* and determine binding kinetics and rates of disassociation.
8.	Determine the mechanisms of regulation and adhesin production.	Study the genetics of well characterized mutant strains.
9.	Determine what factors may prevent adhesion and association in humans (e.g., antifungal drugs, immunization, etc)	After all the above parameters have been well characterized in experimental animals, studies in human volunteers with wild-type and mutant C. *albicans.*

From Kennedy, M. J., in *Current Topics in Medical Mycology,* Vol. 2, McGinnis, M. R., Ed., Springer-Verlag, New York, 1988, 73. With permission.

particular sugars and sugar derivatives into the assay medium resulted in the blockage of adherence of *Candida* cells to human epithelial cells and to synthetic surfaces. Thus, Segal et al.[280] showed that N-acetyl-glucosamine, glucosamine and mannosamine, but not mannan, glucan, glucose, α-methylglucose, mannose, or α-methylmannose could block the adhesion of *C. albicans* to vaginal epithelial cells. The results on the effect of mannose are contradictory. Several authors found it inhibitory to adherence when added into the assay medium. Thus, Centeno et al.[281] showed that mannose, but not glucose, significantly decreased the attachment of *C. albicans* to epithelial cells. Similarly, Ghannoum et al.[282] found that both mannose, and to a lesser extent, N-acetylglucosamine, in addition to inositol, were effective (Table 13). However, several other investigators could not trace any blocking effect for mannose added into

TABLE 13
Effect of Saccharides on the Adherence of *Candida albicans* Cells to Buccal Epithelial Cells

Saccharide (10 m*M*/l)	Mean adherence (± SE)[a]		
	Control	With saccharide	(% Reduction, *p*)
Cellobiose	495 ± 23	238 ± 17	(52, <0.001)
Galactose		508 ± 27	NS
Glucose		486 ± 24	NS
Inositol		312 ± 22	(37, <0.001)
Lactose		227 ± 19	(54, <0.001)
Maltose		288 ± 22	(42, <0.001)
Mannose		295 ± 21	(40, <0.001)
N-acetyl-D-glucosamine		318 ± 22	(36, <0.001)
Sucrose		293 ± 26	(41, <0.001)
Trehalose		219 ± 19	(56, <0.001)

[a] Data are expressed in mean number of yeast cells adhering to 100 buccal epithelial cells. NS, not significant.

Adapted from Ghannoum, M. A., Burns, G. R., Abu Elteen, K. H., and Radwan, S. S., *Infect. Immun.*, 54, 189, 1986; and Ghannoum, M. A., Abu Elteen, K., and Radwan, S. S., *Mykosen,* 30, 371, 1987.

the assay medium.[283-286] A number of disaccharides viz. maltose, sucrose, lactose, cellobiose, and trehalose were effective, but the monosaccharide galactose was not.[27] Most effective are amino-sugars like mannosamine, glucosamine, and galactosamine.[287]

The lectin Con A, which binds to α-linked D-mannose residues, was used by a number of workers to investigate the role of mannoproteins in adherence. Most of these studies indicated that this wall component acts as a *Candida* adhesin.[270,288,289] Tkaez et al.[290] used Con A to locate mannan-rich areas of the cell surface. Their studies were based on the observation that fluorescein-conjugated Con A could stain mannan-containing whole cells of *S. cerevisiae* but not mannan-deficient cells. Whole cells of *C. albicans* were found to stain strongly with fluorescein-conjugated Con A but became resistant to this stain in the presence of lectins such as soybean agglutinin.[291] Using Con A in their experiments, Maisch and Calderone[292,293] showed that cell surface mannan may play an essential role in the adherence of *C. albicans* to the fibrin-platelet matrices which form on the endocardium of heart valves. Sheep erthrocytes coated with an alkali-soluble cell wall extract (72% polysacccharide and less than 1% protein) attached to a fibrin-platelet matrix, whereas nonconjugated erythrocytes did not adhere. Pretreating the erythrocytes by the extract nullified the adhesion-promoting effect.[292,293] Sandin et al.[294] and Sandin and Rogers[295] used various lectins and sugars to investigate the possible role of saccharide-containing moieties on surfaces of *C. albicans* and buccal epithelial cells in the adherence process. These workers found that different lectins showed affinities for various sugar moieties, and therefore such lectins were used to pretreat cells before they were mixed into the adherence assay medium. Their results revealed that Con A inhibited adherence when it was used for the pretreatment of either the yeast or the epithelial cells.[294,295] Adherence was restored by pre-incubating Con A with a mannose derivative, but not with other sugars. Lectins that do not recognize mannose apparently did not affect adherence. Similar confirming results were reported by Rossano and Tufano.[296]

The treatment of *C. albicans* with tunicamycin, an antibiotic that inhibits protein glycosylation and, at low concentrations, inhibits mannoprotein synthesis but not chitin or glucan synthesis, interferred with the adherence of this yeast.[297] Douglas and McCourtie[278] reported that addition of tunicamycin, at early stationary phase, inhibited the formation of the fibrillar layer with a consequential decrease in adherence to buccal epithelial cell (over 60% reduction as compared with control yeast cells).

The role of the protein portion of mannoprotein in adherence has received little attention. However, there is evidence to suggest that this portion may serve as an adhesin, with some reports suggesting that the protein portion of the mannoprotein complex is more important than the carbohydrate moiety in mediating adherence (for review see Kennedy[262]). These suggestions are not unexpected since proteinaceous adhesins represent the majority of bacterial adhesins.[8,12,298] Evidence to support the role of proteins as *Candida* adhesins comes from experiments showing that *C. albicans* cells adherence decreases following their exposure to heat (resulting in protein denaturation) or various proteolytic enzymes.[283,284] Maisch and Calderone[292] showed that pretreatment of *C. albicans* with pronase, trypsin, and chymotrypsin led to a reduction of yeast cells adhesion to fibrin-platelet clots. Further supporting evidence came from the investigation of McCourtie and Douglas,[148] who showed that pretreatment of an extracellular polymer (identified subsequently as mannoprotein) from *C. albicans* with heat, dithiothreitol, or certain proteases, but not α-mannosidase or sodium metaperiodate, destroyed, partly or entirely, its ability to inhibit yeast adhesion to buccal epithelial cells.[270] Also, the binding affinity of *C. albicans* pseudohyphae to neutrophils was reported to be reduced under the influence of a protein component derived from this yeast.[299] Other workers reported that binding of *C. albicans* to fibrinogen was diminished when the fungus was pretreated with 2-mercaptoethanol alone or together with pronase, with α-mannosidase, or with trypsin but not with pronase alone nor with chitinase.[300] These results clearly demonstrate that the protein portion, in collaboration with the polysaccharide moiety, of the mannoprotein layer influences the adhesion process and may partly explain why *Candida* adhesion is not always inhibited by mannose, methyl-α-D-mannoside, or mannoprotein preparations devoid of protein portion.[280,283,284] The studies of Segal and co-workers[280,301,302] suggest a role for chitin in the adhesion of *C. albicans*. (See Chapter 7 for a review of this aspect.) They reported that chitin, its constituent N-acetylglucosamine (NAG) and chitin soluble extract (CSE) inhibited yeast adhesion to vaginal epithelial cells *in vitro*. Furthermore, Lehrer et al.[301] observed that both NAG and CSE also reduced the infection rate in a rat model of vaginitis. While work carried out in our laboratory confirmed the effect of chitin and its derivatives on the adhesion of *Candida* to buccal epithelial cells and intestinal tissues of mice (see Chapter 7), other investigators have not been able to reproduce the results of Segal and her colleagues.[280,301,302] This suggests that more research is necessary to determine unequivocally whether chitin and related components can serve as adhesins, and if so, by what mechanism(s).

2. Lipids

The role of lipids in adherence has so far received limited attention by researchers in spite of the fact that lipids are major constituents of the cell surfaces (see Sections II. B. iv, this chapter). In addition, there are some relevant findings in the literature about the role of lipoteichoic acids (LTA) in the adherence of bacteria to epithelial cells (for review see Jones,[8] Christensen et al.,[298] and Beachey and Simpson[303]) and to teeth (for review see Kessler[304]). LTA is a linear polyglycerolphosphate polymer containing a glycerophosphoryl/diglucosyl/diglyceride moiety at its nonpolar end.[305,306] Such compounds were detected in Gram-positive bacteria,[305,307] and their fatty acid moieties are believed to serve to bind the molecules spontaneously to animal cells.[306,308-311] It is well known that acylated LTA occur mainly in association with the plasma membrane. However, there is experimental evidence that indicates such compounds become exposed on the surfaces of intact bacteria.[306,312-317] It has been suggested that acylated LTA exposed at the cell wall surface may be anchored LTA of sufficient chain length to reach through the wall.[318] It is believed that bacterial adherence to eukaryotic cells is mediated by fatty acid moieties of LTA.[318] Another lipid class, glycosphingolipids, normally detected in small amounts in plasma membranes of animal cells, has recently been reported to be the site at which the fimbriae of Gram-negative bacteria adhere.[319-323]

It has already been mentioned that cell walls of *C. albicans* contain lipids.[27,36,56,111,202] Lipids are also major constituents of the plasma membranes both of *Candida* and of epithelial cells. Epithelial cells lack walls and, therefore, the plasma membrane represents the outermost surface. The plasma membranes of Can*dida* cells, however, are coated with the cell wall. Whether or not lipid moieties of plasma membrane molecules penetrate through the wall and become exposed on the cell surface, as is the case with many bacteria,[298] has not yet been investigated.

Experimental evidence for the involvement of lipids in adherence of *C. albicans* to buccal epithelial cells has been reported.[282] Total lipids extracted from *C. albicans* and other *Candida* spp. significantly blocked *in vitro* adherence of yeast cells to buccal epithelial cells. The constituent lipid classes of *C. albicans* were then isolated in a pure form and tested individually. The results indicated that various phospholipids viz. diacylglycerophosphocholines, diacylglycerophosphoethanolamines, diacylglycerophosphoglycerols, diacylglycerophospho-inositols, and diacylglycerophosphoserines were highly active in blocking adherence (Table 14).[282] Similar results were obtained with sterols and steryl esters, but triacylglycerols and fatty acids did not have any significant effects. Lipid classes found efficient in blocking adherence are those known to be constituents of the plasma membrane. In another study, *C. albicans* glycolipids, namely ceramide monohexosides and ceramide dihexosides from yeast forms and sterylglycoside from mycelial forms, were also found to block adherence *in vitro*.[27] Such glycolipid classes most probably also occur in the plasma membrane. The results of the same study have further shown that yeast wall lipids and lipids from epithelial cells were also significantly effective in inhibiting adherence (Table 15).[27]

On the basis of the above discussion it is possible to speculate that *Candida* possess more than one adhesin. Nine different fimbrial adhesins have been identified in *Escherichia coli*.[12] Also the presence of a component on the surface envelope of the yeast does not necessarily indicate that the chemical plays a role in adhesion. The converse seems also to be true; i.e., the presence of a molecule in the inner layers of the yeast cell wall (e.g., chitin) does not necessarily indicate that the molecule has no role to play in adhesion.

B. RECEPTORS

1. Definition and Characterization

A receptor is a cell membrane constituent that mediates certain cellular activity as a consequence of interaction between it and an extracellular biological agent (e.g., adhesin). It should be able to interact specifically with a ligand and this interaction should lead to a biologically relevant response.[324,325] In the case of microbial adhesion to epithelial tissue this response may be manifested as infection.

Like other cell membrane constituents, receptors undergo dynamic change in their production and expression as a function of age and physiological state of the cell. Furthermore, receptors are subject to modifications under the influence of external (stimuli) e.g., drugs, and proximity to other components present in the outer cell envelope.[252,326-328] These alterations may interfere with the ability of receptors to bind to adhesins.

A number of steps are usually followed in order to present direct evidence for the existence, identification, and characterization of specific receptors. These include:

1. Binding studies which demonstrate the presence of receptor molecules for a particular ligand.[252]
2. Identification studies which include several experimental approaches: inhibition of adhesion by soluble substances, lectins, or enzymatic destruction of the receptor; attachment to inert surfaces coated with the putative receptor; and comparison of genetically resistant and sensitive animal cells.[8,329] Ofek et al.[252] have tabulated procedures commonly used to identify surface receptors (Table 16).
3. Isolation and characterization of receptors. These involve isolation and purification of

TABLE 14
Effects of Constituent Lipid Classes of Total Lipid from *Candida albicans* on Its Adherence to Buccal Epithelial Cells

	Mean adherence (± SE)[a]		
Lipid class	Control	With lipid	(% Reduction, *p*)
Steryl esters	515 ± 28	435 ± 25	(15.5, <0.05)
Steryl glycosides		243 ± 20	(53, <0.001)
Ceramide monohexosides		276 ± 24	(46, <0.001)
Monogalactosyldiacylglycerols		579 ± 25	NS
Digalactosyldiacylglycerols		510 ± 28	NS
Sulfoquinovosyldiacylglycerols		524 ± 24	NS
Triacylglycerols		522 ± 30	NS
Fatty acids		498 ± 22	NS
Sterols		413 ± 25	(20, <0. 01)
Diacylglycerophosphoethanolamines		277 ± 24	(46, <0.001)
Diacylglycerophosphoglycerols		315 ± 26	(39, <0.001)
Diacylglycerophosphocholines		309 ± 25	(40, <0.001)
Diacylglycerophosphoinositols		228 ± 25	(56, <0.001)
Diacylglycerophosphoserines		292 ± 24	(43, <0.001)

[a] Data are expressed as mean numbers of yeast cells adhering to 100 buccal epithelial cells. NS, not significant.

Adapted from Ghannoum, M. A., Burns, G. R., Abu Elteen, K. H, and Radwan, S. S., *Infect. Immun*,., 54, 189, 1986; and Ghannoum, M. A., Abu Elteen, K., and Radwan, S. S., *Mykosen*, 30, 370, 1987.

TABLE 15
Effect of Total Lipids from Whole Cells, Cell Walls, and Cell Membranes of Candida albic*ans* and of Whole Epithelial Cells on the Yeast Adherence to Buccal Epithelial Cells

	Mean adherence (± SE)[a]		
Source of lipids (175 μg/ml of assay mixture)	Without lipids	With lipids	(% Reduction, *p*)
Whole yeast forms	495 ± 23	324 ± 18	(35, <0.001)
Whole mycelial forms	495 ± 23	309 ± 21	(38, <0.001)
Yeast cell walls	495 ± 23	214 ± 19	(57, <0.001)
Yeast cell membranes	495 ± 23	298 ± 23	(40, <0.001)
Epithelial cells	495 ± 23	232 ± 23	(53, <0.001)

[a] Data are expressed in mean number of yeast cells adhering to 100 buccal epithelial cells.

From Ghannoum, M. A., Abu Elteen, K., and Radwan, S. S., *Mykosen*, 30, 371, 1987. With permission.

cell membrane receptors with the aim of elucidating their physicochemical nature.

Such studies are usually followed by investigations giving evidence that these receptors mediate the cellular activity culminating in the appropriate response. Table 17 summarizes the procedures followed for the purification and characterization of cell membrane receptors.[252]

2. Cell Surface Receptors of *Candida* Adhesins

Compared to information available on *Candida* adhesins, reports on its receptors are sparse and the receptors on various tissues to which this yeast adhere are not well characterized.[262]

TABLE 16
Procedures Commonly Used to Identify the Chemical Nature of Recognition Sites of Cell Surface Receptors

Procedure	Materials	Information gained and comments
Hapten inhibition of the interaction of a biological agent with cells by low-molecular-weight compounds. The interaction can be measured either as binding or as biological activity of the agent.	The inhibitors should be soluble, well defined, and free of damaging effects on the cell	Inhibition of the interaction suggests that the structure of the attachment site is closely related to that of the inhibitor. Confirmation of this assumption can be obtained by showing that the agent binds to immobilized inhibitor
Effect of enzymatic and chemical modifications of the cell surface on the interaction of the agent with the cells	Proteolytic enzymes, glycosidases, or chemical reagents which cleave specific covalent bonds	This approach is helpful in determining whether the recognition site is a protein or a carbohydrate. The more specific the treatment for its moleculr target, the more information is gained on the structure of the recognition site
Inhibition of binding of the agent by compounds that bind to known structures on the cell surface	The inhibitors may be lectins of known specificity or antibodies (preferably monoclonals) against specific epitopes	Inhibition suggests that the molecular target site of the inhibitors may be involved in binding of he agent under study
Binding of the agent to tissue culture cells that exhibit alterations in the composition or structure of surface constituents (e.g., tunicamycin, or swainsonine, which affect the biosynthesis of carbohydrate chains of glycoproteins)	Clones selected for resistance to lectin or cells treated with drugs that alter specific oligomeric structure of cell surface	Correlation between decreased binding of the agent and well-defined alterations of cell surface indicates the involvement of the affected structures in binding of the agent

From Ofek, I., Lis, H., and Sharon, N., in *Bacterial Adhesion, Mechanisms and Physiological Significance*, Savage, D. C. and Fletcher, M., Eds., Plenum, New York, 1985, chap. 3. With permission.

Experimental evidence suggests that cell surface receptors of *Candida* adhesins are carbohydrates (including glycoproteins) and lipids.[270,282,286] Chemically, these are of the same nature as receptors suggested for bacterial adhesins.[8,12,298]

Carbohydrates including mannose, L-fucose, neuraminic acid, and galactose have been reported as possible receptors for bacteria (for review see References 8, 252).[330-334] Critchley and Douglas[335] observed that *Candida* adhesion is mediated by more than one receptor. They showed that glycosides containing L-fucose, N-acetyl-D-glucosamine, and possibly D-mannose can all function as epithelial receptors for different strains of *C. albicans*. Evidence for their suggestions was partly derived from experiments utilizing various lectins. Treatment of buccal epithelial cells with winged-pea lectin (which is specific for L-fucose) but not with other lectins (e.g., wheat-germ agglutinin [specific for N-acetyl-β-D-glucosaminyl residues] or peanut lectin [specific for D-galactosyl residues] significantly inhibited *C. albicans* GDH 4346 adhesion. *C. albicans* strain GDH 2023 was most efficiently inhibited by the latter two lectins.[335] These findings suggest that more than one surface component on epithelial cells may serve as receptors, and that adhesion to different sites may be strain specific.[335] Furthermore, they

TABLE 17
Procedures for the Identification and Purification of Cell Membrane Receptors

Procedure	Comments
Overlay	
SDS electrophoresis or thin-layer chromatography solubilized cell membranes, followed by overlay of the gel with a solution of radiolabeled (or fluorescent) ligand and radiography	Reveals membrane constituents capable of of binding the agent
Receptor cross-linking	
Binding of the agent, derivatized with a cleavable photolabile cross-linking reagent, to cells, followed by light-sensitive cross-linking and solubilization of the receptor-agent complex. The receptor can then be released from its bond to the ligand and analyzed.	Leads to the isolation of membrane constituents available to the ligand on the surface of intact cells
Affinity chromatography	
Chromatography of solubilized radiolabeled cell membranes on columns derivatized with the agent; elution with a suitable buffer or specific reagent, e.g., hapten inhibitor	Yields membrane constituents capable of binding the agent
Reconstitution	
Incorporation of the purified receptor into artificial membranes or cells devoid of the receptor, followed by binding studies with the agent and determination of biological response	Reconstitution of the biological response is the best proof that the isolated cell membrane constituent is indeed the receptor for the agent

From Ofek, I., Lis, H., and Sharon, N., in *Bacterial Adhesion, Mechanisms and Physiological Significance*, Savage, D. C. and Fletcher, M., Eds., Plenum, New York, 1985, chap. 3. With permission.

suggested that fucose-type receptors may be most commonly required for *Candida* adhesion.[335] Based on inhibiton studies carried out by a number of workers,[331,332,335] it is suggested that the natural epithelial receptor is not an L-fucose residue as such but, most probably, a larger molecule with a particular stereochemical configuration.

Glucoproteins have been reported to act as receptor molecules for streptococcal adhesins.[4,336,337] The glycoprotein, fibronectin, which is one of the major surface glycoproteins on mammalian cells,[338] has been suggested to function as a ligand to bind bacteria to animal cells (Table 18).[298,339-342] For a review of the role of fibronectin in the adhesion of bacteria to animal tissues the reader is referred to Christensen et al.[298] Rotrosen et al.[355] suggested that the host cell receptor for *Candida* adhesion is fibronectin. Skerl et al.[286] investigated the binding of *C. albicans* to human fibronectin *in vitro*. Their results indicated that around 40% of the yeast cells adhered to fibronectin, with greater adherence achieved at 30°C than at 4°C and with viable cells than heat-killed cells.[286] It was also demonstrated that the adherence process was blocked when yeast cells were pretreated with the enzymes chymotrypsin, pronase, or papain but not with pepsin.[286] They were able to provide experimental evidence for a possible role of fibronectin as a receptor for the binding of *Candida* to epithelial cells. Thus, they demostrated the presence of fibronectin in buccal and vaginal epithelial cells. Furthermore they reported that fibronectin-pretreated yeast cells showed lesser adherence to epithelial cells than nontreated ones.[286]

The most clear demonstration of the role of lipid components as receptors for microbial adhesins came from the work of Sato and co-workers[343,344] (for review see Jones[8]). These researchers reported that sterol-containing liposomes inhibited hemagglutination by *Bordetella pertussis*. Further evidence for the interaction of bacterial adhesins with lipids is provided by investigations by Gibbons and colleagues.[345,346] These workers showed that *Streptococcus salivarius* and *S. pyogenes* adhesins may interact with lipids of the animal cell membranes. Attachment of *S. salivarius* to epithelial cells was reduced in the presence of sphingomyelin,

TABLE 18
Fibronectin Binding by Bacteria

Bacteria which consistently bind fibronectin	Bacteria with low-level or infrequent binding of fibronectin, or bacteria that do not bind fibronectin on the basis of selected strains
Group A streptococci	Group B streptococci
Group C streptococci	Group D streptococci
Group G streptococci	Group M streptococci
Staphylococcus aureus	Group N streptococci
Staphylococcus haemolyticus	Group P streptococci
Bacteria with variable binding of fibronectin, or bacteria demonstrated to bind fibronectin on the basis of selected strains	Group U streptococci
Streptococcus mutans	*Staphylococcus epidermidis*
Streptococcus salivarius	*Staphylococcus capitis*
Streptococcus sanguis	*Staphylococcus xylosus*
Streptococcus pneumoniae	*Staphylococcus cohnii*
Staphylococcus saprophyticus	*Streptococcus mitis*
Staphylococcus warneri	*Pseudomonas aeruginosa*
Staphylococcus hyicus	*Klebsiella pneumoniae*
Staphylococcus simulans	*Escherichia coli*
Staphylococcus hominis	*Neisseria meningitidis*
	Neisseria gonorrhoeae
	Hemophilus influenzae

From Christensen, G. D., Simpson, W., and Beachey, E. H., in *Bacterial Adhesion: Mechanisms and Physiological Significance,* Savage, D. C. and Fletcher, M., Eds., Plenum, New York, 1985, 279. With permission.

lecithin, phosphatidylserine, and phosphatidylethanolamine.[345] Similar results were reported for *S. pyogenes.*[346] Work carried out in our laboratory showed that total lipids extracted from buccal epithelial cells blocked the adherence of *C. albicans* by 53%.[27] Analysis of lipids extracted from buccal epithelial cells showed that they consist of sterols (24.2%), triacylglycerols (16.8%), fatty acids (14.4%), steryl esters (12.3%), diacylglycerophosphocholines (18.4%) and diacylglycerophosphoethanolamines (13.9%).[27] Four of these fractions namely, sterols, steryl esters, and the two phospholipids have already been shown to be highly effective in blocking adherence of *C. albicans in vitro.*[282] These results and those reported on bacteria suggest a role for sterols and phospholipids as receptors for both bacteria and yeast. Based on the available literature, Kennedy[262] tabulated putative surface receptors of *Candida* adhesins (Table 19). Figure 12 brings *C. albicans* adhesins "face to face" with epithelial cell receptors.

IV. MECHANISM(S) OF *CANDIDA* ADHERENCE

The association and subsequent adhesion of microorganisms to host surfaces employ more than one mechanism(s) of binding.[298] Freter[347] commented, "It is apparent that the seemingly simple process of bacterial adherence represents a potentially very complex sequence of host-bacterium interactions."[347] This is illustrated in Table 20 giving examples of bacteria, from different systems, using multiple mechanisms of adherence. Mechanisms of adhesion of bacteria with animal tissues and inert surfaces have been reviewed (see References 2, 8, 250, 298, 347 to 349). It is clear from these reviews that the mechanisms responsible for bacterial adhesion to surfaces are complex-multiple systems, and that no single bacterial species can deploy every one of these mechanisms.[298,347] Furthermore, the interaction of bacteria to surfaces is a complex series of events consisting of at least three consecutive but overlapping phases: phase one — accumulation and adsorption, phase two — adhesion, and phase three — invasion.[350] Different mechanisms are responsible for bringing about different phases. Similar mechanisms have been suggested for *Candida* adhesion (Table 21).

TABLE 19
Possible Surface Receptors of *Candida* Adhesins

Possible receptor moieties	Epithelial cell type	Examples of inhibitors	Ref.
Fibronectin	Buccal	—	286
Fucose	Buccal	L-fucose Lectin from *Lotus tetragonolobus*	270
Lipids	Buccal	Phospholipids, sterols	282
Mannose	Buccal	Con A D-mannose Methyl-α-D-mannoside	294
N-acetyl-D-glucosamine	Buccal	N-acetyl-D-glucosamine Wheat-germ agglutinin	270
Mucins	Intestinal	—	271
Fucose	Vaginal	L-fucose	285

From Kennedy, M. J., in *Current Topics in Medical Mycology,* Vol. 2, McGinnis, M. R., Springer-Verlag, New York, 1988, 73. With permission.

Review of the literature shows that *Candida* adheres to epithelial surfaces through either direct interaction, i.e., yeast cells to epithelial tissues, or indirectly through bridging involving intermediate cells/components, or both.

A. DIRECT INTERACTIONS

Direct *Candida* — Epithelial cell adhesion seems to pass through two phases prior to cell invasion: nonspecific interactions (referred to as loose or reversible)[351] and specific interactions (referred to as irreversible, tight, or intimate)[255,256,351,352] brought about by the molecular interaction of the adhesins and receptors. The reversible phase lasts for about 20 min; thereafter the cells bind irreversibly.[351]

Candida and cell surfaces to which they adhere have a net negative charge (see Section II this chapter, and Sections II and IV Chapter 5) and, therefore, repel one another.[8] Once in close proximity other forces of attraction and repulsion exert their effects (refer to Table 3 Chapter 5 and Reference 353), which counteract electrostatic repulsion and probably initiate cationic bridges as suggested for bacteria.[348] However, without a specific attachment mechanism(s), the association is easily broken by external forces and the yeast removed. This is clear in a system such as the small intestine where the effect of fluid shear, due to peristalsis and bathing action of gastrointestinal secretions, detach yeast cells from tissue surfaces (see Section II Chapter 5 for the role of cell surface hydrophobicity in adhesion processes).

The cationic bridges mentioned probably culminate in the formation of a definite loose association, apparently mediated by a ruthenium red positive matrix.[255] The loose stage is followed by tight or intimate adhesion. Ultrastructural studies on colonization and invasion of oral epithelium carried out by Howlett and Squier[256] showed that the adherence of yeast to the superficial cells of the oral mucosa appeared to involve intimate contact between the epithelial cell surface and the deeper layers of the *C. albicans* cell wall. This close contact is maintained during invasion where a "close seal" between the invading hyphae and the surrounding epithelial cell envelope was observed.[256] Marrie and Costerton[255] observed a tight adhesion between yeast cells and mucosal epithelial cells where no space could be seen between the host and yeast cell. Results reported in these studies do not demonstrate unequivocally that a distinct phase of reversible adsorption need precede adhesion of yeast to host cells. Jones et al.[350] commented that the majority of bacteria that attach to host cells have preformed adhesins, and these may

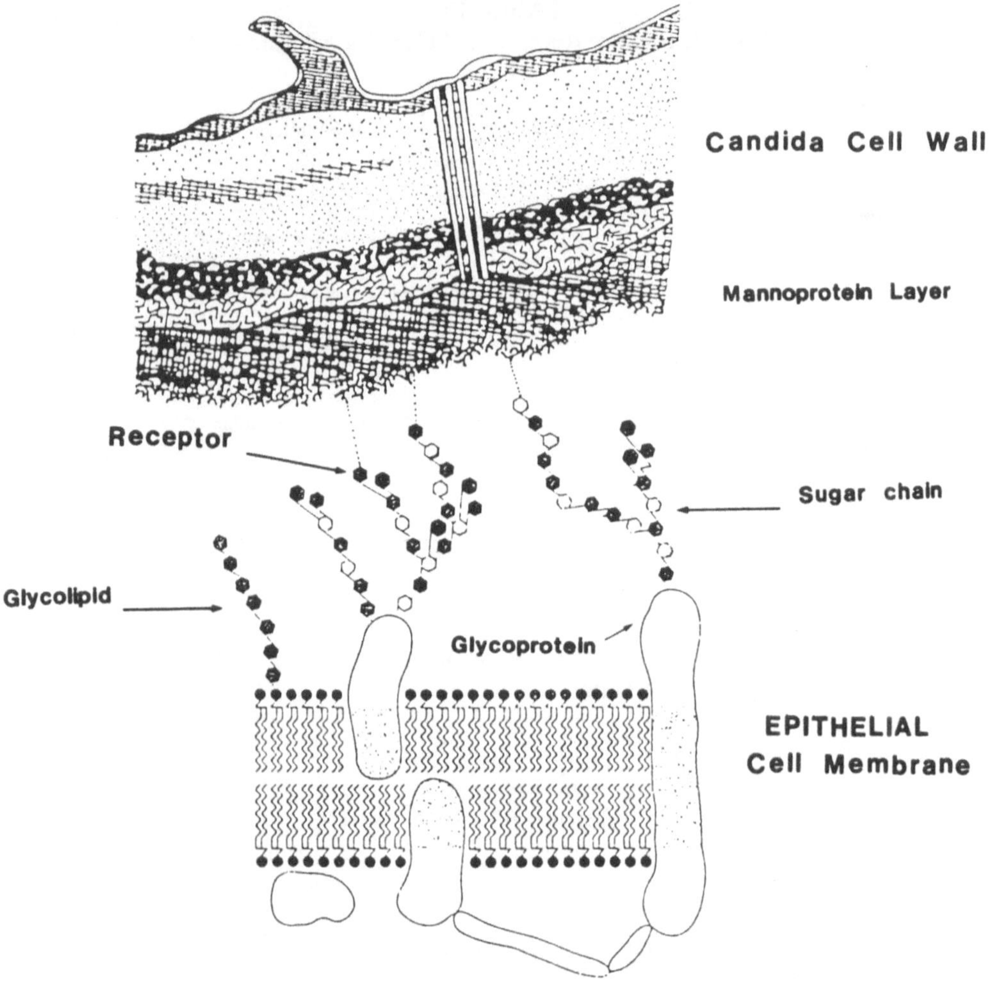

FIGURE 12. Diagram illustrating interaction of *Candida* and mammalian epithelial cell membrane. The outer surface of the yeast cell wall is a mass of mannoprotein fibrils attaching to the polysaccharides extending from the epithelial membrane.

intervene directly in assisting the bacterial cell to adhere rapidly and firmly to the animal cell. It is possible however, that *in vivo* the microorganism may produce adhesins only after they become reversibly adsorbed to the cell surface. In this respect Tronchin et al.[260] showed that *C. albicans* cells undergo ultrastructural modifications in their cell wall coat during the adhesion of the yeast to epithelial cells. Similarily, Howlett and Squier[256] observed structural changes in the floccular layer of the cell walls of *Candida* hyphae invading tissues. This layer appeared to condense or disperse so that contact between the epithelial cell membrane and the inner layers of the cell wall becomes established.[256]

The mechanism(s) responsible for attachment and penetration of vascular endothelium have been studied by a number of investigators. These studies proposed different mechanisms. Klotz et al.,[354] using porcine vascular endothelium, followed microscopically the early progression of association and penetration of these tissues by *C. albicans*. These authors showed that within 15 min yeast could attach tightly to the endothelium with a partial disruption of the endothelial surface.[354] More severe damage to the endothelium was observed by 30 min, and by 60 min *C. albicans* cells had penetrated deep into the porcine endothelial cells and were nearing the

TABLE 20
Examples of Bacteria Using Multiple Mechanisms to Bind to Animal Tissues

Bacteria	Adhesin	Functional importance
Escherichia coli	Type 1, K88, K99 CFA/, CFA/2, 987P, F41, E8775, P fimbriae	Each fimbria binds to particular hosts and tissues; each fimbrial type is associated with a particular infection, gastroenteritis or urinary tract, in a particular host
Neisseria gonorrhoeae	Type II outermembrane protein, fimbriae	Each adhesin binds to different genital tissues or conditions, enabling the organism to persist and infect a variety of microhabitats
Streptococcus agalactiae	LTA, lipid end LTA, charged backbone	One mechanism may mediate binding of *S. agalactiae* to immature tissues; the other mechanism may mediate binding to mature tissues
Streptococcus mutans and other oral microorganisms	LTA, glucosyltrasferase cell surface lectins, fimbriae, fibrillae	The oral microorganisms simultaneously employ a variety of adhesins to bind themselves to each other and to the oral surfaces

From Christensen, G. D., Simpson, W. A., and Beachey, E. H., in *Bacterial Adhesion: Mechanisms and Physiological Significance*, Savage, D. C. and Fletcher, M., Eds., Plenum, New York, 1985, 279. With permission.

TABLE 21
Adhesion and Association Mechanisms of *Candida albicans*

	Nature of mechanism			
Mechanism	Active or passive	Specific or nonspecific	Direct or indirect	Ref.
Adhesin-Receptor interaction	A	S	D, I	271, 280, 284, 287, 294, 301, 382, 410, 448
Nonspecific adhesion	A, P	N	D, I	271, 354, 401, 425, 428, 430, 432
Co-adhesion to adherent organisms	A	S, N	I	259, 271, 281, 470
Entrapment in tissue	P	N	D	271, 361, 367, 369
Germ-Tube penetration	A	N, S	D, I	255, 256, 355, 414

From Kennedy, M. J., in *Current Topics in Medical Mycology*, Vol. 2, McGinnis, M. R., Ed., Springer-Verlag, New York, 1988, 73. With permission.

basement membrane. This penetration occurred with only viable cells, which suggests that it is mediated by enzymatic activity. Heat- or formalin-killed yeasts failed to show evidence of penetration, although they adhered to the endothelium.[354] Although Rotrosen et al.[355] proposed a similar sequence of events for the attachment and penetration of endothelial cells by *Candida* to that proposed by Klotz et al.,[354] (initial attachment of the organism to endothelial cells lining the vessel wall, then evasion of the fungicidal activity of circulating phagocytes, and finally, penetration of and migration through endothelial cells into organ parenchyma),[355] the mechanism(s) of adhesion and penetration proposed was different. These authors suggested that phagocytosis contributes to the initial entry of *Candida* into the endothelial cell, as opposed to a strictly invasive process requiring enzymatic lysis, mechanical rupture, or invagination of the endothelial membrane.[355] It is obvious that further studies are required to more clearly define the nature of these mechanisms.

An alternative mechanism(s) for the adhesion of *C. albicans* to endothelial cells is attachment to damaged endothelium. The predominance of bacteria, particularly *S. aureus*, *S. pyogenes*, and oral streptococci such as *S. mutans* and *S. sanguis*, as the most common cause of bacterial endocarditis is attributed to their ability to adhere selectively to damaged endothelial tissue (for review see Christensen et al.[298] and Thompson et al.[356]). Klotz and Maca[357] simulated endothelial injury by treating bovine vascular endothelial cells with EDTA, urea, or thrombin. These treatments led to a marked, reversible contraction of the endothelial layer resulting in exposure of subendothelial extracellular matrix (ECM). The authors showed that *Candida* cells adhered more to contracted monolayers than to confluent monolayers ($P < 0.01$) but preferentially adhered to the ECM, suggesting that exposure of the ECM *in vivo* may contribute to the establishment of metastatic *Candida* lesions.[357]

Another way for microorganisms to adhere to injured endothelial tissues is through binding to a fibrin-platelet matrix. Investigations using rabbits with experimentally induced cardiac valvular injury showed that bacteria adhered selectively to the fibrin-platelet network overlying the traumatized valvular leaflets.[358-360] Following experimental infection, the fibrin-platelet matrix increased in size with bacterial recruitment of additional fibrin and platelets; these blood elements further enmesh the bacteria in the growing thrombus (vegetation).[298] A similar mechanism(s) may be found in *Candida*. Maisch and Calderone[293] studied the adherence of *C. albicans* to a fibrin-platelet matrix formed *in vitro* These authors reported that viability appears to be essential for adherence to fibrin-platelet clots. Furthermore, *C. albicans* and *C. stellatoidea* adhered to fibrin-platelet matrices to a greater extent than the other species of *Candida* and *S. cerevisiae* tested.[293] This latter finding correlates well with the incidence of endocarditis caused by various *Candida* spp. In another study these authors investigated the surface components of *C. albicans* responsible for adherence. Their data suggested a possible role for surface mannan in the adherence of *C. albicans* to fibrin-platelet clots which form *in vivo* on the endocardium of heart valves.[292] Workers from the same laboratory investigated *Candida* endocarditis using a rabbit model and characterized the development of vegetation formation following intravenous injection of *Candida*.[361,362] *C. albicans* cells adhered to fibrin-platelet-erythrocyte deposits within 30 to 90 min.[361] Phagocytes were abundant within the vegetation, and could be seen with ingested yeast cells. Many of the *Candida* appeared capable of surviving phagocytosis and produced germ-tubes. The infection progressed with *Candida* cells developing into large colonies of abundant pseudohyphae and blastospores. These fungal elements were apparently protected within the vegetation by the continued deposition of host elements.[361] Modification of epithelial cells during the attachment of *Candida* was also observed.[256,260] It is possible that the changes observed in the *Candida* cell wall and host cells would lead to an increase or concentration of adhesins, or an unmasking of different or more adhesins in the yeast with a concomitant increase in number or nature of receptors available at the surfaces of host cells.[262] Since adhesion of *Candida* is manifested in ultrastructural studies to be of either a loose or tight nature it is quite difficult to decide whether these associations represent two distinct adhesive mechanisms or phases of the same mechanism(s). Vudhichamnog et al.[363] showed that formalin-killed yeast cells adhered only loosely, while viable *Candida* bound irreversibly.

B. INDIRECT INTERACTIONS

Microorganisms may adhere to specific tissues by first binding to intermediary microbes, host cells and/or "bridging" ligands derived from host and microorganisms.[298] For bacteria, this is best exemplified in the formation of dental plaque.[364] *C. albicans* interacts with mucosal surfaces from oral, gastrointestinal, and urogenital systems by similar mechanisms including co-adhesion to adherent fungi, co-adhesion to adherent bacteria, adhesion to mucus, and entrapment in the mucous gel overlying the epithelium (see Section VI Chapter 1).[255,271,272,365,366]

Pope and Cole,[367] examining the stomach mucosa of infant mice, revealed that at early times after yeast inoculation, *Candida* cells were adhering to both keratinized squamous epithelia and columnar secreting epithelium. At later stages *Candida* were also seen attached to the secreting epithelial tissues surrounded by mucus, with some cells nearly covered by the mucus layer. In addition *Candida* cells were observed to be associated with lactobacilli adherent to the keratinized epithelium (see Chapter 1 for the role of bacteria in yeast adhesion).[367] These results suggest that adhesion of *C. albicans* to the stomach mucosa is achieved through a number of mechanisms including both direct and indirect attachment (see also Chapter 5 Section V.A).

Similarly, the association of *C. albicans* with small intestinal mucosa has been studied and reported to be mediated by several different mechanisms.[271,367-369] Cole and co-workers[367-370] showed *C. albicans* to be in contact with intestinal microvilli 1 h after oral-intragastric challenge of infant mice (Figure 13). Many yeasts were also seen frequently in association with the mucus, and appeared to be attached to, embedded in, or covered by, a layer of mucus. Yeast cells were capable of progressive extracellular digestion of the intestinal mucus barrier and microvillus layer, followed by intracellular invasion of epithelial cells (Figure 14). Kennedy et al.[271,371] showed that *C. albicans* associates with the large intestinal mucosal surfaces of antibiotic treated mice by similar mechanisms. However, control mice that possessed an indigenous wall-associated bacterial flora did not behave in the same way.[371] Examination of cecal mucosa from antibiotic-treated mice showed that *C. albicans* associated with the mucosal surfaces by at least five distinct mechanisms (Figures 15 to 18).

Examination of attachment of *C. albicans* to vaginal mucosa revealed that yeast cells could attach to the epithelium by several adhesion and association mechanisms. These include *Candida* direct adhesion and co-adhesion to adherent organisms.[255,365,366] In addition to the mechanisms described so far there are several other mechanisms whereby microorganisms could indirectly bind to host tissues. These mechanisms utilize host-derived "bridging" ligands, such as fibronectin and plasma proteins (e.g., fibrinogen), and/or "bridging" cells.[298] The role of fibronectin in the adhesion of bacteria to animal tissues has been reviewed by Christensen et al.[298] Studies reviewed suggest that soluble fibronectin may function as a ligand to bind bacteria to animal cell surfaces.[339-342,372] The role of fibronectin in the adhesion of *C. albicans* has already been mentioned. Other serum proteins were suggested as possible ligands in the adhesion of microorganisms to animal tissues. Work by Mogensen and Dishon[373] demonstrated the ability of the immunoglobulin-protein A complex to function *in vitro* as a ligand and to bind staphylococci to infected tissues. Furthermore, work by Austin and Daniels[374] suggests the strong possibility that staphylococci and streptococci can exploit IgG *in vivo* as a bridging ligand for targeted adhesion to compromised tissues. Fibrinogen acts as a ligand bridging staphylococci to platelets (for review see Hawiger et al.[375]). This binding is brought about by the presence of two 15 amino acid terminals at opposite sides of the fibrinogen molecule. Staphylococci bind to one terminal while the platelets bind to the other.[298] Espersen et al.[376] demonstrated the enhancing effect of human fibrinogen on *S. aureus* virulence in mice. Other serum proteins such as albumin, β2 microglobulin, and haptoglobin are known to be bound by bacteria.[377-380] However, the physiological significance of these interactions and their relevance to microbial colonization is unknown.

Interactions of *Candida* with human plasma proteins are of particular importance in relation to adhesion. This is due to the fact that exposure of *C. albicans* to mammalian serum results in blastospore-hyphal conversion and/or changes in cell surface structure,[43,381] events known to influence adherence.[261,382] A number of studies have recently been published on the binding of human serum proteins to *Candida*. Thus Bouali et al.[383] carried out cytochemical and ultrastructural studies on the binding of human fibrinogen to *C. albicans in vitro*. These workers reported that purified human fibrinogen appeared to bind more strongly to hyphae than to yeast forms.[383] Annaix et al.[384] reported confirmatory results. In a comprehensive study Page and Odds[385] evaluated the propensity of three plasma proteins, albumin, fibrinogen, and transferrin, to bind to *Candida* spp. *in vitro*. Their results showed that all the three proteins tested bound with high avidity to germ-tubes formed by *C. albicans* (with fibrinogen binding more readily than the other

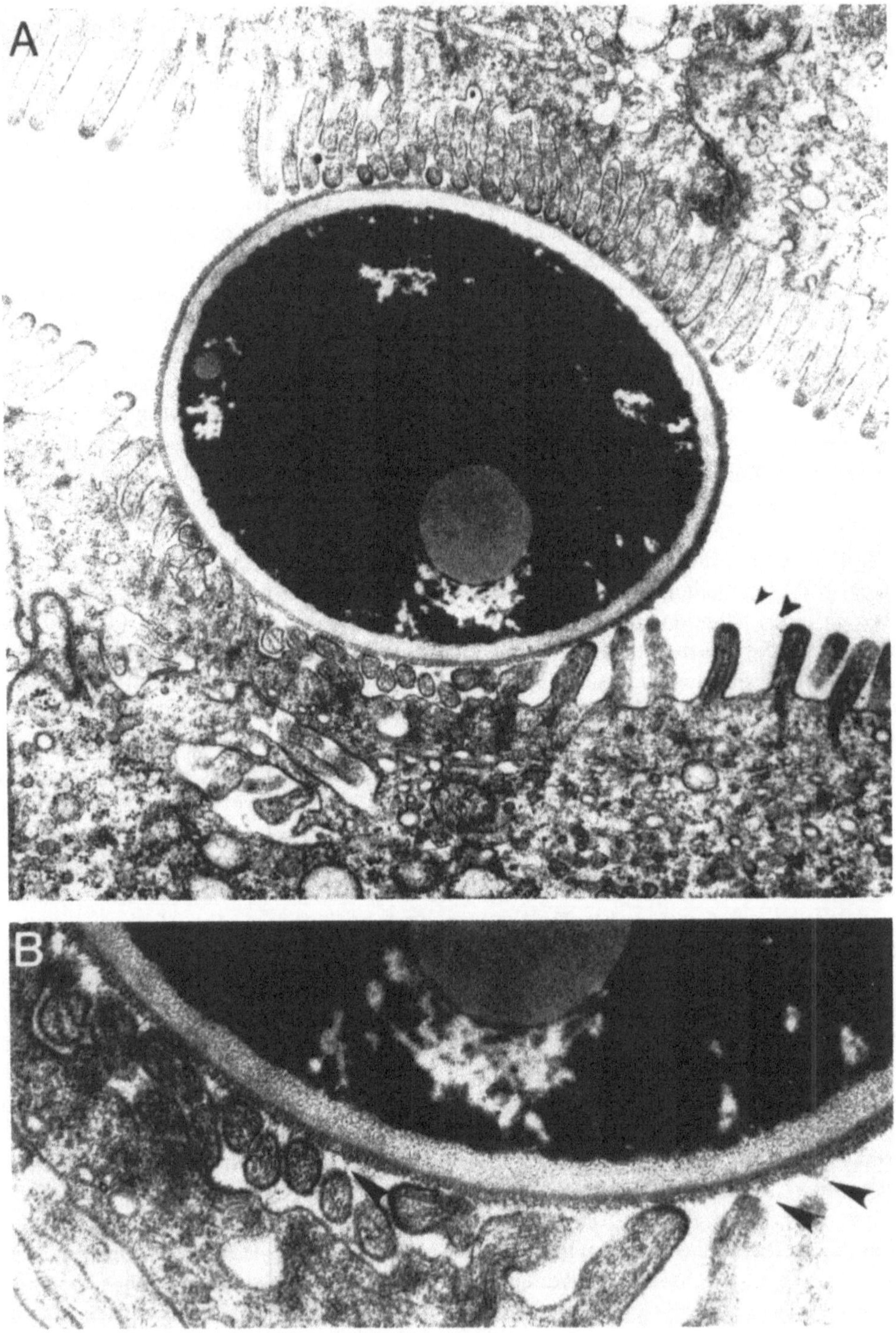

FIGURE 13. Thin sections of yeast in contact with microvilli in the jejunum of an infant mouse 1 h after oral-intragastric inoculation. Arrow heads locate thin fibrils which link tips of microvilli with outer, fibrillar, layer of yeast wall. (Magnification [A] × 27,450 and [B] × 48,000.) (From Cole, G. T., Seshan, K. R., Pope, L. M., and Yancey, R. J., *J. Med. Vet. Mycol.*, 26, 173, 1988. With permission.)

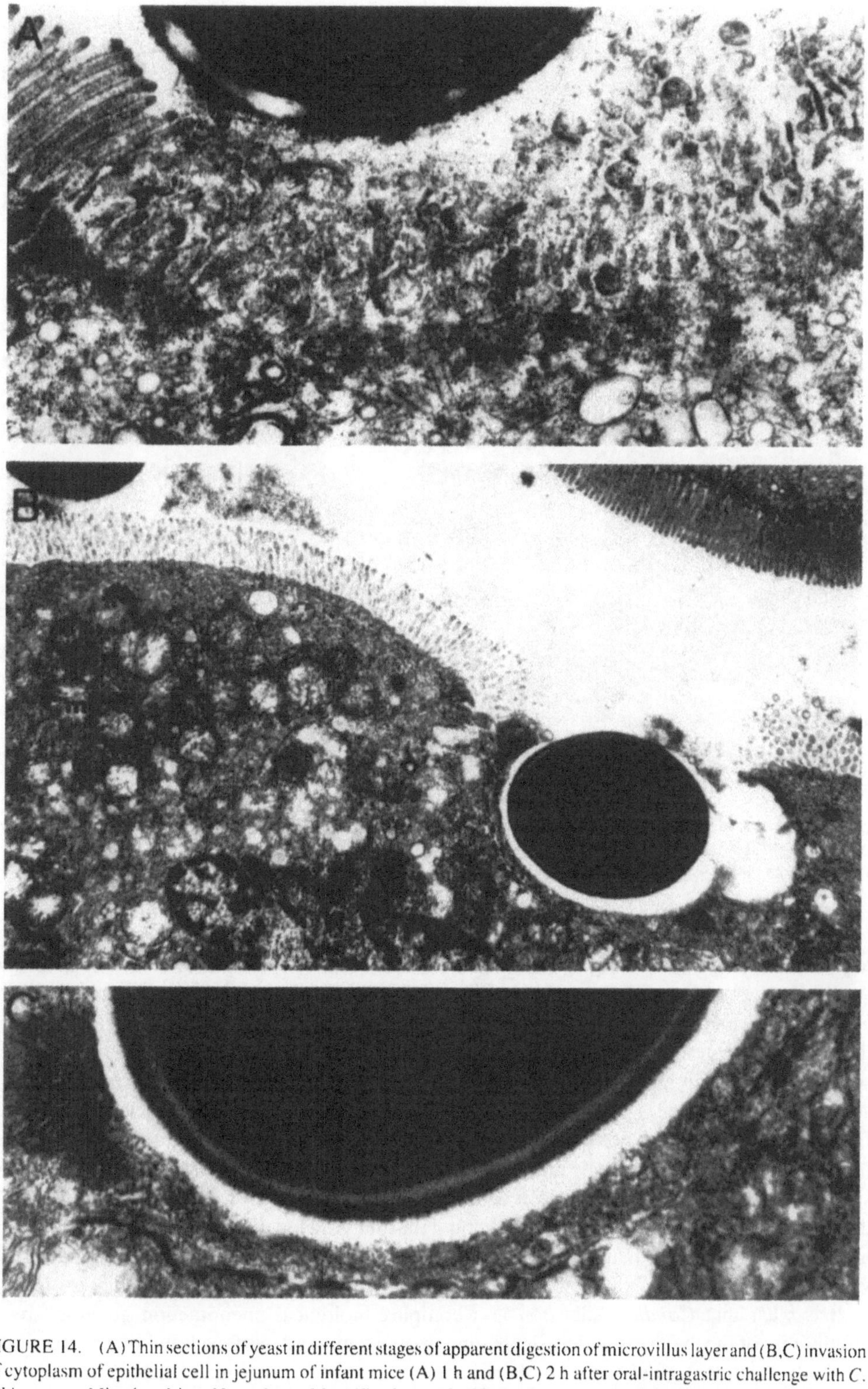

FIGURE 14. (A) Thin sections of yeast in different stages of apparent digestion of microvillus layer and (B,C) invasion of cytoplasm of epithelial cell in jejunum of infant mice (A) 1 h and (B,C) 2 h after oral-intragastric challenge with *C. albicans*. mt, Mitochondrion; N, nucleus. (Magnification × 18,600, × 6900, and × 24,000, respectively.) (From Cole, G. T., Sesan, K. R., Pope, L. M., and Yancey, R. J., *J. Med. Vet. Mycol.*, 26, 173, 1988. With permission.)

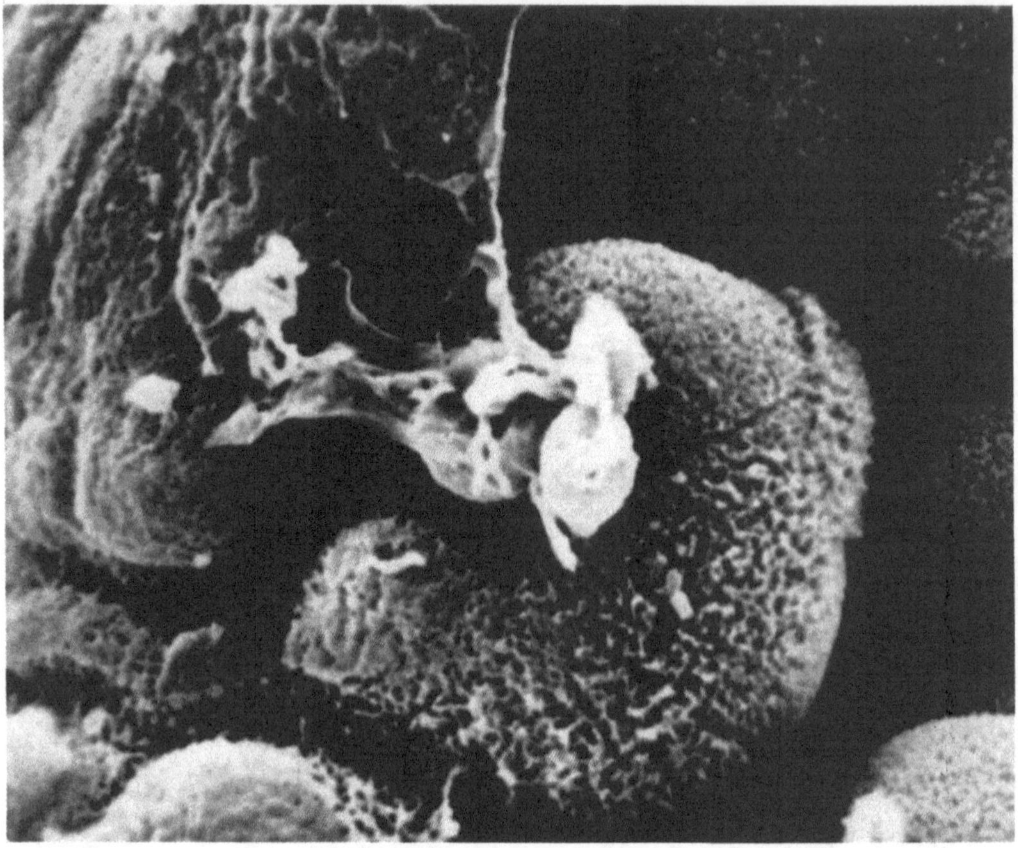

FIGURE 15. Cecal mucosa from a mouse treated with vancomycin showing direct attachment of *C. albicans* to cecal epithelium. (Magnification × 5000.). (From Kennedy, M. J., Volz, P. A., Edwards, C. A., and Yancey, R. J., *J. Med. Microbiol.*, 24, 333, 1987. With permission.)

two proteins) but did not bind to blastospores of *C. albicans* or other pathogenic species, not even to parent blastospores bearing germ-tubes.[385] This confirmed the work of Bouali et al.[383] and Annaix et al.[384] Although Trouchin et al.[386] proposed that fibrinogen bound to germ-tubes acts as a ligand in binding to fibronectin on host cell surfaces and/or as a mechanism(s) for providing the yeast with a protein coat that could protect it from host immune systems, Page and Odds[385] suggested, however, that adhesion via plasma protein intermediates is more likely to be a secondary mechanism(s) than a primary one. These authors based their conclusion on the findings that fibronectin binds directly to blastospores *in vitro*,[387] and that surface mannoproteins on blastospores directly mediate adhesion of blastospores to epithelial tissues, whereas blastospores can not bind purified fibrinogen, albumin, or transferrin *in vitro*.[385] Moreover, the finding that an attenuated *C. albicans* strain could bind plasma proteins more readily than strains with higher lethality suggests that plasma protein binding, by itself, is not an essential character for virulence.[385]

It is clear that *Candida* adhesion is a complex biological phenomenon governed by a multiplicity of mechanisms which, depending upon the mucosal surface involved, are partially or wholly functional. However, a number of points remain to be investigated in order to elucidate clearly the molecular basis of the adhesion mechanisms. The affinity and number of binding sites involved in yeast adhesion to epithelial tissues is essential for a complete and accurate description of the phases of adhesion. Indentification of surface factors that serve as *C. albicans*

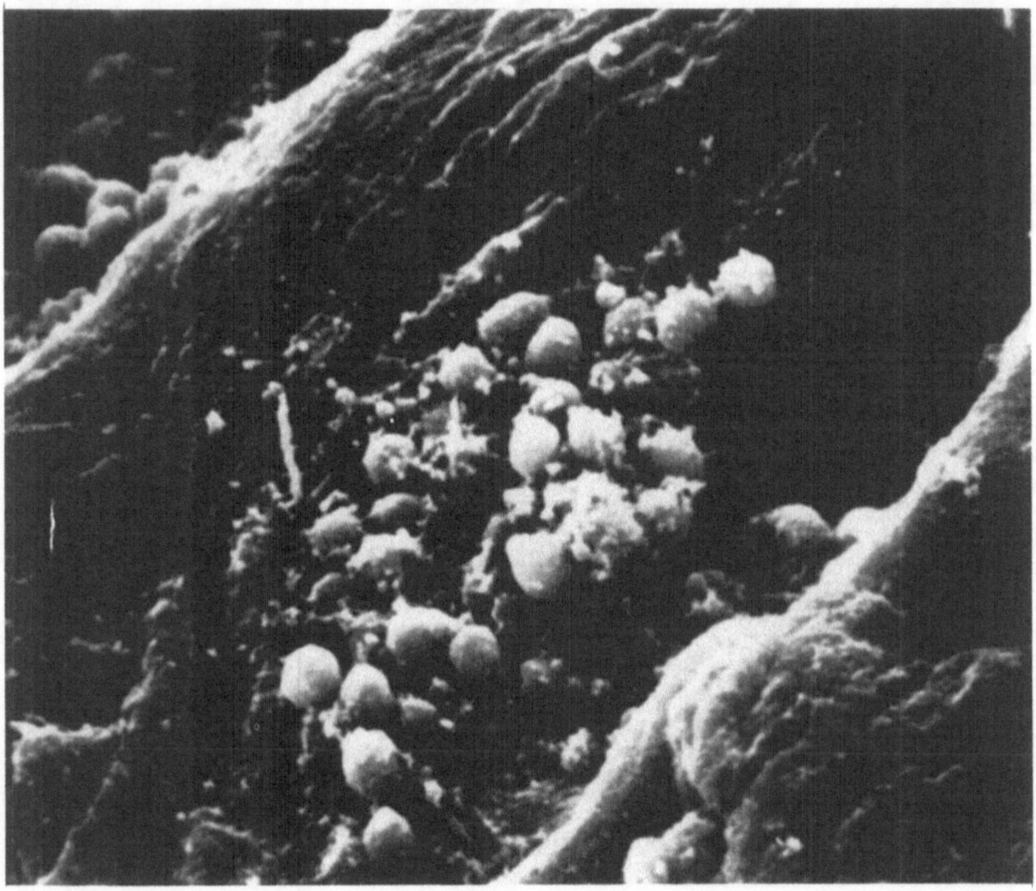

FIGURE 16. Cecal mucosa from a penicillin-treated mouse showing adhesion of *Candida* cells to the epithelium. Note that some yeast cells appear to be located in depressions in the surface of the colonized tissue. (Magnification × 400.). (From Kennedy, M. J., Volz, P. A., Edwards, C. A., and Yancey, R. J., *J. Med. Microbiol.*, 24, 333, 1987. With permission.)

adhesins or mucosal receptors that might be involved in tissue colonization (particularly gastric and endothelial) and the nature of bonding that takes place between these components should be investigated. An understanding of the nature of the adhesive events between *Candida* and epithelial tissues, as well as between this yeast and other bridging intermediates including adherent microorganisms or host-derived ligands is needed.

V. ADHERENCE ASSAYS

A. GENERAL CONSIDERATIONS

Numerous *in vivo* and *in vitro* models have been developed to quantify and characterize *Candida* adhesion to both living and nonliving surfaces. Kennedy,[262] in an excellent review, described the models used *in vitro* and *in vivo* to study *C. albicans* adhesion. Consequently, the reader is referred to his review and Table 22, which summarizes adhesion models described in the literature for *Candida*. These models suffer from many methodological difficulties and drawbacks,[388] which could be responsible for the variations in results coming out from different laboratories and even from the same laboratory. The source of these pitfalls is the fact that a large number of factors have been shown to influence *Candida* adhesion.[389] These include, yeast

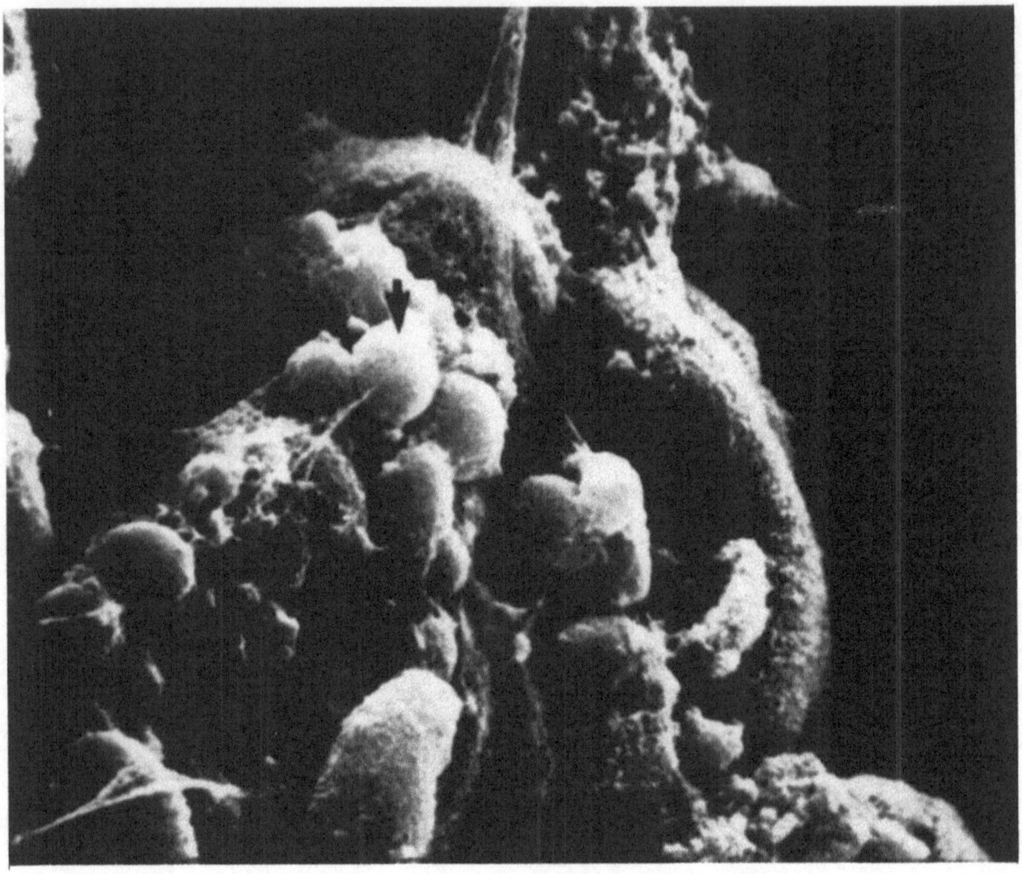

FIGURE 17. Cecal mucosa from a penicillin-treated mouse showing (arrow) yeast cells indirectly associated with the mucosa by attaching to other adherent yeast. (Magnification × 3000.) (From Kennedy, M. J., Volz, P. A., Edwards, C. A., and Yancey, R. J., *J. Med. Microbiol.*, 24, 333, 1987. With permission.)

characters, epithelial cell characters, environmental factors, and characters of inanimate surfaces. Chapter 5 is a review of the influence these parameters exert on *Candida* adhesion. *In vitro* models suffer from further drawbacks in that they cannot simulate adequately and reproducibly the conditions prevailing *in vivo*.[262,390] This makes it difficult to design a single laboratory model to mimic the situation that exists *in vivo*.

Kennedy[262] and, to some extent, Rotrosen et al.[289] have reviewed comprehensively factors complicating adhesion studies. We feel that Kennedy has comprehensively dealt with this aspect and, therefore, we will not attempt to cover this part in our review. His review and the disparities in results reported in the literature emphasize the great need for a standardized adhesion assay. Towards this aim, Kennedy proposed a standard test assay for studying *C. albicans* adherence to buccal epithelial cells and called for its adoption as a step towards establishing a laboratory model that could be adopted in future and which would assist us in our understanding of the molecular basis of the *C. albicans* adhesion processes.[262] Preliminary experiments in our laboratory show that this proposed method is quite simple and easy to follow, thus lending itself to experimental manipulations. In these preliminary experiments, we compared the results obtained using the procedure recommended by Kennedy with those obtained in parallel experiments using the procedure that we had routinely used in our own laboratory up to that time.[282,392] Using the Kennedy procedure we observed a value of 550 ± 23 yeast adhering to 100 buccal epithelial cells (BEC) compared with 585 ± 25 using our own technique. We feel that this

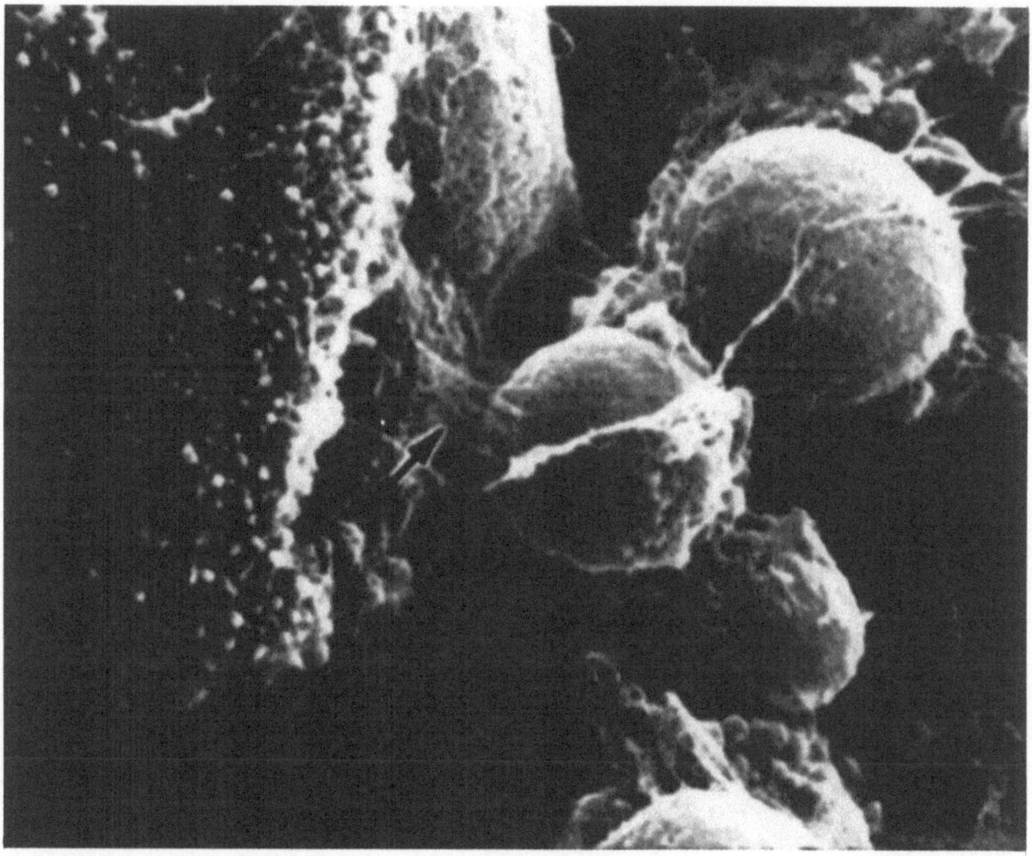

FIGURE 18. Cecal mucosa from a mouse treated with penicillin showing yeast associated with the epithelium, probably by adhesion to the (arrow) epithelial glycocalyx. (Magnification × 10,000.) (From Kennedy, M. J., Volz, P. A., Edwards, C. A., and Yancey, R. J., *J. Med. Microbiol.*, 24, 333, 1987. With permission.)

represents a satisfactory agreement and we recommend general adoption of the procedure outlined by Kennedy.[262] To encourage this process we are suggesting a methodology for the procedure with details which were not included in Kennedy's review and which we feel will be helpful to workers wishing to follow the technique.

B. STANDARDIZED ADHESION ASSAY

1. Preparation of Buccal Epithelial Cells

1. Collect and pool BEC from at least ten healthy adult volunteers, male and female, who are not using antimicrobial agents, by gently rubbing the mucosal surface of the cheek with a sterile tongue depressor.
2. Suspend the epithelial cells in 50 ml of phosphate-buffered saline (PBS) (0.15 *M* NaCl; 0.01 *M* phosphate, pH7.2) and standardize. Ideally these cells should be used shortly after collection, but to avoid problems arising from day to day or donor to donor variation between assays, the cells may be stored at 4°C for up to 2 weeks before use.

2. Preparation of *C. albicans* Cells

In any standardized assay procedure the use of particular test strains is usually recommended. In this case a problem arises in using an established laboratory strain in that its behavior might

TABLE 22
Models for Studying *C. albicans* and Association Mechanisms

Basis of model	Model has been used with	Ref.
In vitro		
Exfoliated Epithelial cells	Buccal epithelial cells	148,255,259,260,264,265, 272,278,279,282,283, 294,295,351,363,366,382, 391,394,396-407
	Corneocytes	287,408
	Urogenital epithelial cells	255,281,407
	Vaginal epithelial cells	280,283-285,296,301,366, 404-407,409-413
Tissue culture	HeLa cells	277,415-417
	Cervical epithelial cells	418
	Vaginal epithelial cells	285,419
Tissue slices, disks, or explants	Endothelial tissue	355,420
	Oral mucosa	256,421,422
	Gastrointestinal mucosa	302,371
	Fibrin-plalelet matrices	292,293
	Intestinal mucus gel	271
Non-biological surfaces	Acrylic	148,261,265,279,398,423,424
	Glass	428
	Plastic	354,429
	Denture base resins	430
Hydrophobicity testing	Phase-partition test	259,271,354,424,431,432
	Hydrophobic interaction chromatography (HIC)	428
	Modified HIC	262
	Contact angle measurements	354,425,430,432
	Salting out procedures	NT
	Replica plate test	NT
In vivo		
Gastrointestinal mucosa	Infant mice	367-369,433-435
	Adult euthymic mice	365,436
	Adult germ-free mice	365,436,437
	Adult conventional and antibiotic-treated hamsters and mice	271,371,438-445
	Adult conventional rats	446
	Dogs	447
Heart	Rabbit	361,448-452
Oral mucosa	Mice	365,436,453
	Hamsters	454
	Rats	455-467
	Monkeys	468,469
Renal endothelium	Rabbit	257
Vaginal mucosa	Mouse	301,410
	Rats	414

Note: NT, not tested

From Kennedy, M. J., Adhesion and association mechanisms of *Candida albicans*, in *Current Topics in Medical Mycology*, Vol. 2, McGinnis, M. R., Ed., Springer-Verlag, New York, 1988, chap. 4. With permission.

not accurately represent that of an immediate clinical isolate. There is little point in recommending a recent clinical isolate, however, as this will not remain so for very long. It seems reasonable, then, to recommend the use of a strain of *C. albicans* which has been shown to be consistent in its behavior and has previously been employed in a number of laboratories. With these considerations in mind we recommend the use of strain ATCC 10231 which we have observed to be consistent in its adhesion behavior over a number of years in our laboratory and has been studied by other workers.[282,392-395]

1. Inoculate 1 ml from an overnight culture of *C. albicans* into 50 ml of Difco® Yeast Nitrogen Base (with amino acids and ammonium sulfate) supplemented with 500 m*M* galactose in a 250 ml erlenmeyer flask.
2. Incubate in a rotary shaker (100 r.p.m.) at 37°C for 12 h (early stationary phase).
3. Yeast cells prepared in this way should be standardized and used immediately.

3. Adherence Assay — Method

1. Take a volume of the yeast culture containing 4×10^7 cells, centrifuge (4°C, 4000 r.p.m., 10 min) and resuspend in 2 ml of assay medium (KCl, 0.05 *M*; phosphate buffer, 1 m*M*, pH 6.0; $CaCl_2$, 1 m*M*; $MgSO_4 . 7H_2O$, 0.01 m*M*).
2. Take a volume of the BEC suspension containing 4×10^5 cells, centrifuge (4°C, 2000 r.p.m., 10 min) and resuspend in 2 ml of the same assay medium.
3. Mix the two suspensions in a McCartney bottle and incubate at 37°C with shaking (100 r.p.m.) for 1 h.
4. Immediately filter through a 20 μm (pore size) filter to remove nonadhering yeast cells.
5. Wash the cells on the filter twice with 5 ml portions of assay medium and suspend them by gently swirling with 1 ml of assay medium.
6. Mount a drop of the suspension on a glass slide, air dry, heat fix by passing the slide quickly through a Bunsen burner flame, and stain with crystal violet for 1 min.
7. Determine adherence microscopically by counting the mean number of yeast cells adhering to 100 BEC. Only those yeast cells directly in contact with BEC membrane should be counted.

4. Statistical Treatment of Results

Each assay should be performed at least in duplicate using the Student t-test to evaluate adherence values. A P value of less than 0.05 would indicate significant difference between the results of two determinations.

5. Deviation from Standard Procedure

Depending upon the particular nature of the investigation being undertaken, the standard procedure outlined above might not always be appropriate. For example, if a study involves a comparison of BEC from individual donors, it would obviously be impossible to pool BEC from several volunteers. Similarly, if the objective is to compare yeast strains or phenotypes for a particular reason the recommended type strain or its processing might not be appropriate.

The protocol proposed here is particularly suitable for studying the mechanism(s) of the adhesion process of *C. albicans* to BEC. Nevertheless, it is desirable that when other objectives are sought that research workers should follow the procedure as closely as is compatible with the objectives of their work. Clearly, if other types of epithelial or yeast cells are to be studied a different protocol might be necessary. That suggested here could form the basis for the development of such alternative assays.

REFERENCES

1. **Ofek, I. and Beachey, E. H.,** Bacterial adherence, in *Advances in Internal Medicine*, Vol. 25, Stollerman, G. H., Ed., Year Book Medical Publishers, Chicago, 1980, 503.
2. **Costerton, J. W., Marrie, T. J., and Cheng, K. J.,** Phenomena of bacterial adhesion, in *Bacterial Adhesion, Mechanisms and Physiological Significance*, Savage, D. C. and Fletcher, M., Eds., Plenum, New York, 1985, chap. 1.
3. **Reed, W. P. and Williams, R. C., Jr.,** Bacterial adherence: first step in pathogenesis of certain infections, *J. Chron. Dis.*, 31, 67, 1978.
4. **Gibbons, R. J. and van Houte, J.,** Bacterial adherence in oral microbial ecology, *Annu. Rev. Microbiol.*, 29, 19, 1975.
5. **Bitton, G. and Marshall, K. C.,** *Adsorption of Microorganisms to Surfaces*, John Wiley & Sons, New York, 1980.
6. **Berkeley, R. C. W., Lynch, J. M., Melling, J., Rutter, P. R., and Vincent, B.,** *Microbial Adhesion to Surfaces*, Ellis Horwood, Chichester, 1980.
7. **O'Connor, M., Whelan, J., and Elliott, K.,** *Adhesion and Microorganism Pathogenicity*, Ciba Found. Symp. 80, Pitman Medical, London, 1981.
8. **Jones, G. W.,** The attachment of bacteria to the surfaces of animal cells, in *Microbial Interactions, Receptors and Recognitian*, Series B, Vol. 3, Reissig, J. L., Ed., Chapman and Hall, London, 1977, 139.
9. **Ellwood, D. C., Melling, J., and Rutter, P.,** *Adhesion of Microorganisms to Surfaces*, Academic Press, London, 1979.
10. **Beachey, E. H.,** *Bacterial Adherence, Receptor and Recognition*, Series B, Vol. 6, Chapman and Hall, London, 1980.
11. **Schlessinger, D.,** *Microbiology — 1982*, American Society for Microbiology, Washington, D.C., 1982, 259.
12. **Jones, G. W. and Isaacson, R. E.,** Proteinaceous bacterial adhesins and their receptors, *Crit. Rev. Microbiol.*, 10, 229, 1983.
13. **Savage, D. C. and Fletcher, M.,** *Bacterial Adhesion, Mechanisms and Physiological Significance*, Plenum, New York, 1985.
14. **Mergenhagen, S. E. and Rosan, B.,** *Molecular Basis of Oral Microbial Adhesion*, American Society for Microbiology, Washington, D.C., 1985.
15. **Hurley, R., de Louvois, J., and Mulhall, A.,** Yeasts as human and animal pathogens, in *The Yeasts*, 2nd ed., Vol. 1, Rose, A. H. and Harrison, J. S., Eds., Academic Press, London, 1986, 207.
16. **Odds, F. C.,** *Candida and Candidosis, A Review and Bibliography*, 2nd ed., Bailliere Tindall, London, 1988.
17. **Fletcher, M., Latham, M. J., Lynch, J. M., and Rutter, P. R.,** The characteristics of interfaces and their role in microbial attachment, in *Microbial Adhesion to Surfaces*, Berkeley, R. C. W., Lynch, J. M., Melling, J., Rutter, P. R., and Vincent, B., Eds., Ellis Horwood, Chichester, 1980, chap.3.
18. **Marshall, K. C. and Bitton, G.,** Microbial adhesion in perspective, in *Adsorption of Microorganisms to Surfaces*, Bitton, G. and Marshall, K. C., Eds., John Wiley & Sons, New York, 1980, chap. 1.
19. **Tadros, Th. F.,** Particle-Surface adhesion, in *Microbial Adhesion to Surfaces*, Berkeley, R. C. W., Lynch, J. M., Melling, J., Rutter, P. R., and Vincent, B., Eds., Ellis Horwood, Chichester, 1980, chap.5.
20. **Rutter, P. R. and Vincent, B.,** The adhesion of microorganisms to surfaces: physico-chemical aspects, in *Microbial Adhesion to Surfaces*, Berkeley, R. C. W., Lynch, J. M., Melling, J., Rutter, P. R., and Vincent, B., Eds., Ellis Horwood, Chichester, 1980, chap. 4.
21. **James, A. M.,** Molecular aspects of biological surfaces, *Chem. Soc. Rev.*, 8, 389, 1979.
22. **Healy, T. W., Homola, A., James, R. O., and Hunter, R. T.,** Coagulation of amphoteric latex colloids, reversibility and specific ion effects, *Discuss. Faraday Soc.*, 65, 156, 1978.
23. **Homola, A. and James, R. O.,** Preparation and characterization of amphoteric polystyrene latexes, *J. Colloid Interface Sci.*, 59, 123, 1977.
24. **Arends, J. and Jongebloed, W. L.,** The enamel substrate — characteristics of the enamel surface, *Swedish Dent. J.*, 1, 215, 1977.
25. **Bretscher, M. S.,** The molecules of the cell membrane, *Sci. Am.*, 253, 100, 1985.
26. **Soll, D. R.,** *Candida albicans*, in *Fungal Dimorphism with Emphasis on Fungi Pathogenic for Humans*, Szaniszlot, P. J. and Harris, J. L., Eds., Plenum, New York, 1985, chap. 7.
27. **Ghannoum, M. A., Abu El Teen, K., and Radwan, S. S.,** Blocking adherence of *Candida albicans* to buccal epithelial cells by yeast glycolipids, yeast wall lipids and lipids from epithelial cells, *Mykosen*, 30, 371, 1987.
28. **Ghannoum, M. A., Mudher, S., and Burns, G.,** Incorporation of dexamethasone by *Candida albicans*, *Microbios*, 42, 103, 1985.
29. **Reiss, E.,** Cell wall composition, in *Fungi Pathogenic for Humans and Animals, Part B: Pathogenicity and Detection*, Howard, D. H., Ed., Marcel Dekker, New York, 1985, 57.
30. **Arnold, W. N.,** Physical aspects of the yeast cell envelope, in *Yeast Cell Envelopes: Biochemistry, Biophysics, and Ultrastructure*, Vol. 1, Arnold, W. N., Ed., CRC Press, Boca Raton, FL, 1981, 25.

31. **Phaff, H. J.**, Structure and biogenesis of the yeast cell envelope, in *The Yeasts*, Vol. 2, 1st ed., Rose, A. H. and Harrison, J. S., Eds., Academic Press, London, 1971, chap. 5.
32. **Sentandreu, R., Herrero, E., Martinez-Garcia, J. P., and Larriba, G.**, Biogenesis of the yeast cell wall, *Subcell Biochem.*, 10, 193, 1984.
33. **Rogers, H. J., Perkins, H. R., and Ward, J. B.**, Biosynthesis of wall components in yeast and filamentous fungi, in *Microbial Cell Wall and Membranes*, Rogers, H. J., Perkins, H. R., and Ward, J. B., Eds., Chapman and Hall, London, 1980, chap. 13.
34. **Houwink, A. L., Kreger, D. R., and Roelofsen, P. A.**, Composition and structure of yeast cell walls, *Nature (London)*, 168, 693, 1951.
35. **Kreger, D. R.**, Observations on cell walls of yeasts and some other fungi by X-ray diffraction and solubility tests, *Biochim. Biophys. Acta*, 13, 1, 1954.
36. **Kessler, G. and Nickerson, W. J.**, Glucomannan-protein complexes from cell walls of yeasts, *J. Biol. Chem.*, 234, 2281, 1959.
37. **San-Blas, G., Ordaz, D., and Yegres, F. J.**, *Histoplasma capsulatum:* chemical variability of the yeast cell wall, *Sabouraudia*, 16, 279, 1978.
38. **Farkas, V.**, Biosynthesis of cell walls of fungi, *Microbiol. Rev.*, 43, 117, 1979.
39. **Bacon, J. C. D.**, Nature and disposition of polysaccharides within the cell envelope, in *Yeast Cell Envelopes: Biochemistry, Biophysics, and Ultrastructure*, Vol.1, Arnold, W. N., Ed., CRC Press, Boca Raton, FL, 1981, 65.
40. **Montes, L. F., Patrick, T. A., Martin, S. A., and Smith, M.**, Ultrastructure of blastospores of *Candida albicans* after permanganate fixation, *J. Invest. Dermatol.*, 45, 227, 1965.
41. **Cassone, A., Simonetti, N., and Strippoli, V.**, Ultrastructural changes in the wall during germ tube formation from blastospores of *Candida albicans*, *J. Gen. Microbiol.*, 77, 417, 1973.
42. **Yamaguchi, H., Kanda, Y., and Osumi, M.**, Dimorphism in *Candida albicans*. II. Comparison of fine structure of yeast-like and filamentous phase growth, *J. Gen. Appl. Microbiol.*, 20, 101, 1974.
43. **Poulain, D., Tronchin, G., Dubremetz, J. F., and Biguet, J.**, Ultrastructure of the cell wall of *Candida albicans* blastospores: study of its constitutive layers by the use of a cytochemical technique revealing polysaccharides, *Ann. Microbiol.*, 129, 10, 1978.
44. **Scherwitz, C., Martin, R., and Ueberberg, H.**, Ultrastructural investigations of the formation of *Candida albicans* germ tubes and septa, *Sabouraudia*, 16, 115, 1978.
45. **Takamiya, H., Vogt, A., Batsford, S., Kuttin, E. S., and Muller, J.**, Further studies on the immunoelectron microscopic localization of polysaccharide antigens on ultra-thin sections of *Candida albicans*, *Mykosen*, 28, 17, 1985.
46. **Soll, D. R., Herman, M. A., and Staebell, M. A.**, The involvement of cell wall expansion in the two modes of mycelium formation of *Candida albicans*, *J. Gen. Microbiol.*, 131, 2367, 1985.
47. **Anderson, J. M. and Soll, D. R.**, Differences in actin localization during the bud and hypha formation in the yeast *Candida albicans*, *J. Gen. Microbiol.*, 132, 2035, 1986.
48. **Brawner, D. L. and Cutler, J. E.**, Ultrastructural and biochemical studies of two dynamically expressed cell surface determinants on *Candida albicans*, *Infect. Immun.*, 51, 327, 1986.
49. **Suzuki, S. and Sunayama, H.**, Studies on the antigenic activities of yeasts. II. Isolation and inhibition assay of the oligosaccharides from acetolysate of mannan in *Candida albicans*, *Jpn. J. Microbiol.*, 12, 413, 1968.
50. **Suzuki, S. and Sunayama, H.**, Studies on the antigenic activities of yeasts. III. Isolation and inhibition assay of the oligosaccharides from acid-hydrolysate of mannan of *Candida albicans*, *Jpn. J. Microbiol.*, 13, 95, 1969.
51. **Venezia, R. and Lachapelle, R.**, The use of ferritin-conjugated antibodies in the study of cell wall components of *Candida albicans*, *Can. J. Microbiol.*, 19, 1445, 1973.
52. **Jones, G. R. and Stewart-Tull, D. E. S.**, Antigenic analysis of yeast cell walls, *Sabouraudia*, 13, 94, 1975.
53. **Suzuki, M. and Fukazawa, Y.**, Immunochemical characterization of *Candida albicans* cell wall antigens: specific determinant of *Candida albicans* serotype A mannan, *Microbiol. Immunol.*, 26, 387, 1982.
54. **Brawner, D. L. and Cutler, J. E.**, Variability in expression of a cell surface determinant on *Candida albicans* as evidenced by an agglutinating monoclonal antibody, *Infect. Immun.*, 43, 966, 1984.
55. **Brawner, D. L. and Cutler, J. E.**, Variability in expression of cell surface antigens of *Candida albicans* during morphogenesis, *Infect. Immun.*, 51, 337, 1986.
56. **Hopwood, V., Poulain, D., Fortier, B., Evans, G., and Vernes, A.**, A monoclonal antibody to a cell wall component of *Candida albicans*, *Infect. Immun.*, 54, 222, 1986.
57. **Tronchin, G., Poulain, D., and Biguet, J.**, Etudes cytochimiques et ultrastructurales de la paroi de *Candida albicans*. I. Localization des mannanes par utilisation de concanavalin A surcoupes ultrafines, *Arch. Microbiol.*, 123, 245, 1979.
58. **Tronchin, G., Poulain, D., and Biguet, J.**, Ultrastructural localization of acid-phosphatase in *Candida albicans*, *Biol. cell*, 38, 147, 1980.
59. **Tronchin, G., Poulain, D., Herbaut, J., and Biguet, J.**, Cytochemical and ultrastructural studies of *Candida albicans*. II. Evidence for a cell wall coat using concanavalin A, *J. Ultrastruct. Res.*, 75, 50, 1981.

60. **Pugh, D. and Cawson, R. A.,** The surface layer of *Candida albicans, Microbios,* 23, 19, 1978.
61. **Chattaway, F. C., Holmes, M. R., and Barlow, A. J. E.,** Cell wall composition of the mycelial and blastospore forms of *Candida albicans, J. Gen. Microbiol.,* 51, 367, 1968.
62. **Northcote, D. H.,** The molecular structure and shape of yeast glycogen, *Biochem. J.,* 53, 348, 1953.
63. **Jones, J., Rasch, S., and Strand, A.,** Specific polysaccharides in *Candida.* I. Precipitation and properties of water soluble fractions, *Acta Pathol. Microbiol. Scand.,* 37, 449, 1955.
64. **Kemp, G. and Solotorovsky, M.,** Localization of antigen in mechanically disrupted cells of certain species of *Candida* and *Torulopsis, J. Immunol.,* 93, 305, 1964.
65. **Pepys, J., Faux, J., Longbottom, J. L., McCarthy, D. S., and Hargreave, F. E.,** *Candida albicans* precipitins in respiratory disease in man, *J. Allergy,* 41, 305, 1968.
66. **Reiss, E., Stone, H., and Hasenclever, H. F.,** Serological and cellular immune activity of peptidoglucomannan fractions of *Candida albicans* cell walls, *Infect. Immun.,* 9, 881, 1974.
67. **Stewart-Tull, D. E. S., Timperley, W. K., and Horne, C. H. W.,** Some immunochemical observations of the cell wall of yeasts, *Sabouraudia,* 5, 104, 1966.
68. **Zamiechowska-Maizga, J.,** Isolation and immunochemical analysis of *Candida albicans* fractions, *Bull. Acad. Pol. Sci. Ser. Sci. Biol.,* 22, 275, 1974.
69. **Master, L., Sikl, D., Bauer, S., and Sandula, J.,** Extracellular polysaccharide-protein complexes produced by selected strains of *Candida albicans, Folia Microbiol.,* 11, 373, 1966.
70. **Odds, F. C. and Hierholzer, J. C.,** Purification and properties of a glycoprotein acid phosphatase from *Candida albicans, J. Bacteriol.,* 114, 257, 1973.
71. **Manner, D. J. and Masson, A. T.,** The structure of two glucans from yeast cell walls, *FEBS Lett.,* 4, 122, 1969.
72. **Manner, D. J., Masson, A. T., and Patterson, J. C.,** The structure of a β-(1,3)-D-glucan from yeast cell walls, *Biochem. J.,* 135, 19, 1973.
73. **Bartnicki-Garcia, S.,** Cell wall chemistry, morphogenesis and taxonomy of fungi, *Annu. Rev. Microbiol.,* 22, 87, 1968.
74. **Bartnicki-Garcia, S.,** Cell wall composition and other biochemical markers in fungal phylogeny, in *Phytochemical Phylogeny,* Hasborne, J. B., Ed., Academic Press, New York, 1970, 81.
75. **Zevenhuizen, L. P. T. M. and Bartnicki-Garcia, S.,** Chemical structure of the insoluble hyphal wall glucan of *Phytophthora cinnamomi, Biochemistry,* 8, 1496, 1969.
76. **Wang, M. C. and Bartnicki-Garcia, S.,** Structure and composition of walls of the yeast form of *Verticillum alboatrum, J. Gen. Microbiol.,* 64, 41, 1970.
77. **Bishop, C. T., Blank, F., and Gardner, P. E.,** The cell wall polysaccharides of *Candida albicans;* glucan, mannan and chitin, *Can. J. Chem.,* 38, 869, 1960.
78. **Yu, R. J., Bishop, C. T., Cooper, F. P., Blank, F., and Hasenclever, H. F.,** Glucans from *Candida albicans* (serotype B) and from *Candida parapsilosis, Can. J. Chem.,* 45, 2264, 1967.
79. **Sullivan, P. A., Chiew, Y. Y., Molloy, C., Templeton, M. D., and Shepherd, M. G.,** An analysis of the metabolism and cell wall composition of *Candida albicans* during germ-tube formation, *Can. J. Microbiol.,* 29, 1514, 1983.
80. **Gopal, P. K., Shepherd, M. G., and Sullivan, P. A.,** Analysis of cell wall glucans from yeast, hyphal and germ-tube forming cells of *Candida albicans, J. Gen. Microbiol.,* 130, 3295, 1984.
81. **Gopal, P., Shepherd, M., and Sullivan, P.,** Glucan synthase in *Candida albicans, Proc. Univ. Otago Med. Sch.,* 60, 32, 1982.
82. **Orlean, P. A. B.,** (1-3)-β-D-glucan synthase from budding and filamentous cultures of the dimorphic fungus *Candida albicans, Eur. J. Biochem.,* 127, 397, 1982.
83. **Gopal, P. K., Sullivan, P. A., and Shepherd, M. G.,** Metabolism of [^{14}C] glucose by regenerating spheroplasts of *Candida albicans, J. Gen. Microbiol.,* 130, 325, 1984.
84. **Shepherd, M. G., Surarit, R., Gopal, P. K., and Sullivan, P. A.,** Cell wall metabolism of *C. albicans:* β(1 → 3) and (1 → 6) glucan synthesis, in *Microbial Cell Wall Synthesis and Autolysis,* Nombela, C., Ed., Elsevier, Amsterdam, 1984, 73.
85. **Elorza, M. V., Ricco, H., Gozalbo, D., and Sentandreu, R.,** Cell wall composition and protoplast regeneration in *Candida albicans, Antonie van Leeuwenhoek J. Microbiol. Serol.,* 49, 457, 1983.
86. **Gopal, P., Sullivan, P. A., and Shepherd, M. G.,** Isolation and structure of glucan from regenerating spheroplasts of *Candida albicans, J. Gen. Microbiol.,* 130, 1217, 1984.
87. **Sullivan, P. A., Yin, C. Y., Molloy, C., Templeton, M. D., and Shepherd, M.,** An analysis of the metabolism and cell wall composition of *Candida albicans* during germ tube formation, *Can. J. Microbiol.,* 29, 1514, 1983.
88. **Poulain, D., Hopwood, V., and Vernes, A.,** Antigen variability of *Candida albicans, Crit. Rev. Microbiol.,* 12, 223, 1985.
89. **Osumi, M., Baba, M., and Yamaguchi, E.,** Electron microscopical study on the biosynthesis and assembly of yeast cell wall components, in *Microbial Cell Wall Synthesis and Autolysis,* Nombela, C., Ed., Elsevier Science, Amsterdam, 1984, 137.
90. **Cabib, E.,** Molecular aspects of yeast morphogenesis, *Annu. Rev. Microbiol.,* 29, 191, 1975.

91. **Wang, M. C. and Bartnicki-Garcia, S.,** Biosynthesis of β-1,3 and β-1,6 linked glucan by *Phytophthora cinnamomi* hyphal walls, *Biochem. Biophys. Res. Commun.*, 24, 832, 1966.
92. **Wang, M. C. and Bartnicki-Gracia, S.,** Synthesis of β-1,3-glucan microfibrils by a cell free extract from *Phytophthora cinnamomi, Arch. Biochem. Biophys.*, 175, 351, 1976.
93. **Nanba, H. and Kuroda, H.,** Studies on fungicides. X. Biosynthesis of β-glucan and chitin-like substance of cell wall from *Cochliobolus miyabeamus* (1), *Chem. Pharm.*, 22, 610, 1974.
94. **Nanba, H. and Kuroda, H.,** Studies on fungicides. XII. Biosynthesis of β-glucan and chitin-like substance of cell wall from *Cochliobolus miyabeamus* (2), *Chem. Pharm.*, 22, 1895, 1974.
95. **Sentandreu, R., Elorza, M. V., and Villanueva, J. R.,** Synthesis of yeast wall glucan, *J. Gen. Microbiol.*, 90, 13, 1975.
96. **Bálint, S., Farkas, V., and Bauer, S.,** Biosynthesis of β-glucans catalyzed by a particulate enzyme preparation from yeast, *FEBS Lett.*, 64, 44, 1976.
97. **López-Romero, E. and Ruiz-Herrera, J.,** Biosynthesis of β-glucans by free extracts from *Saccharomyces cerevisiae, Biochim. Biophys. Acta*, 500, 372, 1977.
98. **Shematek, E. M., Braatz, J., and Cabib, E.,** Synthesis of yeast-cell wall glucan and its regulation, *Fed. Proc. Fed. Am. Soc. Exp. Biol.*, 37, 1394, 1978.
99. **Mishra, N. C. and Tatum, E. L.,** Effect of L-sorbose on polysaccharide synthetase of *Neurospora crassa, Proc. Natl. Acad. Sci. U.S.A.*, 69, 313, 1972.
100. **Johnson, B. F.,** Autoradiographic analysis of regional cell wall growth of yeasts-*Schizosaccharomyces pombe, Exp. Cell Res.*, 39, 613, 1965.
101. **Johnson, B. F. and Gibson, E. J.,** Autoradiographic analysis of regional cell wall growth of yeasts, *Exp. Cell Res.*, 41, 297, 1966.
102. **Johnson, B. F. and Gibson, E. J.,** Autoradiographic analysis of regional cell wall growth of yeasts. III. *Saccharomyces cerevisiae, Exp. Cell Res.*, 41, 580, 1966.
103. **Févre, M.,** Subcellular localization of glucanase and cellulase in *Saprolegnia monoica* (Pringsheim), *J. Gen. Microbiol.*, 103, 287, 1977.
104. **Févre, M. and Dumas, C.,** β-Glucan synthetases from *Saprolegnia monoica* (Pringsheim), *J. Gen. Microbiol.*, 103, 297, 1977.
105. **Scarborough, G. A.,** Isolation and characterization of *Neurospora crassa* plasma membrane, *J. Biol. Chem.*, 250, 1106, 1975.
106. **Shematek, E. M. and Cabib, E.,** Biosynthesis of the yeast-cell wall. II. Regulation of β-(1-3) glucan synthetase by ATP and GTP, *J. Biol. Chem.*, 255, 895, 1980.
107. **Ruiz-Herrera, J. and Bartnicki-Garcia, S.,** Proteolytic activation and inactivation of chitin synthetase from *Mucor rouxii, J. Gen. Microbiol.*, 97, 241, 1976.
108. **Larriba, G., Morales, M., and Ruiz-Herrera, J.,** Biosynthesis of β-glucan microfibrils by cell-free extracts from *Saccharomyces cerevisiae, J. Gen. Microbiol.*, 124, 375, 1981.
109. **Parodi, A. J.,** Biosynthetic mechanisms for cell envelope polysaccharides, in *Yeast Cell Envelopes: Biochemistry, Biophysics and Ultrastructure*, Vol. 2, Arnold, W. N., Ed., CRC Press, Boca Raton, FL, 1981, 47.
110. **Shematek, E. M., Braatz, J. A., and Cabib, E.,** Biosynthesis of the yeast cell wall. I. Preparation and properties of a β(1-3) glucan synthetase, *J. Biol. Chem.*, 255, 888, 1980.
111. **Northcote, D. H. and Horne, R. W.,** The chemical composition and structure of the yeast cell wall, *Biochem. J.*, 51, 232, 1952.
112. **Parodi, A. and Leloir, L. F.,** The role of lipid intermediates in the glycosylation of proteins in the eucaryotic cell, *Biochim. Biophys. Acta*, 559, 1, 1979.
113. **Shepherd, M. G., Poulter, R. T. M., and Sulivan, P. A.,** *Candida albicans:* biology, genetics and pathogenicity, *Annu. Rev. Microbiol.*, 39, 579, 1985.
114. **Elorza, M. V., Murgui, A., and Sentandreu, R.,** Dimorphism in *Candida albicans:* contribution of mannoprotein to the architecture of yeast and mycelial cell wall, *J. Gen. Microbiol.*, 131, 2209, 1985.
115. **Elorza, M. V., Murgui, A., Rico, H., Miragall, F., and Sentandreu, R.,** Formation of new cell wall by protoplasts of *Candida albicans:* effect of papulacandin B, tunicamycin and nikkomycin, *J. Gen. Microbiol.*, 133, 2315, 1987.
116. **Miragall, F., Rico, H., and Sentandreu, R.,** Regeneration of the cell wall in protoplasts of *Candida albicans*. A cytochemical study using wheat germ agglutinin and concanavalin A, *Arch. Microbiol.*, 149, 286, 1988.
117. **Nakajima, T. and Ballou, C. E.,** Microheterogeneity of the inner core region of yeast mannoprotein, *Biochem. Biophys. Res. Commun.*, 66, 870, 1975.
118. **Nakajima, T. and Ballou, C. E.,** Structure of the linkage region between the polysaccharide and protein part of *Saccharomyces cerevisiae* mannan, *J. Biol. Chem.*, 249, 7685, 1974.
119. **Nakajima, T. and Ballou, C. E.,** Characterization of the carbohydrate fragments obtained from *Saccharomyces cerevisiae* mannan by alkaline degradation, *J. Biol. Chem.*, 249, 7679, 1974.
120. **Ballou, C. E.,** Some aspects of the structure, immuno-chemistry and genetic control of yeast mannan, *Adv. Enzymol. Relat. Areas Mol. Biol.*, 40, 239, 1974.

121. **Bacon, J. S. D.,** Nature and disposition of polysaccharides within the cell envelope, in *Yeast Cell Envelopes: Biochemistry, Biophysics, and Ultrastructure,* Vol. 1., Arnold, W. N., Ed., CRC Press, Boca Raton, FL, 1981, chap. 5.
122. **Murgui, A., Elorza, M. V., and Sentandreu, R.,** Tunicamycin and papulacandin B inhibit incorporation of specific mannoprotein into the wall of *Candida albicans* regenerating protoplasts, *Biochim. Biophys. Acta,* 884, 550, 1986.
123. **Ballou, C. E.,** Structure and biosynthesis of the mannan component of the yeast cell envelope, *Adv. Microb. Physiol.,* 14, 93, 1976.
124. **Sentandreu, R. and Northcote, D. H.,** The structure of a glycopeptide isolated from the yeast cell wall, *Biochem. J.,* 109, 419, 1968.
125. **Farkas, V., Kovarik, A., Kosinova, A., and Bauer, S.,** Autoradiographic study of mannan incorporation into growing cell walls of *Saccharomyces cerevisiae, J. Bacteriol.,* 117, 265, 1974.
126. **Farkas, V., Vagabov, V. M., and Bauer, S.,** Biosynthesis of yeast mannan. Independent formation of two carbohydrate moieties, *Biochim. Biophys. Acta,* 428, 583, 1976.
127. **Farkas, V., Vagabov, V. M., and Bauer, S.,** Biosynthesis of yeast mannan. Diversity of mannosyltransferases in the mannan-synthesizing enzyme system from yeast, *Biochim. Biophys. Acta,* 428, 573, 1976.
128. **Douglas, R. H. and Ballou, C. E.,** Isolation of *Kluyveromyces lactis* mannoprotein mutants by fluorescence-activated cell sorting, *J. Biol. Chem.,* 255, 5979, 1980.
129. **Cohen, R. E., Zhang, W., and Ballou, C. E.,** Effects of mannoprotein mutations on *Saccharomyces cerevisiae* core oligosaccharide structure, *J. Biol. Chem.,* 257, 5730, 1982.
130. **Cohen, R. E., Ballou, L., and Ballou, C. E.,** *Saccharomyces cerevisiae* mannoprotein mutants: isolation of the *mnn5* mutant and comparison with the *mnn3* strain, *J. Biol. Chem.,* 255, 7700, 1980.
131. **Slodki, M. E.,** Phosphate linkages in phosphomannans from yeast, *Biochim. Biophys. Acta,* 57, 525, 1962.
132. **Sentandreu, R. and Northcote, D. H.,** The characterization of oligosaccharides attached to threonine and serine in mannan glycopeptides obtained from the cell wall of yeast, *Carbohydrate Res.,* 10, 584, 1969.
133. **Nakajima, T. and Ballou, C. E.,** Yeast mannoprotein biosynthesis: solubilization and selective assay of four mannosyl-transferases, *Proc. Natl. Acad. Sci. U.S.A.,* 72, 3912, 1975.
134. **Rosenfeld, L. and Ballou, C. E.,** Acetolysis of disaccharides: comparative kinetics and mechanism, *Carbohydrate Res.,* 32, 287, 1974.
135. **Saxena, A. and Cihlar, R. L.,** Mannan synthetase activity in three strains of *Candida albicans, FEMS Microbiol. Lett.,* 51, 105, 1988.
136. **Durandy, A., Fischer, A., LeDeist, F., Drouhet, E., and Griscelli, C.,** Mannan-specific and mannan-induced T-cell suppressive activity in patients with chronic mucocutaneous candidiasis, *J. Clin. Immunol.,* 7, 400, 1987.
137. **Garcia-de-Lomas, J., Morales, C., Grau, M. A., and Mir, A.,** Detection of *Candida* spp. mannan antigen by indirect ELISA-inhibition. *In vitro* crossed reactivity among mannans obtained from *C. albicans* A, *C. albicans* B, and *C. tropicalis, Mycopathologia,* 102, 175, 1988.
138. **Ray, T. L., Hanson, A., Ray, L. F., and Wuepper, K. D.,** Purification of a mannan from *Candida albicans* which activates serum complement, *J. Invest. Dermatol.,* 73, 269, 1979.
139. **Sunayama, H. and Suzuki, S.,** Studies on the antigenic activities of yeasts. VI. Analysis of the antigenic determinants of the mannan of *Candida albicans* serotype B-792, *Jpn. J. Microbiol.,* 14, 371, 1970.
140. **Yu, R. J., Bishop, C. T., Cooper, F. P., Hasenclever, H. F., and Blank, F.,** Structural study of mannans from *Candida albicans* (serotype A and B), *C. parapsilosis, C. stellatoidea* and *C. tropicalis, Can. J. Chem.,* 45, 2205, 1967.
141. **Jones, G. H. and Ballou, C. E.,** Studies on the structure of yeast mannan. II. Mode of action of the Arthrobacter-mannosidase on yeast mannan, *J. Biol. Chem.,* 244, 1052, 1969.
142. **Kocourek, J. and Ballou, C. E.,** Method for finger printing yeast cell wall mannans, *J. Bacteriol.,* 100, 1175, 1969.
143. **Siki, D., Masler, L., and Bauer, S.,** Mannan from the extracellular surface of *Candida albicans* (Berkhout), *Experientia,* 20, 456, 1964.
144. **Stewart, T. S. and Ballou, C. E.,** A comparison of yeast mannans and phosphomannans by acetolysis, *Biochemistry,* 7, 1855, 1968.
145. **Suzuki, S. and Usuyama, M.,** Effect of Arthrobacter α-mannosidase digestion on the immunochemical properties of the mannans of two serotypes of *Candida* yeasts, *Tohuku Yakka Daigaku Kenkyu Nempo,* 18, 55, 1971.
146. **Kogan, G., Pavliak, V., and Masler, L.,** Structural studies of mannans from the cell walls of the pathogenic yeasts *Candida albicans* serotypes A and B and *Candida parapsilosis, Carbohydrate Res.,* 172, 243, 1988.
147. **Elorza, M. V., Marcilla, A., and Sentandreu, R.,** Wall mannoproteins of the yeast and mycelial cells of *Candida albicans*: nature of the glycosidic bonds and polydispersity of their mannan moieties, *J. Gen. Microbiol.,* 134, 2393, 1988.
148. **McCourtie, J. and Douglas, L. J.,** Extracellular polymer of *Candida albicans:* isolation, analysis and role in adhesion, *J. Gen. Microbiol.,* 131, 495, 1985.

149. **Abu-Elteen, K., Ghannoum, M., and Stretton, R. J.,** Effects of sub-inhibitory concentrations of antifungal agents on adherence of *Candida* spp. to buccal epithelial cells *in vitro,* FEMS Symp. *Candida* and Candidamycosis, Alanya, Turkey, 1989.
150. **Sentandreu, R., Elorza, M. V., and Martinez, J. P.,** *Candida albicans* cell wall macromolecular entities and their probable contribution to the cellular morphology and physiology, in *10th Cong. Int. Soc. Human Animal Mycol., ISHAM,* Torres-Rodrìgues, J. M., Ed., J. R. Prous Science Publisher, Barcelona, Spain, 988, 177.
151. **Friis, J. and Ottolenghi, P.,** Localization of invertase in a strain of yeast, *C.R. Trav. Lab. Carlsberg,* 31, 259, 1959.
152. **Neumann, N. P. and Lampen, J. O.,** Purification and properties of yeast invertase, *Biochemistry,* 6, 468, 1967.
153. **Neumann, N. P. and Lampen, J. O.,** Glycoprotein structure of yeast invertase, *Biochemistry,* 8, 3552, 1969.
154. **Trimble, R. B. and Maley, F.,** Subunit structure of external invertase from *Saccharomyces cerevisiae, J. Biol. Chem.,* 252, 4409, 1977.
155. **Gascòn, S., Neumann, N. P., and Lampen, J. O.,** Comparative study of properties of purified internal and external invertases from yeast, *J. Biol. Chem.,* 243, 1573, 1968.
156. **Barbariè, S., Kozuliè, B., Ries, B., and Mildner, P.,** Purification and evidence for heterogeneity of acid phosphatase from *Saccharomyces cerevisiae, Biochem. Biophys. Res. Commun.,* 95, 404, 1980.
157. **Boer, P. and Steyn-Parvè, E. P.,** Isolation and purification of an acid phosphatase from bakers yeast (*Saccharomyces cerevisiae*), *Biochim. Biophys. Acta,* 128, 400, 1966.
158. **Elorza, M. V. and Sentandreu, R.,** Effect of cycloheximide on yeast cell wall synthesis, *Biochem Biophys. Res. Commun.,* 36, 741, 1969.
159. **Sentandreu, R. and Lampen, J. O.,** Biosynthesis of yeast mannan: inhibition of synthesis of mannose acceptor by cycloheximide, *FEBS Lett.,* 11, 95, 1970.
160. **Liras, P. and Gascön, S.,** Biosynthesis and secretion of yeast invertase. Effect of cycloheximide and 2-deoxy-D-glucose, *Eur. J. Biochem.,* 23, 160, 1971.
161. **Babczinski, P.,** Evidence against the participation of lipid intermediates in the *in vitro* biosynthesis of serine-(threonine)-N-acetyl-D-galactosamine linkages in submaxillary mucin, *FEBS Lett.,* 117, 207, 1980.
162. **Hanover, J. A., Lennarz, W. J., and Young, J. D.,** Synthesis of N- and O-Linked glycopeptides in oviduct membrane preparations, *J. Biol. Chem.,* 255, 6713, 1980.
163. **Waechter, C. J. and Lennarz, W. J.,** Role of polyprenol-linked sugars in glycoprotein synthesis, *Annu. Rev. Biochem.,* 45, 95, 1976.
164. **Lennarz, W. J.,** Lipid-linked sugars in glycoprotein synthesis, *Science,* 188, 986,1975.
165. **Lehle, L.,** Plant-cells synthesize glucose-containing lipid-linked oligosaccharides similar to those found in animals and yeast, *FEBS Lett.,* 123, 68, 1981.
166. **Lehle, L. and Tanner, W.,** Formation of lipid-bound oligosaccharides in yeast, *Biochim. Biophys. Acta,* 399, 364, 1975.
167. **Lehle, L. and Tanner, W.,** The specific site of tunicamycin inhibition in the formation of dolichol-bound N-acetyl-glucosamine derivatives, *FEBS Lett.,* 71, 167, 1976.
168. **Lehle, L. and Tanner, W.,** Biosynthesis and characterization of large dolichyl-phosphate-linked oligosaccharides in *Saccharomyces cerevisiae, Biochim. Biophys. Acta,* 539, 218, 1978.
169. **Lehle, L. and Tanner, W.,** Glycosyl transfer from dolichyl-phosphate sugars to endogenous and exogenous glycoprotein acceptors in yeast, *Eur. J. Biochem.,* 83, 563, 1978.
170. **Parodi, A. J.,** Synthesis of glucosyl-dolichol derivatives in yeast and their role in protein glycosylation, *Eur. J. Biochem.,* 75, 171, 1977.
171. **Muzzarelli, R. A. A.,** *Chitin,* Pergamon, Oxford, 1977.
172. **Horisberger, M. and Vonlanthen, M.,** Location of mannan and chitin on thin sections of budding yeasts with gold markers, *Arch. Microbiol.,* 115, 1, 1977.
173. **Braun, P. C. and Calderone, R. A.,** Chitin synthesis in *Candida albicans:* comparison of yeast and hyphal forms, *J. Bacteriol.,* 135, 1472, 1978.
174. **Mitchell, L. and Soll, D. R.,** Septation during synchronous mycelium and bud formation in released stationary phase cultures of *Candida albicans, Exp. Mycol.,* 3, 298, 1979.
175. **Cabib, E., Duran, A., and Bowers, B.,** Localized activation of chitin synthetase in the initiation of yeast septum formation, in *Fungal Walls and Hyphal Growth,* Burnett, J. H. and Trinci, A. P., Eds., Cambridge University Press, Cambridge, 1979, 189.
176. **Gibson, R. K., Buckey, C. E., and Peberdy, J. F.,** Wall ultrastructure in regenerating protoplasts of *Aspergillus nidulans, Protoplasma,* 89, 381, 1976.
177. **Elorza, M. V., Rico, H., and Sentandreu, R.,** Calco-fluor white alters the assembly of chitin fibrils in *Saccharomyces cerevisiae* and *Candida albicans* cells, *J. Gen. Microbiol.,* 129, 1577, 1983.
178. **Stagg, C. M. and Fether, M. S.,** The characterization of chitin associated D-glucan from the cell walls of *Aspergillus niger, Biochim. Biophys. Acta,* 320, 64, 1973.
179. **Sietsma, J. H. and Wessels, J. G. H.,** Solubility of (1 → 3)-β-D/(1 → 6)-β-D-glucan in fungal walls: importance of presumed linkage between glucan and chitin, *J. Gen. Microbiol.,* 125, 209, 1981.

180. **Sonnenberg, A. S. M., Sietsma, J. H., and Wessels, J. G. H.,** Biosynthesis of alkali-insoluble cell wall glucan in *Schizophyllum commune* protoplasts, *J. Gen. Microbiol.*, 128, 2667, 1981.
181. **Gow, N. A. R. and Gooday, G. W.,** Infrared and X-ray diffraction data on chitins of variable structure, *Carbohydrate Res.*, 165, 105, 1987.
182. **Shepherd, M. G.,** Cell envelope of *Candida albicans, Crit. Rev. Microbiol.*, 15, 7, 1987.
183. **McMurrough, I., Flores-Carreòn, A., and Bartnicki-Garcia, S.,** Pathway of chitin synthesis and cellular localization of chitin synthetase in *Mucor rouxii, J. Biol. Chem.*, 246, 3999, 1971.
184. **Peberdy, J. F. and Moore, P. M.,** Chitin synthetase in *Mortirella vinacea*, properties, cellular location and synthesis in growing cultures, *J. Gen. Microbiol.*, 90, 228, 1975.
185. **Duràn, A., Bowers, B., and Cabib, E.,** Chitin synthetase zymogen is attached to the yeast plasma membrane, *Proc. Natl. Acad. Sci. U.S.A.*, 72, 3952, 1975.
186. **Bracker, C. E., Ruiz-Herrera, J., and Bartnicki-Garcia, S.,** Structure and transformation of chitin synthetase particles (chitosomes) during microfibril synthesis *in vitro, Proc. Natl. Acad. Sci. U.S.A.*, 73, 4570, 1976.
187. **Ruiz-Herrera, J. and Bartnicki-Garcia, S.,** Synthesis of cell wall microfibrils *in vitro* by a soluble chitin synthetase from *Mucor rouxii, Science*, 186, 357, 1974.
188. **Ruiz-Herrera, J., Sing, V. O., Van der Voude, V. J., and Bartnicki-Garcia, S.,** Microfibril assembly by granules of chitin synthetase, *Proc. Natl. Acad. Sci. U.S.A.*, 72, 2706, 1975.
189. **Ruiz-Herrera, J., López Romero, E., and Bartnicki-Garcia, S.,** Properties of chitin synthetase in isolated chitosomes from yeast cells of *Mucor rouxii, J. Biol. Chem.*, 252, 3338, 1977.
190. **Bartnicki-Garcia, S., Bracker, C. E., Reyes, E., and Ruiz-Herrera, J.,** Isolation of chitosomes from taxonomically diverse fungi and synthesis of chitin microfibrils *in vitro, Exp. Mycol.*, 2, 173, 1978.
191. **Hilenski, L. L., Naider, F., and Becker, J. M.,** Polyoxin D inhibits colloidal gold-wheat germ agglutinin labelling of chitin in dimorphic forms of *Candida albicans, J. Gen. Microbiol.*, 132, 1441, 1986.
192. **Barrett-Bee, K. and Hamilton, M.,** The detection and analysis of chitinase activity from the yeast form of *Candida albicans, J. Gen. Microbiol.*, 130, 1857, 1984.
193. **Bartnicki-Garcia, S.,** Role of chitosomes in the synthesis of fungal cell walls, *Microbiology*, 81, 238, 1981.
194. **Singh, B. and Datta, A.,** Glucose repression of the inducible catabolic pathway for N-acetylglucosamine in yeast, *Biochem. Biophys. Res. Commun.*, 84, 58, 1978.
195. **Singh, B. and Datta, A.,** Induction of N-acetylglucosamine catabolic pathway in spheroplasts of *Candida albicans, Biochem. J.*, 178, 427, 1979.
196. **Singh, B. and Datta, A.,** Regulation of N-acetylglucosamine uptake in yeast, *Biochim. Biophys. Acta*, 557, 248, 1979.
197. **Singh, B., Biswas, M., and Datta, A.,** Inducible N-acetyl-glucosamine binding protein in yeasts, *J. Bacteriol.*, 144, 1, 1979.
198. **Chiew, Y. Y., Shepherd, M. G., and Sullivan, P. A.,** Regulation of chitin synthesis during germ tube formation in *Candida albicans, Arch. Microbiol.*, 125, 97, 1980.
199. **Shepherd, M. G., Yin, C. Y., Ram, S. P., and Sullivan, P. A.,** Germ tube induction in *Candida albicans, Can. J. Microbiol.*, 26, 21, 1980.
200. **Sullivan, P. A. and Shepherd, M. G.,** Gratuitous induction by N-acetylglucosamine of germ tube formation and enzymes for N-acetylglucosamine utilization in *Candida albicans, J. Bacteriol.*, 151, 1118, 1982.
201. **Arnold, W. N.,** Lipids, in *Yeast Cell Envelopes: Biochemistry, Biophysics, and Ultrastructure*, Vol. 1, Arnold, W. N., Ed., CRC Press, Boca Raton, FL, 1981, chap. 7.
202. **Bianchi, D. E.,** The lipid content of cell walls obtained from juvenile yeast-like and filamentous cells of *Candida albicans, Antonie Van Leeuwenhoek J. Microbiol. Serol.*, 33, 324, 1967.
203. **Suomalainen, H. and Nurminen, T.,** The lipid composition of cell wall and plasma membrane of baker's yeast, *Chem. Phys. Lipids*, 4, 247, 1970.
204. **Kulkarni, R. K., Hollingsworth, P. J., and Volz, P. A.,** Variation in cell surface features of *Candida albicans* with respect to carbon sources, *Sabouraudia*, 18, 255, 1980.
205. **Sentandreu, R., Herrero, E., and Elorza, M. V.,** The assembly of wall polymers in yeast, in *Microbial Cell Wall Synthesis and Autolysis*, Nombela, C., Ed., Elsevier Science, Amsterdam, 1984, 51.
206. **Shibata, N., Mizugami, K., Takano, K., and Suzuki, S.,** Isolation of mannan-protein complexes from viable cells of *Saccharomyces cerevisiae* X 2180-1A wild type and *Saccharomyces cerevisiae* X 2180-1A-5 mutant strains by the action of zymolyase-60,000, *J. Bacteriol.*, 156, 552, 1983.
207. **Valentin, E., Herrero, E., Pastor, F. I. J., and Sentandreu, R.,** Solubilization and analysis of mannoprotein molecules from the cell-wall of *Saccharomyces cerevisiae, J. Gen. Microbiol.*, 130, 1419, 1984.
208. **Kidby, D. K. and Davies, R.,** Invertase and disulphide bridges in the yeast wall, *J. Gen. Microbiol.*, 61, 327, 1970.
209. **Thiery, J. P.,** Mise en evidence des polysaccharides sur coupes fines en microscopie electronique, *J. Microbiol. (Paris)*, 6, 987, 1967.

210. **Staebell, M. and Soll, D. R.**, Temporal and spatial differences in cell wall expansion during bud and mycelium formation in *Candida albicans*, *J. Gen. Microbiol.*, 131, 1467, 1985.
211. **Soll, D. R.**, The cell cycle and commitment to alternate cell fates in *Candida albicans*, in *The Microbial Cell Cycle*, Nurse, P. and Streiblova, E., Eds., CRC Press, Boca Raton, FL, 1984, 143.
212. **Ponton, J. and Jones, J.**, Identification of two germ-tube specific cell wall antigens of *Candida albicans*, *Infect. Immun.*, 54, 864, 1986.
213. **Lodish, H. F. and Rothman, J. E.**, The assembly of cell membranes, *Sci. Am.*, 240, 38, 1979.
214. **Robertson, D.**, Membrane structure, *J. Cell Biol.*, 91, 1895, 1981.
215. **Sabatini, D. D., Kreibich, G., Morimoto, K., and Adesnik, M.**, Mechanisms of the incorporation of proteins in membranes and organells, *J. Cell Biol.*, 92, 1, 1982.
216. **Karnovsky, M. J., Kleinfeld, A. M., Hoover, R. L., and Klausner, R. D.**, The concept of lipid domains in membranes, *J. Cell Biol.*, 94, 1, 1982.
217. **Smith, R. and Oldfield, E.**, Dynamic structure of membranes by deuterium NMR, *Science*, 225, 280, 1984.
218. **Dautry-Varsat, A. and Lodish, H. F.**, How receptors bring proteins and particles into cell, *Sci. Am.*, 250, 52, 1984.
219. **Wilschut, J. and Hoekstra, D.**, Membrane fusion: from liposomes to biological membranes, *Trends Biochem. Sci.*, 9, 479, 1984.
220. **Wickner, W. T. and Lodish, H. F.**, Multiple mechanisms of protein insertion to and across membranes, *Science*, 230, 400, 1985.
221. **Sheeler, P. and Bianchi, D. E.**, The plasma membrane and the endoplasmic reticulum, in *Cell and Molecular Biology*, Sheeler, P. and Bianchi, D. E., Eds., John Wiley & Sons, New York, 1987, chap. 15.
222. **Marchant, R. and Smith, D. G.**, Membranous structures in yeasts, *Biol. Rev.*, 43, 459, 1968.
223. **Fox, C. F.**, The structure of Cell membrane, *Sci. Am.*, 226, 31, 1972.
224. **Robertson, J. D.**, The membrane of the living cell, *Sci. Am.*, 206, 65, 1962.
225. **Singer, S. J. and Nicholson, G. L.**, The fluid mosaic model of the structure of cell membrane, *Science*, 175, 720, 1972.
226. **Lehinger, A. L.**, Active transport across membranes, in *Biochemistry, The Molecular Basis of Cell Structure and Function*, 2nd ed., Lehinger, A. L., Ed., Worth Publishers, New York, 1975, chap. 28.
227. **Rose, A. H.**, Molecular architecture, in *Chemical Microbiology, An Introduction to Microbiol Physiology*, 3rd ed., Rose, A. H., Ed., Butter Worths, London, 1976, chap. 2.
228. **Singer, S. J.**, Molecular organization of membranes, *Annu. Rev. Biochem.*, 43, 805, 1974.
229. **Seelig, J. and Seelig, A.**, Lipid conformation in model membranes and biological membranes, *Q. Rev. Biophys.*, 13, 19, 1980.
230. **Papahadjopoulos, D., Poste, G., and Vail, W. J.**, Studies on membrane-fusion with natural and model membranes, *Methods. Membr. Biol.*, 10, 1, 1979.
231. **Edidin, M.**, Rotational and translational diffusion in membrane, *Annu. Rev. Biophys. Bioeng.*, 3, 179, 1974.
232. **Cherry, R. J.**, Rotational and lateral diffusion of membrane proteins, *Biochim. Biophys. Acta*, 559, 289, 1979.
233. **Jain, M. K. and Wagner, R. C.**, *Introduction to Biological Membranes*, Wiley-Interscience, New York, 1980.
234. **Tanford, C.**, *The Hydrophobic Effect: Formation of Micelles and Biological Membranes*, 2nd ed., Wiley-Interscience, New York, 1980.
235. **Sharon, N. and Lis, H.**, Glycoproteins: research booming on long-ignored, ubiquitous compounds, *Chem. Eng. News*, 59, 21, 1981.
236. **Sharon, N. and Lis, H.**, Glycoproteins, in *The Proteins*, Vol. 10, 3rd ed., Neurath, H. and Hill, R. L., Eds., Academic Press, New York, 1982, 1.
237. **Marinetti, G. V. and Crain, R. C.**, Topology of amino-phospholipids in the red cell membrane, *J. Supramol. Struct.*, 8, 191, 1978.
238. **Marchesi, V. T., Furthmayer, H., and Tomita, M.**, The red cell membrane, *Annu. Rev. Biochem.*, 45, 667, 1976.
239. **Tanner, M. J. A.**, Erythrocyte glycoproteins, *Curr. Top. Membr. Transp.*, 11, 279, 1978.
240. **Hunter, K. H. and Rose, A. H.**, Yeast lipids and membranes, in *The Yeasts*, Vol. 2, 1st ed., Rose, A. H. and Harrison, J. S., Eds., Academic Press, London 1971, chap. 6.
241. **Keith, A. D., Wisnieski, B. J., Henry, S., and Williams, J. C.**, Membrane of yeast and neurospora: lipid mutants and physical studies, in *Lipids and Biomembranes of Eukaryotic Microorganisms*, Erwin, J. A., Ed., Academic Press, London, 1973, chap. 6.
242. **Aaronson, S.**, The fungi, in *Chemical Communication at the Microbial Level*, Vol. 2, Aaronson, S., Ed., CRC Press, Boca Raton, FL, 1981, chap. 2.
243. **Kerridge, D., Koh, T. Y., Marriott, M. S., and Gale, E. F.**, The production and properties of protoplasts from the dimorphic yeast *Candida albicans*, in *Microbial and Plant Protoplasts*, Peberdy, J. F., Rose, A. H., Rogers, H. J., and Cocking, E. C., Eds., Academic Press, London, 1976, 23.
244. **Marriott, M. S.**, Isolation and chemical characterization of plasma membranes from the yeast and mycelial forms of *Candida albicans*, *J. Gen. Microbiol.*, 86, 115, 1975.

245. **Miragall, F., Rico, H., and Sentandreu, R.,** Changes in the plasma membrane of regenerating protoplasts of *Candida albicans* as revealed by freeze-fracture electron microscopy, *J. Gen. Microbiol.*, 132, 2845, 1986.
246. **Hubbard, M. J., Surarit, R., Sullivan, P. A., and Shepherd, M. G.,** The isolation of plasma membrane and characterization of the plasma membrane ATPase from the yeast *Candida albicans*, *Eur. J. Biochem.*, 154, 375, 1986.
247. **Longley, R. P., Rose, A. H., and Knights, B. A.,** Composition of the protoplast membranes from *Saccharomyces cerevisiae*, *Biochem. J.*, 108, 401, 1968.
248. **Dube, J., Setterfield, G., Kiss, G., and Lusena, C. V.,** Fate of the plasma membrane of *Saccharomyces cerevisiae* during cell rupture, *Can. J. Microbiol.*, 19, 285, 1973.
249. **Ghannoum, M. A., Janini, G., Khamis, L., and Radwan, S. S.,** Dimorphism-associated variations in the lipid composition of *Candida albicans*, *J. Gen. Microbiol.*, 132, 2367, 1986.
250. **Marshall, K. C.,** Mechanisms of bacterial adhesion at solid-water interfaces, in *Bacterial Adhesion: Mechanisms and Physiological Significance*, Savage, D. C. and Fletcher, M., Eds., Plenum, New York , 1985, 133.
251. **Marshall, K. C., Stout, R., and Mitchell, R.,** Mechanism of initial events in the sorption of marine bacteria to surfaces, *J. Gen. Microbiol.*, 68, 337, 1971.
252. **Ofek, I., Lis, H., and Sharon, N.,** Animal cell surface membranes, in *Bacterial Adhesion: Mechanisms and Physiological Significance*, Savage, D. C. and Fletcher, M., Eds., Plenum, New York, 1985, 71.
253. **Montes, L. F. and Wilborn, W. H.,** Ultrastructural features of host-parasite relationship in oral candidiasis, *J. Bacteriol.*, 96, 1349, 1968.
254. **Mohamed, A.-M. H.,** Ultrastructural aspects of chronic oral candidosis, *J. Oral Pathol.*, 4, 180, 1975.
255. **Marrie, T. J. and Costerton, J. W.,** The ultrastructure of *Candida albicans* infections, *Can. J. Microbiol.*, 27, 1156, 1981.
256. **Howlett, J. A. and Squier, A.,** *Candida albicans* ultrastructure: colonization and invasion of oral epithelium, *Infect. Immun.*, 29, 252, 1980.
257. **Barnes, J. L., Osgood, W., Lee, J. C., King, R. D., and Stein, J. H.,** Host-parasite interactions in the pathogenesis of experimental renal candidiasis, *Lab. Invest.*, 49, 460, 1983.
258. **Lee, J. C. and King, R. D.,** Adherence mechanisms of *Candida albicans*, in *Microbiology-1983*, Schlessinger, D., Ed., Washington, D.C., American Society for Microbiology, 1983, 269.
259. **Kennedy, M. J. and Sandin, R. L.,** Influence of growth conditions on *Candida albicans* adhesion, hydrophobicity and cell wall ultrastructure, *J. Med. Vet. Mycol.*, 26, 79, 1988.
260. **Tronchin, G., Poulain, D., and Vernes, A.,** Cytochemical and ultrastructural studies of *Candida albicans*. III. Evidence for modification of the cell wall coat during adherence to human buccal epithelial cells, *Arch. Microbiol.*, 139, 221, 1984.
261. **McCourtie, J. and Douglas, L. J.,** Relationship between cell surface composition of *Candida albicans* and adherence to acrylic after growth on different carbon sources, *Infect. Immun.*, 32, 1234, 1981.
262. **Kennedy, M. J.,** Adhesion and association mechanisms of *Candida albicans*, in *Current Topics in Medical Mycology*, Vol. 2, McGinnis, M. R., Ed., Academic Press, New York, 1988, 73.
263. **Luft, J. H.,** Ruthenium red and ruthenium violet. II. Fine structural purification, methods for use for electron microscopy and localization in animal tissues, *Anat. Rec.*, 171, 369, 1971.
264. **Douglas, L. J., Houston, J. G., and McCourtie, J.,** Adherence of *Candida albicans* to human buccal epithelial cells after growth on different carbon sources, *FEMS Microbiol. Lett.*, 12, 241, 1981.
265. **McCourtie, J. and Douglas, L. J.,** Relationship between cell surface composition, adherence and virulence of *Candida albicans*, *Infect. Immun.*, 45, 6, 1984.
266. **Gardiner, R., Podgorski, C., and Day, A. W.,** Serological studies on the fimbriae of yeast and yeast like species, *Bot. Gaz.*, 143, 534, 1982.
267. **Tokunaga, M., Niimi, M., and Koike, H.,** Electron microscopic observation of adherence of *Candida albicans* to human buccal epithelial cells using rapid freezing technique, poster No. 122, 10th Cong. Int. Soc. Human Anim. Mycol., Barcelona, 27 June to 1 July, 1988.
268. **Joshi, K. R., Wheeler, E. E., and Gavin, J. R.,** Scanning electron microscopy of colonies of 6 species of *Candida*, *J. Bacteriol.*, 115, 341, 1973.
269. **Joshi, K. R., Gavin, J. R., and Armiger, L. C.,** Intercellular matrix in colonies of *Candida*, *J. Bacteriol.*, 123, 1139, 1975.
270. **Douglas, L. J.,** Adhesion of pathogenic *Candida* species to host surfaces, *Microbiol. Sci.*, 2, 243, 1985.
271. **Kennedy, M. J., Volz, P. A., Edwards, C. A., and Yancey, R. J.,** Mechanisms of association of *Candida albicans* with intestinal mucosa, *J. Med. Microbiol.*, 24, 333, 1987.
272. **Sandin, R. L.,** Studies on cell adhesion and concanavalin A induced agglutination of *Candida albicans* after mannan extraction, *J. Med. Microbiol.*, 24, 145, 1987.
273. **Gibbons, R. J.,** Adherence of bacteria to host tissue, in *Microbiology-1977*, Schlessinger, D., Ed., American Society for Microbiology, Washington, D.C., 1977, 395.
274. **Beachey, E. H., Eisenstein, B. I., and Ofek, I.,** *Bacterial Adherence in Infectious Diseases*, Upjohn Company, Michigan, 1982.

275. **Beachey, E. H.,** Bacterial adherence: adhesion-receptor interaction mediating the attachment of bacteria to mucosal surfaces, *J. Infect. Dis.*, 143, 325, 1981.
276. **Goldstein, I. J. and Hayes, C. W.,** The lectins: carbohydrate-binding proteins of plants and animals, *Adv. Carbohyd. Chem. Biochem.*, 35, 127, 1978.
277. **Samaranayake, L. P. and Mac Farlane, T. W.,** The effect of dietary carbohydrates on the *in vitro* adhesion of *Candida albicans* to epithelial cells, *J. Med. Microbiol.*, 15, 511, 1982.
278. **Douglas, L. J. and McCourtie, J.,** Effect of tunicamycin treatment on the adherence of *Candida albicans* to human buccal epithelial cells, *FEMS Microbiol. Lett.*, 16, 199, 1983.
279. **McCourtie, J. and Douglas, L. J.,** Unequal distribution of adhesins within populations of *Candida albicans, FEMS Microbiol. Lett.*, 27, 111, 1985.
280. **Segal, E., Lehrer, N., and Ofek, I.,** Adherence of *Candida albicans* to human vaginal epithelial cells: inhibition by amino sugars, *Exp. Cell Biol.*, 50, 13, 1982.
281. **Centeno, A., David, C. P., Cohen, M. S., and Warren, M. M.,** Modulation of *Candida albicans* attachment to human epithelial cells by bacteria and carbohydrates, *Infect. Immun.*, 39, 1354, 1983.
282. **Ghannoum, M. A., Burns, G. R., Abu Elteen, K., and Radwan, S. S.,** Experimental evidence for the role of lipids in adherence of *Candida* spp. to human buccal epithelial cells, *Infect. Immun.*, 54, 189, 1986.
283. **Sobel, J. D., Myers, P. G., Kaye, D., and Levison, M. E.,** Adherence of *Candida albicans* to human vaginal and buccal epithelial cells, *J. Infect. Dis.*, 143, 76, 1981.
284. **Lee, J. C. and King, R. D.,** Characterization of *Candida albicans* adherence to human vaginal epithelial cells *in vitro, Infect. Immun.*, 41, 1024, 1983.
285. **Sobel, J. D., Myers, P., Levison, M. E., and Kaye, D.,** Comparison of bacterial and fungal adherence to vaginal exfoliated epithelial cells and human vaginal epithelial tissue cultures, *Infect. Immun.*, 35, 697, 1982.
286. **Skerl, K. G., Calderone, A., Segal, E., Streevalson, T., and Scheld, W. M.,** *In vitro* binding of *Candida albicans* yeast cells to human fibronectin, *Can. J. Microbiol.*, 30, 221, 1984.
287. **Collins Lech, C., Kalbfleisch, J. H., Franson, T. R., and Sohnle, P. G.,** Inhibition by sugars of *Candida albicans* adherence to human buccal mucosal cells and corneocytes *in vitro, Infect. Immun.*, 46, 831, 1984.
288. **Douglas, L. J.,** Surface composition and adhesion of *Candida albicans, Biochem. Soc. Trans.*, 13, 982, 1985.
289. **Rotrosen, D., Calderone, R. A., and Edwards, J. E.,** Adherence of *Candida* species to host tissues and plastic surfaces, *Rev. Infect. Dis.*, 8, 73, 1986.
290. **Tkaez, J. S., Cybulska, E. B., and Lampen, J. O.,** Specific staining of wall mannan in yeast cells with fluorescein-conjugated concanavalin A, *J. Bacteriol.*, 105, 1, 1971.
291. **Stoddard, R. W. and Herbertson, B. M.,** The use of lectins in the detection and identification of human fungal pathogens, *Biochem. Soc. Trans.*, 5, 233, 1977.
292. **Maisch, P. A. and Calderone, R. A.,** Role of surface mannan in the adherence of *Candida albicans* to fibrin-platelet clots formed *in vitro, Infect. Immun.*, 32, 92, 1981.
293. **Maisch, P. A. and Calderone, R. A.,** Adherence of *Candida albicans* to a fibrin-platelet matrix formed *in vitro, Infect. Immun.*, 27, 650, 1980.
294. **Sandin, R. L., Rogers, A. L., Patterson, R. J., and Beneke, E. S.,** Evidence for mannose-mediated adherence of *Candida albicans* to human buccal cells *in vitro, Infect. Immun.*, 35, 79, 1982.
295. **Sandin, R. L. and Rogers, A. L.,** Inhibition of adherence of *Candida albicans* to human epithelial cells, *Mycopathologia*, 77, 23, 1982.
296. **Rossano, F. and Tufano, M. A.,** Effect of concanavalin A on the adhesiveness of *Candida albicans* to vaginal cells, *Rass. Med. Sper.*, 22, 313, 1975.
297. **Kuo, S. C. and Lampen, J. O.,** Tunicamycin an inhibitor of yeast glycoprotein synthesis, *Biochem. Biophys. Res. Commun.*, 58, 287, 1974.
298. **Christensen, G. D., Simpson, A., and Beachey, E. H.,** Adhesion of bacteria to animal tissues-complex mechanisms, in *Bacterial Adhesion: Mechanisms and Physiological Significance*, Savage, D. C. and Fletcher, M., Eds., Plenum, New York, 1985, 279.
299. **Diamond, R. D. and Krzesicki, R.,** Mechanisms of attachment of neutrophils to *Candida albicans* pseudohyphae in the absence of serum, and of subsequent damage to pseudohyphae by microbial processes of neutrophils *in vitro, J. Clin. Invest.*, 61, 360, 1978.
300. **Bouali, A., Robert, R., Tronchin, G., and Senet, J. M.,** Characterization of binding of human fibrinogen to the surface of germ-tubes and mycelium of *Candida albicans, J. Gen. Microbiol.*, 133, 545, 1987.
301. **Lehrer, N., Segal, E., and Barr-Nea, L.,** *In vitro* and *in vivo* adherence of *Candida albicans* to mucosal surfaces, *Annu. Rev. Microbiol.*, 134, 293, 1983.
302. **Segal, E. and Savage, D.,** Adhesion of *Candida albicans* to mouse intestinal mucosa *in vitro:* development of the assay and test of inhibitors, *J. Med. Vet. Mycol.*, 24, 477, 1986.
303. **Beachey, E. H. and Simpson, W. A.,** Interaction of surface polymers of *Streptococcus pyogenes* with animal cell, in *Microbial Adhesion to Surfaces*, Berkeley, R. C. W., Lynch, J. M., Melling, J., Rutter, P. R., and Vincent, B., Eds., Ellis Horwood, Chichester, 1980, chap. 21.
304. **Kessler, R. E.,** Contribution of lipoteichoic acids to dental adhesion and pathogenesis of oral diseases, in *Microbiology-1982*, Schlessinger, D., Ed., American Society for Microbiology, Washington, D.C., 1982, 338.

305. **McCarty, M.,** The occurrence of polyglycerophosphate as an antigenic component of various gram-positive bacterial species, *J. Exp. Med.*, 109, 361, 1959.
306. **Wicken, A. J. and Knox, K. W.,** Lipoteichoic acids: a new class of bacterial antigen, *Science*, 187, 1161, 1974.
307. **Armstrong, J. J., Baddiley, J., Buchanan, J. G., and Carss, B.,** Nucleotides and the bacterial cell wall, *Nature (London)*, 181, 1692, 1958.
308. **Stewart, F. S. and Martin, W. T.,** Adsorption of streptococcal red cell sensitizing antigen to various tissues, *J. Pathol. Bacteriol.*, 84, 251, 1962.
309. **Knox, K. W. and Wicken, A. J.,** Immunological properties of teichoic acids, *Bacteriol. Rev.*, 37, 215, 1973.
310. **Chorpenning, F. W. and Stamper, H. B.,** Spontaneous adsorption of teichoic acid to erythrocytes, *Immunochemistry*, 10, 15, 1973.
311. **Ofek, I., Beachey, E. H., Jefferson, W., and Campbell, G. L.,** Cell membrane binding properties of group A streptococcal lipoteichoic acids, *J. Exp. Med.*, 141, 990, 1975.
312. **McCarty, M.,** The role of D-alanine in the serological specificity of group A steptococcal glycerol teichoic acid, *Proc. Natl. Acad. Sci. U.S.A.*, 52, 259, 1964.
313. **Van Driel, D., Wicken, A. J., Dickson, M. R., and Knox, K. W.,** Cellular location of the lipoteichoic acids of *Lactobacillus fermenti* NCTC 6991 and *Lactobacillus casei* NCTC 6375, *J. Ultrastruc. Res.*, 43, 483, 1973.
314. **Joseph, R. J. and Shockman, G. D.,** Synthesis and excretion of glycerol teichoic acid during growth of two streptococcal species, *Infect. Immun.*, 12, 333, 1975.
315. **Markham, J. L., Knox, K. W., Wicken, A. J., and Hewett, M. J.,** Formation of extracellular lipoteichoic acid by oral streptococci and lactobacilli, *Infect. Immun.*, 12, 378, 1975.
316. **Jacques, N. A., Hardy, L., Knox, K. W., and Wicken, A. J.,** Effect of growth conditions on the formation of extracellular lipoteichoic acid by *Streptococcus mutans* BHT, *Infect. Immun.*, 25, 75, 1979.
317. **Hardy, L., Jacques, N. A., Forester, H., Campbell, L. K., Knox, K. W., and Wicken, A. K.,** Effect of fructose and other carbohydrates on the surface properties, lipoteichoic acid production and extracellular proteins of *Streptococcus mutans* Ingbritt grown in continuous culture, *Infect. Immun.*, 31, 78, 1981.
318. **Wicken, A. J. and Knox, K. W.,** Bacterial cell surface amphiphiles, *Biochim. Biophys. Acta*, 604, 1, 1980.
319. **Hakomori, S.,** Glycosphingolipids, *Sci. Am.*, 254, 32, 1986.
320. **Leffler, H., Svanborg-Eden, C., Schoolnik, G., and Wadstrom, T.,** Glycosphingolipids as receptors for bacterial adhesion: host glycolipid diversity and other selected aspects, in *Attachment of Organisms to the Gut Mucosa*, Vol. 2, Boedeker, E. C., Ed., CRC Press, Boca Raton, FL, 1984, chap. 8.
321. **Karlsson, K. A.,** Glycosphingolipids and surface membrane, in *Biological Membranes*, Vol. 4, Chapman, D., Ed., Academic Press, New York, 1982, 7.
322. **Svanborg-Edèn, C. and Leffler, H.,** Glycosphingolipids of human urinary tract epithelial cells as possible receptors for adhering *Escherichia coli* bacteria, *Symp. Host Parasite Interactions*, June 1979, Umea, Sweden, *Scan. J. Infect. Dis.*, 24, 144, 1979.
323. **Leffler, H. and Svanborg-Edèn, C.,** Chemical identification of a glycosphingolipid receptor for *Escherichia coli* attaching to human urinary tract epithelial cells and agglutinating human erythrocytes, *FEMS Microbiol. Lett.*, 8, 127, 1980.
324. **Cuatre-Casas, P.,** Membrane receptors, *Annu. Rev. Biochem.*, 43, 169, 1974.
325. **Schulster, D. and Levitzki, A.,** *Cellular Receptor for Hormones and Neurotransmitters*, Wiley-Interscience, New York, 1980.
326. **Gahmberg, C. G.,** Cell surface protein: changes during cell growth and malignant transformation, in *Cell Surface Reviews*, Poste, G. and Nicolson, G. L., Eds., Elsevier/North-Holland, Amesterdam, 1977, 371.
327. **Gahmberg, C. G. and Andersson, L. C.,** Surface glycoproteins of malingant cells, *Biochim. Biophys. Acta*, 651, 65, 1982.
328. **Hakomori, S.,** Glycosphingolipids in cellular interaction, differentiation, and oncogenesis, *Annu. Rev. Biochem.*, 50, 733, 1981.
329. **Isaacson, R. E.,** Pilus adhesins, in *Bacterial Adhesion, Mechanism and Physiological Significance*, Savage, D. C. and Fletcher, M., Eds., Plenum, New York, 1985, 307.
330. **Old, D. C.,** Inhibition of the interaction between fimbrial haemagglutinins and erythrocytes by D-mannose and other carbohydrates, *J. Gen. Microbiol.*, 71, 149, 1972.
331. **Jones, G. W. and Freter, R.,** Adhesive properties of *Vibrio cholerae:* nature of the interaction with isolated rabbit brush border membranes and human erythrocytes, *Infect. Immun.*, 14, 240, 1976.
332. **Cinco, M., Banfi, E., Ruaro, E., Crevatin, D., and Crotti, D.,** Evidence for L-fucose (6-deoxy-L-galactopyranose)-mediated adherence of *Campylobacter* spp. to epithelial cells, *FEMS Microbiol. Lett.*, 21, 347, 1984.
333. **Gesner, B. and Thomas, L.,** Sialic acid binding sites-role in hemagglutination by *Mycoplasma gallisepticum*, *Science*, 151, 590, 1965.
334. **Gilboa-Garber, N., Mizrahi, L., and Garber, N.,** Purification of galactose-binding hemagglutinin of *Pseudomonas aeruginosa* by affinity column chromatography using sepharose, *FEBS Lett.*, 28, 93, 1972.
335. **Critchley, I. A. and Douglas, L. J.,** Role of glycosides as epithelial cell receptors for *Candida albicans*, *J. Gen. Microbiol.*, 133, 637, 1987.

336. **Ørstavik, D., Kraus, F. W., and Henshaw, C.,** *In vitro* attachment of streptococci to the tooth surface, *Infect. Immun.*, 9, 794, 1974.
337. **Sellwood, R., Gibbons, R. A., Jones, G. W., and Rutter, J. M.,** Adhesion of enteropathogenic *Escherichia coli* to pig intestinal brush-borders-existence of 2 pig phenotypes, *J. Med. Microbiol.*, 8, 405, 1975.
338. **Yamada, K. M. and Olden, K.,** Fibronectins-adhesive glycoproteins of cell surface and blood, *Nature (London)*, 275, 179, 1978.
339. **Proctor, R. A., Pendergrast, E., and Mosher, D. F.,** Fibronectin mediates attachment of *Staphylococcus aureus* to human neutrophils, *Blood*, 59, 681, 1982.
340. **Lanser, M. E. and Saba, T. M.,** Fibronectin as a cofactor necessary for optimal granulocyte phagocytosis of *Staphylococcus aureus*, *J. Reticoloendothel. Soc.*, 30, 415, 1981.
341. **Simpson, W. A., Ofek, I., and Beachey, E. H.,** Fatty acid binding sites of serum albumin as membrane receptor analogs for streptococcal lipoteichoic acid, *Infect. Immun.*, 29, 119, 1980.
342. **Verbrugh, H. A., Peterson, P. K., Smith, D. E., Nguyen, B.-Y. T., Hoidal, J. R., Wilkinson, B. J., Verhoef, J., and Furcht, L. T.,** Human fibronectin binding to staphylococcal surface protein and its relative inefficiency in promoting phagocytosis by human polymorphonuclear leukocytes, monocytes, and alveolar macrophages, *Infect. Immun.*, 33, 811, 1981.
343. **Sato, Y., Arai, H., and Suzuki, K.,** Leukocytosis-promoting factor of *Bordetella pertussis*. I. Biological properties, *Infect. Immun.*, 7, 992, 1973.
344. **Sato, Y., Arai, H., and Suzuki, K.,** Leukocytosis-promoting factor of *Bordetella pertussis*. II. Its identity with protective antigen, *Infect. Immun.*, 9, 801, 1974.
345. **Gibbons, R. J., Van Houte, J., and Liljemark, W. F.,** Parameters that effect adherence of *Streptococcus salivarius* to oral epithelial surfaces, *J. Dent. Res.*, 51, 424, 1972.
346. **Ellen, R. P. and Gibbons, R. J.,** Parameters affecting the adherence and tissue tropisms of *Streptococcus pyogenes*, *Infect. Immun.*, 9, 85, 1974.
347. **Freter, R.,** Mechanisms of association of bacteria with mucosal surfaces, in *Adhesion and Microorganism Pathogenicity, Ciba Found. Symp. 80*, Pitman Medical, London, 1981, 36.
348. **Walker, P. D. and Nagy, L. K.,** Adhesion of organisms to animal tissues, in *Microbial Adhesion to Surfaces*, Berkeley, R. C. W., Lynch, J. M., Melling, J., Rutter, P. R., and Vincent, B., Eds., Ellis Horwood, Chichester, 1980, chap. 26.
349. **Gibbons, R. J. and Van Houte, J.,** Bacterial adherence and the formation of dental plaques, in *Bacterial Adherence, Receptor and Recognition*, series B, Vol. 6, Beachey, E. H., Ed., Chapman and Hall, London, 1980, 61.
350. **Jones, G. W., Leigh, A., Richardson, A., and Vanden Bosch, J. L.,** Phases in the interaction between bacteria and animal cells, in *Microbial Adhesion to Surfaces*, Berkeley, R. C. W., Lynch, J. M., Melling, J., Rutter, P. R., and Vincent, B., Eds., Ellis Horwood, Chichester, 1980, chap. 11.
351. **Sandin, R. L.,** The attachment to human buccal epithelial cells by *Candida albicans:* an *in vitro* kinetic study using concanavalin A, *Mycopathologia*, 98, 179, 1987.
352. **Jones, G. W.,** Adhesion to animal surfaces, in *Microbial Adhesion and Aggregation*, Marshall, K. C., Ed., Springer-Verlag, New York, 1984, 71.
353. **Marshall, K. C.,** *Microbial Adhesion and Aggregation*, Springer-Verlag, New York, 1984.
354. **Klotz, S. A., Drutz, D. J., and Zajic, J. E.,** Factors governing adherence of *Candida* species to plastic surfaces, *Infect. Immun.*, 50, 97, 1985.
355. **Rotrosen, D., Edwards, J. E., Gibson, T. R., Moore, J. C., Cohen, A. H., and Green, I.,** Adherence of *Candida* to cultured vascular endothelial cells: mechanisms of attachment and endothelial cell penetration, *J. Infect. Dis.*, 152, 1264, 1985.
356. **Thompson, J., Meddens, M J. M., Thorig, L., and Van Furth, R.,** The role of bacterial adherence in the pathogenesis of infective endocarditis, *Infection*, 10, 196, 1982.
357. **Klotz, S. A. and Maca, R. D.,** Endothelial cell contraction increases *Candida* adherence to exposed extracellular matrix, *Infect. Immun.*, 56, 2495, 1988.
358. **Durack, D. T. and Beeson, P. B.,** Experimental bacterial endocarditis. I. Colonization of a sterile vegetation, *Br. J. Exp. Pathol.*, 53, 44, 1972.
359. **Garrison, P. G. and Freedman, L. R.,** Experimental endocarditis. I. Staphylococcal endocarditis in rabbits resulting from placement of a polyethylene catheter in the right side of the heart, *Yale J. Biol. Med.*, 42, 394, 1970.
360. **Perlman, B. B. and Freedman, L. R.,** Experimental endocarditis. II. Staphylococcal infection of the aortic valve following placement of a poly ethylene catheter in the left side of the heart, *Yale J. Biol. Med.*, 44, 206, 1971.
361. **Calderone, R. A., Rotondo, M. F., and Sande, M. A.,** *Candida albicans* endocarditis: ultrastructural studies of vegetation formation, *Infect. Immun.*, 20, 279, 1978.
362. **Sande, M. A., Bowman, C. R., and Calderone, R. A.,** Experimental *Candida albicans* endocarditis; characterization of the disease and response to therapy, *Infect. Immun.*, 17, 140, 1977.

363. **Vudhichamnog, K., Walter, D. M., and Ryley, H. C.,** The effect of secretory immunoglobulin A on the *in vitro* adherence of the yeast *Candida albicans* to human oral epithelial cells, *Arch. Oral Biol.*, 27, 617, 1982.
364. **Stiles, H. M., Loesche, W. J., and O'Brien, T. C.,** *Microbial Aspects Dental Caries, Proc. Workshop Microbial Aspects Dental Caries*, St. Simmons Island, GA, Vol. 1-3, Information Retrieval, Washington, D.C., 1979.
365. **Balish, E., Balish, M. J., Salkowski, C. A., Lee, K. W., and Bartzal, K. F.,** Colonization of congenitally athymic, gnotobiotic mice by *Candida albicans, Appl. Envrion. Microbiol.*, 47, 647, 1984.
366. **Calderone, R. A., Lehrer, N., and Segal, E.,** Adherence of *Candida albicans* to buccal and vaginal epithelial cells: ultrastructural observations, *Can. J. Microbiol.*, 30, 1001, 1984.
367. **Pope, L. M. and Cole, G. T.,** SEM studies of adherence of *Candida albicans* to the gastrointestinal tract of infant mice, *Scan. Electron Microsc.*, 3, 73, 1981.
368. **Field, L. H., Pope, L. M., Cole, G. T., Guentzel, M. N., and Berry, L. J.,** Persistence and spread of *Candida albicans* after intragastric inoculation of infant mice, *Infect. Immun.*, 31, 783, 1981.
369. **Pope, L. M. and Cole, G. T.,** Comparative studies of gastrointestinal colonization and systemic spread by *Candida albicans* and nonlethal yeast in the infant mouse, *Scan. Electron Microsc.*, 4, 1667, 1982.
370. **Cole, G. T., Seshan, K. R., Pope, L. M., and Yancey, R. J.,** Morphological aspects of gastrointestinal tract invasion by *Candida albicans* in the infant mouse, *J. Med. Vet. Mycol.*, 26, 173, 1988.
371. **Kennedy, M. J. and Volz, P. A.,** Ecology of *Candida albicans* gut colonization: inhibition of *Candida* adhesion, colonization, and dissemination from the gastrointestinal tract by bacterial antagonism, *Infect Immun.*, 49, 654, 1985.
372. **Mosher, D. F., and Furcht, L. T.,** Fibronectin: review of its structure and possible functions, *J. Invest. Dermatol.*, 77, 175, 1981.
373. **Mogensen, S. C. and Dishon, T.,** Rapid detection of herpes simplex virus and varicella-zoster virus in clinical specimens by the use of *Staphylococcus aureus* rich in protein A, *Acta Pathol. Microbiol. Scand. Sect. B*, 91, 83, 1983.
374. **Austin, R. M. J. and Daniels, C. A.,** The role of protein A in the attachment of staphylococci to influenza-infected cells, *Lab. Invest.*, 39, 128, 1978.
375. **Hawiger, J., Kloczewiak, M., and Timmons, S.,** Interaction of fibrinogen with staphylococcal clumping factor and with platelets, *Ann. N.Y. Acad. Sci.*, 408, 521, 1983.
376. **Espersen, F., Clemmensen, I., Frimodt-Moller, N., and Jensen, K.,** Effect of human IgG and fibrinogen on *Staphylococcus aureus* intraperitoneal infection in mice, *Acta Pathol. Microbiol. Immunol. Scand.*, 92, 311, 1984.
377. **Ericson, D., Bjorck, L., and Kronvall, G.,** Further characteristics of B_2-microglobulin binding to oral streptococci, *Infect. Immun.*, 30, 117, 1980.
378. **Kohler, W. and Prokop, O.,** Relationship between haptoglobin and *Streptococcus pyogenes* T4 antigens, *Nature (London)*, 271, 373, 1978.
379. **Kronvall, G., Myhre, E. B., Bjorck, L., and Berggard, I.,** Binding of aggregated human B_2-microglobulin to surface protein structure in groups A, C, and G streptococcis, *Infect. Immun.*, 22, 136, 1978.
380. **Myhre, E. B. and Kronvall, G.,** Demonstration of specific binding sites of a human serum albumin in groups C and G streptococci, *Infect. Immun.*, 27, 6, 1980.
381. **Hasenclever, H. F.,** The *in vitro* interaction of *Candida albicans* with non specific serum proteins, *Mycopathologia*, 65, 169, 1978.
382. **Kimura, L. H. and Pearsall, N. N.,** Relationship between germination of *Candida albicans* and increased adherence to human buccal epithelial cells, *Infect. Immun.*, 28, 464, 1980.
383. **Bouali, A., Robert, R., Tronchin, G., and Senet, J.-M.,** Binding of human fibrinogen to *Candida albicans in vitro:* a preliminary study, *J. Med. Vet. Mycol.*, 24, 345, 1986.
384. **Annaix, V., Aubry, J., Robert, R., Bouali, A., Tronchin, G., and Senet, J.-M.,** Fibrinogen binding on *Candida albicans* germ tubes and mycelium, *Protides Biol. Fluids*, 35, 399, 1987.
385. **Page, S. and Odds, F. C.,** Binding of plasma proteins to *Candida* species *in vitro*, *J. Gen. Microbiol.*, 134, 2693, 1988.
386. **Tronchin, G., Robert, R., Bouali, A., and Senet, J.-M.,** Immunocytochemical localization of *in vitro* binding of human fibrinogen to *Candida albicans* germ tube and mycelium, *Ann. Institut Pasteur Microbiol.*, 138, 177, 1987.
387. **Calderone, R. A. and Scheld, W. M.,** Role of fibronectin in the pathogenesis of candidal infections, *Rev. Infect. Dis.*, 9, supp.; 4, S400, 1987.
388. **Freter, R. and Jones, G. W.,** Models for studying the role of bacterial attachment in virulence and pathogenesis, *Rev. Infect. Dis.*, 5, S647, 1983.
389. **Douglas, L. J.,** Adhesion to surfaces, in *The Yeasts*, Vol. 2, 2nd ed., Rose, A. H. and Harrison, J. S., Eds, Academic Press, London, 1987, 239.
390. **Odds, F. C. and Webster, C. E.,** Effects of azole antifungals *in vitro* on host/parasite interactions relevant to *Candida* infections, *J. Antimicrob. Chemother.*, 22, 473, 1988.

391. **Ghannoum, M. A. and Abu Elteen, K.,** Correlative relationship between proteinase production, adherence and pathogenicity of various strains of *Candida albicans, J. Med. Vet. Mycol.,* 24, 407, 1986.
392. **Ghannoum, M. A. and Abu Elteen, K.,** Effect of growth of *Candida* spp. in the presence of various glucocorticoids on the adherence to human buccal epithelial cells, *Mycopathologia,* 98, 171, 1987.
393. **Ghannoum, M. A., Abu Elteen, K. H., and Motawy, M. S.,** Effect of antineoplastic agents and X-irradiation on the adherence of *Candida* spp. to human buccal epithelial cells *in vitro, Mycopathologia,* 104, 171, 1988.
394. **Cox, F.,** Adherence of *Candida albicans* to buccal epithelial cells in children and adults, *J. Lab. Clin. Med.,* 102, 960, 1983.
395. **Bobichon, H. and Bouchet, P.,** Action of chlorhexidine on budding *Candida albicans:* scanning and transmission electron microscopic study, *Mycopathologia,* 100, 27, 1987.
396. **Barrett-Bee, K., Hayes, Y., Wilson, R. G., and Ryley, J. F.,** A comparison of phospholipase activity, cellular adherence and pathogenicity of yeast, *J. Gen. Microbiol.,* 131, 1217, 1985.
397. **Cox, F.,** *Candida albicans* adherence in newborn infants, *J. Med. Vet. Mycol.,* 24, 121, 1986.
398. **Critchley, I. A. and Douglas, L. J.,** Differential adhesion of pathogenic *Candida* species to epithelial and inert surfaces, *FEMS Microbiol. Lett.,* 28, 199, 1985.
399. **Epstein, J. B., Kimura, L. H., Menard, T. W., Truelove, E. L., and Pearsall, N. N.,** Effects of specific antibodies on the interaction between the fungus *Candida albicans* and human oral mucosa, *Arch. Oral Biol.,* 27, 469, 1982.
400. **Gorman, S. P., McCafferty, D. F., and Anderson, L.,** Application of an electronic particle counter to the quantification of bacterial and *Candida* adherence to mucosal epithelial cells, *Lett. Appl. Microbiol.,* 2, 97, 1986.
401. **Karaev, Z. O., Velichko, E. V., and Bykov, V. L.,** Main characteristics of the process of adhesion of *Candida* to human epithelial cells, *Zhurnal. Mikrobiol. Epidemiol. Immunol.,* 7, 59, 1986.
402. **Kearns, M. J., Davies, P., and Smith, H.,** Variability of the adherence of *Candida albicans* strains to human buccal epithelial cells: inconsistency of differences between strains related to virulence, *Sabouraudia,* 21, 93, 1983.
403. **Kimura, L. H. and Pearsall, N. N.,** Adherence of *Candida albicans* to human buccal epithelial cells, *Infect. Immun.,* 21, 64, 1978.
404. **King, R. D., Lee, J. C., and Morris, A. L.,** Adherence of *Candida albicans* and other *Candida* species to mucosal epithelial cells, *Infect. Immun.,* 27, 667, 1980.
405. **Macura, A. B.,** Adherence of *Candida* to mucosal epithelial cells, *Acta Microbiol. Pol.,* 34, 55, 1985.
406. **Macura, A. B., Pawlik, B., and Wita, B.,** *Candida* adherence to mucosal epithelial cells with regard to its pathogenicity, *Zbl. Bakt. Hyg. I Abt. Orig.,* 254, 561, 1983.
407. **Sandin, R. L., Rogers, A. L., Beneke, E. S., and Fernandez, M. I.,** Influence of mucosal cell origin on the *in vitro* adherence of *Candida albicans:* are mucosal cells from different sources equivalent? *Mycopathologia,* 98, 111, 1987.
408. **Ray, T. L., Digre, K. B., and Payne, C. D.,** Adherence of *Candida* species to human epidermal corneocytes and buccal mucosal cells: correlation with cutaneous pathogenicity, *J. Invest. Dermatol.,* 83, 37, 1984.
409. **Anderson, M. L. and Odds, F. C.,** Adherence of *Candida albicans* to vaginal epithelia: significance of morphological form and effect of ketoconazole, *Mykosen,* 28, 531, 1985.
410. **Lehrer, N., Segal, E., Cihlar, R. L., and Calderone, R. A.,** Pathogenesis of vaginal candidiasis: studies with a mutant which has reduced ability to adhere *in vitro, J. Med. Vet. Mycol.,* 24, 127, 1986.
411. **Persi, M. A., Burnham, J. C., and Duhring, J. L.,** Effects of carbon dioxide and pH on adhesion of *Candida albicans* to vaginal epithelial cells, *Infect. Immun.,* 50, 82, 1985.
412. **Segal, E., Soroka, A., and Schechter, A.,** Correlative relationship between adherence of *Candida albicans* to human vaginal epithelial cells *in vitro* and candidal vaginitis, *Sabouraudia,* 22, 191, 1984.
413. **Sobel, J. D. and Obedeanu, N.,** Effects of subinhibitory concentration of ketoconazole *in vitro* adherence of *Candida albicans* to vaginal epithelial cells, *Eur. J. Clin. Microbiol.,* 2, 445, 1983.
414. **Sobel, J. D., Muller, G., and Buckley, H. R.,** Critical role of germ tube formation in the pathogenesis of candidal vaginitis, *Infect. Immun.,* 44, 576, 1984.
415. **Makrides, H. C. and MacFarlane, T. W.,** Effect of commensal bacteria on the adherence of *Candida albicans* to epithelial cells *in vitro, Microbios Lett.,* 21, 55, 1982.
416. **Samaranayake, L. P. and MacFarlane, T. W.,** The adhesion of the yeast *Candida albicans* to epithelial cells of human origin *in vitro, Arch. Oral Biol.,* 26, 815, 1981.
417. **Samaranayake, L. P. and MacFarlane, T. W.,** Factors affecting the *in vitro* adherence of the fungal oral pathogen *Candida albicans* to epithelial cells of human origin, *Arch. Oral Biol.,* 27, 869, 1982.
418. **Farell, S. M., Hawkins, D. F., and Ryder, T. A.,** Scanning electron microscope study of *Candida albicans* invasion of cultured human cervical epithelial cells, *Sabouraudia,* 21, 251, 1983.
419. **Persi, M. A., Whalen, S. A., and Burnham, J. C.,** Use of a controlled environment chamber to study interactions between *Candida albicans* (ATCC 18804) and cultured vaginal cells, *Micron,* 12, 217, 1981.
420. **Klotz, S. A., Drutz, D. J., Harrison, J. L., and Huppert, M.,** Adherence and penetration of vascular endothelium by *Candida* yeast, *Infect. Immun.,* 42, 374, 1983.

421. **Howlett, J. A.,** Candidial infection of the oral mucosa: an *in vitro* model, *Proc. R. Soc. Med.*, 69, 766, 1976.
422. **Howlett, J. A.,** The infection of rat tongue mucosa *in vitro* with five species of *Candida, J. Med. Microbiol.*, 9, 309, 1976.
423. **McCourtie, J., MacFarlane, T. W., and Samaranayake, L. P.,** Effect of chlorhexidine gluconate on the adherence of *Candida* species to denture acrylic, *J. Med. Microbiol.*, 20, 97, 1985.
424. **McCourtie, J., MacFarlane, T. W., and Samaranayake, L. P.,** Effect of saliva and serum on the adherence of *Candida* species to chlorhexidine-treated denture acrylic, *J. Med. Microbiol.*, 21, 209, 1986.
425. **Miyake, Y., Fujita, Y., Minagi, S., and Suginaka, H.,** Surface hydrophobicity and adherence of *Candida* to acrylic surfaces, *Microbios*, 46, 7, 1986.
426. **Samaranayake, L. P. and MacFarlane, T. W.,** An *in vitro* study of the adherence of *Candida albicans* to acrylic surfaces, *Arch. Oral Biol.*, 25, 603, 1980.
427. **Samaranayake, L. P., McCourtie, J., and MacFarlane, T. W.,** Factors affecting the *in vitro* adherence of *Candida albicans* to acrylic surfaces, *Arch. Oral Biol.*, 25, 611, 1980.
428. **Reinhart, I. I., Muller, G., and Sobel, J. D.,** Specificity and mechanism of *in vitro* adherence of *Candida albicans, Annu. Clin. Lab. Sci.*, 15, 406, 1985.
429. **Savage, D. C.,** Association of indigenous microorganisms with gastrointestinal epithelial surfaces, in *Human Intestinal Microflora in Health and Disease*, Hentges, D. J., Ed., Academic Press, New York, 1983, 55.
430. **Minagi, S., Miyake, Y., Inagaki, K., Tsuru, H., and Suginaka, H.,** Hydrophobic interaction in *Candida albicans* and *Candida tropicalis* adherence to various denture base resin materials, *Infect. Immun.*, 47, 11, 1985.
431. **Hazen, K. C., Plotkin, B. J., and Klimas, D. M.,** Influence of growth conditions on cell surface hydrophobicity of *Candida albicans* and *Candida glabrata, Infect. Immun.*, 54, 269, 1986.
432. **Minagi, S., Miyaka, Y., Fugioka, Y., Tsuru, H., and Suginaka, H.,** Cell-surface hydrophobicity of *Candida* species as determined by the contact-angle and hydrocarbon-adherence methods, *J. Gen. Microbiol.*, 132, 1111, 1986.
433. **Hector, R. F. and Domer, J. E.,** Mammary gland contamination as a means of establishing long-term gastrointestinal colonization of infant mice with *Candida albicans, Infect. Immun.*, 38, 788, 1982.
434. **Pope, L. M., Cole, G. T., Guentzel, M. N., and Berry, L. J.,** An experimental model of candidiasis in infant mice: systemic infection following gastrointestinal colonization, in *Human and Animal Mycology*, Kuttin, E. S. and Baum, G. L., Eds., Escerpta Medica, Amsterdam, 1979, 71.
435. **Pope, L. M., Cole, G. T., Guentzel, M. N., and Berry, L. J.,** Systemic and gastrointestinal candidiasis of infant mice after intragastric challenge, *Infect. Immun.*, 25, 702, 1979.
436. **Helstrom, P. B. and Balish, E.,** Effect of oral tetracycline, the microbial flora, and the athymic state on gastrointestinal colonization and infection to BALB/c mice with *Candida albicans, Infect. Immun.*, 23, 764, 1979.
437. **Phillips, A. W. and Balish, E.,** Growth and invasiveness of *Candida albicans* in the germ-free and conventional mouse after oral challenge, *Appl. Microbiol.*, 14, 737, 1966.
438. **Kennedy, M. J. and Volz, P. A.,** Dissemination of yeasts after gastrointestinal inoculation in antibiotic-treated mice, *Sabouraudia*, 21, 27, 1983.
439. **Kennedy, M. J. and Volz, P. A.,** Effect of various antibiotics on gastrointestinal colonization and dissemination by *Candida albicans, J. Med. Vet. Mycol.*, 23, 265, 1985.
440. **Turner, J. R., Butler, T. F., Johnson, M. E., and Gordee, R. S.,** Colonization of the intestinal tract of conventional mice with *Candida albicans* and treatment with antifungal agents, *Antimicrob. Agents Chemother.*, 9, 787, 1976.
441. **Umenai, T.,** Systemic candidasis produced by oral *Candida* administration in mice, *Tokoku J. Exp. Med.*, 126, 173, 1978.
442. **Umenai, T., Konno, S., and Ishida, N.,** Systemic candidiasis from *Candida albicans* colonizing the gastrointestinal tract of mice, *Experientia*, 35, 1331, 1979.
443. **Umenai, T., Konno, S., Yamauchi, A., Limura, Y., and Fugimoto, H.,** Growth of *Candida* in the upper intestinal tract as a possible source of systemic candidiasis in mice, *Tokoku J. Exp. Med.*, 130, 101, 1980.
444. **Wingard, J. R., Dick, J. D., Merz, W. G., Sandford, G. R., Saral, R., and Burns, W. H.,** Pathogenicity of *Candida tropicalis* and *Candida albicans* after gastrointestinal inoculation in mice, *Infect. Immun.*, 29, 808, 1980.
445. **Wingard, J. R., Dick, J. D., Merz, W. G., Sandford, G. R., Saral, R., and Burns, W. H.,** Differences in virulence of clinical isolates of *Candida tropicalis* and *Candida albicans* in mice, *Infect. Immun.*, 37, 833, 1982.
446. **Myerowitz, R. L.,** Gastrointestinal and disseminated candidiasis: an experimental model in the immunosuppressed rat, *Arch. Pathol. Lab. Med.*, 105, 138, 1981.
447. **Stone, H. H., Geheber, C. E., Kolb, L. D., and Litchens, W. R.,** Alimentary tract colonization by *Candida albicans, J. Surg. Res.*, 14, 273, 1973.
448. **Calderone, R. A., Cihlar, R. L., Lee, D. D. S., Hoberg, K., and Scheld, W. M.,** Yeast adhesion in the pathogenesis of endocarditis due to *Candida albicans:* studies with adherence negative mutants, *J. Infect. Dis.*, 152, 710, 1985.

449. **Freedman, L. R. and Johnson, M. L.,** Experimental endocarditis, *Ann. N.Y. Acad. Sci.*, 236, 456, 1974.
450. **Scheld, W. M., Calderone, R. A., Alliegro, G. M., and Sande, M. A.,** Yeast adherence in the pathogenesis of *Candida* endocarditis, *Proc. Soc. Exp. Biol. Med.*, 168, 208, 1981.
451. **Scheld, W. M., Calderone, R. A., Brodeur, J. P., and Sande, M. A.,** Influence of preformed antibody on the pathogenesis of experimental *Candida albicans* endocarditis, *Infect. Immun.*, 40, 950, 1983.
452. **Scheld, W. M., Strunk, R. W., Balian, G., and Calderone, R. A.,** Microbial adhesion to fibronectin *in vitro* correlates with production of endocarditis in rabbits, *Proc. Soc. Exp. Biol. Med.*, 180, 474, 1985.
453. **Sofaer, J. A., Holbrook, W. P., and Southan, J. C.,** Experimental oral infections with the yeast *Candida albicans* in mice with or without inherited iron-deficiency anemia (SLa.), *Arch. Oral Biol.*, 27, 497, 1983.
454. **Franklin, C. D. and Martin, M. W.,** The effects of *Candida albicans* on turpentine-induced hyperplasia of hamster cheek pouch epithelium, *J. Med. Vet. Mycol.*, 24, 281, 1986.
455. **Allen, C. M., Blozis, G. G., Rosen, S., and Bright, J. S.,** Chronic candidiasis of the rat tongue: a possible model for human median rhombid glossitis, *J. Dent. Res.*, 61, 1287, 1982.
456. **Fisker, A. V., Schiott, C. R., and Philipsen, H. P.,** Short-term oral candidiasis in rats, with special reference to the site of infection, *Acta Pathol. Microbiol. Immunol. Scand.*, 90, 49, 1985.
457. **Fisker, A. V., Rindon, C., Schiott, C. R., and Philipsen, H. P.,** Long-term oral candidiasis in rats, *Acta Pathol. Microbiol. Immunol. Scand. Sect. B*, 90, 221, 1982.
458. **Hassan, O. E., Jones, J. H., and Russell, C.,** Experimental oral candidal infection and carriage of oral bacteria in rats subjected to a carbohydrate-rich diet and tetracycline treatment, *J. Med. Microbiol.*, 20, 291, 1985.
459. **Jones, J. H. and Adams, D.,** Experimentally induced acute oral candidiasis in the rat, *Br. J. Dermatol.*, 83, 670, 1970.
460. **Jones, J. H. and Russell, C.,** The histology of chronic candidal infection of the rat's tongue, *J. Pathol.*, 113, 97, 1974.
461. **Jones, J. H., Russell, C., Young, C., and Owen, D.,** Tetracycline and the colonization and infection of the mouths of germ-free and conventionalized rats with *Candida albicans*, *J. Antimicrob. Chemother.*, 2, 247, 1976.
462. **Olsen, I. and Bondevik, O.,** Experimental *Candida* induced denture stomatitis in the wister rat, *Scand. J. Dent. Res.*, 86, 392, 1978.
463. **Russell, C. and Jones, J. H.,** Effects of oral inoculation of *Candida albicans* in tetracycline-treated rats, *J. Med. Microbiol.*, 6, 275, 1973.
464. **Russell, C. and Jones, J. H.,** The effects of oral inoculation of the yeast and mycelial phases of *Candida albicans* in rats fed on normal and carbohydrate-rich diets, *Arch. Oral Biol.*, 18, 409, 1973.
465. **Russell, C. and Jones, J. H.,** The histology of prolonged candidal infection of the rat's tongue, *J. Oral Pathol.*, 4, 330, 1975.
466. **Russell, C., Jones, J. H., and Gibbs, A. C. C.,** The carriage of *Candida albicans* in the mouths of rats treated with tetracycline briefly or for a prolonged period, *Mycopathologia*, 58, 125, 1976.
467. **Shakir, B. S., Martin, M. V., and Smith, C. J.,** Induced palatal candidiasis in the wistar rat, *Arch. Oral Biol.*, 26, 787, 1981.
468. **Budtz-Jorgensen, E.,** Denture stomatitis. IV. An experimental model in monkeys, *Acta Odont. Scand.*, 29, 513, 1971.
469. **Budtz-Jorgensen, E.,** Effects of triamcinolone acetonide on experimental oral candidiasis in monkeys, *Scand. J. Dent. Res.*, 83, 171, 1975.
470. **Bagg, J. and Silverwood, R. W.,** Coagglutination reactions between *Candida albicans* and oral bacteria, *J. Med. Microbiol.*, 22, 165, 1986.

REFERENCE ADDED IN PROOF

471. **Houston, J. G. and Douglas, L. J.,** Interaction of *Candida albicans* with neutrophils: effect of phenotypic changes in yeast cell-surface composition, *J. Gen. Microbial.*, 135, 1885, 1989.

Chapter 5

RESPONSES OF ADHERENCE TO ENVIRONMENTAL VARIABLES

TABLE OF CONTENTS

I. INTRODUCTION

Adherence of microorganisms to living or inanimate surfaces is initiated once these two elements come in close proximity to each other. For this to take place a number of factors conducive to adherence should prevail. The microorganism, the surface involved, and the environmental conditions are likely to affect the adherence process.[1,2] There have been many studies to determine the relative importance of these variables. Based on these studies, Daniels[3] divided parameters which affect microbial adsorption of microorganisms on solid surfaces into three categories that are related to either the microorganism, the adsorbent, or the environment. Similarly, Douglas[4] categorized factors that affect the adhesion of *Candida* spp. to epithelial cells *in vitro* to three comparable categories. Table 1 outlines various parameters affecting adhesion of *Candida* on epithelial cells, plastic surfaces, and denture base resin materials. These parameters are the subject of this chapter.

II. CHARACTER OF YEAST

The relationship to adherence of various *Candida* species and strains within a species as well as germ-tube formation has been discussed previously (Chapter 3, Section II. A).

A. CULTURE MEDIUM

Since adherence of *Candida* to epithelial cells involves the outer cell envelopes of both partners any factors influencing them will affect the adhesion process. Medium composition and temperature of growth are known to affect cell morphology of dimorphic fungi (for review see Mariat[5]) including *Candida albicans* and, therefore, the cell wall composition of the two forms.[6] McMurrough and Rose[7] found that the nature of substrate limitation affected the composition and structure of the yeast cell wall. They reported that the growth of *Saccharomyces cerevisiae* under different conditions of glucose limitation influenced the level of glucan, mannan, protein, and phosphorus in the cell wall. Phosphate limitation alters the carbohydrate content of the walls, by decreasing glucan and the total lipids content.[8] Sterol esters, triacylglycerols, diacylglycerophosphoserines, and glucosamine also diminished, but there was a concomitant increase in the production of diacylglycerophosphoethanolamines, diacylglycerophosphocholines, and protein.[8] It is clear that variation in medium composition has important effects on the cell yeast wall and its adherence properties.

With this in mind, several workers devoted efforts to investigate the adherence and pathogenicity of *C. albicans* grown in media supplemented with different carbohydrates. Many investigations have demonstrated that dietary carbohydrates, including glucose, sucrose, maltose, galactose, and xylitol, enhance candidal adhesion to epithelial cells and acrylic denture surfaces. Samaranayake et al.[9] showed that yeasts incubated in sucrose and glucose had a better adhesion to acrylic strips than controls grown in sugar-free media. Growth in the presence of lactose and xylitol showed no significant difference when compared with the control yeasts.[9]

Pre-incubation of *C. albicans* in the presence of a range of sucrose concentrations (from 50 to 500 m*M*) gave a significant positive correlation between the number of adherent yeasts to acrylic surfaces or epithelial cells and the sucrose concentration.[10,11] Similarly, Douglas and co-workers[12,13] showed that adherence of *C. albicans* to acrylic surfaces, as well as epithelial cells, was enhanced significantly after growth of the yeast to the stationary phase in media containing high concentrations of various sugars as the carbon source. Galactose was the most effective sugar tested and fructose the least.[13] Yeast grown in the presence of 500 m*M* of this sugar showed more than ninefold greater adherence than control *C. albicans* grown in medium with a relatively low concentration (50 m*M*) of glucose.[13]

Although much attention has been paid to the enhancing role carbohydrates play in adherence

TABLE 1
Parameters Affecting Adhesion of *Candida* Species to Epithelial Cells, Plastic Surfaces, and Denture Base Resins

Yeast	Epithelial cells	Inanimate surfaces	Environment
Species, strains, and phenotype (white, opaque)	Cell type	Hydrophobic forces (London-van der Waals forces)	Hydrogen-ion concentrationn and CO_2 level
Culture medium	Variability and viability	Electrostatic forces (surfaces charge)	Temperature and time of contact
Culture age and viability		Surface free energy	Buffer
Culture concentration		Surface roughness and physical presentation	Serum and saliva
Germination		Wettability	Presence of secretions (e.g., hormones, mucus, crevicular fluid)
Cell surface hydrophobicity			Presence of commensal bacteria
			Presence of antibodies
			Age-related alterations

(which is understandable since diets rich in carbohydrates have been identified by many workers as creating an environment conducive to the development of mucosal candidosis), there is a lack of information regarding the relation of other medium constituents to adherence.[14-19] In this respect, we compared the adherence of *C. albicans* cells grown in yeast nitrogen base (YNB) supplemented with 2.5% (w/v) glucose with other yeast cells grown in a minimal medium (MM) (ammonium sulphate, 1.0 g/l and monopotassium sulphate 1.0 g/l, pH adjusted to 6.0) supplemented with glucose. No statistical difference was observed between these two groups (number of yeast adhering to 100 buccal epithelial cells were 445 ± 25 and 457 ± 25, $p > 0.05$, for *C. albicans* grown in YNB and MM, respectively). More research should be directed towards this point. (See Reference 332.)

The mechanism(s) by which carbohydrates seem to enhance adherence appears to be by the production of an additional, fibrillar surface layer. This layer may be analogous to bacterial fimbriae,[11,12] whose importance in adhesion is well established.[20] Samaranayake and MacFarlane[10] suggested that an extracellular metabolic product could be responsible for the enhanced adherence observed for yeast grown in the presence of sucrose. Their conclusion is based on the fact that the increase disappeared when *C. albicans* were killed prior to incubation in sucrose-containing medium. At about the same time McCourtie and Douglas[13] showed that *C. albicans* cells grown in the presence of galactose undergo a change in cell surface composition and develop a fibrillar-floccular layer on the yeast cell surface which makes the yeast cells more adherent and more resistant to spheroplast formation. Further evidence for the development of a fibrillar layer by *C. albicans* under the influence of the environment came from cytochemical and ultrastructural studies carried out by Tronchin et al.[21] They reported that adhering yeasts develop a fibrogranular surface layer visualized by a polysaccharide detection technique. Their results indicated that adherence of *C. albicans* to host cells is associated with the ability of yeast to actively rearrange substructural aspects of their cell wall coat in response to external stimuli.[21] *C. albicans* strains that lack this capability completely, or possess it to a much lower degree, have relatively little adherence.[22] A fibrillar-floccular layer on the surface of *C. albicans* growing *in vivo* has been also demonstrated by electron microscopy of scrapings taken from the tongue and buccal mucosa of patients with oral candidosis.[23-25] Under *in vitro* conditions materials from this fibrillar layer are readily released into the medium.[13] Chemical characterization of this layer was carried out by Douglas and her group.[26,27] It was found to contain carbohydrates, mannose with some glucose, (65 to 82%), proteins (7%), phosphorus (0.5%), and glucosamine (1.5%).

B. CULTURE AGE AND VIABILITY

1. Cell Age

The attachment of microorganisms is known to be influenced by the culture age.[2] Investigations regarding the role of the growth phase on the adherence of yeast gave contradictory results. King et al.[28] showed that the growth phase of *C. albicans* had a marked influence on its ability to attach to mucosal epithelial cells. Stationary-phase yeasts (grown for periods longer than 18 h) were found to adhere to a greater degree than exponential-phase blastospores. By contrast, Segal et al.[29] reported a higher rate of adhesion with exponential-phase yeasts. Work carried out in our laboratory failed to demonstrate any significant difference in the adherence ability of *C. albicans* at different growth phases. Although stationary-phase yeast showed greater adherence than exponential-phase cells (441 ± 23 compared with 397 ± 24 yeast adhering to 100 buccal epithelial cells, respectively), this was not statistically significant ($p > 0.5$).[30] The reasons for these difference are not clear, although it may be attributable to the different growth media used. In addition, both Segal et al.[29] and King et al.[28] used vaginal epithelial cells, while buccal epithelial cells were used in our study.

The influence of culture age on adherence of yeast is probably due to changes in cell surface chemistry and/or hydrophobicity. There is appreciable evidence that production or secretion of adhesive polymers by bacteria is influenced by the age of the bacterial culture.[31] Similarly, McCourtie and Douglas[13] linked changes which facilitate adherence in cell surface composition to the stationary phase. They reported that *C. albicans* produce fibrillar layers on the yeast surfaces which enhance attachment to epithelial and acrylic surfaces *in vitro*.[13,22] These fibrils are released into the surrounding environment upon prolonged incubation of the yeast (5 d).

Recently, Hazen et al.[32] correlated the growth phase with cell surface hydrophobicity (CSH) of *C. albicans* and *C. glabrata*. Exponential-phase cell CSH was significantly lower than that of cells at the stationary phase. The above cell surface alterations in composition and hydrophobicity may explain the observations reported by King et al.[28,33] that adherence of *C. albicans* to vaginal epithelial cells is greater for stationary-phase cells than for exponential-phase cells.

2. Viability

Since adhesion is a physicochemical phenomenon involving interaction between specific chemicals present in both the host (called receptors) and the parasite (called adhesins), then it is expected that adherence between these two systems will take place irrespective of whether they are viable or not, although the biological properties cannot be ignored altogether. This is so, provided the adhesins and the receptors are not interfered with, i.e., the physicochemical nature of the components is not altered and accessibility of one to the other is maintained.

Kimura and Pearsall[34] were the first to study the effect of viability of *C. albicans* on the adherence of this yeast to human buccal epithelial cells. Although they reported that viable yeast cells adhere better than nonviable cells, it was found that this is related to changes in the yeast (particularly to germination) rather than to cell viability. No difference was observed in the adherence of viable and nonviable cells once germination had occurred. Maisch and Calderone[35] reported that heat- and formaldehyde-killed cells did not adhere as well as viable cells. Similarly, Samaranayake and MacFarlane[10,11] found that the enhancement of adhesion of *C. albicans* to HeLa cells and acrylic surfaces disappeared once yeast cells were heat killed prior to incubation in sucrose-containing medium. The work carried out subsequently by Lee and King[33] and Klotz et al.[36] stressed the fact that the method of fungal killing bore a relationship to adherence capacity. Gentle heat killing of the yeast results in some loss of polysaccharides and, presumably, adhesins and, thus, decreased adherence.[37] Severe killing (e.g., heating for prolonged period, over 30 min) will result in considerable damage to the outer surfaces of the yeast cells, with concomitant loss of adhesins and consequently poor adhesions. Table 2 summarizes the effect of various treatments on the viability and adherence of *C. albicans*. It is obvious that heating the yeast cells for only a short period (i.e., gentle killing) affected adherence slightly (5%

TABLE 2
Variation in the Adherence of *C. albicans* Following Various Treatment

Treatment	Viability	Relative adherence
No treatment (control)	100	100
Formalin (18 h, 4°C)		
0.27%	< 0.1	98
2.7%	< 0.1	56
27.0%	< 0.1	19
0.5% Formaldehyde (18 h, 4°C)[a]	< 0.1	50
Heat (65°C)		
5 min	< 0.1	95
15 min	< 0.1	15
30 min	< 0.1	9
2 h[a]	< 0.1	12
U. V. irradiation		
10 min	< 0.1	95
4 h	< 0.1	68
24 h	< 0.1	11
X-irradiation[a]		
400 c-Gy (3.84 min)	< 0.1	45
500 c-GY (4.80 min)	< 0.1	34

Based on data from Lee and King[33] and [a]unpublished data from our laboratory.

reduction in relative adherence compared with the control), while heating for longer periods (15 min and longer) resulted in considerable reduction in the relative adherence (between 88 to 91% reduction in relative adherence compared with the control). This could explain the results reported above by Maisch and Calderone[35] and Samaranayake and MacFarlane.[10,11] Both these groups used severe killing methods when compared with those shown in Table 2 (heating at 63°C for 2 h and at 70°C for the same period, respectively). As has previously been discussed, *C. albicans* produces an extracellular fibrillar layer on the yeast surface which has been shown to be responsible for adherence of this yeast to other surfaces.[26,27] Killing of the yeast using severe methods will no doubt lead to either loss of or damage to this layer and will therefore affect the process of adherence.

C. CULTURE CONCENTRATION

The dependence of microbial adhesion upon culture concentration is well known.[2,38,39] In general, the adherence of microorganisms is reported to increase proportionally with the microbial density.[2,39] Comparison of results originating from different laboratories regarding the influence of cell concentration on adherence is often difficult because various researchers employ different contact times (ranging from 30 min to 2 h), growth media, and incubation temperatures. Furthermore, variation arises from differences inherent to the *C. albicans* strains tested. Although yeast concentrations ranging from 10^2 to 10^9 cells per ml were tested for their influence on adherence, no detectable attachment was observed at concentrations below 10^4 yeasts per ml.[10,29,34] Above this concentration adherence showed a significant positive correlation ($p < 0.001$, $r = 0.95$) between the number of attached yeasts per unit area of HeLa cells monolayer and the yeast concentration in the incubating suspension.[10] Similar observations were reported by other workers.[28,34]

Whether adherence of yeasts to epithelial cells was saturable or not, gave contradictory results. Segal et al.[29] reported that varying the concentrations of both yeasts (10^2 to 10^9 cells per ml) and epithelial cells (10^2 to 10^7 cells per ml) in the assay mixture revealed that adherence of yeast to epithelial cells was saturable.[29] King et al.,[28] on the other hand, reported that saturation of blastospore receptor sites on the surface of epithelial cells was not achieved at the yeast concentrations that were tested (between 5×10^5 and 5×10^8 yeast added to assay medium). However, formation of yeast aggregates through coadherence, i.e., adherence to other yeasts already attached to an epithelial cell, could be responsible for the increases in adherence values noted when high concentrations of yeasts was added to the assay system. Recently, Klotz and Penn[40] showed that upon the addition of high yeast inocula, aggregration of yeasts on monolayers of human intestinal epithelium occurred rather than a random distribution of yeast cells. The greater the yeast inocula size, the greater were the aggregate sizes. This seems to support the evidence that the adherence of yeast to epithelial cells is saturable.[40] In this respect Fletcher[2] showed that saturation of attachment surface was attained at high bacterial cell concentrations.

D. CELL SURFACE HYDROPHOBICITY

Cell surface hydrophobicity seems to play an important role in adherence of microorganism to various surfaces.[41-44] The role of this property in the adherence of *Candida* spp. to inert materials as well as to living cells has been considered in recent years. Macura[45] divided *C. albicans* strains into two groups according to their hydrophobicity. This correlated with their adherence values. The most hydrophobic strains revealed the highest adherence capacity, while less hydrophobic ones were less capable of adhering. Minagi et al.[46] presented evidence that the adherence of *C. albicans* and *C. tropicalis* to denture base resin materials depend on the hydrophobicity of both the microorganisms and the base materials. Particularly, the closer the surface free energy of the substrate surface and the microorganisms, the higher was the probability of adherence. Klotz et al.,[47] studying factors governing adherence of *Candida* spp. to plastic surfaces, showed that the adherence of *Candida* spp. to these surfaces is predominantly controlled by attractive London van der Waals forces (or hydrophobic forces) and electrostatic forces.[48] These authors suggested that the electrical forces are minor to the hydrophobic ones, since adherence to a considerable extent occurs even in the presence of this repulsive force[47] (this is created because the surface charges of eukaryotic and prokaryotic cells are, in sum, both negative).[49] Klotz et al.[47] suggested that adherence of yeast to plastic surfaces is dependent on the interfacial free energy (or interfacial tension).[50-52] Thus, adherence occurs whenever the sum total of interfacial tension (γ_T) i.e., yeast-liquid (γye), yeast-solid (γys), and solid liquid (γse) is reduced.[47]

In addition to hydrophobicity and electrical charges, other factors such as the size and the shape of microorganism may be of importance in the adherence of *Candida* to plastic surfaces.[47,322] Also, environmental factors and the growth phase can variably affect the relative cell surface hydrophobicity. Hazen et al.[32] showed that *C. albicans* cells grown at room temperature were more hydrophobic than cells grown at 37°C. Furthermore, cell surface hydrophobicity of exponential-phase yeast cells was significantly lower than that of cells at the stationary phase. These results may partly explain earlier findings that adherence of *C. albicans* to vaginal epithelial cells is greater for cells grown at room temperature than cells grown at 37°C, and greater for stationary-phase cells than for exponential-phase cells.[28,33]

Two recent studies, in an attempt to elucidate the mechanisms which contribute to the adherence of *Candida* yeasts to living cells, have considered the role of hydrophobicity with other mechanisms. Klotz and Penn[40] showed that the adherence of hydrophobic yeasts was twice that of identical yeasts grown in glucose as a carbon source. (Yeasts grown in hexadecane as the sole carbon source tend to be more hydrophobic than those grown on glucose.[40]) This suggested that this type of hydrophobic interactions as well as electrostatic effects are of some importance in the adherence of yeast to living cells. In contrast, Kennedy et al.[53] suggested that hydrophobic interactions were unlikely to be of significance in the process of adhesion. These authors concluded that it is unlikely that such a nonspecific phenomenon could account for all the observed adhesive activities of *C. albicans* with the intestinal mucosa.[53] We believe that hydrophobicity is one of the mechanisms responsible for the adherence of *Candida* spp. to living and inanimate surfaces and it is by no means the sole factor. The extent of its contribution to this process may be enhanced if the right environmental conditions prevail.[323] (See Reference 333.)

III. CHARACTER OF EPITHELIAL CELLS

Epithelial cells are the second partner involved in the adherence process (with the yeast being the other). Cytoplasmic membranes are the surface where direct contact occurs. In free-living cells the plasma membrane usually has the appearance of a structure that is homogeneous over the entire cell surface. In differentiated cells that are organized into tissues, discrete parts of the membrane may show special modifications such as tight and gap junctions and desmosomes and outward or inward proliferation of the plasma membrane. Such modifications may facilitate the performance of particular functions.[54] As discussed in Chapter 4, biological membranes consist of lipids and proteins, both in their nonglycosylated forms. The relative amounts of each of these groups of compounds vary considerably, as does the nature of the various membrane constituents.[54] In spite of these large variations, the organization of the membranes is remarkably similar. Experimental work on the adhesion of *Candida* made use of a large variety of cell types including buccal,[34,55-59] vaginal,[33,60-63] and uroepithelial cells,[64,65] epidermal corneocytes,[66,67] and gastrointestinal epithelium.[53,68,69] In addition to these cell types, which were obtained from either human volunteers or animals, culture cell lines were also exploited in adhesion studies. Cell lines used include HeLa cells,[10,70,71] kidney epithelial cells,[10] cervical epithelial cells,[72] endothelial cells,[73,74] and intestine cell lines.[40] Buccal and vaginal epithelial cells are by far the most used, reflecting the fact that infections of the mouth and vagina are the most common forms of superficial candidosis.[75] Of the two types buccal cells have been more widely used, possibly due to the ease and convenience in obtaining them.

Microbial-epithelial cell adherence is essentially a surface phenomenon. Thus, differences in the nature of both microbial and epithelial cell surfaces might be expected to affect adherent interactions. Variations in the oral as well as vaginal epithelium are well documented and may explain the variations observed in the adherence of the same type of epithelial cells to microorganisms.[76-81] This may also explain the differences observed in susceptibility to disease in certain population groups.

Comparative studies related to variation in adherence of *Candida* to buccal epithelial cells (BEC) derived from children and adults, newborn infants, and neonates have been performed.[57-59] Cox,[58] studying the adherence of *C. albicans* in children and adults, showed that BEC from healthy adults had no significant daily or individual change in yeast adherence.[58] He also reported that cells from healthy adults adhered *C. albicans* to the same extent as cells from healthy children. This is in disagreement with the findings of Tobgi et al.[82] who found that BEC from children exhibited a threefold higher affinity for yeast, as compared with cells from adult. Davidson et al.[57] studied the age-related variation in the adherence of *C. albicans* to BEC of newborn infants. They found no statistically significant increase in the number of yeasts attached to epithelial cells related to different age groups. Interestingly, studying the same group of infants twice at a mean age of 2.3 and 6.2 d gave significant increase in the adherence of *C. albicans* with age.[57] In a recent study, Cox[59] showed that the adherence of *C. albicans* to BEC from full-term infants was significantly lower than adherence to epithelial cells from premature infants and healthy school-age children. This remained so until full-term infants reached 5 d of age, where the difference evened out.

Explanations put forward for the variations reported above include alterations in cell surface receptors or a change in the fluids that bathe these cells, such as saliva.[59] In this context it is relevant to point to the work of Hayward[83] and Hayward and Hackemann[84] who reported the presence of a glycoprotein surface coat on keratinized epithelial cells. Nonkeratinized cells were devoid of this coat. However, it is obvious that there is little information available on the reasons for these variations and it is felt that more research should be directed towards understanding the nature of the epithelial cell surface, their interactions with salivary components and the mechanism(s) involved.

Variations observed in the adherence of *C. albicans* to vaginal epithelial cells (VEC) is no surprise in view of the known influence exerted by reproductive hormones on the vaginal mucosa. These hormones cause cyclical shedding and proliferation of mucosal cells,[64,80] together with alterations in morphology, cellular composition and function of these cells.[81,85-87] In addition, Botta[64] suggested that the synthesis or accessibility of receptors involved in the adhesion of bacteria to the urogenital tract fluctuates during the menstrual cycle. These changes could account for the variance in the normal microflora observed during the menstrual cycle, pregnancy, labor, and puerperium.[88-92]

A number of studies have attempted to explain the variation in the adherence of *Candida* to VEC. Sobel et al.[63] observed that a variety of microorganisms adhered best to cultured human VEC that were keratinized and in the process of desquamating; however, they did not adhere to newly dividing young cells. Sklavounou and Germaine[56] reported similar results. They found that oral *Streptococcus mitis* and *S. salivarius* exhibited significantly greater adherence to keratinized and nonkeratinized buccal cells, whereas *S. sanguis* and *S. mutans* showed no such preference.[56,83,84] Segal et al.[61] related the increased *C. albicans* adherence to an increase in the number of intermediate epithelial cells which are rich in glycogen as opposed to parabasal cells which had lower adherence capabilities.[93] Conditions which contribute to the increase in the number of such cells include: pregnancy, the first or fourth weeks of the menstrual cycle, and diabetes.[61] Persi et al.[62] went a step further and showed that adhesion of *C. albicans* to the VEC was affected by the side of the vaginal cell exposed (in addition to other factors). These cells have two distinct surfaces: microridge surfaces (which face the vaginal lumen) and nonmicroridge surfaces. They reported that most of the yeast cells attached to the microridge side of the VEC, suggesting the presence of receptors that are recognized by the adhesin of *C. albicans*.[62] This is an interesting finding that should be investigated further.

So far we have discussed the variations observed in the adherence of yeast to the same type of epithelial cells. Few studies exist which compare the relative adherence of two different cell types, particularly BEC and VEC. Sobel et. al.[94] found that the adherence of *C. albicans* to BEC was slightly greater than to VEC. In contrast, King et al.[28] found exactly the opposite effect. He reported that *C. albicans*, *C. tropicalis*, and *C. stellatoidea* adhered in greater numbers to vaginal cells than to buccal cells.[28] The work of Tomsikova et al.[95] and Macura et.al.[96] supports the findings of the latter, i.e., adherence was found to be more pronounced with VEC than with cells of the buccal mucosa. Ryley[97] suggested that these variations could be a reflection of the ability of the yeast to form germ-tubes under the conditions of the assay employed by different workers rather than of the yeasts to adhere as such. Adherence of *Candida* spp. to human epidermal corneocytes and BEC was investigated in two separate studies. Collins-Lech et al.[67] reported that the adherence of yeast was somewhat greater to BEC than to corneocytes. Adherence to the cells of different individuals was variable, but the ratios of values for the two kinds of cells from a single individual were quite constant. Results obtained by Ray et al.[66] confirm that the adherence of various yeast strains to buccal cells was higher than to epidermal corneocytes. Here again the variability could be a reflection of differences in the number or nature of available receptors on the epithelial cell membrane.

It is clear that variability of both host and microbial cells influence the results of adherence studies. Consequently, it is crucial to introduce standard controls in the study of microbial adherence. If the variability of *Candida* is under investigation, the selection of the target host cell must be such that its adhering characteristics are stable and well defined. Conversely, if the variability of the host cell is under study, the yeast must be standardized by assuring that the test strain is grown at the same growth and environmental parameters and that its adherence is assayed at optimal conditions. The ability of *C. albicans* to undergo high frequency switching to various phenotypes, both *in vitro* and *in vivo*[324,325,329] complicates matters further. Thus, as suggested by Kennedy et. al.,[326] the phenotypic state of *Candida*, at the time of assay, should be determined, otherwise, the results, even within the same laboratory, may be difficult to interpret.

IV. CHARACTER OF INANIMATE SURFACES

Modern therapeutic procedures are necessarily invasive, and a very large number of plastic and metal prostheses are implanted into patients in both short-term and long-term applications. When such tubes are left *in situ* for periods of days or even months the likelihood of infections is increased. This is particularly true in the case of infections caused by *C. albicans* and related species (see Chapter 2).[75] Invasion of the human host occurs by microorganisms attaching to plastic surfaces or similar materials that comprise the indwelling component of many intravenous devices, bladder catheters, and other prostheses. It has been shown by a number of investigators that attachment of microbial cells to tissue surfaces or to prosthetic implants is a precursor to pathogenesis in many cases.[98-101] Adherence of *Candida* spp. to plastic intravenous catheters has been reported.[102] Besides catheters, denture base resin materials are an important inanimate substance closely associated with candidosis (chronic atrophic candidosis, also known as denture stomatitis and denture sore mouth). *C. albicans* is recovered more often and in higher numbers from the fitting surface of the denture than from the palate.[103] This suggests that the acrylic denture acts as a reservoir of infection and that adherence of the yeast to the denture surface is a normal prerequisite for colonization of the palate.[103-105]

Surface properties of these nonbiological substances play a vital role in mediating adhesion with the biological systems (such as yeast). Little precise information is available, however, on the properties of inert surfaces and the bulk of our knowledge is derived from studies concerning the attachment of microorganisms in the aquatic environment.[38,106-108] Few studies exist that deal with the properties of nonbiological substrates in relation to *Candida* adhesion.[9,11,13,22,46,47] The majority of these investigations look at the adhesion process as it is affected by parameters related to the yeast with only slight reference to parameters related to nonbiological substrates.

Inanimate surface characteristics reported to be of importance in adherence are (1) interfacial free energy,[47,109,110] (2) critical surface tension (sometimes referred to as wettability),[107,111] (3) hydrophobic forces (attractive London van der Waals forces),[47,107,112,113] (4) surface charges, and (5) surface roughness.[114]

The surface free energy represents the potential bonding energy of a surface. Attachment of substances onto the surface utilizes part of this free energy. The remaining unsatisfied bonding potential is known as the interfacial tension or interfacial free energy.[115] The interfacial free energy influences both the adsorption of dissolved components at the interface, which probably influence the formation of conditioning film, and the accumulation or movement of bacteria at the phase boundary.[116]

Since adhesion of *Candida* to inanimate surfaces occurs in an aqueous environment, the interfacial free energy that we are concerned with is related to free energies at the aqueous interfaces of both the microbial cell and the inanimate surface, i.e., solid/liquid interfacial free energies than to solid/air surface energies. From the thermodynamic point of view adhesion involves the change in free energy of materials as they form interfaces and the underlying premise is that adhesion is energetically favored if the formation of the adhesion interface is accompanied by a decrease in the free energy of the system.[2,107,117] Loeb[110] commented that the general expression for the free energy change of the adhesion processes could be expressed in terms of the surface free energies of the two materials of interest, e.g., 1 and 2, initially in contact with a surrounding fluid phase, 3, and then coming together to form 1-2 interface. Thus,

$$\Delta F^a = \Delta F_{1,2} - \Delta F_{1,3} - \Delta F_{2,3}$$

where ΔF^a is the free energy of adhesion and $\Delta F_{1,2}$, $\Delta F_{1,3}$, and $\Delta F_{2,3}$ are the free energies per unit area of interface between phases 1 and 2, 1 and 3, and 2 and 3, respectively, expressed in erg cm^{-2}. Minagi et al.[46] studied the relationship between the surface free energy of denture base resin plates and adherence of *C. albicans* (a relatively hydrophilic species) and *C. tropicalis* (a hydrophobic species). Close relationships were observed between the surface free energies of resin plates and adherence of these two species. Increasing surface free energy of resin plates resulted in increasing the adherence of *C. albicans*, but decreased the adherence of *C. tropicalis*. In the same study Minagi et al.[46] investigated the change of the free energy accompanying the adherence per unit area of adherence (ΔGa). The ΔGa was obtained from the interfacial free energy between the plate and the fungus, the plate and the liquid medium, and the fungus and the liquid medium. They found that a decrease of ΔGa resulted in increasing adherence of both *C. albicans* and *C. tropicalis*.

Wettability can be represented by the empirical parameter, γcrit, which is the critical surface tension for wetting of the substrate. γ crit is determined by measuring the contact angle, θ between each of a series of liquid droplets of known surface tension and the substrate in question, and plotting the cosine of these angles against the droplet surface tension. The critical surface tension for wetting of the substrate is then defined as the intercept of the best straight line through the data with the $\cos\theta = 1$ axis. Physically, γcrit separates liquids which form contact angles with the substrate of less than about 1° (i.e., spread spontaneously) from those forming higher contact angles and not spreading (see Zisman[118,119] for contact angle techniques). Wettability of the inanimate surfaces is a second parameter found to be associated with attachment of both microbial and mammalian cells to various surfaces.[107,111,120,121] Baier,[121] in his studies of the attachment of cells from human blood to the surfaces of implant materials, has shown that the attachment rate is a function of the initial surface properties of the substrate. He correlated γcrit with the rate of attachment of cells from human blood to candidate metallic and polymeric implant materials and found that there appeared to be a range in critical surface tension from 25 to 30 dyn/cm corresponding to a minimum rate of attachment. This was called the biocompatible range and corresponded to a substrate surface composed mostly of close-packed methyl and hydroxyl groups. Substrates with values of γcrit both above and below the biocompatible range were said to fall in the bioadhesive range. This grouping encompassed nearly all structural materials, and it was shown that such substrates acquired glycoprotein-type conditioning films during the first few minutes of exposure to human blood.[121] Deposition of these conditioning films shifted the value of γ crit toward biocompatible range but always left the surface well within the bioadhesive range. Under such conditions, those microorganisms attaching to a surface after the initial colonization by other bacteria may be adhering to a very different surface to that encountered by the initial group of microorganisms. Thus, the situation is complex, particularly when the formation of a biofilm is complete. The role of wettability on adhesion of marine bacteria to various surfaces has also been studied. Dexter et al.[111] presented data which supported the influence of the critical surface tension of the substrate on the attachment of marine microorganisms, particularly during the first 10 hours of exposure. Similarly, Fletcher and Loeb[107] found a relationship between bacterial attachment and water wettability in a series of polymeric materials.

The remaining two properties of inanimate surfaces found to play a role in adhesion are hydrophobic interactions and electrostatic forces. These two properties are often considered together.[47] Minagi et al.[46] examined the effects of hydrophobicities of substrate surfaces using two species of *Candida*. They showed that hydrophobic interaction takes part in the adherence of both species to hydrophobic surfaces. Klotz et al.[47] showed that hydrophobic forces predominantly control the adherence of *Candida* spp. to plastic surfaces. In the same study they showed that electrostatic interactions occur simultaneously creating a condition of negative

TABLE 3
Forces of Attraction and Repulsion Between Microbial Cells and Adsorbent Surfaces

Forces of Attraction	Forces of Repulsion
Chemical bonding (hydrogen, thio, amide and ester bonds)	Charge repulsion between surfaces of similar charge
Ion-pair formation (-NH^+....OOC-)	Van der Waals forces of repulsion
Ion-triplet formation (-COO...Ca^{2+}...OOC-)	Steric exclusion (hindrance)
Interparticle bridging (polyelectrolytes)	Negative chemotaxis (cellular mobility)
Charge fluctuations	
Charge mosaics	
Charge attraction of opposite signs	
Electrostatic attraction between surfaces of similar charge	
Electrostatic attraction due to image forces	
Surface tension	
Vander Waals forces of attraction	
Electromagnetic forces	
Hydrodynamic forces	
Diffusional forces	
Gravitational forces	
Positive chemotaxis (cellular mobility)	

From Daniels, S. L., *Adsorption of Microorganisms to Surfaces,* Bitton, G. and Marshall, K. C., Eds., John Wiley & Sons, New York, 1980. With permission.

cooperative binding. The electrical forces are minor to hydrophobic forces, since adherence to a considerable extent occurs even in the presence of this repulsive force.[47] Electrostatic interactions were responsible for the strong adherence observed by Zierdt[122] of bacteria, yeasts, erythrocytes, leukocytes, platelets, spores, and polystyrene spheres to membrane filter materials.

Types of forces leading to the destabilization of microbial suspensions and adhesion to surfaces are varied and complex. Table 3 lists possible attractive and repulsive forces that may exist between microbial cells and adsorbent surfaces.[327] The relative importance of each force depends on the types of microbial and adsorbent surfaces.

In addition to the physico-chemical properties of inanimate surfaces other factors affecting the adherence of yeast to plastic could be of importance. One such factor is the surface area of the material, as well as the form in which it is manufactured. We tried to compare the adherence of *C. albicans* to braided and monofilament sutures using scanning electron microscopy.[330] *C. albicans* adhered to braided sutures more than monofilament ones. Spaces between the filaments provided a trapping site for the yeast and other suspended materials (Figure 1) and braided sutures provide higher surface areas. Furthermore, the place at which the sutures were cut could provide an attachment area for the yeast (Figure 1).

Two points are clear: (1) the mechanisms by which microorganisms adhere to inanimate surfaces involve many overlapping phenomena. The adhesion of microbial cells when they are near inanimate surfaces may involve a balance between London-van der Waals forces (hydrophobic forces) and electrostatic forces of electric double layers. (2) More studies on the adherence of *Candida* spp. to inanimate surfaces should be undertaken if information that could be used to affect or alter the adherence of these yeasts to catheter and other protheses is to be obtained.

V. ENVIRONMENTAL PARAMETERS

Candida spp. are encountered most often in the oral cavity, gastrointestinal tract, and the

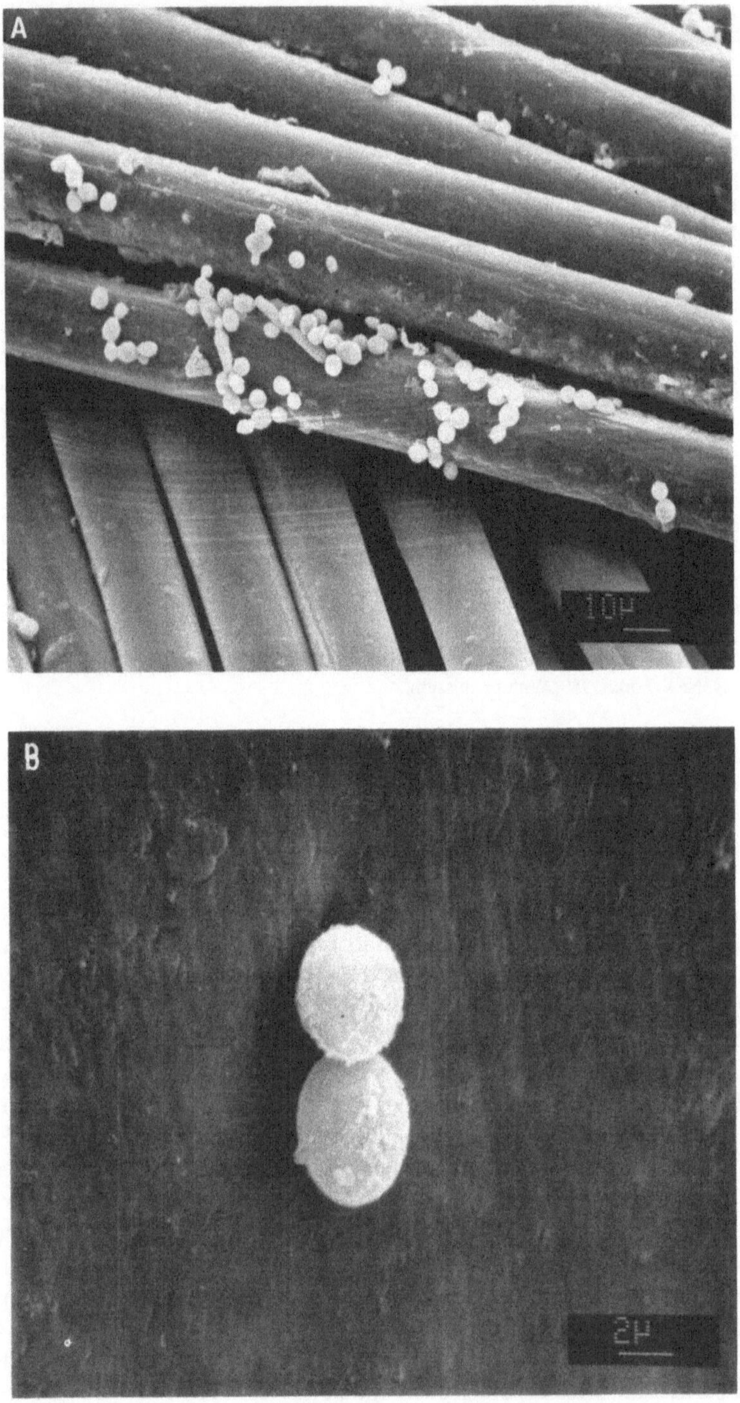

suture (Prolene). The smooth surface did not permit excessive adhesion. (C) and (D) *C. albicans* cells lodging in the suture cut area. (Ghannoum, M. A., Unpublished data.)

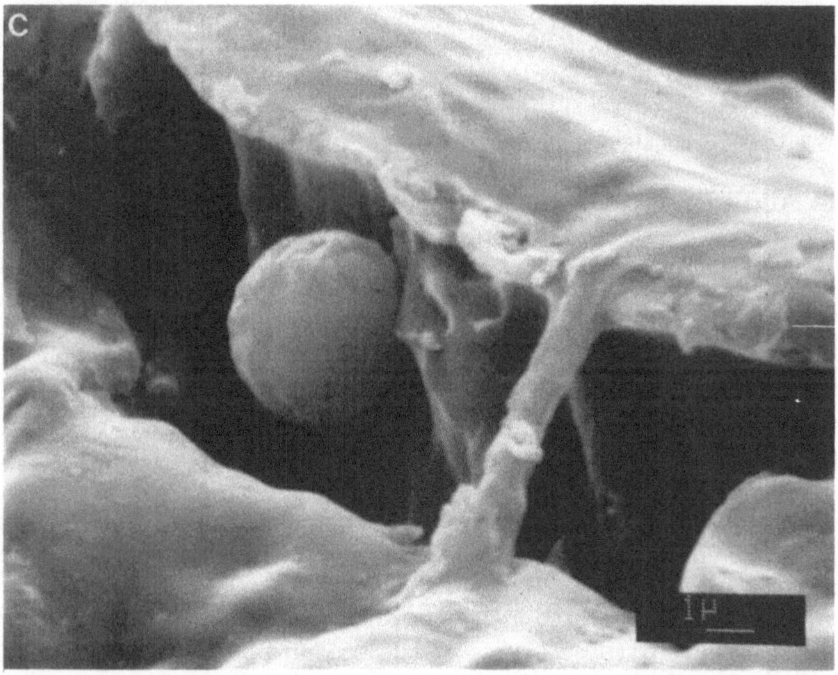

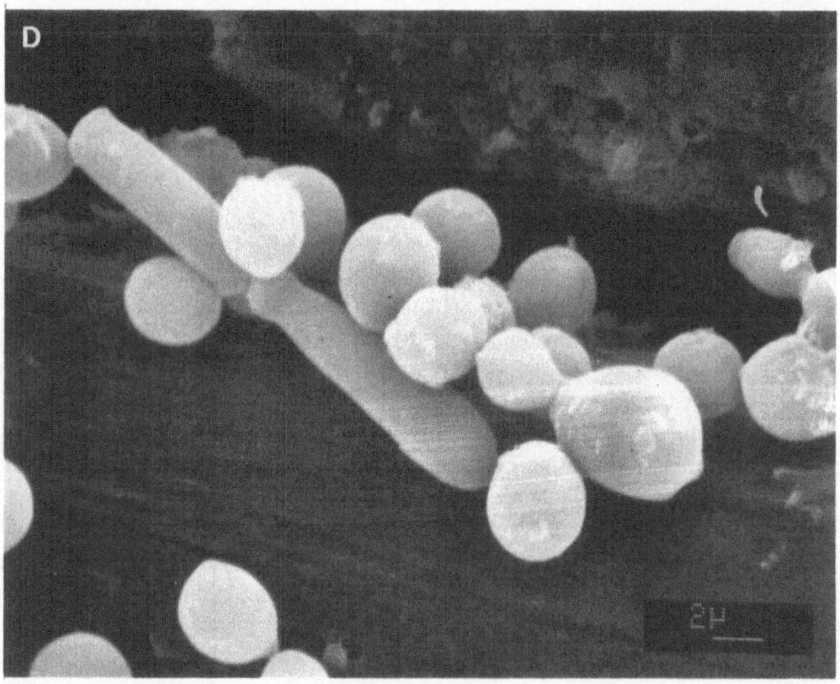

FIGURE 1 C, D.

vagina. Each of these sites has its own environmental conditions that distinguish it from others. Adherence of microorganisms is influenced by the particular environmental conditions prevailing at a particular site. In this section we will discuss the environmental conditions present in the oral cavity, gastrointestinal tract, and vagina with emphasis on the conditions known to influence *Candida* adherence.

A. FEATURES OF THE ORAL CAVITY

The oral cavity is a warm and moist environment which provides various surfaces suitable for microbial colonization. These surfaces include teeth, external gingivae (gums), tongue, and keratinized and nonkeratinized mucosa. The presence of these different surfaces makes the mouth a nonuniform environment comprising several distinct habitats each of which will support the growth of a charactersitic microbial community.[123,124] (See Gibbons and van Houte[78] for distribution of bacteria in the mouth.) In addition, the nature of the various microenvironments within the mouth may change from time to time during the life of an individual. At birth the mouth consists of soft epithelial surfaces which are kept moist by the secretions of the salivary glands. With the development of primary dentition a significant change in the environment occurs which is accompanied by a major change in the oral flora.[125,126] Once teeth are present, the flora appear to be generally similar to that of adults, though there are a few notable exceptions.[127-129]

Apart from the development of primary and permanent dentition, the ecological conditions within the mouth may be affected by various other environmental factors such as diet and nutrition,[130-133] antibiotic therapy,[134-137] radiation,[138,139] and various types of dental treatment including scaling, polishing, fillings, and denture wearing.[104,105,140-143]

The oral cavity is kept moist by oral secretions which include saliva and crevicular fluid. These secretions affect the growth of microorganisms in the oral cavity. Other factors which may influence microbial growth include physical conditions, nutrition, and inhibitory substances.[78] A review of these secretions and parameters is vital because of the role they play in influencing microbial adhesion.

1. Oral Secretions

a. Saliva

Saliva is derived from the secretions of three pairs of large glands, namely the parotid, submandibular, and sublingual, in addition to numerous smaller glands located in different sites of the oral cavity.[144] There are differences in the chemical composition of the secretions from each gland,[145-147] but the complex mixture is termed whole saliva.[123] Saliva flows, unstimulated, continuously at an average rate of 19 ml/h which leads to swallowing.[148] In addition to lubricating the mouth and acting as a transport fluid for oral microorganisms, saliva appears to play an important role in protecting the tooth enamel.[149] This is achieved by the formation of a proteinaceous layer known as enamel pellicle, from the saliva on the enamel surface.[150] This layer ensures that bacteria do not come into contact with the enamel.

Saliva consists of a complex mixture of inorganic and organic substances dispersed or dissolved in water. The main organic substances include proteins, lipids, vitamins, free amino acids, glucose, lactate, citrate, kallikrein, and an array of proteinaceous and glycoprotein components.[151-153] The main electrolytes in human saliva are sodium, potassium, calcium, chloride, bicarbonate, inorganic phosphate, and thiocyanate with small amounts of magnesium, iodide, and fluoride.[154] In addition, whole saliva contains variable amounts of organic and inorganic substances derived from gingival exudate; degenerating leukocytes and epithelial cells and the metabolic products of the oral microbial flora. The composition of saliva varies between individuals and the time of day and rate of collection. Besides differences in the composition of saliva derived from different salivary glands, there is a difference in its physical properties.[149] Other factors which affect the composition of saliva such as flow rate, duration and type of stimulus, diet, and hormones have been reviewed and discussed by Afonsky,[155] Burgen and Emmelin,[156] Jenkins,[157] and Dawes.[158]

Saliva may interfere with the adhesion process in two different ways: either altering the ionic strength of the environment providing an ionic state particularly favorable for adhesion, or interacting between salivary macromolecules and the microbial surfaces, adhesin(s) or epithelial cell receptors, causing a variety of surface modifications that affect adhesion.[159-161]

The adhesiveness of saliva-treated bacteria may be increased, unaltered, or decreased, depending on the bacterial species and on the experimental conditions.[159,162-164] Evidence for the interactions of salivary components with the microbial cells came from work on *Bacteroides gingivalis*. When this bacterium is suspended in a buffer, the cells can attach in large numbers to experimental pellicles, to epithelial cells, and to erythrocytes. However, when the cells are suspended in saliva, few, if any, organisms associate with erythrocytes or epithelial cells, and the numbers which adhere to experimental pellicles are markedly reduced.[165]

The role of saliva on the adherence of *C. albicans* to human epithelial cells and acrylic strips has been studied. McCourtie and Douglas[13] reported that adherence of *C. albicans* to acrylic was inhibited when saliva-treated acrylic strips were used in assays or when the yeasts were suspended in saliva. Studies with oral bacteria carried out by Clark and Gibbons[163] support this finding. Samaranayake and MacFarlane[166] showed that adhesion of *C. albicans* to HeLa cells increased significantly under the influence of salivary pellicle. Kimura and Pearsall[34] showed that viable *C. albicans* cells, preincubated in saliva for 90 min at 37°C before being washed and mixed with epithelial cells in phosphate-buffered saline, adhered better than nonviable yeasts or yeasts pre-incubated in phosphate-buffered saline. In contrast, no difference in the adherence of streptococci to BEC *in vitro* was observed when they were pre-incubated in phosphate-buffered saline or saliva.[167]

Further examination showed that the enhanced adherence of *C. albicans* observed after incubation with saliva is more a reflection of the ability of the yeasts to form germ-tubes, rather than to a requirement for prolonged interaction between fungi and epithelial cells.[34] A salient point is the association between the enhanced *Candida* growth in saliva and glucose availability.[168] Knight and Fletcher[169] demonstrated that saliva devoid of glucose is unable to support candidal growth. These authors concluded that oral candidosis seen in individuals on high-carbohydrate diets or in patients with diabetes mellitus is due to increased salivary glucose concentrations. Other conditions leading to the increase in glucose concentrations, such as antibiotic and steroid treatments, may also lead to similar growth enhancement.[169]

b. Crevicular Fluid

Serum-like crevicular fluid, which is secreted via the gingival crevice, is another secretion of the oral cavity. Compared to the saliva, the volume of crevicular fluid is small.[170] It is considered to have a protective function due to its flushing action and the availability of components for both humoral and cell-mediated immune responses. It is distinct from the saliva in composition and its protein content is approximately 20 to 30-fold higher.[171,172] It contains proteins such as albumin, immunoglobulins IgG, IgA, and IgM (present in about the same proportions and concentrations as in serum), and complement. In addition, cells such as neutrophils, leucocytes (predominantly polymorphonuclear leucocytes), monocytes, B and T lymphocytes, and blast cells are present. Like saliva, crevicular fluids contain low levels of glucose and free amino acids.[78,123] The very close similarity in total protein and immunoglobulins between crevicular fluid and serum led Brandtzaeg[171,173] to consider crevicular fluid an inflammatory exudate, and with the exception of some components, its immunoelectrophoretic protein pattern does indeed mimic that of serum.

Crevicular fluid influences the oral ecology, particularly the gingival crevice, by providing essential growth factors for some microorganisms residing there. For example, supply of hemin for the growth of *Bacteroides melaninogenicus* and α-2-globulin for *Treponema denticola* is attributed to crevicular fluid.[174]

To what extent crevicular fluid interferes with the process of adhesion of microorganisms is unknown. Unlike saliva, studies related to this aspect are lacking. Taking into consideration the role serum and IgA known to exert on adherence (the former enhances yeast adhesion to acrylic denture surfaces, while the latter inhibits candidal adherence to epithelial cells)[9,175] one would expect a role for this fluid in adhesion. Further investigation, however, is necessary.

2. Physical Conditions

Various physical circumstances which operate within host ecosystems such as the mouth influence both the growth and adhesive properties of microorganisms.[78,176] Conditions existing in the mouth include: narrow-temperature range (35 to 36°C), oxidation-reduction potential (which could reach a highly reduced level with E_h levels as low as -300 mV),[123,173,178] pH (the pH of the mouth is controlled mainly by saliva whose average pH is 6.75). The role these factors play on the adherence of microorganisms to surfaces is equivocal. Studies relating temperature and pH to adhesion suggested that variation in the ability of *C. albicans* to adhere to various epithelial surfaces resulted more from the ability of the yeasts to form germ-tubes under conditions of the assay than from the yeast's ability to adhere per se (for review see Ryley[97]). This point was reported by Kimura and Pearsall,[34] who demonstrated that conditions permitting germination (growth in tissue culture medium 199 at 37°C but not at 25°C) enhanced adherence.

Lee and King,[33] using a radiometric adherence assay, found that blastospores cultivated at 25°C adhered to VEC to a significantly greater extent than blastospores grown at 37°C. Their observation that blastospores grown at 25°C arise from budding yeast cells, while blastospores obtained at 37°C originate from pseudohyphal elements is yet another indication that germination and the cell wall modifications which accompany this process are of significance in the adherence process.[33] Optimal pH for adherence of both bacteria and yeast is reported to be in the range 6 to 7.[4,28,179-181] However, Samaranayake and MacFarlane[166] reported adhesion of *C. albicans* to HeLa cells varied with the pH of the test medium. They showed maximal adherence at pH 3, with less under neutral pH conditions.

The presence of dissolved carbon dioxide in saliva and in most human tissues (at concentrations of 5 to 6%)[123,182,183] could be another important environmental factor influencing adherence of *C. albicans* to epithelial cells. This was first suggested by Weld[184] and Mardon et al.[185] Persi et al.[62] investigated the effects of CO_2 and pH on adhesion of *C. albicans* to VEC and gave evidence that, among other factors, the adhesion process is affected by the pH and CO_2 levels present in the adhesion assay.

Denture wearing could provide a suitable environment where CO_2 accumulates from diffusion, through the oral surfaces between the palate and the denture. A similar situation is reported for skin covered with an occlusive dressing. Under these circumstances the concentration of CO_2 under the dressing equilibrates with that of the underlying tissues and reaches 5 to 7% within a few hours.[186]

The mechanism through which CO_2 promotes adherence is not known. However, a number of findings are relevant in this context. First, the ability of individual isolates of seven potentially pathogenic yeast species in the genus *Candida* to grow under elevated concentrations of CO_2.[187] Second, the observation by Sims[188] of the induction of germ-tubes and mycelial growth caused by intracellular accumulation of CO_2. Third, changes in the surface structures of *C. albicans* observed in cultures incubated in ambient air supplemented with 10% CO_2.[62] These observations may suggest that CO_2 enhances adhesion as a result of alterations taking place in the yeast partner.

3. Nutrition

Microorganisms in the oral cavity derive their nutrients from oral secretions (saliva and crevicular fluids), desquamated epithelial cells, and diet. These constitute a rich supply of nutrients which will allow the growth of a microbial community of great diversity and satisfy the requirements of many highly fastidious microorganisms.[78,123] We have already covered the role played by oral secretions in adherence. Studies on the influence of dietary components on the oral flora is limited, with little attention directed towards the role of dietary proteins and lipid.[78] In contrast, carbohydrates have been studied more often because, most probably, of their association with dental caries development and oral candidosis.[14-19] Samaranayake and MacFarlane[189] proposed a unifying hypothesis related to the role of dietary carbohydrates on the enhancement of growth, multiplication, and adherence of *Candida* spp. to oral surfaces. Also, the role of carbohydrates in the modification of candidal outer envelop and the consequence to adherence has been discussed above.

B. FEATURES OF THE GASTROINTESTINAL TRACT

The gastrointestinal (GI) tract is a long tube extending from the mouth to the anus. Associated with the digestive tube are various glands situated outside which deliver their secretions into it by duct systems. Each part of the digestive tube has four layers, their nature and thickness varying with functional requirements in the different regions. These layers are the tunicae mucosa (mucous membrane), submucosa, muscularis, and serosa. The mucous membrane is a wet, surface epithelial membrane that is lubricated by mucus and rests upon a basal lamina. The basal lamina is supported by a layer of connective tissue, the lamina propria, and in many regions by a thin outer layer of smooth muscle, the muscularis mucosae. In most regions, the mucous membrane is irregular and shows finger-like projections, the villi, which greatly increase surface area, and deep epithelial-lined invaginations (intestinal glands or crypts).[190]

The GI environment is dark, moist and almost devoid of oxygen. It contains high concentrations of bicarbonate and a large amount of carbon dioxide.[191-194] The pH level in the GI tract varies between 2.0 (in the stomach) to a pH 6 to 7 (in the lower regions of the gut). This variation is related to gastric acidity, which is neutralized towards the duodenum. The GI environment is subject to a constant cleansing action brought about by the movement of intestinal contents down the digestive tract. The normal peristaltic flow is fastest in the jejunum and upper ileum and slows in the distal ileum. A degree of stagnation proximal to the ileo-cecal valve occurs due to water absorption.[195]

Factors controlling and influencing the microbial presence and distribution in the GI tract are related to host physiology, environmental conditions, and bacterial interactions.[196] Factors related to host physiology include intestinal secretions (gastric acid, enzymes, mucus, and bile salts), intestinal mucosa, immune mechanisms, and intestinal motility. Important environmental conditions include diet and antibacterial drugs. Bacterial interactions are related to the total numbers and the relative frequency of the various species in the normal flora.[196]

C. albicans associates with the GI mucosa by at least five distinct mechanisms.[53] These include adhesion to epithelium, adhesion to mucus (Figure 2), co-adhesion to adherent fungi, co-adhesion to adherent bacteria, and entrapment in the mucous gel overlying the epithelium (Figure 2). The predominant association mechanisms appear to be entrapment in the mucous gel, and adhesion to mucus and the epithelium.[53,68] Consequently, the discussion will concentrate on these.

1. Gastrointestinal Epithelium

The GI epithelium forms an interface between the external environment (the intestinal lumen) and the internal environment (lamina propria). The surface epithelial cells of the stomach are tall and columnar (only one cell type is present), which distinguishes the stomach from all other regions of the digestive tract. The columnar cells are mucin secreting. The mortality rate of the surface epithelial cells is high, and they are replaced by less-differentiated cells situated in the deeper parts of the foveolae and upper regions of the gastric glands. Indeed, there is evidence that the entire surface epithelium is replaced every 3 to 4 d.[190]

The epithelium of the intestinal mucosa is simple columnar in type but differs from the

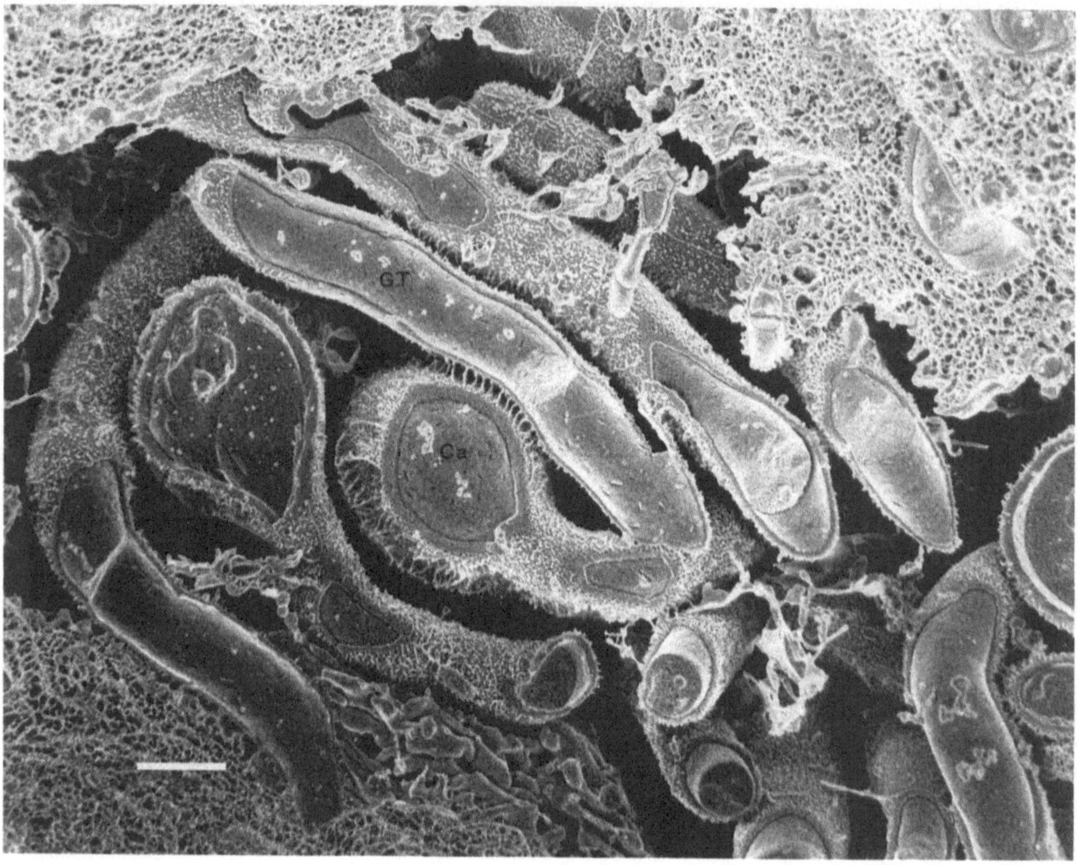

FIGURE 2. Scanning electron micrographs, prepared using rapid freezing and fracturing technique, showing the adhesion of *C. albicans* cells to one another. These adherent cells also mediate the adhesion of *Candida* to epithelial cells. (Courtesy of Drs. Tokunaga, M., Niimi, M., and Koike, H., Department of Microbiology, School of Dentistry, University of Kagoshima, Kagoshima 890, Japan).

surface epithelium of the stomach in that more than one cell type is present. These are secreting-absorbing cells (enterocytes), goblet (mucus) cells, Paneth cells, and others.[190] The enterocytes are found both in intestinal glands and on villi and they are the predominant cell type. As many as five different cell types have been described, and all are spherical or pyramidal in shape, with basal granules. The various types produce gastrin (in the stomach glands only), serotonin, secretin, norepinephrine, and possibly cholecystokinin. The goblet cells are scattered on the intestinal villi between columnar cells, and their numbers increase from duodenum to the terminal ileum. Goblet cells arise by differentiation of some of the undifferentiated cells and contain a prominent Golgi apparatus and a few mucous droplets.[197] They then migrate up the gland and onto a villus, accumulating more and more mucous droplets and acquiring the distended or goblet shape. Goblet cells are then shed from the tips of villi into the lumen. The period for the epithelial cells to be created and shed into the lumen is about 2 d in man.[198]

Paneth cells are found in the bases of intestinal glands and are scattered between undifferentiated cells. They are found throughout the small intestine, but are absent in the large intestine.[190] The Paneth cells contain zinc and secrete peptidase and possibly lysozyme. A number of cells are often observed between intestinal epithelial cells. These include intra-epithelial lymphocytes (may have an important role in the immune function of the intestine),[199] eosinophils, PMN leukocytes, and macrophages.[200,201]

The cytoplasmic membrane on the side of the epithelial cells exposed to the intestinal lumen is ramified into rod-shaped structures called microvilli.[190] These structures increase the intestinal epithelial cell surface area by a factor of 15 to 25. Each microvillous is covered by an extension of the plasma membrane, the outer lamina of which is associated with a network of fine filaments giving a fuzzy or granular appearance, depending on the transmission electron microscopy techniques used (Figure 3).[202,203]

This filamentous coat (surface coat or glycocalyx) occupies the spaces between microvilli and their tips and extends 0.2 μm or more from the tips of the microvilli.[204] It is highly prominant over microvilli of intestinal epithelial cells, while less so over microvilli of goblet and undifferentiated crypt cells.[202] This surface coat is a product of the intestinal epithelial cells which is continuously synthesized in the cell, transported to the cell surface, and lost into the lumen. It is not a layer of mucus and is distinctive from other components of the polysaccharide coat found on the laterial and basal membranes. This surface coat may play a role in preventing the large particles, such as bacteria, from gaining access to the epithelia. However, it does not appear to restrict the movement of colloidal materials, large molecules such as thorium dioxide, or small molecules. These cell surface mucopolysaccharides provide a variety of potential binding sites for bacterial toxins,[205,206] microorganisms, and experimental ligands such as plant lectins.[207-211]

Histological evidence suggest that the surface coat is composed of a sulfated, weak acidic mucopolysaccharide which is resistant to proteolytic and mucolytic agents.[190,202] In addition, a large number of enzymes (25 or more) have been localized in the surface coat and its plasma membrane. These include alkaline phosphatase, adenosine triphosphatase, thiamine triphosphatase, numerous disaccharides, amylase, and aminotransferase.[202]

Direct interaction between organisms including protozoans, spiral-shaped microorganisms, yeasts, fusiform bacteria, and lactobacilli is known to occur.[202,212,213] This association is effected by the adherence of these organisms to the epithelial cells with concomitant alterations in the ultrastructural morphology of the microvillus border of the intestinal epithelial cells. The process of bacterial attachment to intestinal cells was described in an excellent ultrastructural study by Neutra.[214] This author used a Gram-negative bacterium that is able to associate with the host cells with a minimal damage to either participant, thus providing a useful model for studying the adhesion process. Electron microscopy of thin sections revealed that both the bacterium and the host show structural specializations that serve to maintain the attachment.[214] As seen by transmission electron microscopy, the bacteria are located quite close to the surface of the outer layer of the host membrane. The bacterial cell wall is rigid and thickened at the blunt attached end, and the host cell cytoplasm forms a stable, cup-shaped (socket-like) depression lined by the plasma membrane that accommodates the bacterial tip. A little material recognizable as glycocalyx can be seen crossing a 30 to 40 nm gap existing between the two participants.[214] Studies of intestinal-bacterial attachment sites have shown that both spirochetes and flagellated bacteria indent the host cell epical membrane.[215-218] However, the attachment sites of the two organisms differ dramatically, even on a single intestinal cell, suggesting that distinctive bacterial surface proteins may provide the "signals" that induce host cell modifications.[214,219]

Direct association of yeast cells to GI epithelial tissue is known.[220-229] This is not surprising since the GI tract has been regarded as a reservoir for *C. albicans*,[225-227] and an important source in cases of recurrent *Candida* vaginitis or perianal itch.[228] Moreover, the passage of *C. albicans* through the intestinal mucosa is recognized as important since its dissemination from the GI tract was first reported by Krause et al.[229] following oral administration. Another relevant finding is the demonstration that a number of opportunistic fungal pathogens (e.g., *C. albicans*, *C. tropicalis*, *Sporothrix schenckii*, *Cryptococcus neoformans*, and *Trichosporon* species) can spread systematically from the GI tract to invade visceral organs.[230-233]

Two studies show *Candida albicans* in direct association with the GI epithelium. Pope and Cole[68] used scanning electron microscopy to reveal the location of this yeast and its structural

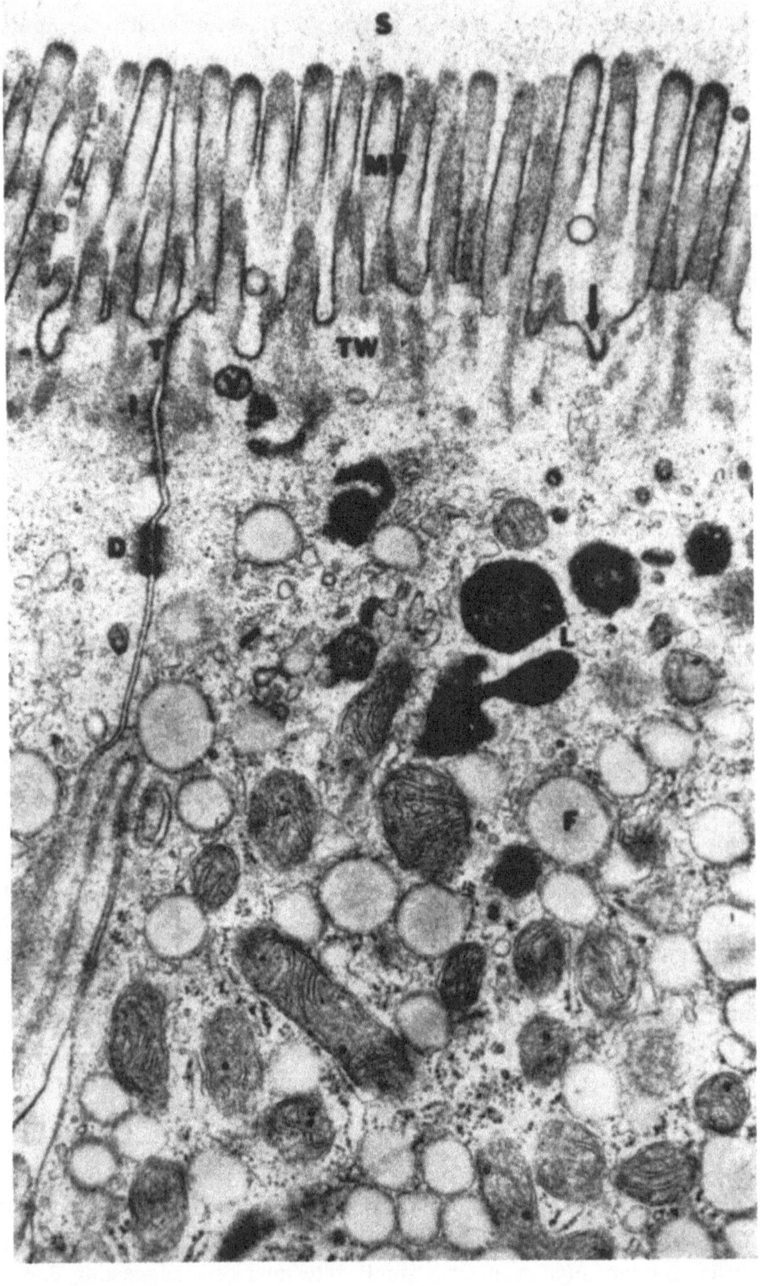

FIGURE 3. Electron micrograph of apical portion of two intestinal absorptive cells in normal man during fat absorption. (MV) Microvilli are closely and regularly spaced and possess a very prominent (S) surface coat. (TW) Microvillous core tubules extend into the terminal web. There is a charactristic junctional complex consisting of (T) tight junction, (I) intermediate junction, and (D) desmosome. Membrane invaginations at the base of microvilli, called apical pits (arrow), have been shown to be involved in ferritin uptake but not in fat uptake. Apical pits eventually contribute their contents to (L) apical lysosomes. Note, (V) the "secretory" vesicle to the right of the junctional complex, an organelle important for transport of secretory IgA. Droplets of (F) resynthesized triglycerides can be seen within profiles of the smooth endoplasmic reticulum. Bicarbonate buffered osmium tetroxide fixation. Double stained. (Magnification × 20,000.) (From Dobbins, W. O., III, *Attachment of Organisms to the Gut Mucosa*, Boedeker, E. C., Ed., CRC Press, Boca Raton, FL, 1984.)

association with the surface of tissues of the gut at early stages after intragastric inoculation of infant mice. The yeast adheres to both keratinized squamous epithelial and columnar secreting epithelial surfaces of the stomach, and some yeast cells were found deep wihin tissue folds. Also it was shown that yeast cells adhered to all areas of the villus surface of the upper intestine, and some *Candida* cells appeared to be embedded in the villus surface.[68] Kennedy et al.,[53] studying the mechanisms of association of *C. albicans* with intestinal mucosa, demonstrated that direct adhesion to epithelium is one of the predominant mechanisms responsible for *Candida*-epithelial association. These authors gave evidence that yeast cells adhere directly to the epithelium, probably by adhesion to the epithelial glycocalyx. Supporting data are also provided by Pope and Cole.[68] They presented an ultrathin section showing a yeast cell adhering to the villus surface. Dark fibrous material (presumably glycocalyx) was observed filling the gap between the two participants.

Recently, Cole et. al.[328] investigated the interaction between *C. albicans* and the GI tract further using the infant mouse model. These authors presented microscopic evidence which suggests that the highest frequency of invasion of the mucosa by yeast cells occurred in the region of the jejunum 1 to 3 h after yeast inoculation. Furthermore, it was shown that *C. albicans* cells are capable of progressive extracellular digestion of the intestinal mucus barrier and microvillus layer, followed by intracellular invasion of columnar epithelial cells without total distruption of cytoplasmic contents (Figure 4).

Depressions in the epithelium similar to those observed under bacteria, were also observed under *Candida* cells, possibly due to enzymic lysis which may allow the yeast to stabilize itself with the epithelium after the initial adhesion.[53] These depressions were observed for *C. albicans* adhering to BEC *in vitro* (Figure 5).[234]

It would appear that yeast interact directly with the GI epithelium in much the same way that bacteria do, i.e., the microbial cell is initially attached to the epithelium aided by the glycocalyx coat (Figure 4). This is followed by the formation of a depression in the epithelium where more intimate association is established in preparation for systemic spread, provided the host conditions are suitable. It is important, however, to consider the size differences between bacteria and yeast. Since bacteria is smaller in size (approximately the same size as a microvilli) it can align itself parallel to the epithelial cell microvilli with only minor displacement of these host structures.[202] Yeast, on the other hand, are larger and intimate attachment to the plasma membrane of an intestinal cell would require more drastic displacement of microvilli. This situation may be similar to that observed with *Crytosporidium* (a protozoan parasite). Dobbins showed *Cryptosporidium* in an intimate attachment to the plasma membrane of an intestinal absorptive cell with obvious displacement of microvilli (Figure 6).[202]

2. Gastrointestinal Mucus and Mucous Gel

Mucus is a unique secretion that occurs as viscous, sticky, water-insoluble glycoprotein gel. Mucus clings to the epithelial surfaces and is found in all internal tracts of the body.[235-237] The mucus blanket is constantly renewed by the secretion of high molecular weight glycoproteins from individual goblet cells throughout the epithelium (for review, see Neutra[238]). In the GI tract, where mucus is the major organic secretion, its role in protecting the stomach and duodenum against the corrosive gastric juices has intrigued physiologists for more than one and a half centuries.[239] Only recently, interest has emerged in the study of the specific interactions between mucus and microorganisms.[240-244]

a. Structure and Chemical Composition

The mucus layer covering the epithelial mucosa of the alimentary, respiratory, and reproductive tracts of mammals consists of glycoproteins, proteins, and lipids in the form of a gel imbibed with water and electrolytes.[245,246]

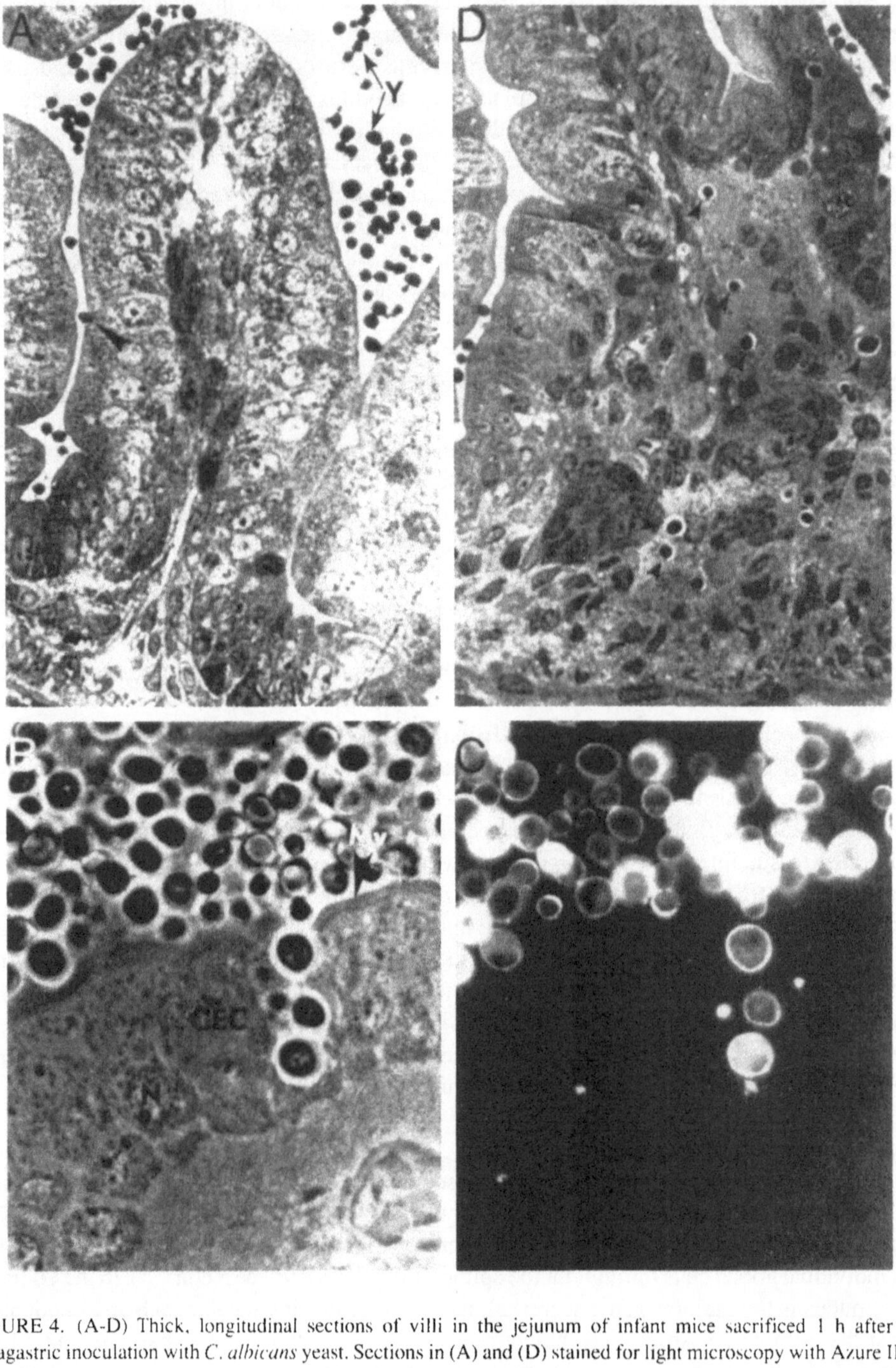

FIGURE 4. (A-D) Thick, longitudinal sections of villi in the jejunum of infant mice sacrificed 1 h after oral intragastric inoculation with *C. albicans* yeast. Sections in (A) and (D) stained for light microscopy with Azure II and methylene blue. In (B) and (C) the same section was examined by phase-contrast microscopy and immunofluorescent microscopy, respectively, and shows three yeast cells which have invaded a columnar epithelial cell (CEC). Arrowhead in (A) locates yeast in early stage of invasion at villus surface. Arrowheads in (D) locate yeast within host tissue, each of which is surrounded by a clear zone. Mv, microvilli; N, nucleus; Y, yeast. (Figure 1 A, D; magnification × 600. Figure 1 B, C; × 4000.) From Cole, G. T., Seshan, K. R., Pope, L. M., and Yancey, R. J., *J. Med. Vet. Mycol.*, 26, 173, 1988. With permission.)

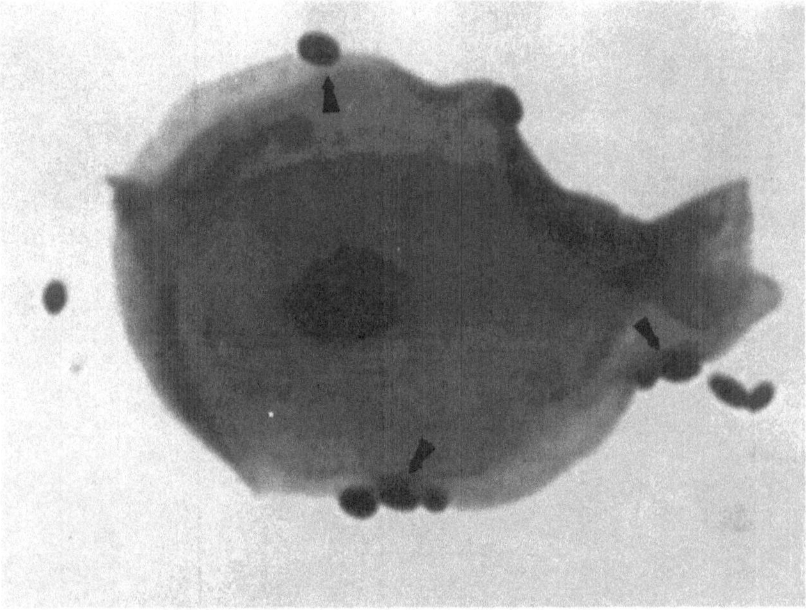

FIGURE 5. A light micrograph of *C. albicans* adhering to (BEC) buccal epithelial cells. Yeast cells are embedded in (arrows) shallow depressions of the BEC.

i. Mucous Glycoproteins or Mucins

Glycoproteins of mucous secretions received considerable recognition because they represent the major organic constituent of mucus and have viscoelastic properties which account for the gelling characteristics of whole mucus.[245,247-256] Mucins are a distinct subgroup of glycoconjugates, which differ structurally and chemically from serum glycoproteins and proteoglycans of connective tissue.[257] They are macromolecules with molecular weights which are usually estimated as greater than 10^6. Mucous glycoproteins are characterized by a predominance of carbohydrate, often over 70 to 80% by weight of the preparation.[257-259] They contain galactose, fucose, N-acetylglucosamine, N-acetylgalactosamine, and sialic acids as the only sugars. Mucins are characterized by the absence of mannose, xylose, glucose, and uronic acid. The presence of these sugars would suggest contamination by serum glycoproteins or connective tissue proteoglycans. The sugars consist of hundreds of oligosaccharides side chains, that can vary in length from between 2 and 5 to 19 and 22 sugars, which are joined to a central protein core.2[57,260] The linkage is an O-glycosidic link joining N-acetylgalactosamine to the hydroxyl of either serine or threonine. The protein core of mucins is atypical when compared to that of globular protein, with a characteristically high content of serine, threonine, and proline and low content of aromatic amino acids.[259] Proline is presumed to be necessary to achieve a conformation that will permit the very close packing of the carbohydrate chain. Viewed in three-dimensional space, mucins appears like a shaggy "bottle brush", where the carbohydrate chains represent the bristles and the proteins core represents the central wire support.

The development of techniques for the isolation of pure, native, undegraded mucin allowed the study of its structure.[257,261] Investigation revealed that native glycoproteins from GI secretions are polymeric structures composed of glycoprotein subunits joined together by disulphide bridges between their protein cores.[245] One of the first and best characterized glycoprotein is that from pig gastric mucus which is a polymer of four glycoprotein subunits (5×10^5 mol. wt.) joined by disulphide bridges.[262,263] The glycoprotein from human gastric mucus (2×10^6 mol. wt.) has a composition very close to that of pig gastric mucus and consists of a

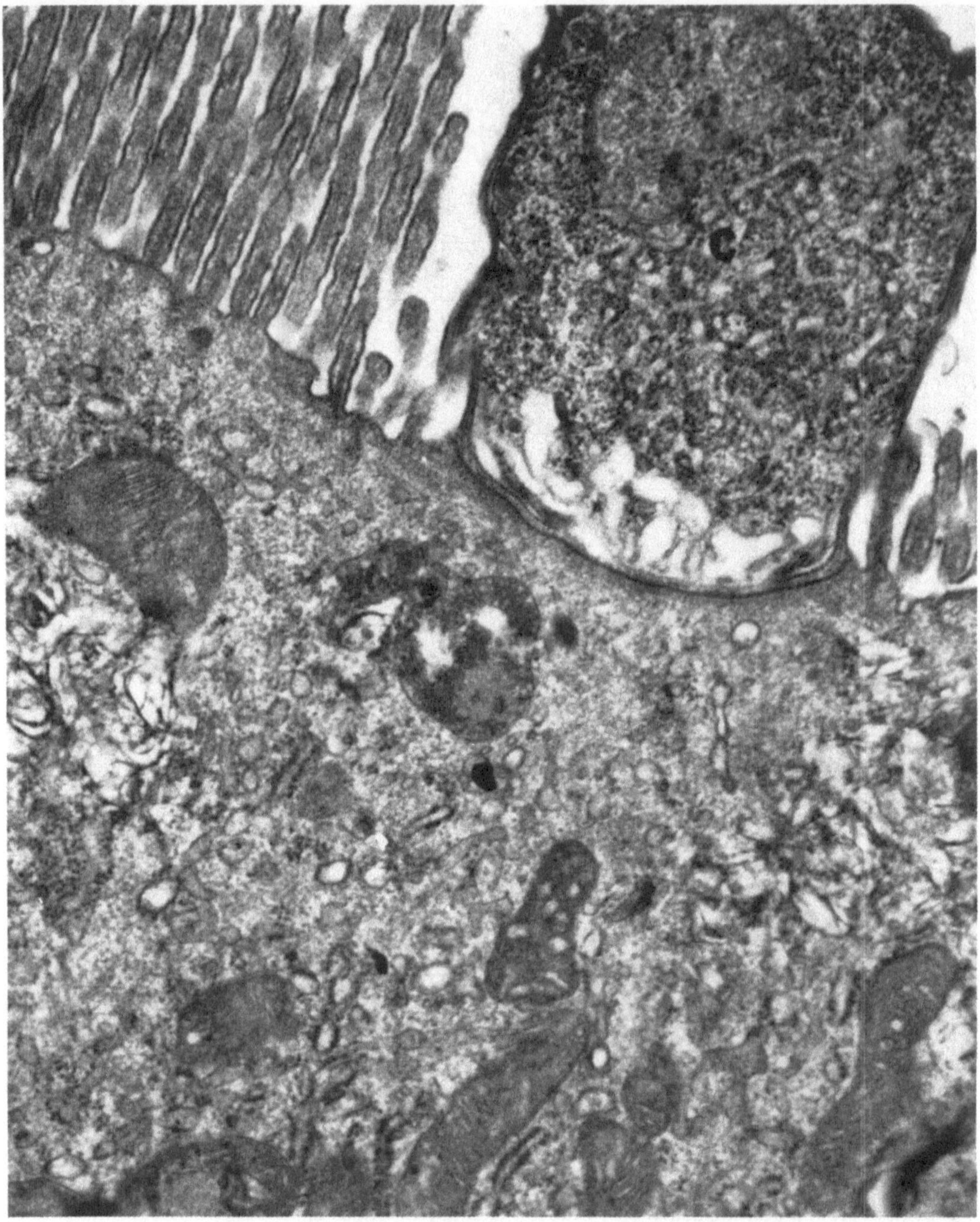

FIGURE 6. Electron microscopic photomicrograph illustrating a (C) cryptosporidial species found in mouse. It is intimately attached to the plasma membrane of an intestinal absorptive cell. Note that the microvilli have been displaced by the organism and that there appears to be little evidence of injury to the epithelial cell. (Fixed in osmium, double stained. Magnification × 25,000.) (From Dobbins, W. O.,III, *Attachment of Organisms to the Gut Mucosa*, Boedeker, E. C., Ed., CRC Press, Boca Raton, FL, 1984.)

similar polymeric structure of glycoprotein subunits joined by disulphide bridges with a 70,000 mol. wt. protein released on reduction.[253,264] Based on dissociation studies involving proteolytic cleavage, employing enzymes, and reduction using mercaptoenthanol, it is possible to represent diagrammatically the structure of gastric mucous protein. Figure 7 shows a mucous glycoprotein

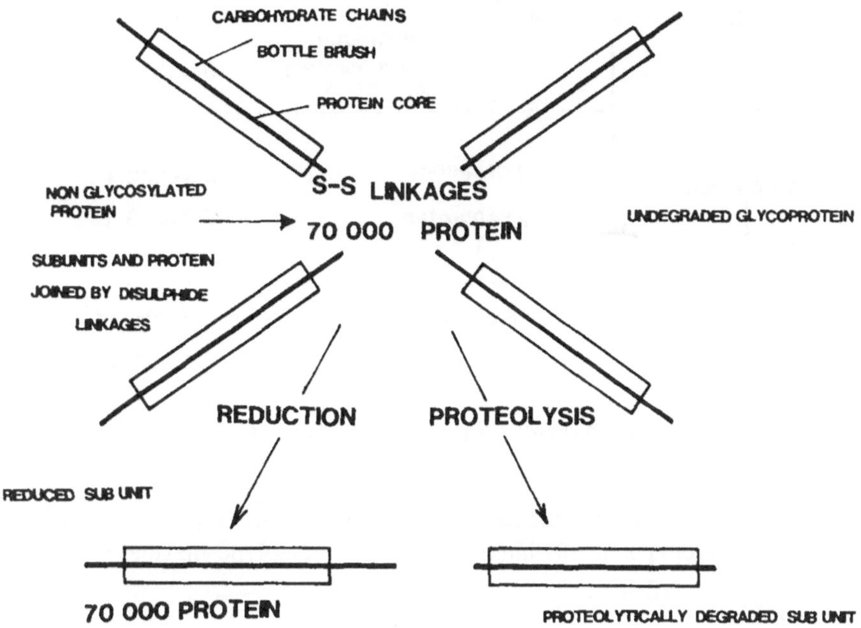

FIGURE 7. Diagrammatic representation of the structure of pig gastric mucous glycoprotein. (Adapted from Allen, A., *Trends Biochem. Sci.*, 8, 169, 1983.)

consisting of four subunits jointed centrally, rather than end to end. This proposed arrangement is based on physical studies showing the molecule to be roughly spherical.[265]

An important property of native mucins in relation to attachment of different pathogens,[206,207] is their negative charge. This negative charge is attributed to the presence of sialic acid residues in terminal positions on the carbohydrate chains, to ester sulphate residues present within the sugar units,[252] and to the presence of a net excess of negatively charged amino acids in the protein core.[259] Since the oligosaccharide composition of human glycoprotein varies greatly from one person to another it is expected that this variation will be reflected in charge variability per mass of sulfoglycoprotein and sialoglycoprotein. Variation is also expected in glycoproteins from different parts of the GI tract of an individual. In general, the further down the GI tract the mucous glycoprotein secretion is, the greater is its negative charge.[259]

ii. Lipids

Compared to mucins, lipids of the mucous secretions have received little attention until recently.[267-269] Lipids are present in the soluble portion of the gastric secretion, in the mucus layer lining the surface of the mucosa and in the preformed intracellular mucus contained within the secretory granules of the epithelial cells.[246,247,270]

Chemical analysis of lipid content of rat gastric mucus showed that they constitute from 20 to 30% of the dry weight of mucus and contain both apolar (50%) and polar (50%) lipids.[246] The apolar lipids consist of fatty acids, cholesterol, cholesteryl esters, and di- and triacylglycerols. The glycolipids are predominantly neutral and sulphated glyceroglucolipids. The major phospholipids are sphingomyelin, diacylglycerophosphoethanolamines, diacylglycerophosphocholines, and acylglycerophosphocholines.[246,271]

Lipids seem to exert some effect on the physicochemical and physiological properties of mucus. Studies on mucous secretions of the GI tract carried out by a number of workers indicate that lipids affect the rheological properties of mucus by decreasing its viscosity and elasticity and increasing the solubility of mucus.[272-274] There are indications that glyceroglucolipids

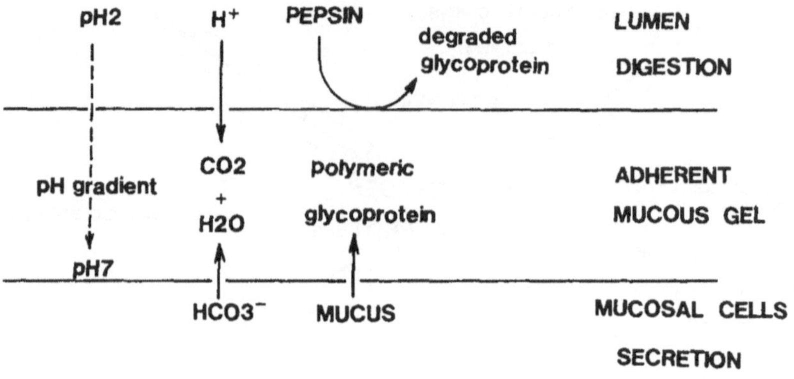

FIGURE 8. Mechanism of protection against gastric acid and pepsin by mucus. (Adapted from Allen, A., *Trends Biochem. Sci.*, 8, 169, 1983.)

enhance the functions of GI mucins such as lubrication, interaction with bacteria, and protection of mucosal surfaces from potentially injurious acids and proteolytic enzymes.[249,251,269,275]

b. Function of Mucus

Mucus is an essential component of the protective mechanisms in the GI, respiratory, and cervical tracts and other areas covered by mucous membranes. The role of mucous gel in protecting the GI tract is already recognized. In 1933 Florey[276] observed that the mucous gel of the small intestine could envelope carbon particles and thereby cause their removal, with the mucus expelled by the pumping action of the intestinal villi. Mucus serves two main functions: (1) it acts as a physical barrier against both large molecules and microorganisms, and (2) it acts as an acid-neutralizing agent. Gastric mucus is permeable to ions and small molecular weight molecules such as glucose, salicilate, and vitamin B_{12}. It is not permeable, however, to molecules, the size of pepsin (mol. wt. 35,000).[245,259] By preventing the passage of pepsin, the mucus protects the epithelial cells from digestion by this enzyme.[277] However, it should be emphasized that the interactions of mucus and proteolytic enzymes, such as pepsin, are more complex and depend on other factors, beside permeability. These include (1) the amount or thickness of adherent mucus covering the epithelial surface and (2) the structure of the mucous gel.[257,259]

The acid-neutralizing ability of mucus was first suggested by Heatley,[278] who proposed that gastric acids (HCl), which can diffuse through the mucus layer, are neutralized by mucosal bicarbonate secretion. Credence for this hypothesis came from detailed studies on (1) mucosal bicarbonate secretion,[279-281] and (2) on the structure and properties of mucus.[245,282,283] Current concepts of GI mucosal protection against both gastric acid and pepsin are represented diagrammatically in Figure 8 which was proposed by Allen.[257] In this system mucous gel provides an unstirred layer and acts primarily as a mixing barrier. The mucosal bicarbonate secretion is confined within the mucous gel matrix and prevented from rapidly mixing with the much larger amount of acid in the lumen.

In addition to the protective effect shown above, mucus plays a role in the protection against invading microorganisms.[284,285] In this respect, Lake et al.[331] have shown that soluble antigen-antibody complexes (or antigen alone in immunized animals) can stimulate the release of mucus gel from intestinal goblet cells. If such a reaction were to be triggered by microorganisms in the immune host, it would constitute a potentially powerful deterrent to mucosal colonization by pathogens.[285]

c. Role in Adherence

Prior to adhesion and penetration of the mucosal epithelial surfaces, microorganisms must first establish contact with the surface of the mucous gel. This is followed by penetration and trapping in the gel, which in certain cases may represent the only manner that pathogenic microorganisms colonize the mucosa.[53] This association with the mucus gel may be quite prolonged, and therefore may be of considerable importance as an ecological and pathogenicity determinant.[284] Studies on the bacterial association with the mucous gel system of the GI tract showed that mucus plays a dual contrasting role in relation to adherence: (1) as an inhibitor of bacterial adhesion and (2) as a receptor for bacterial adhesion. In the first case components of the mucous gel may act as competitive inhibitors of bacterial adhesion to epithelial cells. Mucin possesses oligosaccharide binding sites for bacteria and their toxins which mimic those of the intestinal wall and therefore inhibit mucosal attachment.[260] An example for this type of interaction is the competition of gastric mucin and intestinal epithelium for cholera toxin.[286] In the second case the mucous gel may be invaded by bacteria and, under suitable conditions, may furnish a matrix in which the bacteria can proliferate and associate with the mucosa. Freter[285] proposed that in such cases large bacterial populations can build up within the mucous gel when the rate of bacterial invasion of the mucous gel and/or the rate of bacterial multiplication within the mucous gel exceeds the rate by which mucus is removed from the epithelial surface. In effect, mucus is acting as an adhesion site. An example for this interaction is the colonization of cystic fibrosis patients with *Staphylococcus, Pseudomonas*, or *Hemophilus influenzae*.[287]

Association between GI mucus and *Candida* and its relationship to adherence has been the subject of few investigations. Cole et al.[68,328] frequently observed yeast cells in association with the mucus layer throughout the GI tract (Figure 9). Kennedy et al.,[53] in an effort to determine if *C. albicans* could associate with the mucous gel, mixed mucous gel obtained from antibiotic-treated mice with *C. albicans* in an *in vitro* adhesion assay. They observed that large numbers of yeast associated with the mucous gel within 5 min and became embedded and attached to it.[53] In the same study, these authors cultured mucous gel obtained from penicillin-treated mice 24 h after oral challenge with *C. albicans*. This gel contained about 19% of the total *Candida* isolated from the cecal epithelium, cecal contents, and mucous gel. Solutions of mucins (0.5%) reduced the adherence of *C. albicans* to monolayers of human intestinal epithelium by 83%.[40] This inhibition of *Candida* adherence by mucin has been demonstrated in a study by Butrus and Klotz[288] with plastic surfaces. These studies demonstrate that mucus plays a role in the first step in mucosal association by *C. albicans*. The initial step, i.e., contact with the surface of the mucous gel is random while the second step, i.e., penetration or trapping in the mucus is more specific and, at least with bacteria, has been shown to involve specific mucus receptor.[289] However, further research is needed to isolate and characterize the exact nature of the mucus receptor(s) specific for *Candida* adhesion.

C. FEATURES OF THE VAGINA

The vagina is a fibromuscular canal which is flattened from anterior to posterior. The interior vaginal wall is normally 7 to 10 cm long, and the posterior wall 13 to 15 cm long.[290] The vagina is the final part of the internal female genital and the parturient canal. Simultaneously, it serves as the passage for outflow of secretions from the cervical crypts and for menstrual flow.[291]

Scanning electron microscopy examination shows that the surface of the vagina consists of large, flat polygonal cells with interdigitating borders. In the process of exfoliation, the cells show slightly elevated and detached cell borders, exposing similar-appearing layers of cells beneath them.[292] A fine interlacing network of microridges can be seen at higher magnifications.[293] Parakkal[294] suggested that these microridges may be important for holding and distributing secretory products. Others suggested that the microridges may serve to interlock or hold the epithelial cell layers together.[295] The possible role of the microridges in the adhesion of *C. albicans* to vaginal cells has been mentioned above (see section on Character of Epithelial Cells).

The vagina is lined by a stratified squamous epithelium made of five distinct cell layers: basal,

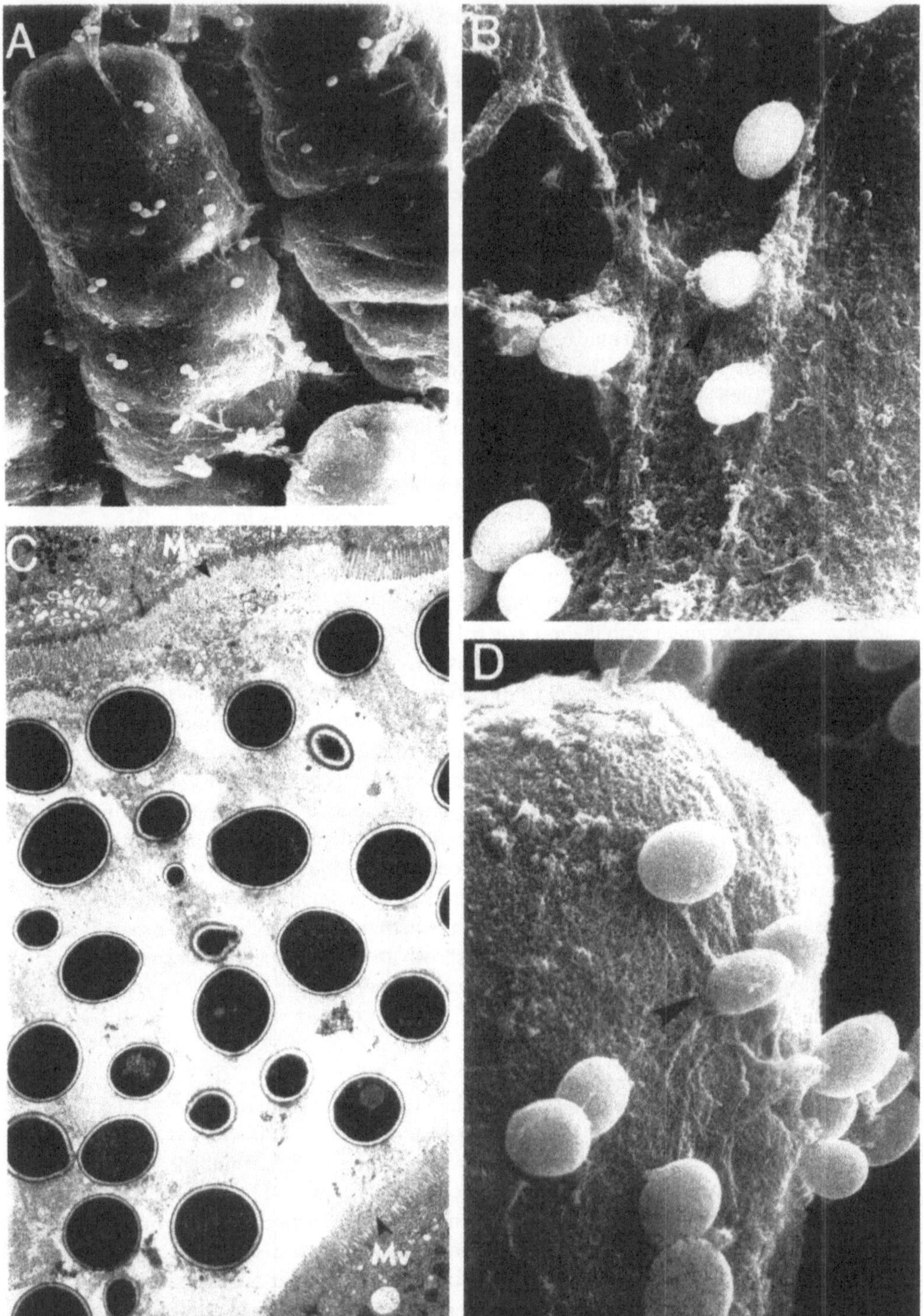

FIGURE 9. (A-D) Yeast cells associated with mucus layer on surface and between villi in jejunum of infant mice (C) 2 h and (A, B, D) 3 h after oral-intragastric challenge with *C. albicans*. The SEM preparations (A, B, D) show yeast cells partially embedded in mucus layer (arrowheads). The (C) TEM preparation reveals clear regions in the mucus sheet adjacent to yeast cells. MV, microvilli. (Figure 2 A to D; magnification × 470, × 2100, × 3000, and × 2300, respectively. (From Cole, G. T., Seshan, K. R., Pope, L. M., and Yancey, R. J., *J. Med. Vet. Mycol.*, 26, 173, 1988. With permission.)

parabasal, intermediate, transitional, and superficial. These layers are difficult to recognize at the ultrastructural level because each layer has ill-defined limits and shows a gradual change in structure.[296]

The basal and parabasal layers are the most active proliferative compartments or germinal beds of the epithelium.[297] The basal layer (approximately 10 μm thick) has a single row of cuboidal cells overlying the basement membrane. The parabasal layer (approximately 14 μm thick) is made of several layers of polyhedral cells with distinct nuclei. The intermediate layer, which may be up to 100 μm thick, has about 10 rows of larger, flatter cells with nuclei. The transitional layer, approximately 80 μm in thickness, has 10 rows of cells which change gradually from an oval elongated type into a squamous type. Finally, the most superficial layer is composed of several layers of large, flat cells with pyknotic nuclei.[293,296]

Glycogen is abundant in all but the deepest epithelial layers throughout the menstrual cycle, but especially in the follicular and ovulatory periods.[293] No glandular elements have been identified in the normal human vagina.[298,299] In spite of this, vaginal epithelium is normally covered with a surface film of moisture similar to a mucus secretion.[300] In the nonmenstruating female, this is a mixture of fluids from a number of sources. Its major components are cervical fluid, exfoliated epithelial cells from the vaginal wall, vaginal fluid per se, with small amounts of secretion from Bartholin's glands. Some secretions from higher up the tract may reach the vagina via uterine fluid, oviductal fluid, follicular fluid at ovulation, and perhaps even peritoneal fluid. Finally, small amounts of urine may enter and contaminate the vagina.[300] Attempts to quantitate the volume of vaginal fluids report values of 0.76 ± 0.04 g (mean ± SD, $n = 113$).[301] These authors determined that the weight increased significantly 13 to 16 days before mensturation due to increases in cervical secretion around ovulation time.

Chemical characterization of vaginal fluid suggests that it contains carbohydrate from the epithelium glycogen, amino acids, aliphatic acids, and proteins.[302-307] (For review on the chemical nature of vaginal secretions, see Preti and Huggins[308] and Bonsall and Michael.[309]) In addition to these components, a special type of IgA is known to be present in the vaginal secretions.[310]

The pH of a healthy mature woman's vagina is usually acidic.[302,311,312] This acidity is supposedly produced mainly by a lactobacillus through fermentation of glycogen, which is deposited in the cells of the vaginal epithelium under the influence of estrogen, to aliphatic acids, primarily lactic acid.[300,313,314] It is also possible that glycogen is fermented to simpler carbohydrates by vaginal enzymes and then reduced by bacteria to lactic acid. Other possibilities are that glycogen is converted to lactic acid by enzymes alone or that bacteria other than lactobacillus are capable of fermenting carbohydrates. Changes in the acidity of the human vagina have been reported.[302] The vagina of the newborn has a pH of about 5.7. Lactobacillus cells appear about the fourth day and the pH falls to about 4.8. This is followed by a rise beginning on the eighth day to neutrality, and neutral values persist until the onset of puberty. Cyclic changes in acidity is known to occur with pH values varying between 4.0 and 5.0 depending on the stage of the menstrual cycle.[302] Lowest values have been found at ovulation and premenstrually.[315] This parallels the curves of estrogen excretion for the normal cycle. There is also a pH gradient, depending on site sampled in the vagina. Thus, high values are found near the vestibule, intermediate ones in the mid-vagina, and the lowest values near the anterior fornix.[315] During pregnancy the pH ranges between 3.8 and 4.4, while in the postmenopausal women neutral or alkaline values may be seen.[315]

1. Changes in Vaginal Physiology and Implications for Microorganisms

The physiological condition of the vagina changes dramatically at different ages, during the menstrual cycle, and pregnancy. Accompanying the physiological changes are variations in the epithelial mucosa, pH and hormonal secretions, which cause major alterations in the microorganisms residing in the vagina.[86,316]

Age-related conditions affecting the vagina are divided into four periods: (1) neonatal, (2) premenarchal, (3) menarchal, and (4) postmenarchal (Figure 10). At 2 to 3 weeks following delivery, the neonate has a high level of estrogen,[317] which is sufficient to cause anatomical and physiological changes in the newborn's vagina to such an extent that it resembles the vagina of a woman in the menarchal stage of life. During this transitional period the microorganisms present in the newborn are similar to those found in the mother. However, once the maternal-derived estrogens are metabolized and excreted, the vaginal epithelium exfoliates to a much thinner form of several layers of stratified epithelial cells. In addition, a marked decrease in glycogen content occurs. Consequently, a corresponding increase in the vaginal pH is observed until the pH becomes neutral or even alkaline (pH 6 to 8). Thus, the acidophilic microorganisms no longer have an advantage. This situation persists throughout the premenarchal period. With the onset of pubescence there is an increase in estrogens, and the vaginal mucosa begins to thicken, with glycogen-rich cells forming multilayered, stratified squamous epithelium. As expected, this results in a new environment which favors microorganisms which can best survive under these conditions. At pubescence the pH becomes acidic (pH range from 3.7 to 6.3) as a result of breakdown/fermentation of glycogen and production of aliphatic acids.[318]

In the postmenarchal period the vaginal histological and physiological changes results in a vaginal environment similar to premenarchal period. Thus, acidophilic organisms no longer predominate.

During pregnancy marked changes occur in the vagina due to increased vascularity and venous stasis, both resulting in copious secretion and a characteristic blue-violet color. The vaginal wall undergoes a series of modifications to prepare it for the distention incident to labor.[315] Extension of the whole vaginal surface, including extension within the individual cells takes place during pregnancy and the epithelial layer is greatly thickened. Microridges seem to be less per cell and form an extended course because of their elasticity.[319] Another important change is the glycogen content of cells which increases above the levels obtained during menarchal period. These conditions create an environment conducive to the growth of acidophilic organisms. Pregnancy, together with diabetes, are the only two factors which were strongly and reproducibly supported by scientific evidence as factors which favor the onset of vaginal thrush.[320]

Changes occurring during the menstrual cycle are controlled by the hypothalamus, which regulates the menstrual cycle through the hormones of the anterior pituitary and the ovaries.[298] Rhythmic changes in the types and levels of hormones excreted accompany the menstrual cycle and early pregnancy. Also variation in the pH with the stages of the menstrual cycle is known to occur with a higher pH when menstrual flow is greatest.[321] In addition, lowest values were reported at ovulation and premenstrually. This parallels the curves of estrogen excretion for the normal cycle.[315] The changing cellular content of the vaginal tract may have relevance to the observed cyclic changes in the microbial population of the vagina.[81]

2. Vaginal Factors Studied in Relation to Adherence

Workers dealing with *Candida* adherence and its relation to the vaginal physiology directed their efforts towards investigating whether a correlation exists between certain hormonal states, pregnancy, environmental conditions (such as pH), menstrual cycle, type of epithelial cells, and the adherence of *Candida*.[61-64,97]

The possible role of hormones on the adhesion of *C. albicans* to vaginal epithelial cells was investigated in some detail by Botta.[64] He showed that adhesion of *C. albicans*, group B streptococci and *E. coli* was greatly enhanced when incubation mixtures were prepared with vaginal and uroepithelial cells collected during the third week of the menstrual cycle. The receptiveness of these epithelial cells was significantly affected by changes during the menstrual cycle.[64] Segal et al.[61] investigating the adherence of *C. albicans* through the 4 weeks of the menstrual cycle arrived at different results. These authors showed that *C. albicans* adhered in

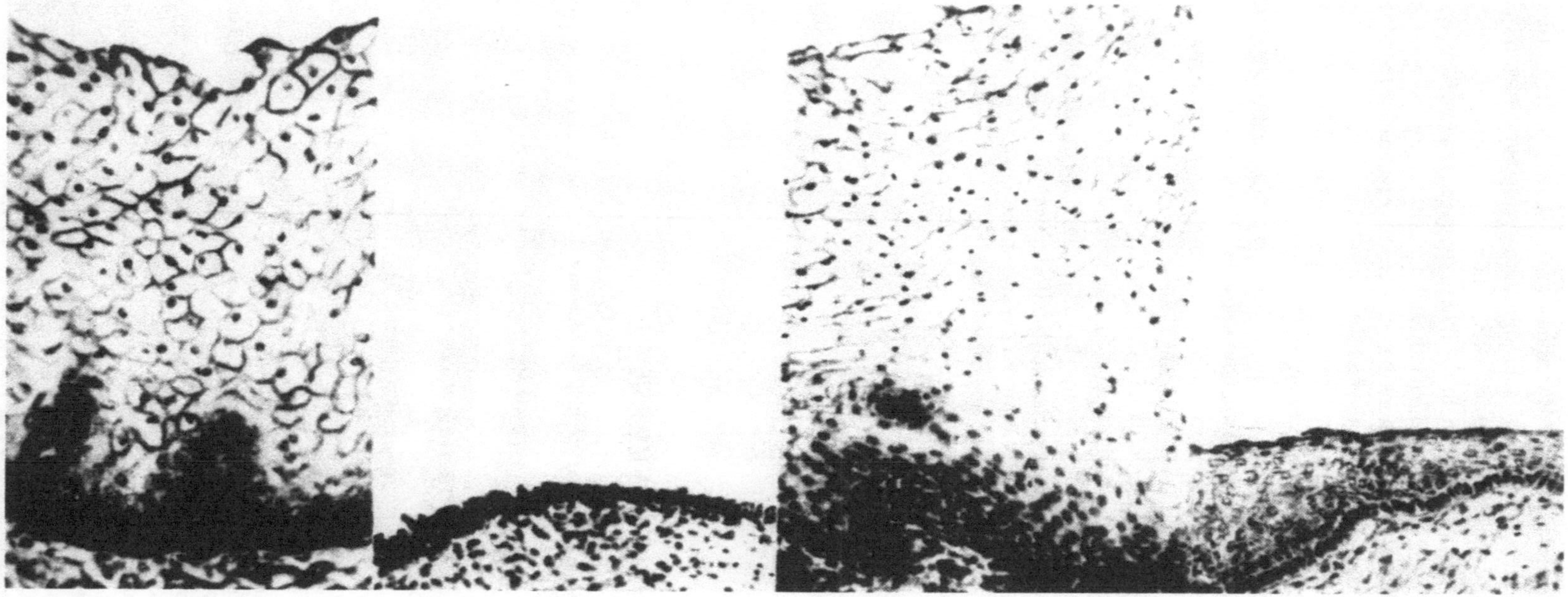

FIGURE 10. Tissue sections of the vaginal epithelium during the life cycle of a woman. From left to right: newborn, child, adult woman, menopausal woman. (From Schmidt, E. H. and Beller, F.K., Biochemistry of the vagina, in *Human Reproductive Medicine, the Human Vagina*, Hafez, E. S. E. and Evans, T. N., Eds., Vol. 2, Elsrier North-Holland Biomedical Press, Amsterdam.

higher numbers to vaginal cells collected during week one and four of the cycle.[61] In the same study Segal et al.[61] showed that various groups of women known to have a high predisposition to vaginitis, and diabetic or pregnant women, had significantly higher values of vaginal epithelial adhesiveness than did the control group. Adherence of cells from postmenarchal women was lower than cells from fertile women. In all of these situations, an inverse correlation between adhesion and Karyopyknotic Index (which indicates a higher proportion of intermediate cells) was observed. Thus, hormone fluctuations may play an important role in the adherence process. If this is the case one might ask what is the molecular mechanism that is responsible for this? Botta[64] suggested that hormones influence this process by acting specifically on cell receptiveness. Evidence for direct interaction between *C. albicans* and a number of steroid hormones have been reviewed by Ryley.[97] *C. albicans* appears to possess both components of a hormone-receptor system, i.e., a ligand and a binding protein. Whether a direct correlation exists between this hormone-receptor system in *Candida* and adherence remains to be investigated.

Persi et al.[62] studied the effect of carbon dioxide and pH, which are present in the vagina *in vivo*, on adhesion of *C. albicans*. These authors showed that adhesion of *C. albicans* to VEC surface was affected by pH and CO_2 levels present in the adhesion assay medium. In the same study, Persi et al.[62] demonstrated that the presence of microridges on the surface of the VEC appears to be necessary for adhesion of *C. albicans*. Whether the change in vaginal cells microridges that occurs during pregnancy (see above) is related to the adhesion process, and consequently the higher incidence in pregnant women, has not been investigated.

REFERENCES

1. **Ellen, R. P. and Gibbons, R. J.,** Parameters affecting the adherence and tissue tropisms of *Streptococcus pyogenes*, *Infect. Immun.*, 9, 85, 1974.
2. **Fletcher, M.,** The effects of culture concentration and age, time, and temperature on bacterial attachment to polystyrene, *Can. J. Microbiol.*, 23, 1, 1977.
3. **Daniels, S. L.,** The adsorption of microorganisms onto surfaces: a review, *Dev. Ind. Microbiol.*, 13, 211, 1972.
4. **Douglas, L. J.,** Adhesion to surfaces, in *The Yeasts*, Rose, A. H. and Harrison, J. S., Eds., Vol. 2, 2nd ed., Academic Press, London, 1987, 239.
5. **Mariat, F.,** Saprophytic and parasitic morphology of pathogenic fungi, *Symp. Soc. Gen. Microbiol.*,14, 85, 1964.
6. **Soll, D. R.,** *Candida albicans*, in *Fungal Dimorphism, with Emphasis on Fungi Pathogenic for Humans*, Szaniszlo, P. J., Ed., Plenum, New York, 1985, chap. 7.
7. **McMurrough, I. and Rose, A. H.,** Effect of growth rate and substrate limitation on the composition and structure of the cell wall of *Saccharomyces cerevisiae*, *Biochem. J.*, 105, 189, 1967.
8. **Ramsay, A. M. and Douglas, L. J.,** Effects of phosphate limitation of growth on the cell wall and lipid composition of *Saccharomyces cerevisiae*, *J. Gen. Microbiol.*, 110, 185, 1979.
9. **Samaranayake, L. P., McCourtie, J., and MacFarlane, T. W.,** Factors affecting the *in vitro* adherence of *Candida albicans* to acrylic surfaces, *Arch. Oral Biol.*, 25, 611, 1980.
10. **Samaranayake, L. P. and MacFarlane, T. W.,** The adhesion of the yeast *Candida albicans* to epithelial cells of human origin *in vitro*, *Arch. Oral Biol.*, 26, 815, 1981.
11. **Samaranayake, L. P. and MacFarlane, T. W.,** An *in vitro* study of the adherence of *Candida albicans* to acrylic surfaces, *Arch. Oral Biol.*, 25, 603, 1980.
12. **Douglas, L. J., Houston, J. G., and McCourtie, J.,** Adherence of *Candida albicans* to human buccal epithelial cells after growth on different carbon sources, *FEMS Microbiol. Lett.*, 12, 241, 1981.

13. **McCourtie, J. and Douglas, L. J.,** Relationship between cell surface compostion of *Candida albicans* and adherence to acrylic after growth on different carbon sources, *Infect. Immun.*, 32, 1234, 1981.
14. **Shuttleworth, C. W. and Gibbs, F. J.,** The aetiological significance of *Candida albicans* in chronic angular cheilitis and its treatment with nystatin, *Br. Dent. J.*, 108, 354, 1960.
15. **Neill, D. J.,** Symposium on denture sore mouth. I. An aetiological review, *Dent. Pract. (Bristol)*, 16, 135, 1965.
16. **Ritchie, G. M., Fletcher, A. M., Main, D. M. G., and Prophet, A. S.,** Etiology, exfoliative cytology and treatment of denture stomatitis, *J. Prosth. Dent.*, 22, 185, 1969.
17. **Gentles, J. C. and LaTouche, C. J.,** Yeasts as human and animal pathogens, in *The Yeasts,* Rose, A. H. and Harrison, J. S., Eds., Vol. 1, 1st ed., Academic Press, London, 1969, 108.
18. **Kaufman, R. H. and Gardner, H. L.,** *Benign Diseases of the Vulva and Vagina,* C. V. Mosby, St. Louis, 1969, 149.
19. **Van Reene, J. F.,** Microbiologic studies on denture stomatitis, *J. Prosth. Dent.*, 30, 493, 1973.
20. **Beachey, E. H.,** *Bacterial Adherence, Receptors and Recognition,* Series B, Vol. 6, Chapman and Hall, London, 1980.
21. **Tronchin, G., Poulain, D., and Vernes, A.,** Cytochemical and ultrastructural studies of *Candida albicans*. III. Evidence for modifications of the cell wall coat during adherence to human buccal epithelial cells, *Arch. Microbiol.*, 139, 221, 1984.
22. **McCourtie, J. and Douglas, L. J.,** Relationship between cell surface composition, adherence and virulence of *Candida albicans*, *Infect. Immun.*, 45, 6, 1984.
23. **Marrie, T. J. and Costerton, J. W.,** The ultrastructure of *Candida albicans* infections, *Can. J. Microbiol.*, 27, 1156, 1981.
24. **Mohamed, A.-M. H.,** Ultrastructural aspects of chronic oral candidosis, *J. Oral Pathol.*, 4, 180, 1975.
25. **Montes, L. F. and Wilborn, W. H.,** Ultrastructural features of host-parasite relationship in oral candidosis, *J. Bacteriol.*, 96, 1349, 1968.
26. **McCourtie, J. and Douglas, L. J.,** Extracellular polymer of *Candida albicans:* isolation, analysis and role in adhesion, *J. Gen. Microbiol.*, 131, 495, 1985.
27. **Critchley, I. A. and Douglas, L. J.,** Isolation and partial characterization of an adhesin from *Candida albicans*, *J. Gen. Microbiol.*, 133, 629, 1987.
28. **King, R. D., Lee, J. C., and Morris, A.,** Adherence of *Candida albicans* and other *Candida* species to mucosal epithelial cells, *Infect. Immun.*, 27, 667, 1980.
29. **Segal, E., Lehrer, N., and Ofek, I.,** Adherence of *Candida albicans* to human vaginal epithelial cells: inhibition by amino sugars, *Exp. Cell Biol.*, 50, 13, 1982.
30. **Ghannoum, M. A. and Abu Elteen, K.,** Effect of growth of *Candida* spp. in the presence of various glucocorticoids on the adherence to human buccal epithelial cells, *Mycopathologia*, 98, 171, 1987.
31. **Harris, R. H. and Mitchell, R.,** The role of polymers in microbial aggregation, *Annu. Rev. Microbiol.*, 27, 27, 1973.
32. **Hazen, K. C., Plotkin, B. J., and Klimas, D. M.,** Influence of growth conditions on cell surface hydrophobicity of *Candida albicans* and *Candida glabrata*, *Infect. Immun.*, 54, 269, 1986.
33. **Lee, J. C. and King, R. D.,** Characterization of *Candida albicans* adherence to human vaginal epithelial cells *in vitro*, *Infect. Immun.*, 41, 1024, 1983.
34. **Kimura, L. H. and Pearsall, N. N.,** Adherence of *Candida albicans* to human buccal epithelial cells, *Infect. Immun.*, 21, 64, 1978.
35. **Maisch, P. A. and Calderone, R. A.,** Adherence of *Candida albicans* to a fibrin-platelet matrix formed *in vitro*, *Infect. Immun.*, 27, 650, 1980.
36. **Klotz, S. A., Drutz, D. J., Harrison, J. L., and Huppert, M.,** Adherence and penetration of vascular endothelium by *Candida* yeasts, *Infect. Immun.*, 42, 374, 1983.
37. **Cassone, A., Mattia, E., and Boldrini, L.,** Agglutination of blastospores of *Candida albicans* by conconavalin A and its relationship with the distribution of mannan polymers and the ultrastructure of the cell wall, *J. Gen. Microbiol.*, 105, 263, 1978.
38. **Floodgate, G. D.,** Factors affecting the settlement of a marine bacterium, Veroeff, *Inst. Meeresforsch. Bremerh-Aven*, 2, 265, 1966.
39. **Mardh, P. A. and Westrom, L.,** Adherence of bacteria to vaginal epithelial cells, *Infect. Immun.*, 13, 661, 1976.
40. **Klotz, S. A. and Penn, R. L.,** Multiple mechanisms may contribute to the adherence of *Candida* yeasts to living cells, *Curr. Microbiol.*, 16, 119, 1987.
41. **Rosenberg, M.,** Bacterial adherence to hydrocarbons: a useful technique for studying cell surface hydrophobicity, *FEMS Microbiol. Lett.*, 22, 289, 1984.
42. **Smyth, C. J., Jonsson, P., Olsson, E., Soderlind, O., Rosengren, J., Hjerten, S., and Wadstrom, T.,** Differences in hydrophobic surface characteristics of porcine enteropathogenic *Escherichia coli* with or without K88 antigen, as revealed by hydrophobic interaction chromatography, *Infect. Immun.*, 22, 462, 1978.
43. **Sobel, J. D. and Obedeanu, N.,** Role of hydrophobicity in adherence of Gram-negative bacteria to epithelial cells, *Ann. Clin. Lab. Sci.*, 15, 216, 1984.

44. **Tylewska, S.,** The effect of culture conditions on the surface hydrophobic properties and adherence of Streptococci, *Med. Dosw. Mikrobiol.,* 35, 175, 1983.
45. **Macura, A. B.,** Hydrophobicity of *Candida albicans* related o their adherence to mucosal epithelial cells, *Zbl. Bakt. Hyg. A.,* 266, 491, 1987.
46. **Minagi, S., Miyake, Y., Inagaki, K. Tsuru, H., and Suginaka, H.,** Hydrophobic interaction in *Candida albicans* and *Candida tropicalis* adherence to various denture base resin materials, *Infect. Immun.,* 47, 11, 1985.
47. **Klotz, S. A., Drutz, D. J., and Zagic, J. E.,** Factors governing adherence of *Candida* species to plastic surfaces, *Infect. Immun.,* 50, 97, 1985.
48. **Van Oss, C. J., Absolom, D. R., and Neumann, A. W.,** The hydrophobic effect essentially a Van der Waals interaction, *Colloid Polymer Sci.,* 258, 424, 1980.
49. **Ofek, I. and Beachey, E. H.,** General concepts and principles of bacterial adherence, in *Bacterial Adherence, Receptors and Recognition,* Series B, Vol. 6, Beachey, E. H., Ed., Chapman and Hall, London, 1980, 1.
50. **Absolom, D. R., Van Oss, C. J., Genco, R. J., Francis, D. W., and Neumann, A. W.,** Surface thermodynamics of normal and pathological human granulocytes, *Cell Biophys.,* 2, 113, 1980.
51. **Andrade, J. D.,** Interfacial phenomena and biomaterials, *Med. Instrum.,* 7, 110, 1973.
52. **Neumann, A. W., Absolom, D. R., Van Oss, C. J., and Zingg, W.,** Surface thermodynamics of leukocyte and platelet adhesion to polymer surfaces, *Cell Biophys.,* 1, 79, 1979.
53. **Kennedy, M. J., Volz, P. A., Edwards, C. A., and Yancey, R. J.,** Mechanisms of association of *Candida albicans* with intestinal mucosa, *J. Med. Microbiol.,* 24, 333, 1987.
54. **Ofek, I., Lis, H., and Sharon, N.,** Animal cell surface membranes, in *Bacterial Adhesion, Mechanisms and Physiological Significance,* Savage, D. C. and Fletcher, M., Eds., Plenum, New York, 1985, 71.
55. **Kimura, L. H. and Pearsall, N. N.,** Relationship between germination of *Candida albicans* and increased adherence to human buccal epithelial cells, *Infect. Immun.,* 28, 464, 1980.
56. **Sklavounou, A. and Germaine, G. R.,** Adherence of oral streptococci to keratinized and non-keratinized human oral epithelial cells, *Infect. Immun.,* 27, 686, 1980.
57. **Davidson, S., Brish, M., and Rubinstein, E.,** Adherence of *Candida albicans* to buccal epithelial cells of neonates, *Mycopathologia,* 85, 171, 1984.
58. **Cox, F.,** Adherence of *Candida albicans* to buccal epithelial cells in children and adults, *J. Lab. Clin. Med.,* 102, 960, 1983.
59. **Cox, F.,** *Candida albicans* adherence in newborn infants, *J. Med. Vet. Mycol.,* 24, 121, 1986.
60. **Botta, G. A.,** Hormonal and type-dependent adhesion of group B streptococci to human vaginal cells, *Infect. Immun.,* 25, 1084, 1979.
61. **Segal, E., Soroka, A., and Schechter, A.,** Correlative relationship between adherence of *Candida albicans* to human vaginal epithelial cells *in vitro* and candidal vaginitis, *Sabouraudia,* 22, 191, 1984.
62. **Persi, M. A., Burnham, J. C., and Duhring, J. L.,** Effect of carbon dioxide and pH on adhesion of *Candida albicans* to vaginal epithelial cells, *Infect. Immun.,* 50, 82, 1985.
63. **Sobel, J. D., Myers, P., Levison, M. E., and Kaye, D.,** Comparison of bacterial and fungal adherence to vaginal exfoliated epithelial cells and human vaginal epithelial tissue culture cells, *Infect. Immun.,* 35, 697, 1982.
64. **Botta, G. A.,** Possible role of hormones in the observed changes in adhesion of several microorganisms to epithelial cells from different body sites, *FEMS Microbiol. Lett.,* 11, 69, 1981.
65. **Centeno, A., Davis, C. P., Cohen, M. S., and Warren, M. M.,** Modulation of *Candida albicans* attachment to human epithelial cells by bacteria and carbohydrates, *Infect. Immun.,* 39, 1354, 1983.
66. **Ray, T. L., Digre, K. B., and Payne, C. D.,** Adherence of *Candida* species to human epidermal corneocytes and buccal mucosal cells: correlation with cutaneous pathogenicity, *J. Invest. Dermatol.,* 83, 37, 1984.
67. **Collins-Lech, C., Kalbfleisch, J. H., Franson, T. R., and Sohnle, P. G.,** Inhibition by sugars of *Candida albicans* adherence to human buccal mucosal cells and corneocytes *in vitro, Infect. Immun.,* 46, 831, 1984.
68. **Pope, L. M. and Cole, G. T.,** SEM studies of adherence of *Candida albicans* to the gastrointestinal tract of infant mice, *Scanning Electron Micros.,* 3, 73, 1981.
69. **Segal, E. and Savage, D. C.,** Adhesion of *Candida albicans* to mouse intestinal mucosa *in vitro:* development of the assay and test of inhibitors, *J. Med. Vet. Mycol.,* 24, 477, 1986.
70. **Makrides, H. C. and MacFarlane, T. W.,** An investigation of the factors involved in increased adherence of *C. albicans* to epithelial cells mediated by *E. coli, Microbios,* 38, 177, 1983.
71. **Makrides, H. C. and MacFarlane, T. W.,** Effects of commensal bacteria on the adherence of *Candida albicans* to epithelial cells *in vitro, Microbios Lett.,* 21, 55, 1982.
72. **Farrell, S. M., Hawkins, D. F., and Ryder, T. A.,** Scanning electron microscopy study of *Candida albicans* invasion of cultured human cervical epithelial cells, *Sabouraudia,* 21, 251, 1983.
73. **Klotz, S. A.,** The adherence of *Candida* yeasts to human and bovine vascular endothelium and subendothelial extracellular matrix, *FEMS Microbiol. Lett.,* 48, 201, 1987.
74. **Rotrosen, D., Edwards, J. E., Jr., Gibson, T. R., Moore, J. C., Cohen, A. H., and Green, I.,** Adherence of *Candida* to cultured vascular endothelial cells: mechanisms of attachment and endothelial cell penetration, *J. Infect. Dis.,* 152, 1264, 1985.

75. **Odds, F. C.,** *Candida and Candidosis,* University Park Press, Baltimore, 1979.
76. **Alvares, O. F. and Meyer, J.,** Variable and regional differences in oral epithelium, in *Current Concepts of the Histology of Oral Mucosa,* Squier, C. A. and Meyer, J., Ed., Charles C. Thomas, Springfield, IL, 1971, 80.
77. **Feldman, R. S. and Szabo, G.,** Comparative microanatomy of skin and oral mucosa, in *Textbook of Oral Biology,* Shaw, H. J., Sweeny, E. A., Cappuccino, C. C., and Meller, S. M., Eds., W.B. Saunders, Philadelphia, 1978, 143.
78. **Gibbons, R. J. and van Houte, J.,** Bacterial adherence in oral microbial ecology, *Ann. Rev. Microbiol.,* 29, 19, 1975.
79. **Squier, C. A., Johnson, N. W., and Hackemann, M.,** Structure and function of normal human oral mucosa, in *Oral Mucosa in Health and Disease,* Dolby, A. E., Ed., Blackwell Scientific, Oxford, 1975, 1.
80. **Papanicolaou, G. N.,** The sexual cycle in human female as revealed by vaginal smears, *Am. J. Anat.,* 52, 519, 1933.
81. **Larsen, B., Markovetz, A. J., and Galask, R. P.,** Relationship of vaginal cytology to alterations of the vaginal microflora of rats during the estrous cycle, *Appl. Environ. Microbiol.,* 33, 556, 1977.
82. **Tobgi, R. S., Samaranayake, L. P., and MacFarlane, T. W.,** Adhesion of *Candida albicans* to buccal epithelial cells exposed to chlorhexidine gluconate, *J. Med. Vet. Mycol.,* 25, 335, 1987.
83. **Hayward, A. F.,** Electron microscopic observations on cell coat and membrane-coating granules of the epithelium of the hard and soft palate in the rat, *Arch. Oral Biol.,* 18, 67, 1973.
84. **Hayward, A. F. and Hackemann, M.,** Electron microscopy of membrane-coating granules and a cell surface coat of keratinized and non-keratinized oral epithelium, *J. Ultrastruct. Res.,* 43, 205, 1973.
85. **Sobel, J. D., Schneider, J., Kaye, D., and Levison, M. E.,** Adherence of bacterial to vaginal epithelial cells at various times in the menstrual cycle, *Infect. Immun.,* 32, 194, 1981.
86. **Larsen, B., Markovetz, A. J., and Galask, R. P.,** Spatial relationship of the genital microflora to the vaginal epithelium of female rats: transmission electron microscopy, *Appl. Environ. Microbiol.,* 35, 444, 1978.
87. **Larsen, B., Markovetz, A. J., and Galask, R. P.,** Spatial relationship of the vaginal microflora to the vaginal epithelium of female rats: scanning electron microscopy, *Appl. Environ. Microbiol.,* 34, 80, 1977.
88. **Barlett, J. G., Onderdonk, A. B., Drude, E., Goldstein, C., Anderka, M., Alpert, S., and McCormack, W. M.,** Quantitative bacteriology of the vaginal flora, *J. Infect. Dis.,* 136, 271, 1977.
89. **Goplerud, C. P., Ohm, M. J., and Galask, R. P.,** Aerobic and anaerobic flora of the cervix during pregnancy and the puerperium, *Am. J. Obstet. Gynecol.,* 126, 858, 1976.
90. **Levison, M. E., Corman, L. C., Carrington, E. R., and Kaye, D.,** Quantitative microflora of the vagina, *Am. J. Obstet. Gynecol.,* 127, 80, 1977.
91. **Saigh, J. H., Sanders, C. C., and Sanders, W. E., Jr.,** Inhibition of *Neisseria gonorrhoeae* by aerobic and facultative anaerobic components of the endocervical flora: evidence for a protective effect against infection, *Infect. Immun.,* 19, 704, 1978.
92. **Thadepalli, H., Chan, W. H., Maidman, J. E., and Davidson, E. C., Jr.,** Microflora of the cervix during normal labour and puerperium, *J. Infect. Dis.,* 137, 568, 1978.
93. **Gold, J. J.,** *Textbook of Gynaecologic Endocrinology,* Harper and Row, New York, 1968, 133.
94. **Sobel, J. D., Myers, P. G., Kaye, D., and Levison, M. E.,** Adherence of *Candida albicans* to human vaginal and buccal epithelial cells, *J. Infect. Dis.,* 143, 76, 1981.
95. **Tomsikova, A., Kotal, L., and Novackova, D.,** The role of adherence of *Candida* species in the pathogenesis of vaginal and oral candidiasis, *Mykosen,* 30, 74, 1987.
96. **Macura, A. B., Pawlik, B., and Wita, B.,** *Candida* adherence to mucosal epithelial cells with regard to its pathogenicity, *Zbt. Bakt. Hyg. I. Abt. Orig. A,* 254, 561, 1983.
97. **Ryley, J. F.,** Pathogenicity of *Candida albicans* with particular reference to the vagina, *J. Med. Vet. Mycol.,* 24, 5, 1986.
98. **Costerton, S. W., Geesey, G. G., and Cheng, K. J.,** How bacteria stick, *Sci. Am.,* 238, 86, 1978.
99. **Reed, W. P. and Williams, R. C.,** Bacterial adherence — 1st step in pathogenesis of certain infections, *J. Chron. Dis.,* 31, 67, 1978.
100. **Schoenbaum, S. C., Gardner, P., and Shillito, J.,** Infections of cerebrospinal fluid shunts: epidemiology, clinical manifestations, and therapy, *J. Infect. Dis.,* 131, 543, 1975.
101. **Gould, K., Ramirez-Ronda, C. H., Holmes, R. K., and Sanford, J. P.,** Adherence of bacteria to heart valves *in vitro, J. Clin. Invest.,* 56, 1364, 1975.
102. **Rotrosen, D., Gibson, T. R., and Edwards, J. E.,** Adherence of *Candida* species to intravenous catheters, *J. Infect. Dis.,* 147, 599, 1983.
103. **Davenport, J. C.,** The oral distribution of *Candida* in denture stomatitis, *Br. Dent. J.,* 129, 151, 1970.
104. **Budtz-Jörgensen, E.,** The significance of *Candida albicans* in denture stomatitis, *Scand. J. Dent. Res.,* 82, 151, 1974.
105. **Olsen, I.,** Denture stomatitis, occurence and distribution of fungi, *Acta Odont. Scand.,* 32, 329, 1974.
106. **Corpe, W. A.,** Attachment of marine bacteria to solid surfaces, in *Adhesion in Biological Systems,* Manly, R. S., Ed., Academic Press, New York, 1970, 73.

107. **Fletcher, M. and Loeb, G. I.,** Influence of substratum characteristics on the attachment of a marine pseudomonad to solid surfaces, *Appl. Environ. Microbiol.*, 37, 67, 1979.
108. **Marshall, K. C., Stout, R., and Mitchell, R.,** Mechanism of the initial events in the sorption of marine bacteria to surfaces, *J. Gen. Microbiol.*, 68, 337, 1971.
109. **Gerson, D. F. and Scheer, D.,** Cell surface energy, contact angles and phase partition. III. Adhesion of bacterial cells to hydrophobic surfaces, *Biochem. Biophys. Acta*, 602, 506, 1980.
110. **Loeb, G. I.,** The properties of nonbiological surfaces and their characterization, in *Bacterial Adherence, Mechanisms and Physiological Significance*, Savage, D. C. and Fletcher, M., Eds., Plenum, New York, 1985, chap. 5.
111. **Dexter, S. C., Sullivan, J. D., Jr., Williams, J., III, and Watson, S. W.,** Influence of substrate wettability on the attachment of marine bacteria to various surfaces, *Appl. Microbiol.*, 30, 298, 1975.
112. **Gerson, D. F. and Zajic, J. E.,** The biophysics of cellular adhesion, in *Immobilized Microbial Cells*, Venkatsubramanian, K., Ed., American Chemical Society, Washington, D.C., 1979, 29.
113. **Daniels, S. L.,** The adsorption of microorganisms onto solid surfaces: a review, *Dev. Ind. Microbiol.*, 13, 211, 1972.
114. **Davenport, J. C.,** The denture surface, *Br. Dent. J.*, 133, 101, 1972.
115. **Fletcher, M.,** Adherence of marine micro-organisms to smooth surfaces, in *Bacterial Adherence, Receptors and Recognition*, Series B, Vol. 6, Beachey, E. H., Ed., Chapman and Hall, London, 1980, 345.
116. **Marshall, K. C.,** *Interfaces in Microbial Ecology*, Harvard University Press, Cambridge, MA, 1976.
117. **Fletcher, M. and Loeb, G. I.,** The influence of substratum surface properties on the attachment of a marine bacterium, in *Colloid and Interface Science*, Vol. III, Kerker, M., Ed., Academic Press, New York, 1976, 459.
118. **Zisman, W. A.,** Relation of the equilibrium contact angle to liquid and solid constitution, in *Contact Angle, Wettability and Adhesion*, Advances in Chemistry, Series 43, American Chemical Society, Washington, D.C., 1964.
119. **Zisman, W. A.,** Surface energetics of wetting, spreading and adhesion, *J. Paint Technol.*, 44, 41, 1972.
120. **Baier, R. E.,** Surface properties influencing biological adhesion, in *Adhesion in Biological Systems*, Manly, R. S., Ed., Academic Press, New York, 1970, 15.
121. **Baier, R. E.,** Influence of the initial surface condition of materials on bioadhesion, in *Proc. 3rd Int. Cong. Marine Corrosion Fouling*, National Bureau of Standards, Gaithersburg, MD, 1972, 633.
122. **Zierdt, C. H.,** Adherence of bacteria, yeast, blood cells and latex spheres to large-porosity membrane filters, *Appl. Environ. Microbiol.*, 38, 1166, 1979.
123. **Marsh, P.,** *Oral Microbiology*, Thomas Nelson and Sons, London, 1980.
124. **Hardie, J. M. and Bowden, G. H.,** The normal microbial flora of the mouth, in *The Normal Microbial Flora of Man*, Skinner, F. A. and Carr, J. G., Eds., Academic Press, London, 1974, 47.
125. **Berkowitz, R. J., Jordan, H. V., and White, G.,** Early establishment of *Streptococcus mutans* in mouth of infants, *Arch. Oral Biol.*, 20, 171, 1975.
126. **Carlsson, J.,** Nutritional requirement of *Streptococcus sanguis*, *Arch. Oral Biol.*, 17, 1327, 1972.
127. **Bailit, H. L., Baldwin, D. C., and Hunt, E. E.,** The increasing prevalence of gingival *Bacteriodes melaninogenicus* with age in children, *Arch. Oral Biol.*, 9, 435, 1964.
128. **de Araujo, W. C. and MacDonald, J. B.,** The gingival cervice microbiota in five pre school children, *Arch. Oral Biol.*, 9, 227, 1964.
129. **Kelstrup, J.,** Incidence of *Bacteroides melaninogenicus* in human gingival sulci and its prevalence in oral cavity at different ages, *Periodontics*, 4, 14, 1966.
130. **Carlsson, J. and Sundstrom, B.,** Variation in composition of early dental plaque following ingestion of sucrose and glucose, *Odontol. Revy*, 19, 161, 1968.
131. **Bowen, W. H.,** Effects of foods on oral bacterial populations in man and animals, *J. Dent. Res.*, 49, 1276, 1970.
132. **Carlsson, J. and Egelberg, J.,** Effect of diet on early plaque formation in man, *Odontol. Revy*, 16, 112, 1965.
133. **Littleton, N. W., McCabe, R. M., and Carter, C. H.,** Studies of oral health on persons nourished by stomach tube. II. Acidogenic properties and selected bacterial components of plaque material, *Arch. Oral Biol.*, 12, 601, 1967.
134. **Handelman, S. L. and Hawes, R. R.,** The effect of long-term systemic antibiotic therapy on the antibiotic resistance of the salivary flora, *J. Oral Ther.*, 1, 23, 1964.
135. **Handelman, S. L. and Hawes, R. R.,** The effect of long-term antibiotic administration on the numbers of salivary organisms, *Arch. Oral Biol.*, 10, 353, 1965.
136. **Lobene, R. L., Brion, M., and Socransky, S. S.,** Effect of erythromycin on dental plaque and plaque-forming microorganisms of man, *J. Periodont.*, 40, 287, 1969.
137. **Bibby, B. G.,** Antibiotics and dental caries, in *Dietary Chemicals vs. Dental Caries*, Advances in Chemistry, series 94, American Chemical Society, Washington, D.C., 1970.
138. **Chen, T. Y. and Webster, J. H.,** Oral monilia study on patients with head and neck cancer during radiotherapy, *Cancer*, 34, 246, 1974.

139. **Llory, H., Dammron, A., Gioanni, M., and Frank, R. M.,** Some population changes in oral anaerobic microorganisms, *Streptococcus mutans* and yeasts following irradiation of the salivary glands, *Caries Res.*, 6, 298, 1972.
140. **Shklair, I. L., Englander, H. R., Stein, L. M., and Kesel, R. G.,** Preliminary report on the effect of complete mouth rehabilitation on oral lactobacilli counts, *J. Am. Dent. Assoc.*, 53, 155, 1956.
141. **Kesel, R. G., Shklair, I. L., Green, G. H., and Englander, H. R.,** Further studies on lactobacilli counts after elimination of carious lesions, *J. Dent. Res.*, 37, 50, 1958.
142. **Shklair, I. L. and Mazzarella, M. A.,** Effects of full-mouth extraction on oral microbiota, *Dent. Prog.*, 1, 275, 1961.
143. **Sakamaki, S. T. and Bahn, A. N.,** Effect of orthodontic banding on localized oral lactobacilli, *J. Dent. Res.*, 47, 275, 1968.
144. **Sicher, H. and Bhaskar, S. R.,** *Orban's Oral Histology and Embryology*, 7th ed., C.V. Mosby, St. Louis, 1972, 269.
145. **Dawes, C.,** Some characteristics of parotid and submandibular salivary proteins, *Arch. Oral Biol.*, 10, 269, 1965.
146. **Dawes, C. and Wood, C. M.,** The composition of human lip mucous gland secretions, *Arch. Oral Biol.*, 18, 343, 1973.
147. **Gron, P.,** The state of calcium and inorganic orthophosphate in human saliva, *Arch. Oral Biol.*, 18, 1365, 1973.
148. **Becks, H. and Wainwright, W. W.,** Human saliva. XIII. Rate of flow of resting saliva of healthy individuals, *J. Dent. Res.*, 22, 391, 1943.
149. **Rutter, P.,** The accumulation of organisms on the teeth, in *Adhesion of Microorganisms to Surfaces*, Ellwood, D. C., Melling, J., and Rutter, P., Eds., Academic Press, London, 1979, 139.
150. **Meckel, A.,** The formation and properties of organic films on teeth, *Arch. Oral Biol.*, 10, 585, 1965.
151. **Dreyfus, P. M., Levy, H. L., and Efron, M. L.,** Concerning amino acids in saliva, *Experimenta*, 24, 447, 1968.
152. **Kauffman, S. L., Kasai, G. S., and Koser, S. A.,** The amounts of folic acid and vitamin B_6 in saliva, *J. Dent. Res.*, 32, 840, 1953.
153. **Prout, R. E.,** Human saliva lipid, composition and its possible relation to dental calculus formation, *J. Dent. Res.*, 55, 134, 1976.
154. **MacFarlane, T. W. and Mason, D. K.,** The physiological responsiveness of the oral mucosa: the role of saliva, in *Oral Mucosa in Health and Disease*, Dolby, A. E., Ed., Blackwell Scientific, Oxford, 1975, 113.
155. **Afonsky, D.,** *Saliva and its Relation to Oral Health*, University of Alabama Press, Birmingham, 1961.
156. **Burgen, A. S. V. and Emmelin, N. G.,** *Physiology of the Salivary Gland*, Edward Arnold, London, 1961, 140.
157. **Jenkins, G. N.,** *The Physiology of the Mouth*, 3rd ed., Blackwell Scientific, Oxford, 1966, 305.
158. **Dawes, C.,** Effect of diet on salivary secretion and composition, *J. Dent. Res.*, 49, 1263, 1970.
159. **Ørstavik, D., Kraus, F. W., and Henshaw, L. C.,** *In vitro* attachment of streptococci to the tooth surface, *Infect. Immun.*, 9, 794, 1974.
160. **Gibbons, R. J. and van Houte, J.,** Bacterial adherence and the formation of dental plaques, in *Bacterial Adherence, Receptors and Recognition*, Series B, Vol. 6, Beachey, E. H., Ed., Chapman and Hall, London, 1980, chap. 3.
161. **Gibbons, R. J.,** Adherence of the oral flora, in *Attachment of Organisms to the Gut Mucosa*, Vol. 1, Boedeker, E. C., Ed., CRC Press, Boca Raton, FL, 1984, 11.
162. **Ørstavik, D.,** Sorption of streptococci to glass: effects of macromolecular solutes, *Acta Pathol. Microbiol. Scand. B*, 85, 47, 1977.
163. **Clark, W. B. and Gibbons, R. J.,** Influence of salivary components and extracellular polysaccharide synthesis from sucrose on the attachment of *Streptococcus mutans* 6715 to hydroxyapatite surfaces, *Infect. Immun.*, 18, 514, 1977.
164. **Ørstavik, J. and Ørstavik, D.,** An *in vitro* model for studying adsorption of oral streptococci to crown and bridge cements, *J. Oral Rehabil.*, 7, 43, 1980.
165. **Slots, J. and Gibbons, R. J.,** Attachment of *Bacteroids melaninogenicus* subsp. *asaccharolyticus* to oral surfaces and its possible role in colonization of the mouth and of periodontal pockets, *Infect. Immun.*, 19, 254, 1978.
166. **Samaranayake, L. P. and MacFarlane, T. W.,** Factors affecting the *in vitro* adherence of the fungal oral pathogen *Candida albicans* to epithelial cells of human origin, *Arch. Oral Biol.*, 27, 869, 1982.
167. **Gibbons, R. J. and van Houte, J.,** Bacterial adherence to oral epithelial surfaces and its role as an ecological determinant, *Infect. Immun.*, 3, 567, 1975.
168. **Samaranayake, L. P., Hughes, A., and MacFarlane, T. W.,** The proteolytic potential of *Candida albicans* in human saliva supplemental with glucose, *J. Med. Microbiol.*, 17, 13, 1984.
169. **Knight, L. and Fletcher, J.,** Growth of *Candida albicans* in saliva: stimulation by glucose associated with antibiotics, cortico steroids, and diabetes mellitus, *J. Infect. Dis.*, 123, 371, 1971.
170. **Egelberg, J.,** Gingival exudate measurements for evaluation of inflammatory changes of the gingivae, *Odontol. Revy*, 15, 381, 1964.

171. **Brandtzaeg, P.,** Immunochemical comparison of proteins in human gingival pocket fluid, serum and saliva, *Arch. Oral Biol.*, 10, 795, 1965.
172. **Mann, W. V. and Stoffer, H. R.,** The identification of protein components in fluid from gingival pockets, *Periodontics*, 2, 263, 1964.
173. **Brandtzaeg, P.,** Immunoglobulin systems of oral mucosa and saliva, in *Oral Mucosa in Health and Disease*, Dolby, A. E., Ed., Blackwell Scientific, Oxford, 1975, 137.
174. **Cimasoni, G.,** The crevicular fluid, in *Monographs in Oral Science*, Vol. 3, Myers, H. M., Ed., S. Karger, Basel, 1974.
175. **Vudhichamnong, K., Walker, D. M., and Ryley, H. C.,** The effect of secretory immunoglobulin A on the *in vitro* adherence of the yeast *Candida albicans* to human oral epithelial cells, *Arch. Oral Biol.*, 27, 617, 1982.
176. **Gibbons, R. J.,** Adhesion of bacteria to surfaces of the mouth, in *Microbial Adhesion to Surfaces*, Berkeley, R. C. W., Lynch, J. M., Melling, J., Rutter, P. R., and Vincent, B., Eds., Soc. Chem. Indust., Ellis Harwood Ltd., Chichester, England, 1980, 351.
177. **Kennedy, E. B. and Ash, M. M.,** Oxidation reduction potential of developing plaque, periodontal pockets and gingival sulci, *J. Periodontol.*, 40, 630, 1969.
178. **Onisi, M. and Kondo, W.,** Ecological aspects of human parotid saliva. I. The existence of constant rH values, *J. Dent. Res.*, 39, 433, 1960.
179. **Hillman, J. D., van Houte, J., and Gibbons, R. J.,** Sorption of bacteria to human enamel powder, *Arch. Oral Biol.*, 15, 899, 1970.
180. **Liljemark, W. F. and Schauer, S. V.,** Studies on the bacterial components which bind *Streptococcus sanguis* and *Streptococcus mutans* to hydroxyapatite, *Arch. Oral Biol.*, 20, 609, 1975.
181. **Ørstavik, D.,** Sorption of *Streptococcus faecium* to glass, *Acta Pathol. Microbiol. Scand. B.*, 85, 38, 1979.
182. **Ham, R. G. and McKeehan, W. L.,** Media and growth requirements, *Methods Enzymol.*, 58, 44, 1979.
183. **Tuller, M. A.,** Acid-base homeostasis and its disorders: use of arterial blood gas analysis, in *Diagnosis and Therapy*, Medical Examination Publishing, Flushing, NY, 1971, 19.
184. **Weld, J. T.,** *Candida albicans:* rapid identification in pure cultures with carbon dioxide on modified eosin-methylene blue medium, *Arch. Dermatol. Syphiol.*, 66, 691, 1952.
185. **Mardon, D. N., Balish, E., and Phillips, A. W.,** Control of dimorphism in a biochemical variant of *Candida albicans*, *J. Bacteriol.*, 100, 701, 1969.
186. **Aly, R. and Maibach, H.,** Susceptibility to skin infection, in *Recent Advances in Dermatology*, No. 6, Rook, A. J. and Maibach, H., Eds., Churchill Livingstone, Edinburgh, 1983, 75.
187. **Webster, C. E. and Odds, F. C.,** Growth of pathogenic *Candida* isolates anaerobically and under elevated concentrations of CO_2 in air, *J. Med. Vet. Mycol.*, 25, 47, 1986.
188. **Sims, W.,** Effect of carbon dioxide on the growth and form of *Candida albicans*, *J. Med. Microbiol.*, 22, 203, 1986.
189. **Samaranayake, L. P. and MacFarlane, T. W.,** Hypothesis: on the role of dietary carbohydrates in the pathogenesis of oral candidosis, *FEMS Microbiol. Lett.*, 27, 1, 1985.
190. **Leeson, R. and Leeson, T.,** The digestive tract, in *Histology*, 3rd ed., W. B. Saunders, Philadelphia, 1976, 318.
191. **Dehority, B. A.,** Carbon dioxide requirement of various species of rumen bacteria, *J. Bacteriol.*, 105, 70, 1971.
192. **Hendrix, T. R. and Bayless, T. M.,** Digestion-Intestinal secretion, *Annu. Rev. Physiol.*, 32, 139, 1970.
193. **Hungate, R. E.,** *The Rumen and its Microbes*, Academic Press, New York, 1966, 1.
194. **Levitt, M. D. and Bond, J. H.,** Volume, composition and source of intestinal gas, *Gastroenterology*, 59, 921, 1970.
195. **Drasar, B. S. and Barrow, P. A.,** The bacterial flora of the normal intestine, in *Intestinal Microbiology*, Van Nostrand Reinhold, London, 1985, 19.
196. **Drasar, B. S. and Hill, M. J.,** Factors controlling and influencing the flora, in *Human Intestinal Flora*, Academic Press, London, 1974, 3.
197. **Bloom, W. and Fawcett, D. W.,** *A Textbook of Histology*, 9th ed., W. B. Saunders, Philadelphia, 1968, 560.
198. **Trier, J. S. and Browning, T. H.,** Epithelial-cell renewal in cultured duodenal biopsies in celiac sprue, *N. Engl. J. Med.*, 283, 1245, 1970.
199. **Selby, W. S., Janssy, G., and Jewell, D. P.,** Immunohistological characterization of intra epithelial lymphocytes of the gastrointestinal tract, *Gut*, 22, 169, 1981.
200. **Austin, L. L. and Dobbins, W. O., III,** Intra epithelial leucocytes of the intestinal mucosa in normal man and in Whipple's disease: a light and electron microscopic study, *Digest. Dis. Sci.*, 27, 311, 1982.
201. **Dobbins, W. O., III,** Human intestinal epithelium as a biological membrane, in *Pathology of Cell Membranes*, Trump, B. F., Ed., Academic Press, New York, 1975, 429.
202. **Dobbins, W. O., III,** Alteration in intestinal structure associated with microorganisms, in *Attachment of Organisms to the Gut Mucosa*, Boedeker, E. C., Ed., CRC Press, Boca Raton, FL, 1984, 161.
203. **Savage, D. C.,** Colonization by and survival of pathogenic bacteria on intestinal mucosa, in *Adsorption of Microorganisms to Surfaces*, Bitton, G. and Marshall, K. C., Eds., John Wiley & Sons, New York, 1980, 175.
204. **Ito, S.,** Structure and function of the glycocalyx, *Fed. Proc. Fed. Am. Soc. Biol.*, 28, 12, 1969.

205. **Hollenberg, M. D., Fishman, P. H., Bennett, V., and Cuatrecasas, P.,** Cholera toxin and cell growth; role of membrane gangliosides, *Proc. Natl. Acad. Sci. U.S.A.*, 71, 4224, 1974.
206. **Draper, R. K., Chin, D., and Simon, M. I.,** Diphtheria toxin has the properties of a lectin, *Proc. Natl. Acad. Sci. U.S.A.*, 75, 261, 1978.
207. **Gottschalk, A., Belyavin, G., and Biddle, F.,** Glycoproteins as influenza virus hemagglutinin inhibitors and as cellular virus receptors, in *Glycoproteins, Their Composition, Structure and Function*, Gottschalk, A., Ed., Elsevier, Amsterdam, 1972, 1082.
208. **Gibbons, R.I.,** Adherence of bacteria to host tissue, in *Microbiology-1977*, Schlessinger, D., Ed., American Society for Microbiology, Washington, D.C., 1977, 395.
209. **Smith, H.,** Microbial surfaces in relation to pathogenicity, *Bacteriol. Rev.*, 41, 475, 1977.
210. **Sharon, N., Eshdat, Y., Silverblatt, J., and Ofek, I.,** Bacterial adherence to cell surface sugars, in *Adhesion and Microorganism Pathogenicity*, Ciba Found. Symp. 80, Pitman Medical, London, 1981, 119.
211. **Etzler, M. E. and Branstrator, M. L.,** Differential localization of cell surface and secretory components in rat intestinal epithelium by use of lectins, *J. Cell Biol.*, 62, 329, 1974.
212. **Savage, D. C.,** Microbial ecology of gastrointestinal tract, *Annu. Rev. Microbiol.*, 31, 107, 1977.
213. **Savage, D. C.,** Electron microscopy of bacteria adherent to epithelia in the murine intestinal canal, in *Microbiology-1977*, Schlessinger, D., Ed., American Society for Microbiology, Washington, D.C., 1977, 422.
214. **Neutra, M. R.,** Ultrastructural studies of the interaction of bacteria with intestinal cell surfaces, in *Attachment of Organisms to the Gut Mucosa*, Vol. I, Boedeker, E. C., Ed., CRC Press, Boca Raton, FL, 1984, 173.
215. **Harland, W. A. and Lee, F. D.,** Intestinal spirochetosis, *Br. Med. J.*, 3, 718, 1967.
216. **Gear, E. V. and Dobbins, W. O.,** Rectal biopsy, *Gastroenterology*, 55, 522, 1968.
217. **Lee, F. D., Kraszewski, A., Gordon, J., Howie, J. G. R., McSenney, D., and Harland, W. A.,** Intestinal spirochetosis, *Gut*, 12, 126, 1971.
218. **Takeuchi, A. and Zeller, J. A.,** Ultrastructural idenification of spirochetes and flagellated microbes at the brush border of the large intestinal epithelium of the rhesus monkey, *Infect. Immun.*, 6, 1008, 1972.
219. **Neutra, M. R.,** Prokaryotic-eukaryotic cell junctions: attachment of spirochetes and flagellated bacteria to primate large intestinal cells, *J. Ultrastruct. Res.*, 70, 186, 1980.
220. **Balish, E., Balish, M. J., Salkowski, C. A., Lee, K. W., and Bartizal, K. F.,** Colonization of congenitally athymic, gnotobiotic mice by *Candida albicans*, *Appl. Environ. Microbiol.*, 47, 647, 1984.
221. **Field, H., Pope, L. M., Cole, G. T., Guentzel, M. N., and Berry, L. J.,** Persistence and spread of *Candida albicans* after intragastric inoculation of infant mice, *Infect. Immun.*, 31, 783, 1981.
222. **Helstrom, P. B. and Balish, E.,** Effect of oral tetracycline, the microbial flora and the athymic state on gastrointestinal colonization and infection of BALB/c mice with *Candida albicans*, *Infect. Immun.*, 23, 764, 1979.
223. **Lee, K. W. and Balish, E.,** Systemic candidosis in germ free, flora-defined and conventional nude and thymus-bearing mice, *J. Reticuloendothel. Soc.*, 29, 71, 1981.
224. **Pope, L. M. and Cole, G. T.,** Comparative studies of gastrointestinal colonization and systemic spread by *Candida albicans* and non-lethal yeast in the infant mouse, *Scanning Electron Micros.*, 4, 1667, 1982.
225. **Myerowitz, R. L., Pazin, G. J., and Allen, C. M.,** Disseminated candidiasis: changes in incidence, underlying diseases and pathology, *Am. J. Clin. Pathol.*, 68, 29, 1977.
226. **Stone, H. H., Geheber, C. E., Kolb, L. D., and Kitchen, W. R.,** Alimentary tract colonization by *Candida albicans*, *J. Surg. Res.*, 14, 273, 1973.
227. **Stone, H. H., Kolb, L. D., Currie, C. A., Geheber, C. E., and Cuzzell, J. F.,** *Candida* sepsis: pathogenesis and principles of treatment, *Ann. Surg.*, 179, 697, 1974.
228. **Miles, M. R., Olsen, L., and Rogers, A.,** Recurrent vaginal candidiasis. Importance of an intestinal reservoir, *JAMA*, 238, 1836, 1977.
229. **Krause, W., Matheis, H., and Wulf, K.,** Fungaemia and funguria after oral administration of *Candida albicans*, *Lancet*, 1, 598, 1969.
230. **Green, J. P. and Bulmer, G. S.,** Gastrointestinal inoculation of *Cryptococcus neoformans* in mice, *Sabouraudia*, 17, 233, 1979.
231. **Haupt, H. M., Merz, W. G., Beschorner, W. E., Vaughan, W. P., and Saral, R.,** Colonization and infection with *Trichosporon* species in the immunosuppressed host, *J. Infect. Dis.*, 147, 199, 1983.
232. **Kennedy, M. J., Bajwa, P. S., and Volz, P. A.,** Gastrointestinal inoculation of *Sporothrix schenckii* in mice, *Mycopathologia*, 78, 141, 1982.
233. **Wingard, J. R., Dick, J. D., Merz, W. G., Sandford, G. R., Saral, R., and Burns, W. H.,** Differences in virulence of clinical isolates of *Candida tropicalis* and *Candida albicans* in mice, *Infect. Immun.*, 37, 833, 1982.
234. **Ghannoum, M. A., Burns, G. R., Abu Elteen, K., and Radwan, S. S.,** Experimental evidence for the role of lipids in adherence of *Candida* spp. to human buccal epithelial cells, *Infect. Immun.*, 54, 189, 1986.
235. **Clamp, J. R., Allen, A., Gibbons, R. A., and Roberts, G. P.,** Chemical aspects of mucus, *Brit. Med. Bull.*, 34, 1, 1978.

236. **Elstein, M. and Parke, D. V.,** *Mucus in Health and Disease*, Plenum, New York, 1977.
237. **Chantler, E., Elder, J., and Elstein, M.,** *Mucus in Health and Disease II*, Plenum, New York, 1982.
238. **Neutra, M. R.,** The mechanism of intestinal mucous secretion, in *Attachment of Organisms to the Gut Mucosa*, Vol. II, Boedeker, E. C., Ed., CRC Press, Boca Raton, FL, 1984, 33.
239. **Hollander, F.,** The two-component mucous barrier; its activity in protecting the gastroduodenal mucosa against peptic ulceration *Arch. Intern. Med.*, 93, 107, 1954.
240. **Freter, R.,** Mechanism of association of bacteria with mucosal surfaces, *Ciba Found. Symp.*, 80, 36, 1981.
241. **Freter, R., Allweis, B., O'Brien, P. C. M., Halstead, S. A., and Macsai, M. S.,** Role of chemotaxis in the association of motile bacteria with intestinal mucosa: *in vitro* studies, *Infect. Immun.*, 34, 241, 1981.
242. **Freter, R., O'Brien, P. C. M., and Halstead, S. A.,** Adhesion and chemotaxis as determinants of bacterial association with mucosal surfaces, *Adv. Exp. Med.*, 107, 429, 1978.
243. **Hartley, C. L., Neumann, C. S., and Richmond, M. H.,** Adhesion of commensal bacteria to the large intestine wall in humans, *Infect. Immun.*, 23, 128, 1979.
244. **Laux, D. C., McSweegan, E. F., and Cohen, P. S.,** Adhesion of enterotoxigenic *Escherichia coli* to immobilized intestinal mucosal preparations: a model for adhesion to mucosal surface components, *J. Microbiol. Methods*, 2, 27, 1984.
245. **Allen, A.,** Structure and function of gastrointestinal mucus, in *Physiology of the Gastrointestinal Tract*, Johnson, L. R., Ed., Raven, New York, 1981, 617.
246. **Slomiany, A., Yano, S., Slomiany, B. L., and Glass, G. B. J.,** Lipid composition of the gastric mucous barrier in the rat, *J. Biol. Chem.*, 253, 3785, 1978.
247. **Slomiany, B. L. and Slomiany, A.,** Lipids of mucous secretions of the alimentary tract, in *Attachment of Organisms to the Gut Mucosa*, Vol. II, Boedeker, E. C., Ed., CRC Press, Boca Raton, FL, 1984, 23.
248. **Glass, G. B. J. and Slomiany, B. L.,** Derangements of the production and secretion of mucus in gastrointestinal diseases, in *Mucus in Health and Disease*, Elstein, M. and Parke, D. V., Eds., Plenum, New York, 1977, 311.
249. **Forstner, J. F.,** Intestinal mucins in health and disease, *Digestion*, 17, 234, 1978.
250. **Boat, T. F. and Cheng, P. W.,** Biochemistry of airway mucus secretions, *Fed. Proc. Fed. Am. Soc. Exp. Biol.*, 39, 3067, 1980.
251. **Beachey, E. H.,** Bacterial adherence: adhesion-receptor mediating the attachment of bacteria to mucosal surfaces, *J. Infect. Dis.*, 143, 325, 1981.
252. **Slomiany, B. L. and Meyer, K.,** Isolation and structural studies of sulfated glycoproteins of hog gastric mucosa, *J. Biol. Chem.*, 247, 5062, 1972.
253. **Pearson, J. P., Allen, A., and Venables, C.,** Gastric mucus: isolation and polymeric structure of the under graded glycoprotein: its breakdown by pepsin, *Gastroenterology*, 78, 709, 1980.
254. **Slomiany, A. and Slomiany, B. L.,** Structures of the acidic oligosaccharides isolated from rat sublingual glycoprotein, *J. Biol. Chem.*, 253, 7301, 1978.
255. **Slomiany, B. L., Murty, V. L. N., and Slomiany, A.,** Isolation and characterization of oligosaccharides from rat colonic mucus glycoprotein, *J. Biol. Chem.*, 255, 9719, 1980.
256. **Slomiany, B. L., Banas-Gruszka, Z., Kojima, K., Herp, A., and Slomiany, A.,** The glycoprotein nature of Forssman antigen in dog gastric mucus, *FEBS Lett.*, 130, 201, 1981.
257. **Allen, A.,** Mucus — a protective secretion of complexity, *Trends Biochem. Sci.*, 8, 169, 1983.
258. **Jabbal, I., Kells, D. I. C., Forstner, G., and Forstner, J.,** Human intestinal goblet cell mucin, *Can. J. Biochem.*, 54, 707, 1976.
259. **Allen, A.,** The structure and function of gastrointestinal mucus, in *Attachment of Organisms to the Gut Mucosa*, Vol. II, Boedeker, E. C., Ed., CRC Press, Boca Raton, FL, 1984, 3.
260. **Forstner, G., Sherman, P., and Forstner, J.,** Mucus: Function and Structure, in *Attachment of Organisms to the Gut Mucosa*, Vol. II, Boedeker, E. C., Ed., CRC Press, Boca Raton, FL, 1984, 13.
261. **Kabat, A.,** *Blood Group Substances*, Academic Press, New York, 1956.
262. **Snary, D., Allen, A., and Pain, R. H.,** Structural studies on gastric mucoproteins lowering of molecular weight after reduction with 2-mercaptoethanol, *Biochem. Biophys. Res. Commun.*, 40, 844, 1970.
263. **Scawen, M. and Allen, A.,** The action of proteolytic enzymes on the glycoprotein from pig gastric mucus, *Biochem. J.*, 163, 363, 1977.
264. **Pearson, J. P., Allen, A., and Parry, S.,** A 70,000 molecular weight protein isolated from purified pig gastric mucus glycoprotein by reduction of disulphide bridges and its implication in the polymeric structure, *Biochem. J.*, 197, 155, 1981.
265. **Kishimoto, A., Kajiwara, N., Shiota, M., and Nishizuka, Y.,** Proteolytic activation of calcium-activated, phospholipid-dependent protein-kinase by calcium-dependent neutral protease, *J. Biol. Chem.*, 258, 1156, 1983.
266. **di Girolamo, R., Liston, J., and Matches, J.,** Ionic binding — the mechanism of viral uptake by shellfish mucus, *Appl. Environ. Microbiol.*, 33, 19, 1977.
267. **Slomiany, B. L. and Slomiany, A.,** Glycolipid constituents of dog and rat gastric secretion, *J. Am. Oil Chem. Soc.*, 55, 239A, 1978.

268. **Slomiany, A., Slomiany, B. L., and Glass, G. B. J.,** Effect of duodenal reflux on glyceroglucolipid component of gastric mucous barrier, *J. Appl. Biochem.*, 2, 336, 1980.
269. **Slomiany, B. L., Banas-Gruszka, Z., and Slomiany, A.,** Association of glyceroglucolipids with glycoproteins of gastric mucus, *IRCS Med. Sci.*, 9, 757, 1981.
270. **Slomiany, A., Patkowska, M. J., Slomiany, B. L., and Glass, G. B. J.,** The effect of ethanol on the constituents of gastric mucous barrier, *Int. J. Biol. Macromol.*, 1, 165, 1979.
271. **Slomiany, A., Galicki, N. I., Kojima, K., Banas-Gruszka, Z., and Slomiany, B. L.,** Glyceroglucolipids of the mucous barrier of dog stomach, *Biochem. Biophys. Acta,* 665, 88, 1981.
272. **Kojima, K., Banas-Gruszka, Z., Galicki, N. I., Slomiany, A., and Slomiany, B. L.,** The effect of lysolecithin on the protein and glyceroglucolipid components of the gastric mucous barrier, *IRCS Med. Sci.*, 9, 12, 1981.
273. **Slomiany, B. L., Kojima, K., Banas-Gruszka, Z., and Slomiany, A.,** Effect of lysolecithin on protein and glycoprotein components of the gastric mucous barrier, *J. Appl. Biochem.*, 2, 448, 1980.
274. **Martin, G. P., Marriot, C., and Kellaway, I. W.,** Direct effect of bile salts and phospholipids on the physical properties of mucus, *Gut,* 19, 103, 1978.
275. **Takagaki, Y. M. and Hotta, K.,** Characterization of peptic inhibitory activity associated with sulfated glycoprotein isolated from gastric mucosa, *Biochem. Biophys. Acta,* 584, 288, 1979.
276. **Florey, H. W.,** Observations on the functions of mucus and the early stages of bacterial invasion of the intestinal mucosa, *J. Pathol. Bacteriol.*, 37, 283, 1933.
277. **Allen, A.,** The structure and function of gastrointestinal mucus, in *Basic Mechanisms of Gastrointestinal Cell Injury and Protection,* Harmon, J. W., Ed., Williams & Wilkins, Baltimore, 1981, 351.
278. **Heatley, N. G.,** Mucosubstance as a barrier to diffusion, *Gastroenterology,* 37, 313, 1959.
279. **Flemstrom, G.,** Gastric secretion of bicarbonate, in *Physiology of the Gastrointestinal Tract,* Johnson, L. R., Ed., Raven, New York, 1981, 603.
280. **Flemstro, G. and Garner, A.,** Gastroduodenal HCO_3^- transport, characteristics and proposed role in acidity regulation and mucosal protection, *Am. J. Physiol.*, 242, G183, 1982.
281. **Flemstrom, G. and Garner, A.,** Gastroduodenal HCO_3^--transport characteristics and proposed role in acidity regulation and mucosal protection, *Am. J. Physiol.*, 242, G193, 1982.
282. **Allen, A., Pain, R. H., and Robson, T.,** Model for the structure of the gastric mucus gel, *Nature (London),* 264, 88, 1976.
283. **Rees, W. D. W. and Turnberg, L. A.,** Mechanisms of gastric-mucosal protection. A role for the mucus-bicarbonate barrier, *Clin. Sci.*, 63, 343, 1982.
284. **Freter, R.,** Bacterial association with the mucus gel system of the gut, in *Microbiology-1982,* Schlessinger, D., Ed., American Society for Microbiology, Washington, D.C., 1982, 278.
285. **Freter, R.,** Factors involved in the penetration of the mucous layer in experimental cholera, in *Attachment of Organisms to the Gut Mucosa,* Boedeker, E. C., Ed., CRC Press, Boca Raton, FL, 1984, 43.
286. **Strombeck, D. R. and Harrold, D.,** Binding of cholera toxin to mucins and inhibition by gastric mucin, *Infect. Immun.*, 10, 1266, 1974.
287. **Sturgess, J. M.,** Perspectives in Cystic Fibrosis, Canadian Cystic Fibrosis Foundation, Toronto, 1980.
288. **Butrus, S. I. and Klotz, S. A.,** Blocking *Candida* adherence to contact lenses, *Curr. Eye Res.*, 5, 745, 1986.
289. **Cohen, P. S., Arruda, J. C., Williams, T. J., and Laux, D. C.,** Adhesion of a human fecal *Escherichia coli* strain to mouse colonic mucus, *Infect. Immun.*, 48, 139, 1985.
290. **Barnes, J.,** Essential gynaecological anatomy, in *Lecture Notes on Gynaecology,* 4th ed., Blackwell Scientific, Oxford, 1980, 1.
291. **Platzer, W., Poisel, S., and Hafez, E. S. E.,** Functional anatomy of the human vagina, in *Human Reproductive Medicine: The Human Vagina,* Vol. 2, Hafez, E. S. E. and Evans, T. N., Eds., Elsevier/North-Holland, Amsterdam, 1978, chap. 3.
292. **Sherman, A. I., Steger, R. W., Hafez, E. S. E., and Brown, S.,** Human vaginal adenosis and adenocarcinoma, in *Symposium of Scanning Electron Microscopy of Fertility and Infertility,* 8th World Cong. Fertility Sterility, *J. Reprod. Med.*, 14, 221, 1975.
293. **Steger, R. W. and Hafez, E. S. E.,** Age-associated changes in the vagina, in *Human Reproductive Medicine: The Human Vagina,* Vol. 2, Hafez, E. S. E. and Evans, T. N., Eds., Elsevier/North-Holland, Amsterdam, 1978, chap. 6.
294. **Parakkal, P. F.,** Cyclic changes in the vaginal epithelium of the rat seen by scanning electron microscopy, *Anat. Rec.*, 178, 529, 1974.
295. **Ludwig, H. and Metzger, H.,** *The Human Female Reproductive Tract: A Scanning Electron Microscopic Atlas,* Springer-Verlag, Berlin, 1976.
296. **Burgos, M. H. and Roig de Vargas-Linares, C. E.,** Ultrastructure of the vaginal mucosa, in *Human Reproductive Medicine: The Human Vagina,* Vol. 2, Hafex, E. S. E. and Evans, T. N., Eds., Elsevier/North-Holland, Amsterdam, 1978, chap. 5.
297. **Averette, H. E., Weinstein, G. D., and Frost, P.,** Autoradiographic analysis of cell proliferation kinetics in human genital tissues. I. Normal cervix and vagina, *Am. J. Obstet. Gynecol.*, 108, 8, 1970.

298. **Clayton, S. G., Lewis, T. L. T., and Pinker, G.,** *Gynaecology,* 13th ed., Edward Arnold, London, 1980.
299. **Vaerman, J. P. and Fèrin, J.,** Local immunological response in the vagina, cervix and endometrium, in *Immunological Approaches to Fertility Control,* Diczfalusy, E., Ed., Karolinska Institutet, Stockholm, 1974, 281.
300. **Wagner, G. and Levin, R. J.,** Vaginal fluid, in *Human Reproductive Medicine: The Human Vagina,* Vol. 2, Hafez, E. S. E. and Evans, T. N., Eds., Elsevier/North-Holland, Amsterdam, 1978, chap. 8.
301. **Stone, A. and Gamble, C. J.,** The quantity of vaginal fluid, *Am. J. Obstet. Gynecol.,* 78, 279, 1959.
302. **Rakoff, A. E., Feo, L. G., and Goldstein, L.,** The biologic characteristics of the normal vagina, *Am. J. Obstet. Gynecol.,* 47, 467, 1944.
303. **Hunter, C. A. and Nicholas, H. J.,** A study of vaginal acids, *Am. J. Obstet. Gynecol.,* 78, 282, 1959.
304. **Dusitin, N., Gregoire, A. T., Johnson, W. D., and Rakoff, A. E.,** Histidine in human vaginal fluid, *Obstet. Gynecol.,* 29, 125, 1967.
305. **Michael, R. P.,** Hormonal studies and sexual communication in primates, *J. Steroid Biochem.,* 6, 161, 1975.
306. **Goldfoot, D. A., Kravetz, M. A., Goy, R. W., and Freeman, S. K.,** Lack of effect of vaginal lavages and aliphatic acids on ejaculatory responses in rhesus monkeys: behavioural and chemical analysis, *Horm. Behav.,* 7, 1, 1976.
307. **Abdallah, M. A. and Roig de Vargas-Linares, C. E.,** Polyacrylamide gel electrophoresis of normal human vaginal fluid, *Acta Physiol. Lat. Am.,* 20, 445, 1970.
308. **Preti, G. and Huggins, G. R.,** Organic constituents of vaginal secretions, in *Human Reproductive Medicine: The Human Vagina,* Vol. 2, Hafez, E. S. E. and Evans, T. N., Eds., Elsevier/North-Holland, Amsterdam, 1978, chap. 10.
309. **Bonsall, R. W. and Michael, R. P.,** Volatile odoriferous acids in vaginal fluid, in *Human Reproductive Medicine,* Vol. 2, Hafez, E. S. E. and Evan, T. N., Eds., Elsevier/North-Holland, Amsterdam, 1978, chap. 11.
310. **Roig de Vargas-Linares, C. E.,** The vagina as a source of immunoglobulins, in *Human Reproductive Medicine: The Human Vagina,* Vol. 2, Hafez, E. S. E. and Evans, T. N., Eds., Elsevier/North-Holland, Amsterdam, 1978, chap. 13.
311. **Oberst, F. W. and Plass, E. D.,** The hydrogen ion concentration of human vaginal discharge, *Am. J. Obstet. Gynecol.,* 32, 22, 1936.
312. **Lang, W. R.,** Vaginal acidity and pH — a review, *Obstet. Gynecol. Surv.,* 10, 546, 1955.
313. **Cruickshank, R. and Sharman, A.,** The biology of the vagina in the human subject. I. Glycogen in the vaginal epithelium and its relation to ovarian activity. *J. Obstet. Gynecol. Br. Empire,* 41, 190, 1934.
314. **Cruickshank, R. and Sharman, A.,** The biology of the vagina in the human subject. II. The bacterial flora and secretion of the vagina at various age-periods and their relation to glycogen in the vaginal epithelium, *J. Obstet. Gynecol. Br. Empire,* 41, 208, 1934.
315. **Kistner, R. W.,** Physiology of the vagina, in *Human Reproductive Medicine: The Human Vagina,* Vol. 2, Hafez, E. S. E. and Evans, T. N., Eds., North-Holland, Amsterdam, 1978, chap. 7.
316. **Brown, W. J.,** Microbial ecology of the normal vagina, in *Human Reproductive Medicine: The Human Vagina,* Vol. 2, Hafez, E. S. E. and Evans, T. N., Eds., North-Holland, Amsterdam, 1978, chap. 29.
317. **Kotcher, E.,** Microbiological considerations in pediatric and adolescent gynecology, *Ann. N.Y. Acad. Sci.,* 142, 586, 1967.
318. **Hunter, C. A. and Long, K. R.,** A study of the microbiological flora of the vagina, *Am. J. Obstet. Gynecol.,* 75, 865, 1958.
319. **Walz, K. A., Metzger, H., and Ludwig, H.,** Surface ultrastructure of the vagina, in *Human Reproductive Medicine: The Human Vagina,* Vol. 2, Hafez, E. S. E. and Evans, T. N., Eds., North-Holland, Amsterdam, 1978, chap. 4.
320. **Odds, F. C.,** *Candida and Candidiosis,* 2nd ed., Baillière Tindall, London, 1988.
321. **Wagner, G. and Ottesen, B.,** Vaginal physiology during menstruation, *Ann. Internal Med.,* 96, 921, 1982.
322. **Rotrosen, D., Calderon, R. A., and Edwards, J. E., Jr.,** Adherence of *Candida* species to host tissues and plastic surfaces, *Rev. Infect. Dis.,* 8, 73, 1986.
323. **Kennedy, M. J., Rogers, A. L., Beneke, E. S., and Yancey, R. J.,** Influence of growth and physiochemical parameters on adsorption of *Candida albicans* to plastics, presented at 10th Cong. Int. Soc. Human Anim. Mycol., Barcelona, 1988, Abstract no. P-124.
324. **Anderson, J. M. and Soll, D. R.,** Unique phenotye of opaque cells in the white-opaque transition of *Candida albicans, J. Bacteriol.,* 169, 5579, 1987.
325. **Slutsky, B., Staebell, M., Anderson, J., Risen, L., Pfaller, M., and Soll, D. R.,** "White-opaque transition": a second high-frequency switching system in *Candida albicans, J. Bacteriol.,* 169, 189, 1987.
326. **Kennedy, M. J., Rogers, A. L., Hanselmen, L. R., Soll, D. R., and Yancey, R. J., Jr.,** Variation in adhesion and cell surface hydrophobicity in *Candida albicans* white and opaque phenotype, *Mycopathologia,* 102, 149, 1988.
327. **Daniels, S. L.,** Mechanisms involved in sorption of microorganisms to solid surfaces, in *Adsorption of Microorganisms to Surfaces,* Bitton, G. and Marshall, K. C., Eds., John Wiley & Sons, New York, 1980, 7.

328. **Cole, G. T., Seshan, K. R., Pope, L. M., and Yancey, R. J.**, Morphological aspects of gastrointestinal tract invasion by *Candida albicans* in the infant mouse, *J. Med. Vet. Mycol.*, 26, 173, 1988.
329. **Soll, D. R.**, Personal communication.
330. **Ghannoum, M. A. and Abu Elteen, K.**, Unpublished data.
331. **Lake, A. M., Bloch, K. J., Neutra, M. R., and Walker, W. A.**, Intestinal goblet cell mucus release. II. *In vitro* stimulation by antigen in the immunized rat, *J. Immunol.*, 122, 834, 1979.

REFERENCES ADDED IN PROOF

332. **Kennedy, M. J. and Sandin, R. L.**, Influence of growth conditions on *Candida albicans* adhesion, hydrophobicity and cell wall ultrastructure, *J. Am. Vet. Mycol.*, 26, 79, 198.
333. **Hazen, K. C.**, Participation of yeast cell surface hydrophobicity in adherence of *Candida albicans* to human epithelial cells, *Infect. Immun.*, 57, 1894, 1989.

Chapter 6

EFFECT OF THERAPEUTIC AGENTS ON ADHERENCE OF CANDIDA

TABLE OF CONTENTS

I. INTRODUCTION

Candidosis is a disease of the debilitated and occurs as a consequence of some breakdown in the host immune defenses, which is brought about by an underlying disease.[1] Patients receiving chemotherapy for their primary illness increase their chances of contracting infection by *Candida*. Exposure to antibiotics, antifungals, immunocompromising compounds, such as steroids and antineoplastic agents, endogenous and exogenous hormones, as well as X-irradiation may ultimately be shown to influence receptors and/or adhesins. Modulations of the microbial ligands and complementary host-cell receptors including expression, synthesis, steric changes, surface potentials, or hydrophobicity by various therapeutic modalities may have an important effect on the adhesion process and consequently on pathogenicity. Identification of these therapeutically induced modifications, that make a person more prone to candidosis is an important area of research which has not received much attention.[2]

Therapeutic agents that require investigation in relation to their effect on adherence may be divided into two categories (1) drugs that are used for the prevention and control of yeast infection, namely, antifungals and antibacterials, and (2) therapeutic regimes which are generally associated with potentiation of candidosis such as steroids, antibiotics, antineoplastic agents, and X-irradiation. Studies involving drugs in the first category are vital because an understanding of the mechanism(s) involved may aid in the use of these agents, hopefully at low levels, in the prophylaxis of patients at high risk for candidosis. Thus, the toxic side effects of such agents can be minimized. In contrast, investigations involving agents in the second category are aimed at considering the possibility that these agents might promote infection, at least in part, by affecting the adherence of the microorganisms to host tissues.

II. DRUGS USED FOR PREVENTION AND CONTROL OF CANDIDOSIS

A. ANTIFUNGAL AGENTS

Interest on the influence of antifungal agents on the adherence of *Candida* to epithelial cells was heightened following reports showing that adherence of bacteria to epithelial cells *in vitro* and *in vivo* was reduced after pre-exposure of bacteria to subinhibitory concentrations of clinically used antibiotics.[3-8,159] Antibiotics at such concentrations may affect the outer surface structures known to mediate adherence, such as alteration in the nature, excretion, or loss of adhesins.[9] A number of investigations have shown that subinhibitory concentrations of antibiotics selectively interfere with the expression of synthesis of certain excreted and nonexcreted components of bacteria without affecting total protein synthesis.[3,6,9-16,159] The work by Al-Bassam et al.[17] showed that subinhibitory levels of amphotericin B led to *Candida albicans* cell wall modification which involved peptidomannans.[17] These components are concerned with the outer layers of the yeast cell in contact with the environment,[18] and have been suggested by, McCourtie and Douglas[19,20] to be responsible for adhesion of *C. albicans* to epithelial cells, with confirmation arising from several laboratories. Al-Bassam et al.[17] reported a reduction in the level of peptidomannans in amphotericin-B-treated yeast. This decrease was mainly due to a fall in the absolute and relative values of mannans. In contrast, the amount of amino acids increased. Such modifications are bound to affect the adhesion process.[21-28]

The effect of antifungal agents such as imidazoles, polyenes, 5-fluorocytosine, and chlorhexidine on the adherence of *Candida* spp. to epithelial cells has also been investigated. Sobel and Obedeanu[29] were the first to study the effect of subinhibitory concentration of antifungal agents on the ability of fungi to adhere to cells. These authors showed that exposure of logarithmic or stationary-phase *C. albicans* cells to subinhibitory concentrations of ketoconazole failed to influence adherence. However, this treatment led to a change in shape with morphological dystrophy, formation of large aggregates due to clumping, and inhibition of germ-tube formation. The overall effect of these alterations was a reduction in the number of individual blastospores directly adhering to vaginal epithelial cells. The mechanism responsible for this reduction was attributed to an indirect effect on germination and clumping rather than a direct effect on adhesin synthesis or expression.[29] Subsequent work by Anderson and Odds[30] confirmed the reported reduction, brought about by ketoconazole, in the attachment of *C. albicans* to vaginal epithelial cells. They showed that inclusion of high ketoconazole concentrations in an adherence assay led to a reduction in numbers of *C. albicans* germ-tubes adhering per epithelial cell. The effect of ketoconazole on the adherence of *Candida* to buccal epithelial cells was investigated by workers studying the influence of a number of antifungal agents on adherence. Brenciaglia et al.[31] showed that pre-incubation of *C. albicans* with ketoconazole at minimal inhibitory concentration (MIC) and subinhibitory levels did not affect either the adherence of yeast nor germ-tube formation. Effects on adherence were observed only at concentrations above MIC values.[31] Macura,[32] however, showed ketoconazole to be effective both at therapeutic and subinhibitory concentrations. Other imidazole derivatives investigated for their effect on adhesion of yeast are miconazole (MN) and clotrimazole, the first discovered imidazole drug. Brenciaglia et al.[31] showed MN to have similar effects on adherence as those reported above for ketoconazole. In the study by Macura,[32] clotrimazole showed a strong adherence-inhibition action. The effect of this azole on germ-tube formation was not investigated.

Recently, Odds and Webster[158] studied the effect of two imidazole (clotrimazole and ketoconazole) and one triazole (itraconazole) derivatives on the adherence of *C. albicans* to vaginal epithelial cells. None of clotrimazole, itraconazole, or ketoconazole had any effect on the adherence of *C. albicans* germ-tubes to vaginal epithelial cells, *in vitro,* regardless of whether the antifungals were used to pretreat the yeast or the epithelial cells or were added to the yeast/epithelial cell mixture. These results contradict the findings of Sobel and Obedeanue[29] and Anderson and Odds.[30] This disparity was attributed to differences in methodologies employed.[158]

Polyenes investigated for their effect on adherence are amphotericin B (AMB) and nystatin (NY). Subinhibitory concentrations of AMB inhibited germination and *Candida* adherence to both serum-coated plastic surfaces and fibrin matrices.[33] In addition, peritoneal clearance of *C. albicans* was accelerated by pretreatment with AMB.[33] Pre-incubation of *C. albicans* with either subinhibitory, MIC, or above MIC levels significantly reduced the adherence of this yeast to buccal epithelial cells.[31] Germ-tube formation was also significantly reduced. These effects were significant irrespective of whether the yeast cells were exposed to AMB for short (3 h) or long (72 h) periods.[31] Compared with AMB, nystatin is reported to have less effect on the inhibition of adhesion of yeast to epithelial cells. Brenciaglia et al.[31] showed that preincubation of *C. albicans* with NY at MIC and sub-MIC levels did not have a significant effect on adherence. However, upon exposure of the yeast to NY at above MIC concentrations a slight improvement in adherence inhibition was noticed. No significant effect on germination was reported for NY pretreatment.[31] In contrast, Macura[32] observed that NY pretreatment with both subinhibitory and therapeutic levels considerably inhibited yeast adhesion, although this polyene had the lowest efficacy when compared to other antifungal agents.

Results reported for the effect of 5-fluorocytosine (5FC) a derivative of fluoropyrimidine, on the adhesion of *Candida* are contradictory. Brenciaglia et al.[31] reported that 5FC interfered with yeast adherence only after long pre-treatment periods (72 h) and only at MIC, or above MIC, values. Although 5FC did not interfere with adherence at sub-inhibitory concentration, it influenced germination at all the concentrations tested, particularly after 72 h exposure.[31] In contrast to these findings, Macura[32] found 5FC, together with ketoconazole, to be the most effective inhibitor of *C. albicans* adherence when compared to other antifungal agents tested in her study.

It is clear that the number of publications regarding the effect of various antifungal agents on

TABLE 1
Effect of Pre-Incubation of Various *Candida* spp. in Subinhibitory Concentrations of Amphotericin B on Adherence to Buccal Epithelial Cells

Mean number of adherent yeasts per 100 epithelial cells (1 h Pre-Incubation)

Organisms	Control	1/4 MIC	P	1/8 MIC	P	1/16 MIC	P
C. albicans	468	286	<0.001	328	<0.001	410	N.S
ATCC 10231	(27)	(19)		(21)		(27)	
C. albicans	490	359	<0.001	387	<0.001	440	N.S
KCCC 14172	(27)	(25)		(24)		(22)	
C. tropicalis	386	216	<0.001	255	<0.001	342	N.S
KCCC 13605	(23)	(18)		(18)		(21)	
C. kefyr[a]	264	189	<0.001	192	<0.001	214	N.S
KCCC 13709	(14)	(14)		(13)		(13)	

Mean number of adherent yeasts per 100 epithelial cells (24 h Pre-Incubation)

Organisms	Control	1/4 MIC	P	1/8 MIC	P	1/16 MIC	P
C. albicans	510	300	<0.001	323	<0.001	386	<0.001
ATCC 10231	(17)	(13)		(12)		(14)	
C. albicans	536	297	<0.001	312	<0.001	349	<0.001
KCCC 14172	(20)	(12)		(12)		(13)	
C. tropicalis	360	179	<0.001	198	<0.001	276	<0.05
KCCC 13605	(15)	(11)		(13)		(12)	
C. kefyr	295	194	<0.001	210	<0.05	234	<0.05
KCCC 13709	(11)	(12)		(13)		(13)	

Note: Values in parentheses are ± SE. *P* values compared with control. N.S, not significant.

[a]*C. kefyr* is the current name for *C. pseudotropicalis*.[1]

the adherence of *Candida* to cells is limited and contradictory. This prompted us to carry out a comprehensive investigation into the effect of AMB, NY, 5FC, and MN nitrate on the adherence of various *Candida* spp. to buccal epithelial cells.[34] Our results showed that treatment of *C. albicans* (two strains), *C. tropicalis,* and *C. pseudotropicalis* with these antifungal agents affected the adherence of yeasts to buccal epithelial cells (Tables 1 to 4). In general, pretreatment of *Candida* for a short period (1 h) had less effect on adhesion than incubation for a long period (24 h). Pretreatment of *C. albicans* with a combination of AMB and 5FC, both at 1/8 MIC, led to stronger adherence inhibition than that obtained for yeast pretreatment with either one alone at higher concentration (1/4 MIC). This applies for both short and long exposure periods (Table 5).

Comparison of adherence-inhibiting activity of antifungals used in this study gave the following pattern of activity: NY > MN > 5FC > AMB (Table 6). This pattern holds for yeasts pretreated with antifungals for a long period (24 h) and applied for the two *C. albicans* strains tested (ATCC 10231 and KCCC 14172). However, a different pattern of activity emerged when yeast cells were pretreated with antifungals for a short period (1 h). A difference also existed between the two strains tested with a 1 h pretreatment. This may provide partial explanation for the difference in adherence-inhibiting activity of antifungal agents reported by different laboratories and points out the need for standardization of methods.

Since *C. albicans* adherence is suggested to be governed by extracellular mannoproteins and germ-tubes which are associated with enhanced adherence to epithelial surfaces,[28,35,36] we chose

TABLE 2
Effect of Pre-Incubation of Various *Candida* spp. in Subinhibitory Concentrations of Nystatin on Adherence to Buccal Epithelial Cells

	Mean number of adherent yeasts per 100 epithelial cells (1 h Pre-Incubation)						
Organisms	**Control**	**1/4 MIC**	***P***	**1/8 MIC**	***P***	**1/16 MIC**	***P***
C. albicans	510	343	<0.001	390	<0.05	451	N.S
ATCC 10231	(28)	(29)		(29)		(35)	
C. albicans	563	315	<0.001	361	<0.001	416	<0.05
KCCC 14172	(37)	(27)		(30)		(39)	
C. tropicalis	368	249	<0.001	270	<0.05	292	<0.05
KCCC 13605	(24)	(19)		(24)		(17)	
C. kefyr	272	160	<0.001	171	<0.001	196	<0.05
KCCC 13709	(27)	(20)		(17)		(19)	

	Mean number of adherent yeasts per 100 epithelial cells (24 h Pre-Incubation)						
Organisms	**Control**	**1/4 MIC**	***P***	**1/8 MIC**	***P***	**1/16 MIC**	***P***
C. albicans	505	144	<0.001	322	<0.001	466	N.S
ATCC 10231	(39)	(16)		(25)		(27)	
C. albicans	507	106	<0.001	252	<0.001	347	<0.05
KCCC 14172	(33)	(13)		(25)		(31)	
C. tropicalis	377	217	<0.001	262	<0.001	295	<0.05
KCCC 13605	(21)	(16)		(21)		(21)	
C. kefyr	246	123	<0.001	138	<0.001	167	<0.05
KCCC 13709	(24)	(13)		(13)		(14)	

Note: Values in parentheses are ± SE. *P* values compared with control. N.S, not significant.

to study the effect of antifungal agents on both germination and secretion of extracellular polymers. All the antifungal agents tested affected the germination of the two strains of *C. albicans* to varying degrees (Figure 1 A and B). The pattern of germ-tube inhibition was NY > AMB > MN > 5FC. Although NY showed the highest activity for both adherence and germination, the two patterns do not exactly match.

Exposure of *C. albicans* KCCC 14172 to subinhibitory concentration of antifungals led to an increase in the excretion of extracellular polymers (Table 7). These polymers were extracted according to the method of McCourtie and Douglas.[19] A chemical analysis of the polymer obtained in our study showed that they are mannoproteins. The loss of extracellular polymers was greatest under the influence of 5FC followed by MN > AMB >NY. Yet again these drugs showed a difference between this pattern and those obtained for both adherence and germination.[34]

Such differences are not unexpected since the antifungals that have been studied do not share the same mode of action. The antifungal activity of polyenes is related to their binding to sterols, particularly ergosterol, which is the primary sterol in the membrane of sensitive fungi.[37-39] This interaction results in the formation of pores in the fungal membrane altering its permeability, leading to loss of cytoplasmic material, and eventual death.[1] By binding with yeast membrane sterols, polyenes lead to a decrease in hydrophobicity of the fungi which probably contribute to the observed reduction in adherence following treatment with these drugs.[40] A plausable explanation of the discrepancy of results between AMB and NY, even though they are structurally similar, could be the heterogeneity of the mechanisms of polyene action.[41]

TABLE 3
The Effect of Pre-Incubation of Various *Candida* spp. in Subinhibitory Concentrations of Miconazole Nitrate on Adherence to Buccal Epithelial Cells

Organisms	Mean number of adherent yeasts per 100 epithelial cells (1h Pre-Incubation)						
	Control	1/4 MIC	*P*	1/8 MIC	*P*	1/16 MIC	*P*
C. albicans	484	268	<0.001	299	<0.001	346	<0.001
ATCC 10231	(20)	(24)		(22)		(23)	
C. albicans	508	251	<0.001	285	<0.001	315	<0.001
KCCC 14172	(24)	(18)		(22)		(27)	
C. tropicalis	363	241	<0.001	276	<0.05	294	N.S
KCCC 13605	(29)	(14)		(18)		(18)	
C. kefyr	263	169	<0.001	201	N.S	222	N.S
KCCC 13709	(26)	(15)		(18)		(20)	

Organisms	Mean number of adherent yeasts per 100 epithelial cells (24 h Pre-Incubation)						
	Control	1/4 MIC	P	1/8 MIC	*P*	1/16 MIC	*P*
C. albicans	515	235	<0.001	265	<0.001	304	<0.001
ATCC 10231	(35)	(25)		(33)		(32)	
C. albicans	520	198	<0.001	234	<0.001	286	<0.001
KCCC 14172	(31)	(17)		(15)		(25)	
C. tropicalis	360	201	<0.001	228	<0.001	275	<0.05
KCCC 13605	(27)	(16)		(12)		(17)	
C. kefyr	250	148	<0.001	187	N.S	207	N.S
KCCC 13709	(22)	(16)		(23)		(20)	

Note: Values in parentheses are ±SE. *P* values compared with control. N.S, not significant.

Azoles antifungals are characterized by heterogeneity of action mechanisms.[1] However, the mode of action shared by all azoles involves the inhibition of ergosterol synthesis by a selective interaction with cytochrome P-450-dependent 14 α–demethylase.[42] Another important effect which is common to most azoles is their ability to prevent or greatly perturb hyphal growth of *C. albicans.*[1] These drugs interfere with germ-tube formation, which may account for the reduction of adherence of *Candida* to epithelial cells.[43] Relevant to adherence is the effect of azoles on cell wall structure. Subinhibitory concentrations of azoles, too low to cause direct damage of the yeast membranes, can cause alterations in cell wall structure.[44-48] These changes could be a consequence of the effect of azoles on the yeast cell membranes and cell wall synthesizing enzymes that are bound to membranes.[49]

5-Fluorocytosine affects both protein and DNA synthesis.[50-55] It enters the fungal cell by the enzyme cytosinepermease. Inside the fungal cell 5FC is deaminated by cytosinedeaminase to 5-fluorouracil which, in turn, is phosphorylated to 5-fluorouridine monophosphate. The latter is phosphorylated further and incorporated into the yeast RNA with eventual disruption of protein synthesis.[51-54] Inhibition of DNA synthesis is brought about by the formation of 5-fluorodeoxyuridine monophosphate, an inhibitor of thymidylate synthetase.[50] Of relevence to adhesion are the changes reported by Polak and Wain[56,57] in the proportion of cell wall polysaccharides, as well as enlargement of cell volumes as a consequence of 5FC action.

The presence of multiple mechanisms that may act simultaneously in the process of adherence of *Candida* to epithelial cells may provide yet another reason for the differences observed between antifungal agents.[58,59] These agents may influence different mechanisms, or

TABLE 4
Effect of Pre-Incubation of Various *Candida* spp. in Subinhibitory Concentrations of 5-Fluorocytosine on Adherence to Buccal Epithelial Cells

Mean number of adherent yeasts per 100 epithelial cells (1h Pre-Incubation)

Organisms	Control	1/4 MIC	*P*	1/8 MIC	*P*	1/16 MIC	*P*
C. albicans	447	269	<0.001	343	<0.001	362	<0.001
ATCC 10231	(22)	(17)		(19)		(19)	
C. albicans	504	282	<0.001	342	<0.001	438	N.S
KCCC 14172	(20)	(15)		(15)		(16)	
C. tropicalis	371	262	<0.001	286	<0.05	345	N.S
KCCC 13605	(15)	(15)		(17)		(19)	
C. kefyr	300	230	<0.05	274	N.S	286	N.S
KCCC 13709	(14)	(13)		(14)		(13)	

Mean number of adherent yeasts per 100 epithelial cells (24 h Pre-Incubation)

Organisms	Control	1/4 MIC	*P*	1/8 MIC	*P*	1/16 MIC	*P*
C. albicans	509	285	<0.001	312	<0.001	360	<0.001
ATCC 10231	(17)	(13)		(14)		(13)	
C. albicans	524	284	<0.001	311	<0.001	320	<0.001
KCCC 14172	(25)	(18)		(16)		(16)	
C. tropicalis	340	232	<0.001	255	<0.001	262	<0.001
KCCC 13605	(16)	(14)		(15)		(14)	
C. kefyr	285	189	<0.001	214	<0.05	250	N.S
KCCC 13709	(13)	(14)		(17)		(12)	

Note: Values in parentheses are ± SE. *P* values compared with control. N.S, not significant.

one particular mechanism to varying degrees. Thus, it is quite difficult to compare the effect of various antifungal agents on the adherence of *Candida* per se without considering these points.

B. CHLORHEXIDINE

Chlorhexidine is one of a family of N^1,- N^5-substituted biguanides which has a wide spectrum of antibacterial activity against both Gram-positive and Gram-negative bacteria.[60] Chlorhexidine is used mainly in medical and veterinary antisepsis and is available as a dihydrochloride, diacetate, and gluconate. Interest in chlorhexidine in relation to *Candida* adhesion followed its successful use (as 0.2% chlorhexidine gluconate) as a mouth rinse in the treatment of chronic atrophic and acute pseudomembranous candidosis.[61-64] In addition, chlorhexidine has been recommended for selected cases as a prophylactic agent in seriously diseased patients to improve oral hygiene and to prevent the spread of infection from the oral cavity.[64] Specifically, chlorhexidine mouth rinses have been tested in a group of leukemic patients as an anti-*Candida* prophylactic agent.[65]

Information on the effect of chlorhexidine on the adherence of *Candida* to denture acrylic and buccal epithelial cells comes from the experimentation of Samaranayake and his colleagues[66] in Scotland. These workers found that pretreatment of acrylic denture strips with chlorhexidine gluconate substantially reduced the adherence of seven *C. albicans* isolates to the strips. In a subsequent study, the chlorhexidine-mediated inhibition of *Candida* adherence to denture acrylic was unaffected by the presence of mixed saliva or serum.[67] Based on these findings it was suggested that soaking dentures in 2.0% chlorhexidine for 15 to 30 min would be an effective regimen for reducing yeast adherence and for killing any attached yeasts, hence treatment of

TABLE 5
Effect of Pre-Incubation of *C. Albicans* KCCC 14172 in a Combination of Amphotericin B Plus 5-Fluorocytosine on Adherence to Buccal Epithelial Cells

Antifungal agents	1 h Pre-Incubation: Yeast adhering to 100 BEC (mean ± SE)	P	Red.	24 h Pre-Incubation: Yeast adhering to 100 BEC (mean ± SE)	*P*	Red.
Control	509 ± 22			532 ± 20		
Amphotericin B (1/4 MIC)	407 ± 16	<0.001	20	324 ± 15	<0.001	39
5-Fluorocytosine (1/4 MIC)	352 ± 16	<0.001	31	318 ± 16	<0.001	40
Amphotericin B plus 5-Fluorocytosine (both at 1/8 MIC)	265 ± 15	<0.001	48	251 ± 16	<0.001	53

Note: *P* values compared with control. Red., percent reduction.

TABLE 6
Percent Reduction of Adherence of *Candida* spp. to Buccal Epithelial Cells Following Pre-Incubation in 1/4 MIC of Antifungal Agents

		% Reduction of adherence			
Antifungals	*Pre-Incubation time*	*C. albicans ATCC 10231*	*C. albicans KCCC 14172*	*C. tropicalis KCCC 13605*	*C. kefyr KCCC 13709*
Amphotericin B	1 h	39	26	44	28
	24 h	41	44	50	34
Nystatin	1 h	33	44	32	41
	24 h	71	79	42	50
Miconazole nitrate	1 h	45	50	34	36
	24 h	54	62	44	41
5-Fluorocytosine	1 h	40	44	29	23
	24 h	44	46	32	34

chronic atrophic candidosis.[66,67] This regimen was found to be superior to using nystatin or amphotericin B.[68]

Only one study exists regarding the effect of chlorhexidine on the adhesion of *C. albicans* to buccal epithelial cells.[69] Oral rinse with chlorhexidine gluconate reduced the adhesion of *C. albicans* to buccal epithelial cells collected immediately after the rinse. This reduction was observed following delay in collection of these cells.[69] Exposure of buccal epithelial cells to chlorhexidine *in vitro* also reduced yeast adhesion.[69] Thus, in addition to soaking of dentures in a solution of chlorhexidine, oral rinses with this antiseptic could be useful in the prevention and treatment of colonization of the oral cavity.

Studies into the mode of action of chlorhexidine against bacteria were carried out in the 1960s.[70-72] These investigations demonstrated that this antiseptic, at low concentrations, inhibits membrane enzymes and promotes leakage of cellular constituents. A more drastic effect is observed upon an increase in the concentration, where cytoplasmic constituents are coagulated

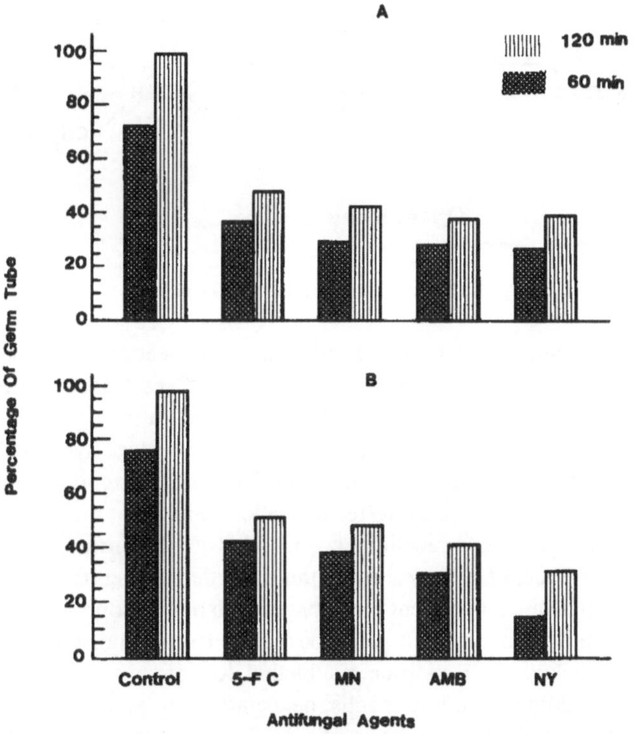

FIGURE 1. Effect of various antifungal agents (each at 1/4 MIC) on the germ-tube formation of (A) *C. albicans* KCCC 14172 and (B) *C. albicans* ATCC 10231.

TABLE 7

Yield of Extracellular Polymer (EP) Obtained from Culture Supernatants of C. *albicans* KCCC 14172 Following Growth with and without Subinhibitory Concentration Of Various Antifungal Agents Singly (1/4 MIC) or in Combination of Amphotericin B Plus 5-Fluorocytosine (Both At 1/8 MIC)

	Antifungal Agents					
	Control	AMB	NY	MN	5-FC	AMB + 5-FC
Dry weight[a]	3.2 ± 0.26	2.0 ± 0.03	2.8 ± 0.05	2.3 ± 0.08	1.2 ± 0.02	0.42 ± 0.001
Percentage yield of EP[b]	2.0 ± 0.41	3.55 ± 0.52	2.8 ± 0.2	5.1 ± 0.3	7.3 ± 0.8	9.2 ± 0.9
% Hexoses from EP[c]	90	91.5	92	98	98	80
Absolute value of hexoses[d]	58 ± 3	65 ± 5	75 ± 4	117 ± 10	87 ± 6	31 ± 1
% Proteins from EP[e]	10.0	8.5	7.8	1.6	1.9	20.1
Absolute value of proteins[f]	6.5 ± 0.9	6.0 ± 0.2	6.3 ± 0.7	1.8 ± 0.1	1.6 ± 0.1	7.8 ± 1.0

[a] Expressed in mg/ml of culture.
[b] E.P. expressed as % from yeast dry weight.
[c] Obtained by Dubois method and expressed as % from EP.
[d] Obtained by Dubois method, absolute values in mg/ml culture.
[e] Obtained by Lowry's method and expressed as % from EP.
[f] Obtained by Lowry's method and absolute value in mg/ml culture.

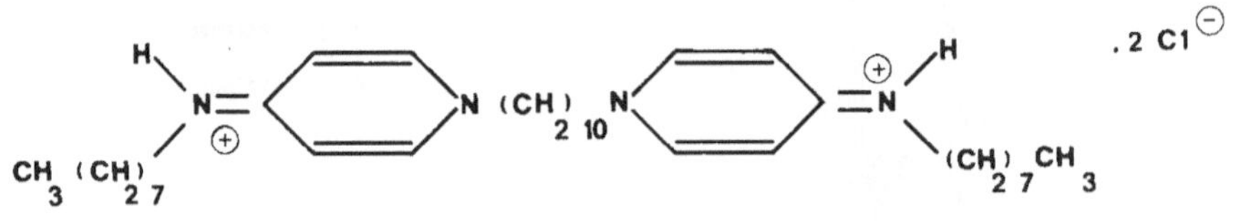

FIGURE 2. Structural formula of octenidine.

with concomitant alterations of the bacterial cell wall.[73,74] In an attempt to elucidate the action of chlorhexidine responsible for the reduction of adherence of yeast, McCourtie et al.[68] reported that *Candida* cells grown in the presence of sublethal concentrations of this agent were more sensitive to zymolase. The sensitivity to zymolase is known to be directly related to the presence of fibrillar material on the outer surface of the yeast.[19] The fibrillar component is vital for the adhesion of *Candida*.[25,28,75,76] Thus, it is inferred that chlorhexidine interferes with the adherence of yeast through an effect on this extracellular layer.[68] This was confirmed by the ultrastructural study of Bobichon and Bouchet.[77] They showed that sublethal concentrations of chlorhexidine effected a loss of cytoplasmic components and coagulation of nucleoprotein of *C. albicans* and caused the typical organization of the cell wall to be replaced by a homogenous, grayish structure with irregularities in thickness.[77] In addition, a sublethal dose of chlorhexidine led to a decrease in the percentage of budding *C. albicans* cells, hence affecting germ-tube formation.[77] Thus, chlorhexidine interferes with the adherence of *Candida* through an effect on the outer cell envelope and germination.

C. OCTENIDINE

Octenidine (Figure 2), like chlorhexidine, has both antibacterial and antimycotic activity. However, the antimycotic activity of octenidine is apparently many times more effective than that of chlorhexidine and is well tolerated by patients in preliminary clinical trials (Whittaker[78]). Octenidine is being developed as a surgical scrub, topical antimicrobial agent, and dental antiplaque mouth wash. These findings prompted us to test the effect of this drug on the adherence, germination, and ultrastructure of *C. albicans*. Pre-incubation of various species of *Candida* in the presence of subinhibitory concentrations of octenidine for 1 and 24 h reduced the adherence of these yeasts to buccal epithelial cells (Table 8). However, octenidine interfered with adherence of *C. tropicalis* and *C. pseudotropicalis* only after 24 h when 1/4 MIC concentrations were used (Table 8).[79] Furthermore, the addition of octenidine to *C. albicans* cells during culture led to a decrease in the percentage of budding cells even after 1 h (Figure 3). Germ-tube formation was similarly reduced by this drug (Figure 4). These effects were dose dependent, i.e., higher concentrations effected higher decreases in budding and germination.

Transmission electron microscopy revealed that octenidine affected the outer envelopes of *C. albicans*. The cell wall was altered, with electron transparent areas appearing between it and the cytoplasm.[160] Cytoplasmic disturbances were also noticed (Figure 5A, B and C). It is clear from these results that octenidine affects germination, budding, and the morphology of *C. albicans* in a manner similar to chlorhexidine.[79,160] Thus, octenidine, like chlorhexidine, reduces the adherence of yeasts to epithelial cells by a similar mechanism.

D. ALLIUM SATIVUM (GARLIC)

Garlic has been used as a spice, food, and folk medicine since ancient times.[80] Its extract has been reported to possess an *in vitro* growth inhibition effect against a large number of fungi including yeasts, and to have a protective effect against *in vivo* experimental fungal infec-

TABLE 8
Effect of Pre-Incubation of Various *Candida* spp. in Subinhibitory Concentrations of Octenidine on Adherence to Buccal Epithelial Cells

	Mean number of adherent yeasts per 100 epithelial cells ± SE (1 h Pre-Incubation)					
Organisms	**Control**	**1/2 MIC**	***P***	**Red.**	**1/4 MIC**	***P***	**Red.**
C. albicans KCCC 14172	551 ± 36	338 ± 26	<0.001	39	374 ± 27	<0.001	32
C. albicans ATCC 10231	509 ± 25	346 ± 26	<0.001	32	386 ± 28	<0.001	24
C. tropicalis KCCC 13605	336 ± 18	222 ± 16	<0.001	34	290 ± 21	N.S	
C. kefyr KCCC 13709	276 ± 24	202 ± 16	<0.05	27	217 ± 18	N.S	

	Mean number of adherent yeasts per 100 epithelial cells ± SE (24 h Pre-Incubation)						
Organisms	**Control**	**1/2 MIC**	***P***	**Red.**	**1/4 MIC**	***P***	**Red.**
C. albicans KCCC 14172	608 ± 42	269 ± 25	<0.001	56	367 ± 28	<0.001	40
C. albicans ATCC 10231	607 ± 46	332 ± 32	<0.001	45	362 ± 28	<0.001	40
C. tropicalis KCCC 13605	374 ± 27	171 ± 16	<0.001	54	285 ± 18	<0.05	24
C. kefyr KCCC 13709	284 ± 23	181 ± 13	<0.001	36	199 ± 16	<0.001	30

Note: *P* values compared with control. Red., percent reduction. N.S, not significant.

tions.[81-83] Kabelik[84] demonstrated that garlic extract was more effective against pathogenic yeasts, especially *C. albicans,* than nystatin, gentian violet, or methylene blue.[84] Clayton,[85] in her book on thrush, described the use of garlic poultices for the treatment of vaginal candidosis, and Odds,[1] in his recent monograph listed garlic as one of the unorthodox treatments of candidosis and called for more concerted efforts towards studying allicin, the active ingredient of garlic, and its derivatives with a view to clinical application.

Studies on the anticandidal mode of action of garlic have been undertaken recently by Ghannoum.[86] This investigation showed garlic exerts its effect by the oxidation of thiol groups present in the essential proteins, causing inactivation of enzymes and subsequent microbial growth inhibition.[86] Scanning electron microscopy and cell leakage studies showed that garlic treatment affected the structure and integrity of the outer surfaces of the yeast cells (Figure 6).[86]

Our findings and the observation of Prasad and Sharma[87] that garlic treatment in feed protected chicks from experimental candidosis and that infection induced by oral inoculation of chicks with *C. albicans* was successfully cured by providing a ration containing 5% garlic, directed us towards an investigation concerning the influence of garlic on the adherence of *Candida* to epithelial cells.[88] Our experiments showed that the adhesion of various *Candida* species to buccal epithelial cells is significantly reduced after both short (1 h) and long (24 h) time periods of yeast exposure to aqueous garlic extract (AGE) (Table 9). A similar adherence inhibition was observed upon pre-incubation of buccal epithelial cells with AGE (Table 10). In addition, germ-tube formation was suppressed when *C. albicans* cells were pretreated with AGE (Figure 7). In the same study, we have shown that there is a significant reduction in the adherence

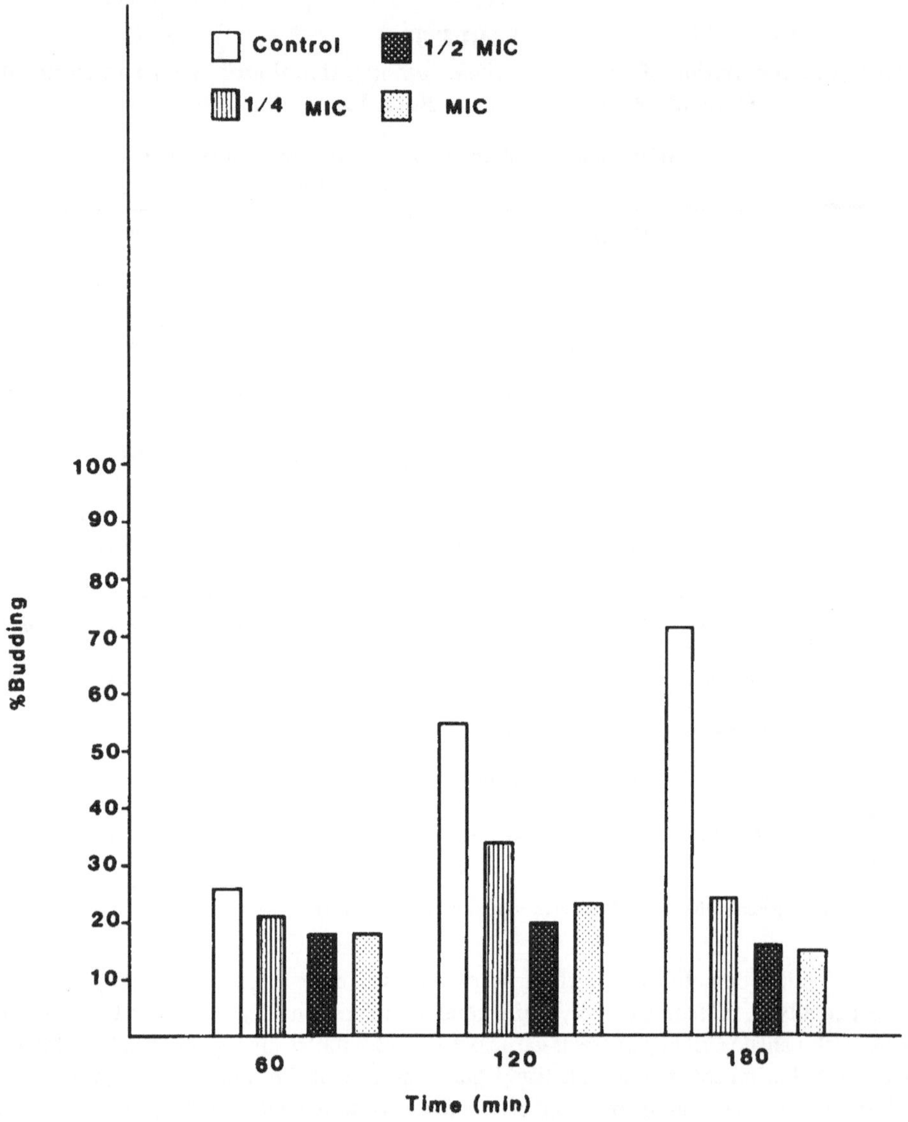

FIGURE 3. Effect of various concentrations of octenidine on the percentage of budding cells of *C. albicans* KCCC 14172.

of yeasts to buccal epithelial cells collected immediately after an oral rinse with AGE (Table 11). These findings may have clinical relevance in the treatment of oral candidosis.

The mechanism(s) responsible for the reduced adhesion of *Candida* following exposure to garlic may result from an interference with the synthesis or expression of adhesins or receptors involved in the attachment of yeast to epithelial cells. Another possibility is the prevention of attachment by mechanical distortion of the adhesins already present in the outer envelopes. The inhibition of germination of AGE-treated yeast cells may also indirectly inhibit the adherence of *Candida* to host cells.[29]

E. FORMALDEHYDE-RELEASING AGENTS

Formaldehyde-releasing compounds are broad spectrum, nonantibiotic antimicrobial agents

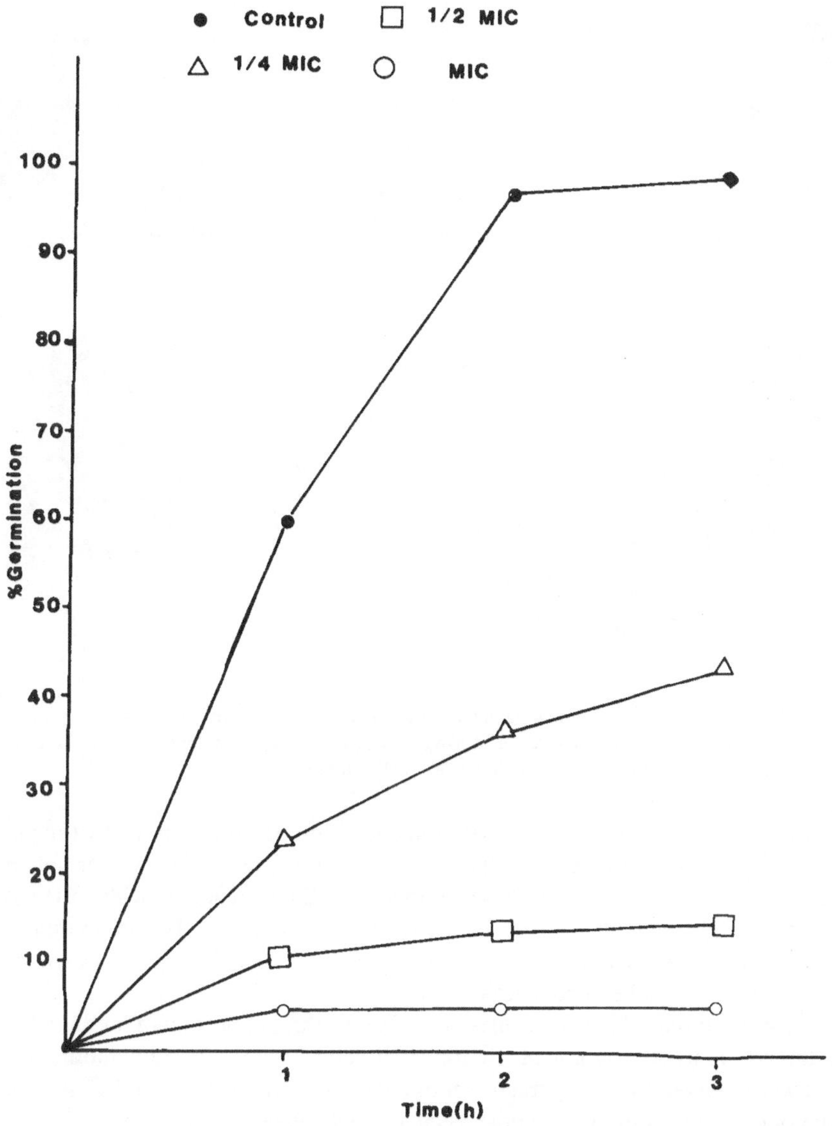

FIGURE 4. Effect of various concentrations of octenidine on the germ-tube formation of *C. albicans* KCCC 14172.

that are used widely in prophylaxis and therapy of infections of body surfaces and accessible body cavities, particularly in the treatment of peritonitis and bladder irrigation.[60,89] Interest in these compounds on the adherence of *C. albicans* followed the findings that noxythiolin has a capacity for reducing microbial adherence.[90]

The yeast anti-adherence properties of three different formaldehyde-releasing agents were studied by Gorman and his co-workers.[91-93] These are noxythiolin, polynoxylin, and taurolin (Figure 8). Noxythiolin reduced the adherence of *C. albicans, Escherichia coli*, and *Staphylococcus saprophyticus* to uroepithelial cells.[92] This reduction was true regardless of whether the drug was used to pretreat the microorganisms, or the epithelial cells, or both. The magnitude of the effect depended on species, growth phase, drug concentration, and contact time.[92] Furthermore, adherence of both the blastospore and pseudohyphal forms of *C. albicans* was reduced following treatment with this agent.[92]

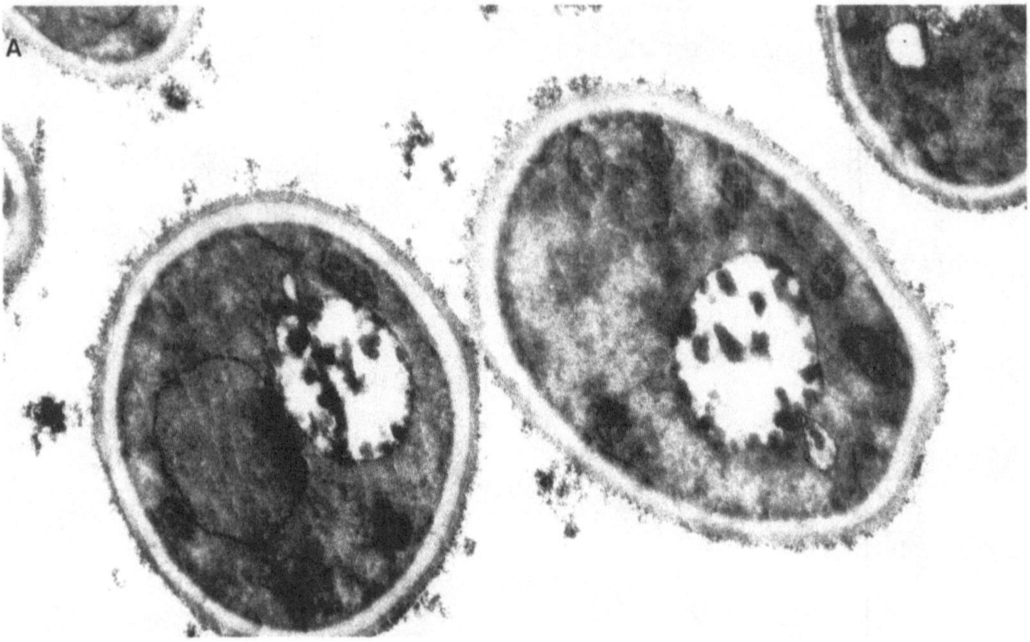

FIGURE 5. Electron microscopic photomicrograph illustrating the effect of octenidine. Cells were grown for 24h at 370°C (A) in the absence of drug, (B and C) or at MIC value. Micrograph shows alterations in the cell wall and cytoplasmic distrubances. (Magnification × 24,000, 8,300 and 10,800.)

The similarity of polynoxylin to noxythiolin and its proposed use in the treatment of oral candidosis,[94,95] prompted Gorman and associates[93] to investigate its effect on the adherence of *C. albicans* to buccal epithelial cells These authors showed that treatment with polynoxylin led to a significant reduction in the adherence of yeast to epithelial cells. This inhibition was time and dose dependent. Unlike noxythiolin, treatment of the epithelial cells with polynoxylin did not result in significant adherence inhibition.[93]

Taurolin, a condensate of two molecules of taurine and three molecules of formaldehyde, also reduced the adherence of *C. albicans* to epithelial cells.[96] Similar to noxythiolin, adherence inhibition was observed following treatment of either yeast or epithelial cells.[96] It is clear that this group of compounds shows potential as an anti-adherence agent.

The effectiveness of these compounds is apparently attributed to the release of formaldehyde.[89,97,98] Noxythiolin, for example, is reported to undergo the following reaction:

$$CH_3\text{–NH–CS–NH–}CH_2OH \rightarrow CH3\text{–NHCS–}NH_2 + H\text{–CHO}$$

This mechanism however, is no longer tenable for taurolin. Myers et al.[99] and Allwood and Myers,[100] suggested that the activity of taurolin is not due entirely to bacterial adsorption of free formaldehyde but also to a reaction with a masked formaldehyde. In a subsequent study Gidley and Sanders[101] showed that the antimicrobial activity of taurolin is due to the delivery of biologically active methylol taurinamide components which react via methylene iminium ions with cell wall components of bacteria. This could be related to the mechanism responsible for adherence inhibition observed by Gorman and co-workers.[91,93] Inhibition of germination and hyphal development of *C. albicans* by polynoxylin and noxythiolin could be another factor responsible for the reduced adherability of yeast treated by such agents.[91,93]

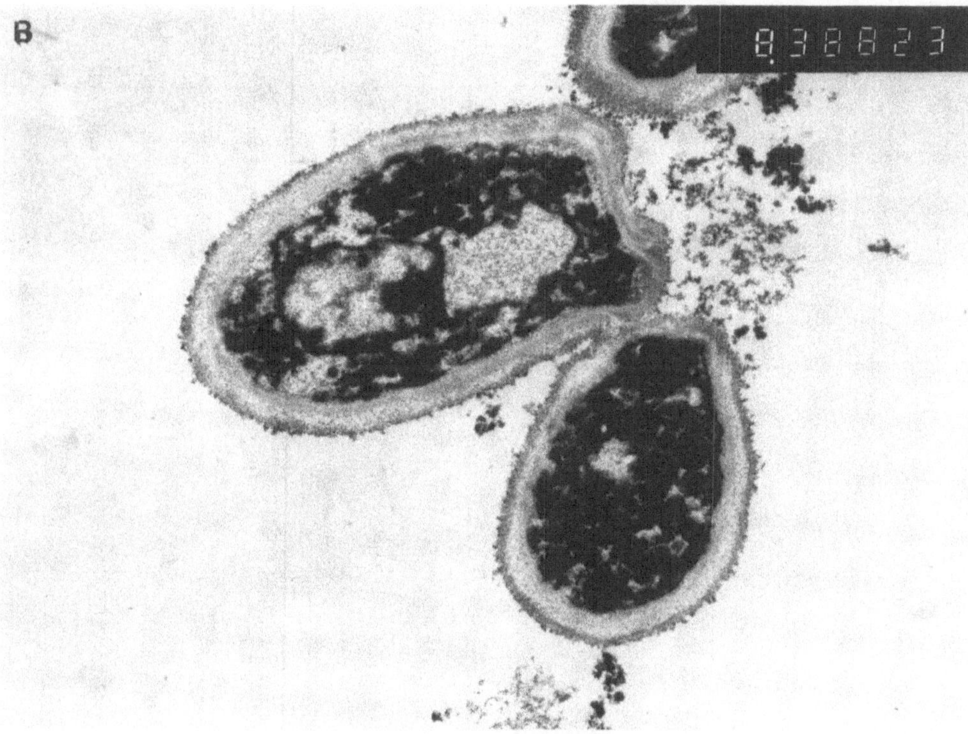

FIGURE 5B.

III. THERAPEUTIC AGENTS ASSOCIATED WITH POTENTIATION OF CANDIDOSIS

The introduction of newer therapeutic modalities into clinical medicine has been responsible for the dramatic change in the incidence of candidosis. Of these regimens, probably the most important is the introduction of antibiotics, steroids, and antineoplastic agents. The precise mechanisms responsible for this susceptibility to infection are not clear. Mechanisms suggested in literature have concentrated almost entirely on the influence of therapeutic agents on the host with little attention directed towards the effects exerted on the parasite (see Chapter 2). We therefore focused our efforts on studying the effect of steroids and antineoplastic agents on *Candida*. Since in Chapter 2, we have covered their effects on growth, morphology, metabolism and tissue colonization of *Candida*, here we will confine the disussion to the influence of steroids, antineoplastic agents, and X-irradiation on the adherence of yeast to epithelial cells.

A. GLUCOCORTICOIDS

Glucocorticoids are known to enhance pathogenicity of opportunistic fungi,[102,103] and a number of case studies suggest that steroid therapy was the main factor, predisposing patients to superficial or systemic candidosis.[103-112] Several mechanisms of action have been suggested for the enhancement of fungal infection by steroids. These may have a direct effect on the organism,[113] or on its host,[114,117] or on both (for review see Ghannoum[118]). Previous studies in our laboratories have shown that unaltered dexamethasone is incorporated primarily into the cell wall and membrane of *C. albicans*.[119] This suggested that steroids might promote infection, at least in part, by affecting the adherence of the yeast to host tissues. To test this hypothesis we

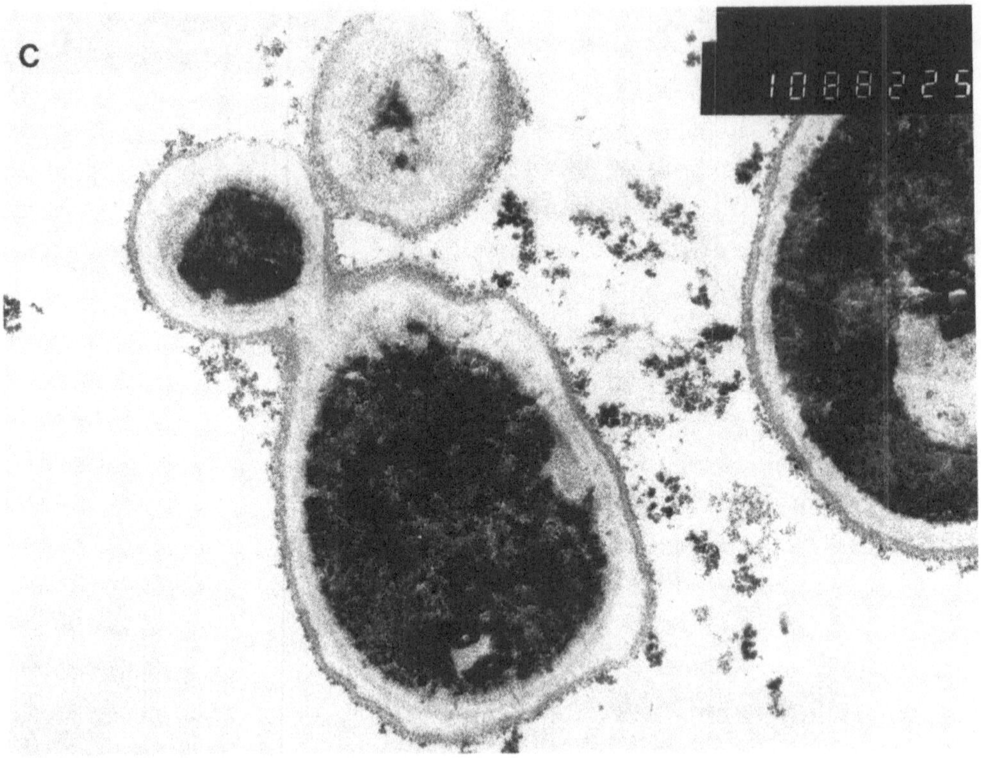

FIGURE 5C.

chose a number of steroids known to predispose to candidosis,[114,116,117,120-122] and investigated their effect on the adherence of various *Candida* species to buccal epithelial cells.

Investigations revealed that *in vitro* culturing of three different yeast species with a number of glucocorticoids altered their adherence ability in two ways: in general, growth with dexamethasone and triamcinolone acetonide promoted the adherence (the increase in adherence ranged between 17 and 44%), while growth in the presence of cortisone acetate or hydrocortisone blocked the adherence (inhibition ranged from 16 to 32%) (Table 12). In addition, pretreatment for 30 min of either yeast or epithelial cells led to a substantial blockage of adherence (Table 13).[123]

Variation in adherence of these yeast strains in the presence of various steroids may be due to: (1) differences in the molecular configurations of the steroids used or (2) to differences in interaction of these compounds with the yeast. Frenkel,[124] in a study of the role of corticosteroids in potentiation of fungal diseases, discussed the criteria for the structural modification required to enhance acute infection. He suggested that 11-hydroxylation was of primary importance in this respect, 17-hydroxylation of secondary importance, and 21-hydroxylation of tertiary importance. The infection-enhancing effects of steroids can be potentiated further by a double bond between 1 and 2, by 6- or 9- fluorination, by 6-methylation, by 16-hydroxylation or methylation, and by certain other modifications. Glucocorticoids tested, in relation to adherence, possess some of these requirements (Figure 9). Furthermore, dexamethasone and triamcinolone acetonide on the one hand and cortisone acetate and hydrocortisone acetate on the other possess common structural features. The structural differences between these two groups may account for the dissimilarities observed in adherence. Differences in the interaction between different glucocorticoids and *C. albicans* have been shown by Loose et al.[125] These workers showed that corticosterone and progesterone were specifically bound by cytosol

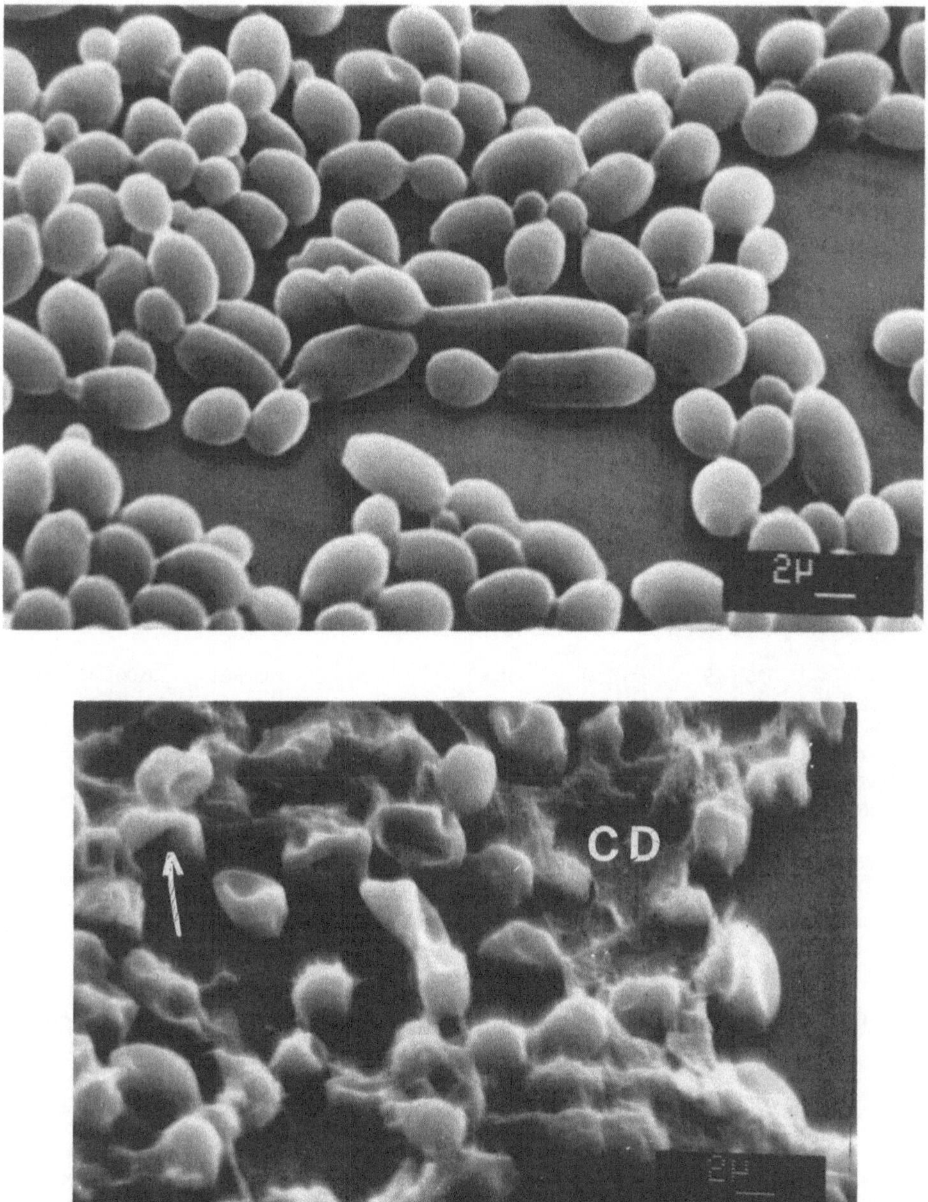

FIGURE 6. Scanning electron micrograph of *C. albicans* KCCC 14172 grown in the absence (top) or presence (bottom) of aqueous garlic extract. Arrow points to cell collapse; CD, cytoplasmic debris. (Magnification × 3500). (From Ghannoum, M. A., *J. Gen. Microbiol.*, 134, 2717, 1988. With permission.)

preparations from *C. albicans*, while dexamethasone failed to bind. Since corticosterone, hydrocortisone, and cortisone acetate are closely related structurally (Figure 9) it is anticipated that they behave in a similar manner, i.e., to interact with the yeast through binding to a specific receptor protein in the cytoplasm, in a similar fashion to mammalian hormones.[126] Relevant to this is the work by Powel et al.,[127] who showed that steroids which act via a receptor mechanism are capable of direct stimulation of protein synthesis and germ-tube formation in *C. albicans*.[127] This interaction may trigger certain biological responses, which in turn, may interfere in the synthesis or expression of adhesins subsequently blocking the adhesion process. The mecha-

TABLE 9
Adherence of *Candida* spp. to Human Buccal Epithelial Cells After Incubation of Yeast in Media Containing Various Concentrations of Age

	Mean number of adherent yeasts per 100 epithelial cells ± SE 1 h Pre-Incubation						
Organisms	**Control**	**MIC**	***P***	**Red.**	**1/2 MIC**	***P***	**Red.**
C. albicans ATCC 10231	517 ± 32	285 ± 20	<0.001	45	377 ± 26	<0.05	27
C. albicans KCCC 14172	556 ± 31	267 ± 22	<0.001	52	303 ± 22	<0.001	45
C. tropicalis KCCC 13605	376 ± 24	217 ± 19	<0.001	42	277 ± 21	<0.05	26
C. kefyr KCCC 13709	289 ± 23	194 ± 17	<0.001	33	238 ± 21	N.S	

	Mean number of adherent yeasts per 100 epithelial cells ± SE 24 h Pre-Incubation						
Organisms	**Control**	**MIC**	***P***	**Red.**	**1/2 MIC**	***P***	**Red.**
C. albicans ATCC 10231	556 ± 29	193 ± 21	<0.001	65	285 ± 21	<0.001	49
C. albicans KCCC 14172	557 ± 34	228 ± 16	<0.001	59	304 ± 22	<0.001	45
C. tropicalis KCCC 13605	395 ± 24	194 ± 14	<0.001	51	247 ± 18	<0.001	37
C. kefyr KCCC 13709	293 ± 18	125 ± 11	<0.001	57	160 ± 14	<0.001	45

Note: *P* values compared with control. Red., percent reduction. N.S, not significant.

TABLE 10
Effect of Pre-Incubation (5 Min.) of Buccal Epithelial Cells with Various Concentrations of AGE on Adherence of *C. albicans* KCCC 14172

AGE Concentration (%)	Adherent yeasts to 100 BEC (mean ± SE)	*P*	Reduction (%)
0.0 (Control)[a]	557 ± 34		
0.04	329 ± 24	< 0.001	41
0.1	314 ± 20	< 0.001	44
0.2	279 ± 23	< 0.001	50
0.3	252 ± 23	< 0.001	55

Note: *C. albicans* was not exposed to AGE. *P* values compared to control.

[a] Buccal epithelial cells not exposed to AGE.

nism responsible for enhancement of *Candida* adhesion by dexamethasone and triamcinolone acetonide is unknown. However, as dexamethasone is incorporated into the outer surfaces of the yeast it could promote adherence by a surface receptor mechanism. However, this requires further investigation.

Since adherence is considered as the first stage of *Candida* infection and since dexam-

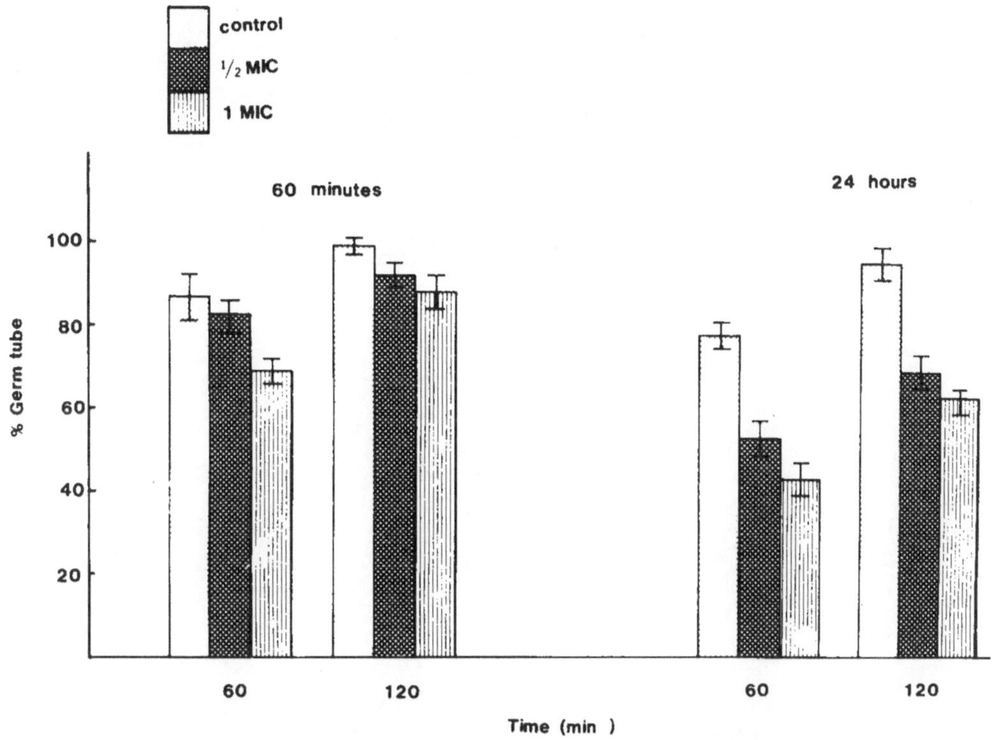

FIGURE 7. Effect of aqueous garlic extract on the germ-tube formation of *C. albicans* KCCC 14172.

TABLE 11
Adherence of *C. Albicans* KCCC 14172 to Buccal Epithelial Cells Collected After Oral Rinses with Various Concentrations of AGE for 1 Min

AGE conc. (%)	Adherent yeasts to 100 BEC (mean ± SE)	*P*	Reduction (%)
0.0 (Control)[a]	558 ± 37		
0.04	526 ± 37	N.S	—
0.1	409 ± 24	<0.001	27
0.2	317 ± 22	<0.001	43
0.3	305 ± 21	<0.001	45

Note: *P* values compared with control. N. S, not significant.

[a] BEC not exposed to AGE.

ethasone and triamcinolone acetonide are found to promote adhesion, while cortisone acetate and hydrocortisone are shown to inhibit adhesion, then it could be concluded that the former are pathogenicity enhancers while the latter are inhibitors. This is, however, a simplistic view since *Candida* virulence is decided by a multitude of determinants, none of which is predominant.[1]

B. ANTINEOPLASTIC AGENTS AND X-IRRADIATION

A combination of radiation therapy and surgery have been advocated in the treatment for certain advanced carcinomas.[128] However, the general dissatisfaction with the results of treatment of carcinomas of the head and neck after radical surgery and extensive radiotherapy

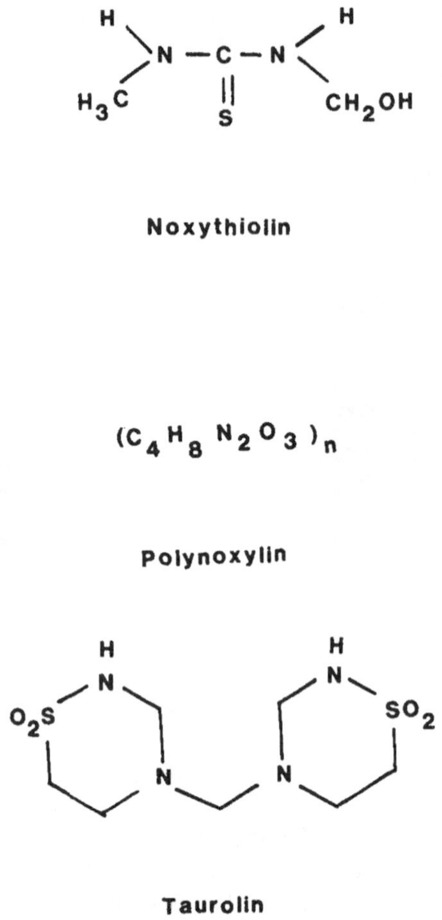

FIGURE 8. Structural formulae of three formaldehyde-releasing compounds.

led to the use of a triad of chemotherapy, radiotherapy, and surgery for cancer treatment.[129,130] Each of these modalities is known to predispose the cancer patient to infection, particularly those due to *Candida* species.[1,131,132] *C. albicans*, *C. tropicalis*, *C. glabrata*, *C. parapsilosis*, *C. krusei*, *C. guilliermondi*, and *C. pseudotropicalis* are the most frequently isolated species from these patients.[133,134] This situation has led to a number of investigations concerning the effect of ionizing radiation and chemotherapy on the host as well as on the infective organisms.[135-138] The effect of these therapeutic modalities on the immunological parameters and resistance to experimental infection in the host has been studied.[139,140] The influence of these regimens on the growth, morphology, genetics and pathogenicity of yeast has been investigated also.[136,141,142] However, the effect of radiation and/or chemotherapy on the adherence of *Candida* species to buccal epithelal cells, has not been studied. It is known that the adhesive properties of microorganisms are greatly influenced by various circumstances operating within the host's ecosystems.[143] In addition, chemotherapy with immunosuppressive agents, as well as, X-irradiation are known to lead to a multitude of alterations in the oral mucosa of patients.[144,145] Therefore, we decided to examine the role of antineoplastic agents, X-irradiation, and a combination of both on the adherence of *Candida* species to buccal epithelial cells *in vitro*.

Growth of three *Candida* species in the presence of 11 antineoplastic agents (Table 14) led to a significant reduction of adherence of these yeasts to buccal epithelial cells, between 30 and 61% of the control value. This effect was observed whether exponential or stationary phase

TABLE 12
Effect of Growth of Various Candida spp. in Glucocorticoids-Containing Media on Adherence to Buccal Epithelial Cells

Candida spp.	Control (0.00)	Yeast grown in the presence of: Dexamethasone (0.01%)	*P*	Triamcinolone acetonide (0.01%)	*P*
C. albicans ATCC 10231	918 ± 35 (100)	1131 ± 35 (123)	<0.001	1083 ± 34 (118)	<0.05
C. albicans KCCC 13878	844 ± 30 (100)	965 ± 30 (114)	<0.05	984 ± 31 (117)	<0.05
C. albicans KCCC 14091	792 ± 34 (100)	958 ± 38 (121)	<0.05	924 ± 36 (117)	<0.05
C.tropicalis KCCC 13065	915 ± 38 (100)	1316 ± 46 (144)	<0.001	1089 ± 46 (119)	<0.001
C. kefyr KCCC 13709	757 ± 30 (100)	868 ± 29 (115)	<0.05	775 ± 27 (102)	N.S

Candida spp.	Yeast grown in the presence of: Cortisone acetate (0.01%)	*P*	Hydrocortisone (0.01%)	*P*
C. albicans ATCC 10231	749 ± 30 (81)	<0.001	715 ± 30 (78)	<0.001
C. albicans KCCC 13878	676 ± 34 (80)	<0.001	626 ± 31 (74)	<0.001
C. albicans KCCC 14091	562 ± 25 (71)	<0.001	577 ± 25 (73)	<0.001
C. tropicalis KCCC 13605	765 ± 34 (83)	<0.05	625 ± 25 (68)	<0.001
C. kefyr KCCC 13709	636 ± 26 (84)	<0.05	558 ± 27 (74)	<0.001

Note: Values are number of yeast cells adhering to 100 BEC (mean ± SE). Values in parentheses are percent attachment. *P* values compared with control. N.S, not significant.

From Ghannoum, M. A. and Abu ElTeen, K., *Mycopathologia*, 98, 171, 1987. With permission.

Candida cells were used (Figure 10). Adherence was not affected, however, when endoxan-asta, thiotepa, and methylprednisolone were used (Table 15). Exposure of *C. albicans* to various doses of X-irradiation also led to a reduction in adherence which was dose dependent (Figure 11). Similar results were obtained when buccal epithelial cells were exposed to radiation. This was accentuated when both yeast and buccal epithelial cells were irradiated simultaneously (Table 16).[146]

Table 17 shows that treating *C. albicans* with a combination of chemotherapy and X-irradiation led to a higher reduction in adherence of yeast to buccal epithelial cells compared to when the yeast was treated by either chemotherapy or radiation alone (reduction between 63 to 74% as compared with control). The adherence capability of buccal epithelial cells obtained from cancer patients was compared with that of epithelial cell obtained from healthy volunteers. Our results showed that a significant difference existed in the binding of *C. albicans* to these cells ($P < 0.001$). *C. albicans* adhered to epithelial cells from healthy subjects to a greater extent than from cancer patients (446 ± 22 compared with 270 ± 21 yeasts adhering to 100 buccal epithelial cells).

TABLE 13
Effect of Pre-Treatment of Cells With Dexamethasone or Cortisone Acetate on the Adherence of *C. Albicans* (KCCC 14091) to Buccal Epithelial Cells

Pretreatment		Dexamethasone			Cortisone acetate		
Yeast cells	Epithelial cells	Number of yeast adhering to 100 BEC (mean ± SE)	Red.	*P*	Number of yeast adhering to 100 BEC (mean ± SE)	Red.	*P*
–	– (control)	764 ± 30			764 ± 30		
+	–	571 ± 30	25	<0.001	500 ± 25	35	<0.001
–	+	560 ± 27	27	<0.001	53 ± 25	30	<0.001

Note: BEC, buccal epithelial cells. Red., percent reduction of adherence from control values. *P* values compared with control.

From Ghannoum, M. A. and Abu ElTeen, K., *Mycopathologia*, 98, 171, 1987. With permission.

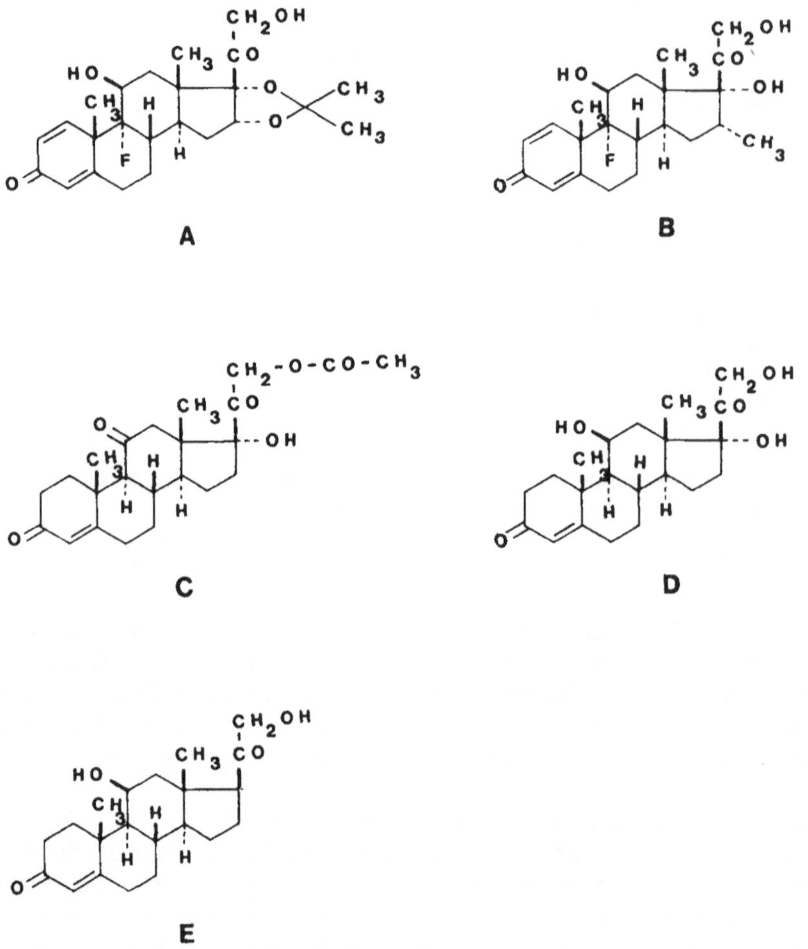

FIGURE 9. Structural formulae of glucocorticoids (A) Triamcinolone acetonide; (B) Dexamethasone; (C) Cortisone acetate; (D) Hydrocortisone; and (E) Corticosterone. (From Ghannoum, M. A. and Abu El Teen, K., *Mycopathologia*, 98, 171, 1987. With permission.)

TABLE 14
Antineoplastic Agents Used With Their MIC Values (μg/ml) Against Various *Candida* spp.

Antineoplastic agents	Supplier	Abbreviation	MIC[a]
Bleomycin hydrochloride	Nippon Kayaku Co. Ltd. Tokyo, Japan	BLM	500
Methotrexate	Cyanamid International Corporation, Switzerland	MTX	750
Methyl prednisolone sodium succinate	Upjohn, Puurs, Belgium	PRN	>1500
Endoxan-asta	Asta-werke AG chemische Fabrik, Germany	CYT	1500
5-Fluorouracil	Hoffmann-La Roche and Co. Ltd., Basel, Switzerland	5-FU	1250
Vincristine sulphate	Eli Lilly and Company Ltd., U.K.	VCR	1000
Cytarabine	Upjohn, Puurs, Belgium	Anac	>1500
Dactinomycin	Merck Sharp & Dohme International Division of Merck & Co. Inc., New Jersy, U.S.A.	DAC	500
Mustine hydrochloride	The Boots Company Nottingham, U.K.	M.CHI	500
Vinblastine sulphate	Eli Lilly & Co. Ltd., U.K.	VELBE	>1500
Thiotepa	American Cyanamid Co., NewYork	THT	500

[a] Minimal variation was observed in the sensitivity of *C. albicans* KCCC 14172, *C. albicans* ATCC 10231, *C. tropicalis* KCCC 13605 and *C. kefyr* KCCC 13709 towards the antineoplastic agents tested.

From Ghannoum, M. A., Abu ElTeen, K. H., and Motawy, M. S., *Mycopathologia*, 104, 171, 1988. With permission.

The mechanism(s) responsible for blocking the adherence of *Candida* to epithelial cells by antineoplastic agents is unknown. Possibly two mechanisms are functioning. The first involves the uptake of antineoplastic agents by *Candida* species. Once inside, these agents may induce certain cellular activities similar to what is known to occur in bacteria.[147,148] It is possible that these changes are reflected by diversified modification (depending on the agent) of the outer surface structures known to mediate adherence (such as alteration in the nature of adhesins). It is also possible that these agents may cause an inhibition of synthesis and expression of yeast adhesins at the concentrations used. The second mechanism may be that these agents are not transported by the yeasts but are adsorbed at the cell surface masking the adhesins and causing an adherence blockage. Further investigations are required to determine the mechanism(s) responsible.

As with chemotherapy, the mechanism responsible for the blockage of adherence by X-irratiation is unknown. However, our results suggest that radiation, somehow, modified *Candida* adhesins and/or epithelial cells receptors. In this investigation low doses of radiation (between 50 to 250 c-Gy) did not affect the viability of the microorganisms, whereas higher doses (>250 c-Gy) did. Since adherence was significantly reduced at all radiation doses (Table 16), this would imply that the effect on adherence is due to reasons other than viability per se. However, cell viability could have played a partial role in the high doses employed (500 c-GY) since it has been shown that nonviable yeasts do not adhere as well as viable cells to fibrin-platelet matrix formed *in vitro*.[149] On the other hand, Lee and King,[150] studying the adherence of

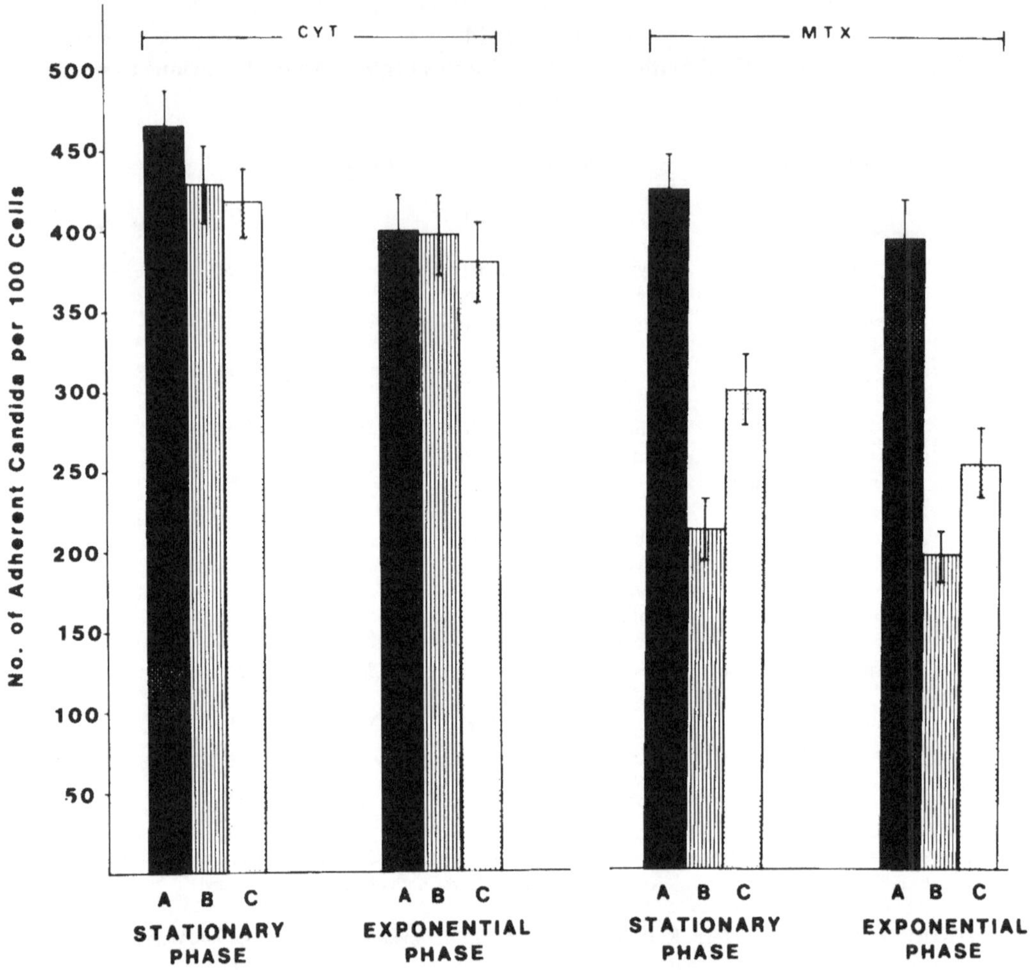

FIGURE 10. Effect of Endoxan-asta (CYT) and Methotrexate (MTX)on the adherence of *C. albicans* KCCC 14172 to buccal epithelial cells. (A) Yeast grown for 24 h (stationary phase) or 12 h (exponential phase) without drug (control). (B) Yeast grown for 24 h or 12h in the presence of CYT or MTX (100 μg/ml). (C) CYT or MTX (100 μg/ml final concentration) included in the assay medium. (From Ghannoum, M. A., Abu ElTeen, K. H., and Motawy, M. S., *Mycopathologia*, i, 104, 171, 1988.With permission.)

C. albicans to human vaginal epithelial cells, showed that viability was not essential for adherence, but severe methods used to kill the blastospores did reduce their attachment. To what extent *Candida* viability is vital to adherence to epithelial cells after various treatments should be examined in greater detail.

The combined action of chemotherapy and radiation on the adherence of *C. albicans* to buccal epithelial cells showed two classes of combined effects (Table 17). In the first, both ionizing radiation and the other agent produced reduction in adherence. This is true in the case of methotrexate and various doses of radiation. The combined effect enhanced the reduction to between 14 to 25% of the values obtained when radiation alone was used. Bleomycin showed similar effects. The second class includes the combination where ionizing radiation produces an effect of which the amount was modified by the other agent which by itself is inactive. This was true in the case of radiation and endoxan-asta which by itself showed no significant effect on adherence ($P > 0.05$). However, the reduction of adherence due to radiation was enhanced between 5 and 14% of the values obtained for radiation alone. Interaction between antineoplastic

TABLE 15
Effect of Growth (24 h) of Various *Candida* spp. in the Presence and Absence (Control) Of Antineoplastic Agents

	Yeast grown in the presence of			
	Control[a]	BLM[b] (50)		
***Candida* spp.**	**Adherence**	**Adherence**	**Red.**	***P***
C. albicans ATCC 10231	439 ± 24	289 ± 24	34	<0.001
C. albicans KCCC 14172	445 ± 25	294 ± 20	34	<0.001
C. tropicalis KCCC 13605	353 ± 23	230 ± 17	35	<0.001
C. kefyr KCCC 13709	280 ± 17	191 ± 15	32	<0.001

	Yeast grown in the presence of					
	MTX[c] (100)			CYT[d] (100)		
***Candida* spp.**	**Adherence**	***Red.***	***P***	**Adherence**	**Red.**	***P***
C. albicans ATCC 10231	222 ± 22	49	<0.001	432 ± 27	—	N.S
C. albicans KCCC 14172	247 ± 17	45	<0.001	390 ± 20	—	N.S
C. tropicalis KCCC 13605	221 ± 17	37	<0.001	294 ± 21	—	N.S
C. kefyr KCCC 13709	162 ± 14	42	<0.001	269 ± 18	—	N.S

Note: Data are presented in numbers of yeast cells adhering to 100 buccal epithelial cells (mean ± SE). Values in parentheses represent levels (mg/ml) of antineoplastic agents used. P values compared with control. Red., percent reduction. N.S, not significant.

[a] The mean of control adhesion assays carried out with each batch of test assays.
[b] Similar results were obtained with DAC, VELBE, and M. HCl.
[c] Similar results were obtained with 5FU, VCR, and Anac.
[d] Similar results were obtained with PRN and THT.

From Ghannoum, M. A., Abu ElTeen, K. H., and Motawy, M. S., *Mycopathologia*, 104, 171, 1988. With permission.

agents and radiation is known and potentiating effects of radiation by antibiotics and/or antineoplastic agents were reported.[151-155] However, the molecular mechanism(s) involved in the sensitization of yeasts to radiation by these agents is not known.

Our findings that *C. albicans* adhered less efficiently to buccal epithelial cells collected from cancer patients than to cells from healthy individuals could be due to changes in the epithelial cell's surface caused by the therapy which could have a detrimental effect on the receptor sites. Chemotherapy with immunosuppressive agents leads to a number of modifications in patients, including mucosal changes in the mouth.[144] Radiation causes acute and chronic xerostomia with altered or absent saliva.[145] Since the adhesive properties of microorganisms are greatly influenced by various circumstances operating within host ecosystems such as the mouth,[143,156,157] the effects of such therapeutic regimes could be of vital importance and will require further study.

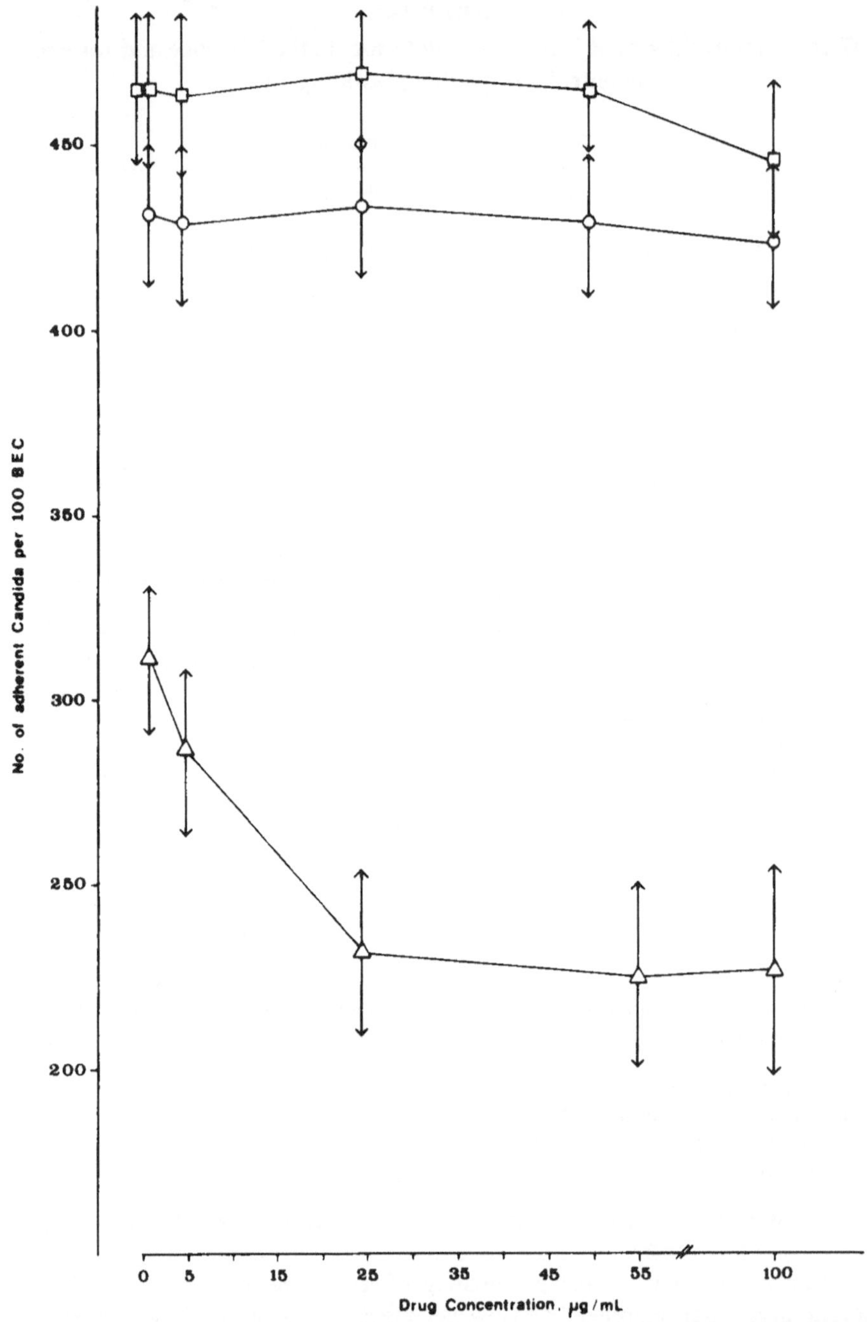

FIGURE 11. Effect of drug concentration, included in the growth medium, on the adherence of *C. albicans* KCCC 14172 to buccal epithelial cells. (□) Control, (Δ) MTX, (O) CYT. (From Ghannoum, M. A., Abu ElTeen, K. H., and Motawy, M. S., *Mycopathologia,* 104, 171, 1988. With permission.)

The findings of this investigation regarding the blockage of adherence by antineoplastic agents and irradiation therapy may have interesting clinical implications. These findings suggest that aggressive cancer therapy may have an adverse effect on the adherence of *Candida*. However, the relevance of our *in vitro* studies to the *in vivo* situation is difficult to assess and necessitates further investigation.

TABLE 16
Effect of X-Irradiation of *C. Albicans* KCCC 14172 (Group 1), Buccal Epithelial Cells (Group 2), and a Mixture of Both (Group 3) on the Adherence of Yeast to Buccal Epithelial Cells

	Type of cells exposed to radiation[a]								
Radiation dose (c-GY)	Yeast	(Red.)	*(P)*	BEC	(Red.)	*(P)*	Yeast and BEC	(Red.)	*(P)*
O (control)	434 ± 25			422 ± 25			432 ± 24		
50	301 ± 19	31	<0.001	280 ± 21	34	<0.001	235 ± 19	46	<0.001
100	287 ± 18	34	<0.001	226 ± 17	46	<0.001	185 ± 1957	57	<0.001
200	248 ± 19	43	<0.001	222 ± 17	47	<0.001	173 ± 1	60	<0.001
250	229 ± 19	47	<0.001	220 ± 20	48	<0.001	160 ± 16	63	<0.001
500	202 ± 19	53	<0.001	218 ± 18	48	<0.001	156 ± 15	64	<0.001

Note: Comparison of significance between the three groups studied showed that no statistical significance exists between group 1 and 2 (*P* > 0.05) while statistical significance existed between group 3 on the one hand and groups 1 and 2 on the other. *P* values are compared with control. Red., percent reduction.

[a] Values given are number of yeast cells adhering to 100 buccal epithelial cells BEC (mean ± SE).

From Ghannoum, M. A., Abu ElTeen, K. H., and Motawy, M. S., *Mycopathologia*, 104, 171, 1988.With permission.

TABLE 17
Effect of Combination Treatment on The Adherence of *C. Albicans* KCCC 14172 to Buccal Epithelial Cells

Doses used to irradiate *C. albicans* (C-Gy)	C. albicans grown in the presence of 10 μg/ml methotrexate			C. albicans grown in the presence of 10 μg/ml Endoxan-asta			C. albicans grown in the presence of 5 μg/ml Bleomycin hydrochloride		
	Cells[a]	Red.	P	Cells[a]	Red.	P	Cells[a]	Red.	P
O (control)	434 ± 25[b]			434 ± 25[b]			430 ± 23[b]		
0	222 ± 22[c]	49	<0.001	432 ± 21	N.S.	<0.001	285 ± 25	34	<0.001
50	162 ± 17	63	<0.001	257 ± 17	41	<0.001	228 ± 24	53	<0.001
100	150 ± 16	65	<0.001	216 ± 17	50	<0.001	172 ± 20	60	<0.001
200	140 ± 15	68	<0.001	207 ± 20	52	<0.001	163 ± 19	62	<0.001
250	123 ± 15	72	<0.001	183 ± 18	58	<0.001	151 ± 17	65	<0.001
500	112 ± 14	74	<0.001	174 ± 18	60	<0.001	133 ± 14	69	<0.001

Note: Red., percent reduction. P values are compared with control. N.S, not significant.

[a] Values given are number of yeast cells adhering to 100 buccal epithelial cells (mean ± SE).
[b] This value refers to the adherence of control *C. albicans* not exposed to antieoplastic agent or radiation.
[c] Compared with control (antieoplastic agent only).

From Ghannoum, M. A., Abu ElTeen, K. H., and Motawy, M. S., *Mycopathologia*, 104, 171, 1988.With permission

The literature and data presented in this chapter indicate that, *in vitro,* therapeutic agents interfere with the adherence of *Candida* to epithelial cells. This suggests a possible role for these compounds in the control and prevention of infections due to these yeasts. However, more systematic screening of various chemotherapeutic agents, antibiotics, immunosuppresive agents, and irradition should be undertaken. In addition, the role of these agents *in vivo* deserves particular attention, since it is difficult to extrapolate results from one situation to the other.

ACKNOWLEDGMENT

This investigation was supported by Kuwait University, Research Council, and Kuwait Foundation for the Advancement of Science, Grant number SO 040/87-07-27.

REFERENCES

1. **Odds, F. C.,** *Candida* and *Candidosis,* A review and Bibliography, Baillière Tindall, London, 1988.
2. **Rotrosen, D., Calderone, R. A., and Edwards, J. E. Jr.,** Adherence of *Candida* species to host tissues and plastic surfaces, *Rev. Infect. Dis.,* 8, 73, 1986.
3. **Alkan, M. L. and Beachey, E. H.,** Excretion of lipoteichoic acid by group A streptococci: influence of penicillin on excretion and loss of ability to adhere to human oral mucosal cells, *J. Clin. Invest.,* 61, 671, 1978.
4. **Bernard, J. P., Francioli, P., and Glauser, M. P.,** Vancomycin prophylaxis of experimental *Streptococcal sanguis* inhibition of bacterial adherence rather than bacterial Killing, *J. Clin. Invest.,* 68, 1113, 1982.
5. **Eisenstein, B. I., Ofek, I., and Beachey, E. H.,** Loss of lectinlike activity in aberrant type 1 fimbriae of *Escherichia coli, Infect Immun.,* 31, 792, 1981.
6. **Eisenstein, B. I., Beachey, E. H., and Ofek, I.,** Influence of sublethal concentrations of antibiotics on the expression of the mannose-specific ligand of *Escherichia coli, Infect. Immun.,* 28, 154, 1980.
7. **Scheld, W. M., Zak, O., Vosbeck, K., and Sande, M. A.,** Bacterial adhesion in the pathogenesis of infective endocarditis. Effect of subinhibitory antibiotic concentrations on streptococcal adhesion *in vitro* and the development of endocarditis in rabbits, *J. Clin Invest.,* 68, 1381, 1981.
8. **Vosbeck, K., Handschin, H., Menge, E. E., and Zak, O.,** Effect of subminimal inhibitory concentrations of antibiotics on adhesiveness of *Escherichia coli in vitro, Rev. Infect. Dis.,* 1, 845, 1979.
9. **Eisenstein, B. I., Ofek, I., and Beachey, E. H.,** Interference with the mannose binding and epithelial cell adherence of *Escherichia coli* by sublethal concentrations of streptomycin, *J. Clin. Invest.,* 63, 1219, 1979.
10. **Boethling, R. S.,** Regulation of extracellular protease secretion in *Pseudomonas maltophilia, J. Bacteriol.,* 123, 954, 1975.
11. **Piovant, M., Varenne, S., Pages, J. M., and Lazdunski, C.,** Preferential sensitivity of synthesis of exported proteins to translation inhibitors of low polarity in *Escherichia coli, Mol. Gen. Genet.,* 164, 265, 1978.
12. **Shibl, A. M. and Al-Sowaygh, I. A.,** Differential inhibition and bacterial growth and hemolysin production by lincosamide antibiotics, *J. Bacteriol.,* 137, 1022, 1979.
13. **Hirashima, A., Childs, G., and Inouye, M.,** Differential inhibitory effects of antibiotics on the biosynthesis of envelope proteins of *Escherichia coli, J. Mol. Biol.,* 79, 373, 1973.
14. **Horne, D. and Tomasz, A.,** Tolerant response of *Streptococcus sanguis* to beta-Lactams and other cell wall inhibitors, *Antimicrob. Agents Chemother.,* 11, 888, 1977.
15. **Horne, D. and Tomasz, A.,** Release of lipoteichoic acid from *Streptococcus sanguis:* stimulation of release during penicillin treatment, *J. Bacteriol.,* 137, 1180, 1979.
16. **Horne, D., Hackenback, R., and Tomasz, A.,** Secretion of lipids induced by inhibition of peptidoglycan synthesis in *streptococci, J. Bacteriol.,* 132, 704, 1977.
17. **Al-Bassam, T., Poulain, D., Giummelly, P., Lematre, J., and Bonaly, R.,** Chemical and antigenic alterations of *Candida albicans* cell walls related to the action of amphotericin B sub-inhibitory doses, *J. Antimicrob. Chemother.,* 15, 263, 1985.
18. **Poulain, D., Tronchin, G., Jouvert, S., Herbaut, J., and Biguet, J.,** Architecture parietale des blastospores de *Candida albicans:* localisation de composants chimiqueset antigeniques, *Ann. Microbiol. (Inst. Pasteur),* 132A, 219, 1981.
19. **McCourtie, J. and Douglas, L. J.,** Extracellular polymer of *Candida albicans:* isolation, analysis and role in adhesion, *J. Gen Microbiol.,* 131, 495, 1985.
20. **Douglas, L. J. and McCourtie, J.,** Effect of tunicamycin treatment on the adherence of *Candida albicans* to human buccal epithelial cells, *FEMS Microbiol Lett.,* 16, 199, 1983.
21. **Barnes, J. L., Osgood, R. W., Lee, J. C., King, R. D., and Stein, J. H.,** Host-parasite interaction in the pathogenesis of experimental rental candidosis, *Lab. Invest.,* 49, 460, 1983.
22. **Calderone, R. A., Lehrer, N., and Segal, E.,** Adherence of *Candida albicans* to buccal and vaginal epithelial cells: ultrastructural observations, *Can. J. Microbiol.,* 30, 1001, 1984.

23. **Farrell, S. M., Hawkins, D. F., and Ryder, T. A.**, Scanning electron microscope study of *Candida albicans* invasion of cultured human cervical epithelial cells, *Sabouraudia*, 21, 251, 1983.
24. **Marrie, T. J., Cooper, J. H., and Costerton, J. W.**, Ultrastructure of *Candida parapsilosis* endocarditis, *Infect. Immun.*, 45, 390, 1984.
25. **Marrie, T. J. and Costerton, J. W.**, The ultrastructure of *Candida albicans* infections, *Can. J. Microbiol.*, 27, 1156, 1981.
26. **Persi, M. A. and Burnham, J. C.**, *Gallus domestica* yolk (vitelline) membrane use for *in vitro* microbial adherence studies, *Appl. Environ. Microbiol.*, 42, 364, 1981.
27. **Pope, L. M. and Cole, G. T.**, Scanning electron microscope studies of adherence of *Candida albicans* to the gastrointestinal tract of infant mice, *Scanning Electron Microsc.*, 1981 (III), 73, 1981.
28. **Tronchin, G., Poulain, D., and Vernes, A.**, Cytochemical and ultrastructural studies of *Candida albicans*. III. Evidence for modification of the cell wall coat during adherence to human buccal epithelial cells, *Arch. Microbiol.*, 139, 221, 1984.
29. **Sobel, J. D. and Obedeanu, N.**, Effects of subinhibitory concentrations of ketoconazole on *in vitro* adherence of *Candida albicans* to vaginal epithelial cells, *Eur. J. Clin. Microbiol.*, 2, 445, 1983.
30. **Anderson, M. L. and Odds, F. C.**, Adherence of *Candida albicans* to vaginal epithelia: significance of morphological form and effect of ketoconazole, *Mykosen*, 28, 531, 1985.
31. **Brenciaglia, M. I., Ghezzi, M. C., Cipriani, P., Mancini, C., and Trancassini, M.**, The influence of antifungal drugs on adherence of *Candida albicans* to buccal epithelial cells, *Chemioterapia*, 5, 200, 1986.
32. **Macura, A. B.**, The influence of some antifungal drugs on *in vitro* adherence of *Candida albicans* to human buccal epithelial cells, *Mycoses*, 31, 371, 1988.
33. **Nugent, K. M. and Couchot, K. R.**, Effects of sublethal concentrations of amphotericin B on *Candida albicans*, *J. Infect. Dis.*, *1*54, 665, 1986.
34. **Abu-Elteen, K. H., Ghannoum, M. A., and Stretton, R. J.**, Effects of subinhibitory concentrations of antifungal agents on *in vitro* adherence of *Candida* spp. to buccal epithelial cells, FEMS Symp. *Candida* Candidamycosis, Alanya, Turkey, 1989.
35. **McCourtie, J. and Douglas, L. J.**, Relationship between cell surface composition of *Candida albicans* and adherence to acrylic after growth on different carbon sources, *Infect. Immun.*, 32, 1234, 1981.
36. **Kimura, L. H. and Pearsall, N. N.**, Relationship between germination of *Candida albicans* and increased adherence to human buccal epithelial cells, *Infect. Immun.*, 28, 464, 1980.
37. **Drouhet, E. and Dupont, B.**, Evolution of antifungal agents: past, present, and future. *Rev. Infect. Dis.*, 9, S4, 1987.
38. **Marriott, M. S.**, Isolation and chemical characterization of plasma membranes from the yeast and mycelial forms of *Candida albicans*, *J. Gen. Microbiol.*, 86, 115, 1975.
39. **Marriott, M. S.**, Inhibition of sterol biosynthesis in *Candida albicans* by imidazole containing antifungals, *J. Gen. Microbiol.*, 117, 253, 1980.
40. **Polak, A.**, Synergism of polyene antibiotics with 5-Fluorocytosine, *Chemotherapy*, 24, 2, 1978.
41. **Strom, P., Crito, C., and Bozzi, A.**, A comparative study of the hemolytic effect of polyenic antibiotics and of other cholesterol-binding agents, *Biochem. Pharmacol.*, 28, 2427, 1979.
42. **Loose, D. S., Kan, P. B., Hirst, M. A., Marcus, R. A., and Feldman, D.**, Ketoconazole blocks adrenal steroidogenesis by inhibiting cytochrome P 450-dependent enzymes, *J. Clin. Invest.*, 71, 1495, 1983.
43. **Van den Bossche, H.**, Biochemical targets for antifungal azole derivatives: hypothesis on the mode of action, in *Current Topics in Medical Mycology*, Vol. 1, McGinnis, M. R., Ed., Springer Verlag, New York, 1985, 313.
44. **Barug, D. and Degroot, K.**, Effect of the imidazole derivative lombazole on the ultrastructure of *Staphylococcus epidermidis* and *Candida albicans*, *Antimicrob. Agents Chemother.*, 28, 643, 1985.
45. **De Nollins, S. and Borgers, M.**, The ultrastructure of *Candida albicans* after *in vitro* treatment with miconazole, *Sabouraudia*, 12, 341, 1974.
46. **Iwata, K., Kanday, Y., Yamaguchi, H., and Osumi, M.**, Electron microscopic studies on the mechanisms of action of clotrimazole on *Candida albicans*, *Sabouraudia*, 11, 205, 1973.
47. **Pesti, M., Becher, D., and Bartsch, G.**, The effect of miconazole on ergosterol-less mutant of *Candida albicans*, *Acta Microbiol. Hung.*, 30, 25, 1983.
48. **Preusser, H. J.**, Effects of *in vitro* treatment with econazole on the ultrastructure of *Candida albicans*, *Mykosen*, 19, 304, 1976.
49. **Surarit, R. and Shepherd, M. G.**, The effects of azole and polyene antifungals on the plasma membrane enzymes of *Candida albicans*, *J. Med. Vet. Mycol.*, 25, 403, 1987.
50. **Diasio, R. B., Bennett, J. E., and Myers, C. E.**, Mode of action of 5-fluorocytosine, *Biochem. Pharmacol.*, 27, 703, 1978.
51. **Polak, A.**, Effect of 5-fluorocytosine on the internal amino acid pool and protein synthesis in *Candida albicans*, *Pathol. Microbiol.*, 40, 132, 1974.
52. **Polak, A.**, Effects of 5-fluorocytosine on protein synthesis and amino acid pool in *Candida albicans*, *Sabouraudia*, 12, 309, 1974.

53. **Polak, A. and Scholer, H. J.**, Fungistatic activity, uptake and incorporation of 5-fluorocytosine in *Candida albicans*, as influenced by pyrimidines and purines. II. Studies on distribution and incorporation, *Pathol. Microbiol.*, 39, 334, 1973.
54. **Polak, A. and Scholer, H. J.**, Mode of action of 5-fluorocytosine and mechanisms of resistance, *Chemotherapia*, 21, 113, 1975.
55. **Polak, A.**, 5-Fluorocytosine—current status with special references to mode of action and drug resistance, *Contrib. Microbiol. Immunol.*, 4, 158, 1977.
56. **Polak, A. and Wain, W. H.**, The influence of 5-fluorocytosine on nucleic acid synthesis in *Candida albicans*, *Cryptococcus neoformans*, and *Aspergillus fumigatus*, *Chemotherapy*, 23, 243, 1977.
57. **Polak, A. and Wain, W. H.**, The effect of 5-fluorocytosine on the blastospores and hyphae of *Candida albicans*, *J. Med. Microbiol.*, 12, 83, 1979.
58. **Klotz, S. A. and Penn, R. L.**, Multiple mechanisms may contribute to the adherence of *Candida* yeasts to living cells, *Curr. Microbiol.*, 16, 119, 1987.
59. **Kennedy, M. J., Volz, P. A., Edwards, C. A., and Yancey, R. J.**, Mechanisms of association of *Candida albicans* with intestinal mucosa, *J. Med. Microbiol.*, 24, 333, 1987.
60. **Hugo, W. B. and Russell, A. D.**, Types of antimicrobial agents, in *Principles and Practice of Disinfection, Preservation and Sterilization*, Russell, A. D., Hugo, W. B., and Ayliffe, G. A., Eds., Blackwell Scientific, Oxford, 1982, chap. 2.
61. **Budtz-Jorgensen, E. and Loe, H.**, Chlorhexidine as a denture disinfectant in the treatment of denture stomatitis, *Scand. J. Dent. Res.*, 80, 457, 1972.
62. **Olsen, I.**, Denture stomatitis. Effects of chlorhexidine and amphotericin B on the mycotic flora, *Acta Odontol. Scand.*, 33, 41, 1975.
63. **Olsen, I.**, Denture stomatitis. The clinical effects of chlorhexidine and amphotericin B, *Acta Odontol. Scand.*, 33, 47, 1975.
64. **Langslet, A., Olsen, I., Lie, S. O., and Lokken, P.**, Chlorhexidine treatment of oral candidiasis in seriously diseased children, *Acta Paediat. Scand.*, 63, 809, 1974.
65. **Sharon, A., Berdicevsky, I., Ben-Aryeh, H., and Gutman, D.**, The effect of chlorhexidine mouth rinses on oral *Candida* in a group of leukemic patients, *Oral Surg.*, 44, 201, 1977.
66. **McCourtie, J., MacFarlane, T. W., and Samaranayake, L. P.**, Effect of Chlorhexidine gluconate on the adherence of *Candida* species to denture acrylic, *J. Med. Microbiol.*, 20, 97, 1985.
67. **McCourtie, J., MacFarlane, T. W., and Samaranayake, L. P.**, Effect of saliva and serum on the adherence of *Candida* species to chlorhexidine-treated denture acrylic, *J. Med. Microbiol.*, 21, 209, 1986.
68. **McCourtie, J., MacFarlane, T. W., and Samaranayake, L. P.**, A comparison of the effects of chlorhexidine gluconate, amphotericin B and nystatin on the adherence of *Candida* species to denture acrylic, *J. Antimicrob. Chemother.*, 17, 575, 1986.
69. **Tobgi, R. S., Samaranayake, L. P., and MacFarlane, T. W.**, Adhesion of *Candida albicans* to buccal epithelial cells exposed to chlorhexidine gluconate, *J. Med. Vet. Mycol.*, 25, 335, 1987.
70. **Hugo, W. B. and Longworth, A. R.**, Some aspects of the mode of action of chlorhexidine, *J. Pharm. Pharmacol.*, 16, 655, 1964.
71. **Hugo, W. B. and Longworth, A. R.**, Effect of chlorhexidine on protoplasts and spheroplasts of *Escherichia coli*, protoplasts of *Bacillus megaterium* and the Gram staining reaction of *Staphylococcus aureus*, *J. Pharm. Pharmacol.*, 16, 751, 1964.
72. **Hugo, W. B. and Longworth, A. R.**, The effect of chlorhexidine on the electrophoretic mobility, cytoplasmic constituents, dehydrogenase activity and cell wall of *Escherichia coli* and *Staphylococcus aureus*, *J. Pharm. Pharamacol.*, 18, 569, 1966.
73. **Davies, A., Bentley, M., and Field, B. S.**, Comparison of the action of vantocil, centrimide and chlorhexidine of *Escherichia coli* and its spheroplasts and the protoplasts of Gram positive bacteria, *J. Appl. Bacteriol.*, 31, 448, 1968.
74. **Kaneshima, S., Koga, S., and Tanaka, K.**, A simple method for cellular identification of free cells by light microscopy and scanning electron microscopy, *J. Electron Microsc.*, 26, 355, 1977.
75. **Montes, L. F. and Wilborn, W. H.**, Ultrastructural features of host-parasite relationship in oral candidasis, *J. Bacteriol.*, 96, 1349, 1968.
76. **Mohamad, A. M. H.**, Ultrastructural aspects of chronic oral candidosis, *J. Oral Pathol.*, 4, 180, 1975.
77. **Bobichon, H. and Bouchet, P.**, Action of chlorhexidine on budding *Candida albicans:* scanning and transmission electron microscopic study, *Mycopathologia*, 100, 27, 1987.
78. **Whittaker, P.**, Unpublished data, 1988.
79. **Ghannoum, M. A., Abu Elteen, K. H., and Whittaker, P.**, Unpublished data, 1988.
80. **Block, E.**, The chemistry of garlic and onions, *Sci. Am.*, 252, 114, 1985.
81. **Appleton, J. A. and Tansey, M. R.**, Inhibition of growth of zoopathogenic fungi by garlic extract, *Mycologia*, 67, 882, 1975.

82. **Barone, F. E. and Tansey, M. R.,** Isolation, purification, identification, synthesis, and kinetics of activity of the anticandidal component of *Allium sativum,* and a hypothesis for its mode of action, *Mycologia,* 69, 793, 1977.
83. **Prasad, G., Sharma, V. D., and Kumar, A.,** Efficacy of garlic *(Allium sativum L.)* therapy against experimental dermatophytosis in rabbits, *Indian J. Med. Res.,* 75, 465, 1982.
84. **Kabelik, J. C.,** Antimikrobialle eigenschaften des knoblauchs, *Pharmazie,* 25, 266, 1970.
85. **Clayton, C.,** Thrush, Sheldon press, London, 1984.
86. **Ghannoum, M. A.,** Studies on the anticandidal mode of action of *Allium sativum* (Garlic), *J. Gen. Microbiol.,* 134, 2917, 1988.
87. **Prasad, G. and Sharma, V. D.,** Efficacy of garlic *(Allium sativum)* treatment against experimental candidiasis in ckicks, *Br. Vet. J.,* 136, 448, 1980.
88. **Ghannoum, M. A.,** Inhibition of *Candida* adhesion to buccal epithelial cells by *Allium sativum* (Garlic), FEMS Symp. Candida Candidamycosis, Alanya, Turkey, 1989.
89. **Pickard, R. G.,** Treatment of peritonitis with pre-and post-operative irrigation of the peritoneal cavity with noxythiolin solution, *Br. J. Surgery,* 59, 642, 1972.
90. **Anderson, L., Gorman, S. P., McCafferty, D. F., and Woolfson, A. D.,** Clinical implications of the microbial anti-adherence properties of noxythiolin, *J. Pharm., Pharmacol.,* 37, 645, 1985.
91. **Gorman, S. P., McCafferty, D. F., Woolfson, A. D., and Anderson, L.,** Inhibition of hyphal development and kill of *Candida albicans* blastospores by noxythiolin *in vitro,* J. *Appl. Bacteriol.,* 60, 319, 1986.
92. **Gorman, S. P., McCafferty, D. F., Woolfson, A. D., and Anderson, L.,** Decrease in adherence of bacteria and yeasts to human mucosal epithelial cells by noxythiolin *in vitro, J. Appl. Bacteriol.,* 60, 311, 1986.
93. **Gorman, S. P., McCafferty, D. F., Woolfson, A. D., and Anderson, L.,** Reduced adherence and inhibition of hyphal development of *Candida albicans* by the antimicrobial agent polynoxylin, *Lett. Appl. Microbiol.,* 6, 145, 1988.
94. **Kennet, S. and Cohn, L.,** Anaflex lozenges in the treatment of oral candidosis, *Br. Med. J.,* 2, 1200, 1966.
95. **Phillips, H. I. B.,** Treatment of denture stomatitis with polynoxylin, *Br. Dent. J.,* 128, 78, 1970.
96. **Gorman, S. P., McCafferty, D. F., Woolfson, A. D., and Jones, D. S.,** Reduced adherence of micro-organisms to human mucosal epithelial cells following treatment with taurolin, a novel antimicrobial agent, *J. Appl. Bacteriol.,* 62, 315, 1987.
97. **Kingston, D.,** Release of formaldehyde from polynoxylin and noxythiolin, *J. Clin. Pathol.,* 18, 666, 1965.
98. **Gucklhorn, I. R.,** Antimicrobials in cosmetics, parts 1 to 7, *Manuf. Chemists Aerosol News,* Vol. 41, (6) 44, (7) 51, (8) 28, (10) 49, (11) 48, (12) 50.
99. **Myers, J. A., Allwood, M. C., Gidley, M. J., and Sanders, J. K. M.,** The relationship between structure and activity of taurolin, *J. Appl. Bacteriol.,* 48, 89, 1980.
100. **Allwood, M. C. and Myers, J. A.,** Formaldehyde releasing agents, *Society for Applied Bacteriology Technical Series,* 16, Academic Press, London, 1981, 69.
101. **Gidley, M. J. and Sanders, J. K. M.,** Mechanisms of antibacterial formaldehyde delivery from noxythiolin and other 'masked formaldehyde' compounds, *J. Pharm Pharmacol.,* 35, 712, 1983.
102. **Borman, M. C. and Pittelkow, R. B.,** Hypercortisonism, moniliasis and arthritis, *Wis. Med. J.,* 66, 240, 1967.
103. **Lehner, T. and Ward, R. C.,** Iatrogenic oral candidosis, *Br. J. Dermatol.,* 83, 161, 1970.
104. **Bradford, L. G. and Montes, L. F.,** Perioral dermatitis and *Candida albicans, Arch. Dermatol.,* 105, 892, 1972
105. **Gale, A. E.,** Beclomethasone thrush, *Med. J. Austr.,* 1, 58, 1982.
106. **Gill, K. A. Jr., Katz, H. I., and Baxter, D. L.,** Fungus infections occurring under occlusive dressings, *Arch. Dermatol.,* 88, 348, 1963.
107. **Katoh, T. and Takahashi, S.,** Epidemiological and clinical survey of cutaneous candidiasis for 15 years (1958 - 1972), *Tohoku J. Exp. Med.,* 114, 359, 1974.
108. **Lynch, P. J., Minkin, W., and Smith, E. B.,** Ecology of *Candida albicans* in candidiasis of the groin, *Arch. Dermatol.,* 99, 154, 1969.
109. **Schechtman, R. L., Archard, H. O., and Cox, D.,** Oropharyngeal candidiasis associated with steroid-containing inhalers, *N. Z. State Dent. J.,* 52, 24, 1986.
110. **Zegarelli, E. V. and Kutscher, A. H.,** Oral moniliasis following intraoral topical corticosteroid therapy, *J. Oral. Ther.,* 1, 304, 1964.
111. **Date, A., Chandy, M., and Pulimood, B. M.,** Filarial chyluria with opportunistic infections, *Trans. R. Soc. Trop. Med. Hyg.,* 77, 112, 1983.
112. **Robertson, D. M., Riley, F. C., and Hermans, P. E.,** Endogenous *Candida* oculomycosis. Report of two patients treated with flucytosine, *Arch. Ophthamol.,* 91, 33, 1974.
113. **Hasany, S. M., Basu, P. K., and Kazdan, J. J.,** Production of corneal ulcer by opportunistic and saprophytic fungi. I. Effect of pretreatment of fungi with steroid, *Can. J. Opthamol.,* 8, 119, 1973.
114. **Seligmann, E.,** Virulence enhancement of *Candida albicans* by antibiotics and cortisone, *Proc. Soc. Exp. Biol. Med.,* 83, 778, 1953.

115. **Mountain, I. M.,** Antibody production by spleen *in vitro*. I. Influence of cortisone and other chemicals, *J. Immunol.*, 74, 270, 1955.
116. **Kass, E. H., Kendrick, M. I., and Finland, M.,** Effect of certain corticosteroids and of growth hormones on nucleoproteins of lymph nodes, *Ann. N.Y. Acad. Sci*, 56, 737, 1953.
117. **Stern, K. and Davidsohn, I.,** Effect of estrogen and cortisone on immune hemoantibiotics in mice of inbred strains, *J. Immunol.*, 74, 479, 1955.
118. **Ghannoum, M. A.,** Mechanisms potentiating Candida infections (a review), *Mycoses*, 1988, 31, 543, 1988.
119. **Ghannoum, M. A., Mudher, S., and Burns, G.,** Incorporation of dexamethasone by *Candida albicans*, *Microbios*, 42, 103, 1985.
120. **Pingleton, W., Bone, R., Kerby, G., and Ruth, W.,** Orpharyngeal candidiasis in patients treated with triamcinolone acetonide aerosol, *J. Allergy Clin. Immunol.*, 60, 254, 1977.
121. **Raab, W.,** Effects of local corticosteroids in skin infections, *Dermatologica*, 152, 67, 1976.
122. **Winter, W. and Foley, G.,** Enhancement of *Candida* infection: differential distribution of renal lesions in mice treated with aureomycin, *J. Infect. Dis.*, 98, 150, 1956.
123. **Ghannoum, M. A. and Abu Elteen, K.,** Effect of growth of *Candida* spp. in the presence of various glucocorticoids on the adherence to human buccal epithelial cells, *Mycopathologia*, 98, 171, 1987.
124. **Frenkel, J. K.,** Evaluation of infection-enhancing activity of modified corticoids, *Proc. Soc. Exp. Biol. Med.*, *103, 552, 1960.*
125. **Loose, D. S., Schurman, D. J. and Feldman, D.,** A corticosteroid binding protein and endogenous ligand in *Candida albicans* indicating a possible steroid-receptor system, *Nature (London)*, 293, 477, 1981.
126. **Raspe, G.,** *Advances in the Biosciences*, Vol. 7, Pergamon-Vieweg, Brauschweig, Germany, 1971, 417.
127. **Powel, B. L., Frey, C. L., and Drutz, D. J.,** Estrogen receptor in *Candida albicans:* a possible explanation for hormonal influences in vaginal candidiasis, Abst. No. 75. Interscience Conf. Antimicrobial Agents Chemotherapy, Las Vegas, 1983.
128. **Lindberg, R. D.,** Present day role of radiation therapy in the treatment of head and neck cancer, in *Cancer of the Head and Neck*, Chambers, R. G., Janssen de Limpens, A. M. P., Jaques, D. A., and Routledge, R. T., Eds., *Proc. Int. Symp.*, American Elsevier, New York, 1975, 7.
129. **Richards, G. J. and Chambers, R. G.,** Combined therapy in the treatment of carcinomas of the head and neck, in *Cancer of the Head and Neck*, Chambers, R. G., Janssen de Limpens, A. M. P., Jaques, D. A., and Routledge, R. T., Eds., *Proc. Int. Symp.*, American Elsevier, New York, 1975, 23.
130. **Shanta, V. and Krishnamurthi, S.,** The combined therapy of oral cancer, in *Fundamental and Clinical Studies of Bleomycin*, Carter, S., Ichikawa, T., Mathe, G., and Umezawa, H., Eds., GANN monograph on cancer research, No. 19, University Park Press, Baltimore, 1976, 159.
131. **Cho, S. Y. and Choi, H. Y.,** Opportunistic fungal infection among cancer patients, *Am. J. Clin. Pathol*, 72, 617, 1979.
132. **Gentles, J. C. and La Touche, C. J.,** Yeasts as human and animal pathogens, in *The Yeasts*, Vol. 1, Rose, A. H. and Harrison, J. S., Eds., Academic Press, London, 1969, 107.
133. **Ghannoum, M. A., Motawy, M. S., Mubarak, A. L., and Al-Awadhi, H. A.,** Incidence and susceptibility of yeasts isolated from cancer patients in Kuwait, *Mykosen*, 28, 619, 1985.
134. **Kiehn, T. E., Edwards, F. F., and Armstrong, D.,** The prevalence of yeasts in clinical specimens from cancer patients, *Am. J. Clin. Pathol.*, 73, 518, 1980.
135. **Gordee, R. S. and Simpson, R. J.,** Relationships of X-irradiation to the enhancement of *Candida albicans* infections, *J. Bacteriol.*, 94, 6, 1967.
136. **Ghannoum, M. A. and Al-Khars, A.,** Effect of antineoplastic agents on the growth and ultrastructure of *Candida albicans*, *Mykosen*, 27, 452, 1984.
137. **Moore, C. W.,** Responses of radiation-sensitive mutants of *Saccharomyces cerevisiae* to lethal effects to bleomycin, *Mutation Res.*, 51, 165, 1978.
138. **Tattersall, M. H.,** Aggressive cancer treatment and its role in predisposing to infection, *Eur. J. Cancer*, *11*, 9, 1975.
139. **Hersh, E. M. and Freireich, E. J.,** Host defense mechanisms and their modifications by cancer chemotherapy, in *Methods in Cancer*, Vol. 4, Busch, H., Ed., New York Research, Academic Press, New York, 1968, 355.
140. **Campbell, A. C., Hersey, P., MacLennan, I. C. M., Kay, H. E. M., Pike, M. C., and the Medical Research Council's Working Party on Leukaemia in Childhood,** Immunosuppressive consequence of radiotherapy and chemotherapy in patients with acute lymphoblastic leukaemia, *Br. Med. J.*, 2, 385, 1973.
141. **Ghannoum, M. A., Khattar, M., Kayed, F., and Shridar, P.,** The influence of antineoplastic agents on experimental candidosis in mice, *Mykosen*, 28, 369, 1985.
142. **Kondo, S.,** Radiation genetics in microorganisms and evolutionary considerations, *Genetics*, 78, 149, 1974.
143. **Gibbons, R. J.,** Adhesion of bacteria to surfaces of the mouth, in *Microbial Adhesion to Surfaces*, Berkely, R. C. W., Lynch, J. M., Melling, J. M., Rutter, P. R., and Vincent, B., Eds., Horwood, Chichester, 1980, 351.
144. **Newman, K., Schimpff, S. C., and Wade, J. C.,** Antibiotic prophylaxis of infection for patients with granulocytopenia, in *Infection in the Immunocompromised Host—Pathogenesis, Prevention and Therapy*, Verhoef, J., Peterson, P. K., and Duie, P. G., Eds., Elsevier/North-Holland, Amsterdam, 1980, 187.

145. **Silverman, S., Luangjarmekorn, L., and Greenspan, D.,** Occurrence of oral *Candida* in irradiated head and neck cancer patients, *J. Oral. Med.*, 39, 194, 1984.
146. **Ghannoum, M. A., Abu-Elteen, K. H., and Motawy, M. S.,** Effect of antineoplastic agents and X-irradiation on the adherence of *Candida* spp. to human buccal epithelial cells—*in vitro, Mycopathologia*, 104, 171, 1988.
147. **Radman, M.,** SOS repair hypothesis: phenomenology of an inducible DNA repair which is accompanied by mutagenesis, in *Molecular Mechanisms for Repair DNA,* Hanawalt, P and Setlow, R. B., Eds., Plenum, New York, 1975, 355.
148. **Witkin, E. M.,** Ultraviolet mutagenesis and inducible DNA repair in *Escherichia coli, Bacteriol. Rev.*, 40, 869, 1976.
149. **Maisch, P. A. and Calderone, R. A.,** Role of surface mannan in the adherence of *Candida albicans* to fibrin-platelet clots formed *in vitro, Infect. Immun.*, 32, 92, 1981.
150. **Lee, J. C. and King, R. D.,** Characterization of *Candida albicans* adherence to human vaginal epithelial cells *in vitro, Infect. Immun.*, 41, 1024, 1983.
151. **Elkind, M. M., Whitmore, G. F., and Alescio, T.,** Actinomycin D: suppression of recovery in X-irradiated mammalian cells, *Science,* 143, 1454, 1964.
152. United Nations: *Ionizing Radiation: Sources and Biological Effects.* United Nations Scientific Committee on the Effects of Atomic Radiation, Report to the General Assembly, New York, 1982, 727.
153. **Bases, R.,** Enhancement of X-ray damage in Hela cells by exposure to lucanthone (Miracyl D) following radiation, *Cancer Res.*, 30, 2007, 1970.
154. **Henderson, S. D. and Kimler, B. F.,** Response of cultured 9L cells to bleomycin and X-ray. Abst. 28th ann. meet. Radiation Res. Soc., New Orleans, LA, *Radiation Res.*, 83, 372, 1980.
155. **Kimler, B. F.,** Interaction of radiation and DHAQ: effect on cell survival. Abst. 28th ann. meet. Radiation Res. Soc., New Orleans, LA, *Radiation Res.*, 83, 372, 1980.
156. **Kimura, L. and Pearsall, N.,** Adherence of *Candida albicans* to human buccal epithelial cells, *Infect. Immun.*, 21, 64, 1978.
157. **Samaranayake, L. and MacFarlane, T.,** Factors affecting the *in vitro* adherence of the fungal oral pathogen *Candida albicans* to epithelial cells of human origin, *Arch. Oral Biol.*, 27, 869, 1982.
158. **Odds, F. C. and Webster, C. E.,** Effects of azole antifungals *in vitro* on host/parasite interactions relevant to candida inpections, *J. Antimicrob. Chemother.*, 22, 473, 1988.
159. **Shibl, A. M.,** Effect of antibiotics on adherence of microorganisms to epithelial cell surfaces, *Rev. Infect. Dis.*, 7, 51, 1985.
160. **Ghannoum, M. A., Abu Elteen, K., Ellabib, M., and Whittaker, P. A.,** Antimycotic effects of octenidine and pirtenidine, *J. Antimicrob. Chemother.*, in press.

Chapter 7

ADHERENCE BLOCKAGE IN THE CONTROL OF CANDIDOSIS

TABLE OF CONTENTS

I. INTRODUCTION

Fungal infections, including candidosis, now pose a serious threat to certain groups of patients particularly the immunocompromised and transplant patients.[1-3] Although there is a relatively large number of compounds available for the treatment of these cases,[2] only a few of the antifungal compounds developed reach the clinical stage; this could be due to poor tissue penetration or inactivation by induced enzymes.[4] Further, the use of antifungals in clinical situations is limited because they are toxic towards the host.[2,4] Considerable efforts have been directed towards finding new antifungals with reduced toxicity. Although azole antifungals have been shown to be less toxic than polyenes, the former still exhibit toxicity particularly in long-term therapy.[4] This situation is accentuated by the fact that patients prone to candidosis, are treated simultaneously with other drugs (known for their toxicity) to combat a primary illness.[5-7] Therefore, there is considerable interest in seeking alternative approaches for the control and prevention of candidosis. One possibility that has shown considerable promise in recent years involves interference with the adhesion process. Since adhesion of microorganisms to host cells is considered to be the initial stage of colonization and infection,[8,9] intervention at this early stage of the infectious process represents a potentially viable means of prevention and control of infectious diseases. Researchers have carried out extensive studies aimed at the elucidation of mechanisms that mediate attachment of both bacteria and *Candida*.[10-18] As a result of these investigations, adherence to host surfaces has been generally accepted as an important step in the pathogenesis of both bacteria and fungi.[18-20] Furthermore, knowledge of bacterial attachment mechanisms has been translated into approaches for the modification or prevention of infection in experimental and farm animals and, to a lesser extent, humans.[13,21] By comparison, knowledge of fungal adherence mechanisms has not been evaluated in animal models.[4] Ryley[22] concluded that, "Adherence of *C. albicans* to epithelial cells has become a very fashionable topic during the last decade, yet we have only slender experimental evidence that the extensive studies carried out *in vitro* have any relevance to the infective process *in vivo*." The situation is still the same. Apart from initial studies correlating adherence with virulence,[23,25] only a few studies, using experimental animals suggest a possible clinical application of this approach.[26] In this chapter we describe the work carried out by these authors and present results from our own *in vivo* studies as a stepping stone towards a more detailed evaluation and correlation of *in vitro* and *in vivo* effects. It is hoped that these studies will be ultimately of practical value in attempts to control and prevent candidosis.

II. PREVENTION OF MURINE CANDIDAL VAGINITIS

Work on the inhibition of candidal adherence to murine vaginal mucosa *in vivo* was initiated by Segal et al.[26] These workers demonstrated that treatment of the vaginas of mice before yeast inoculation with either the *Candida* cell wall component chitin; or its constituent N-acetyl-glucosamine (NAG); or its derivative, (designated chitin-soluble extract, CSE), blocked the attachment of *C. albicans* to the vaginal mucosal surface and thereby prevented vaginal infection.[26] Treatment of mice with CSE or NAG after inoculation of the yeast did not prevent infection.[26] These authors used normal cycling mice rather than estrogen-treated animals. One drawback of this animal model is that the induced infection is short lived and unstable. Consequently, the effect of the inhibitors used could not be fully appreciated. To avoid this shortcoming, subsequent studies used estradiol-treated mice in which a persistent candidal vaginitis was induced.[27] The results obtained confirmed earlier findings in that CSE was found to be an effective measure for preventing candidal vaginitis.[27]

Attempts to characterize the CSE showed that it consists of two fractions: FI and FII, of which only FI exhibited inhibitory activity. Chemical analysis of CSE and its two fractions showed that CSE contains over 70% of proteins, concentrated mostly in FII. In addition, 3% of amino-sugars

TABLE 1
Attachment Of Three *Candida Albicans* Strains To Intestinal Disk

Strain	No. of tissue disks	Mean no. of adhering yeasts ± SD (x 10^5)
C. albicans ATCC 10231	8	26 ± 1
C. albicans KCCC 14172	8	103 ± 6
C. albicans KCCC 13878	8	91 ± 4

were detected both in the whole CSE and in FI, and lipids in the unfractioned CSE and both fractions.[28]

III. PROTECTION AGAINST *CANDIDA* GASTROINTESTINAL COLONIZATION AND SYSTEMIC SPREAD

The passage of viable *C. albicans* through the gastrointestinal (GI) mucosa into the bloodstream is considered to be an important mechanism leading to systemic candidosis.[29-32] Association of the yeast with intestinal mucosa is believed to be the very first stage in the colonization and subsequent dissemination from the GI tract.[33-36] Consequently, an understanding of the mechanism of interaction between the yeast and the GI mucosa could shed some light on how the organisms colonize the GI tract and act as the primary settings for disseminated candidosis.[2] Furthermore, investigations related to blockage of yeast adherence to the GI mucosa are of obvious clinical connotations.

Work on the adherence of *Candida* to the GI tract and prevention of its colonization by *Candida* cell wall components *in vivo* is lacking. Therefore, we explored the effect of *Candida* cell wall components (CSE, NAG, and mannose) known to block adherence of yeast to epithelial cells *in vitro*[26,35,37-39] on the GI colonization and dissemination and as a possible means for the prevention of infection. CSE was prepared by suspending chitin (Sigma) (2%, w/v) in phosphate buffered saline (PBS) by shaking at room temperature for 5 h. The supernatant designated as CSE was then removed and processed further as described by Lehrer et al.[26]

A. INHIBITION OF CANDIDA ADHESION TO INTESTINAL TISSUES BY SACCHARIDE DERIVATIVES

Prior to our *in vivo* experimentations we investigated the attachment of *C. albicans* to murine intestinal tissues *in vitro* using a modification of the technique suggested by Clancy and Savage.[40] This represented an intermediate stage between *in vitro* experimentation using epithelial cells and *in vivo* animal models. Briefly, the technique involved obtaining disks of duodenal tissue from female mice (strain MFI) of an average weight of 20 g. These disks were exposed for 30 min *in vitro* to suspensions of blastospores of *C. albicans*, 10^8 cells/ml. Three different strains were assayed: ATCC 10231, KCCC 14172, and KCCC 13878.[41,42] The disks were then washed three times with sterile distilled water to remove nonadherent yeasts and homogenized. The number of adhering yeasts per tissue was determined by plating a portion of the homogenate using pour plate technique. We observed a variation in the attachment of different strains of *C. albicans* (Table 1). This difference was pronounced particularly between the laboratory strain (ATCC 10231) and the two clinical strains (KCCC 14172 and KCCC 13878). This is expected, however.

To evaluate the effect of *Candida* cell wall components on the adhesion of yeast to intestinal tissues we performed the following experiments:

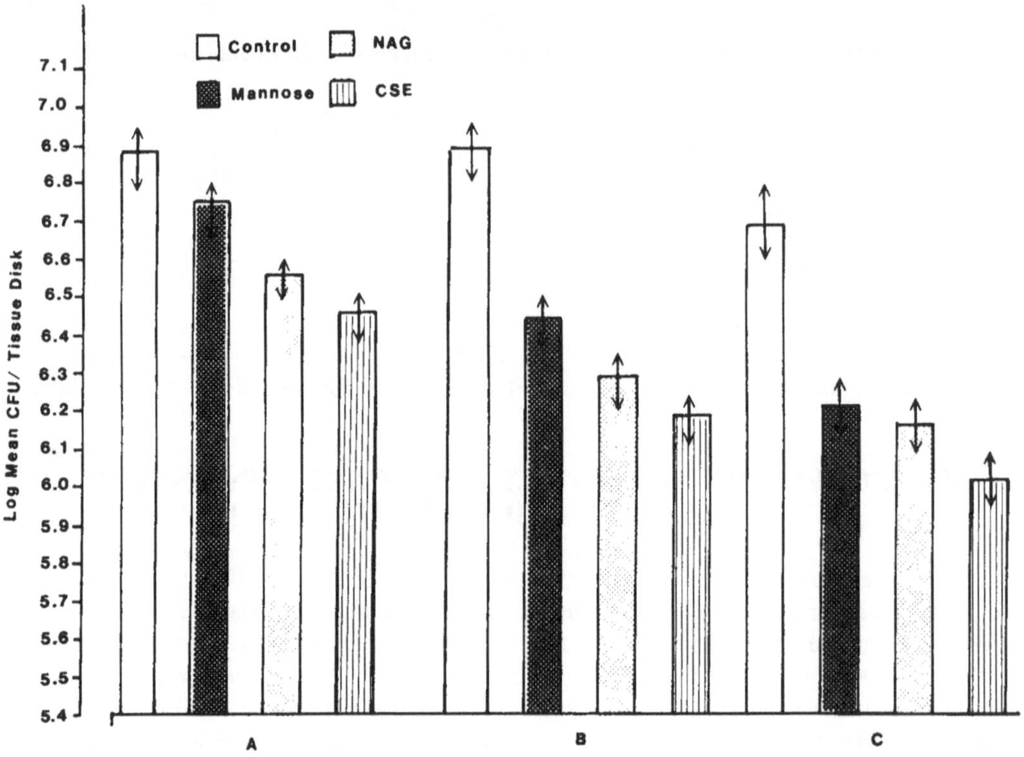

FIGURE 1. Effect of *Candida* cell wall components on the adhesion of yeast to intestinal tissues. (A) Inclusion of test compounds in the adhesion assay. Controls consisted of assays with no test compound. (B) Posttreatment of duodenal disks with candida blastospores for 30 min. (C) Pretreatment of duodenal disks with 2% test compounds for 30 min.

1. Inclusion of test compounds in the adhesion assay. No test compounds were added to the assay mixtures in the case of controls.
2. Pretreatment of duodenal disks with 2% test compounds for 30 min. These disks were then assayed for adhesion. Control disks were pretreated with sterile distilled water containing no test compound.
3. To test the effect of posttreatment, i.e., treatment with saccharides following exposure to *Candida* blastospores, disks were suspended in PBS containing 10^8 yeast per ml for 30 min, at 37°C, in an orbital shaker. This was followed by washing twice with PBS to remove nonadherant yeasts. Disks were then suspended for another 30 min in buffer containing 2% test compound and re-incubated. Controls consisted of suspension of yeast-treated disks in PBS containing no test compound. The number of attached *C. albicans* was then determined as mentioned above.

To monitor the effect of test compounds microscopically, separate disks for control and tests were prepared for scanning electron microscopy by the method of Ghannoum et al.[43] Our results are summarized in Figure 1. Inclusion of all the three compounds tested in the adhesion mixture led to blockage of adherence of *C. albicans* to intestinal tissues to varying degrees. CSE was the most active inhibitor, followed by NAG, and mannose was the least active (Figures 1 and 2). These results agree with the findings of Segal and Savage[44] who showed inhibition of *Candida* adherence to intestinal tissues under the influence of CSE. These authors did not, however, test the influence of either NAG or mannose.

Pre- and posttreatment enhanced the inhibition of adherence, with pretreatment resulting in a higher magnitude of inhibition (Figure 1, B and C).

FIGURE 2. Scanning electron microscopy of intestinal tissues from (A) control tissue, (B) CSE-treated, and (C) NAG-treated tissue. Note the direct attachment of *C. albicans* to duodenal epithelium. (Magnification × 3500, 5000, and 3500, respectively.)

B. INHIBITION OF CANDIDA ATTACHMENT TO THE GI TRACT AND SYSTEMIC SPREAD BY YEAST SACCHARIDES *IN VIVO*

We have used infant mice, 5 to 6 d old, to study the possiblity of preventing GI colonization and dissemination by *C. albicans* by treating the animals with saccharides.[45]

White mice (MFI strain) were used to establish a breeding colony in the animal house of the department of zoology, Kuwait University. Offsprings of these animals were used in all experiments. We used infant mice rather than adults because oral-intragastric inoculation of infant mice with *C. albicans* leads to GI and systemic candidosis.[45,46,55] Adult mice are resistant to colonization unless they are pretreated with antibiotics.[32]

For oral challenge of mice, an overnight culture of *C. albicans* KCCC 14172 was grown at 37°C in Yeast Nitrogen Base (Difco®) supplemented with 2.5% (w/v) glucose. Flasks containing 50 ml of the same medium were inoculated with 1 ml of the overnight culture and grown for 24 h in a shaking water bath at 37°C. The cells were harvested by centrifugation (1200 g, 10 min) and washed twice with sterile PBS, and resuspended to the estimated inoculum size. This estimate was based on repeated dilution counts of similarly prepared suspensions, and the actual challenge dose was determined by a dilution count at the time of challenge.

1. Animal Preparation and Inoculation

At 2 d prior to challenge with *C. albicans,* infant mice were treated with various saccharides

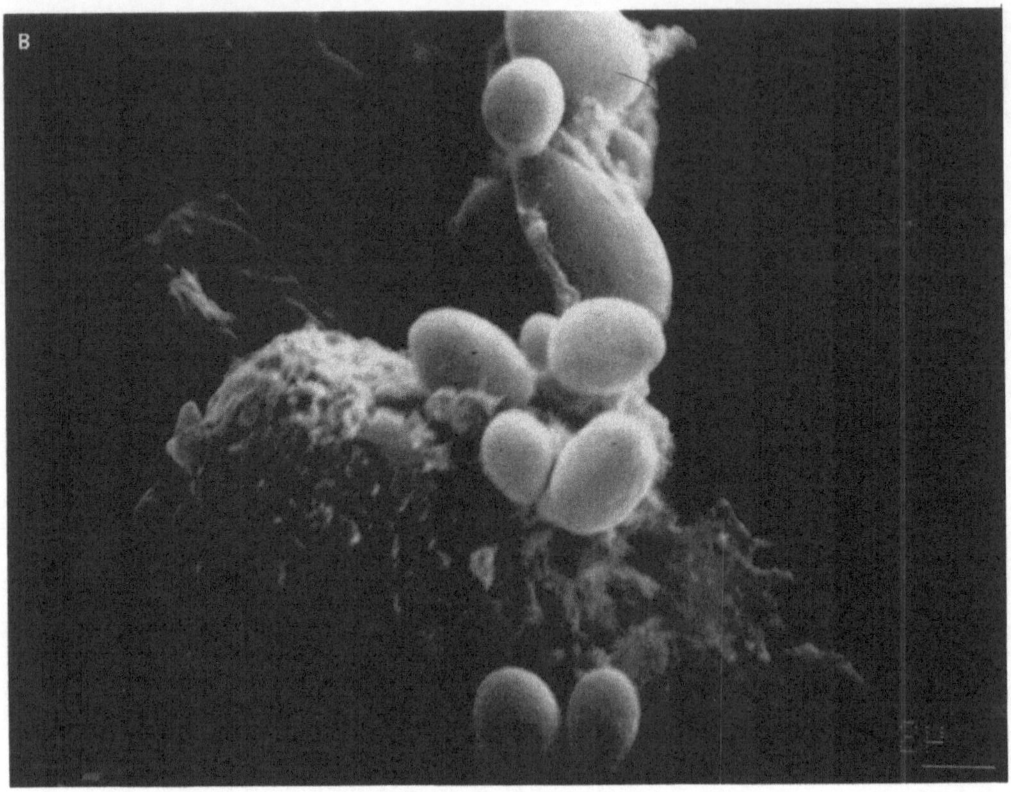

FIGURE 2B.

by orally introducing 20 μl of a 2% solution of either mannose, NAG, or CSE. At 6 h prior to the challenge infant mice were removed from their mothers. Immediately before inoculation a further 20 μl of the inhibitors were given. The yeast inoculum, containing approximately 1×10^8 cells in 20 μl, was delivered orally by using Eppendorf® tips. Treatment with inhibitors was continued on a daily basis. Control animals were treated similarly with sterile distilled water. After challenge, the animals were returned to mothers without concern for parentage. To check if the timing of the saccharide inoculations are important to prevent *C. albicans* colonization and dissemination, we carried out other experiments where a group of mice were given a saccharide dose 30 min prior to *Candida* inoculation. The influence of saccharide treatment post-*Candida* inoculation was also tested.

The dose-response effect for CSE was carried out with a single time point, 48 h. In this group the mice were treated with different concentrations of CSE 2 d prior to challenge with *C. albicans*. CSE treatment was continued for a further 48 h, when the mice were sacrificed and the yeast colony forming units (CFU) determined for the whole intestine and the stomach.

2. Enumeration of Organisms

The levels of *C. albicans* in the GI tract and selected body organs were determined by

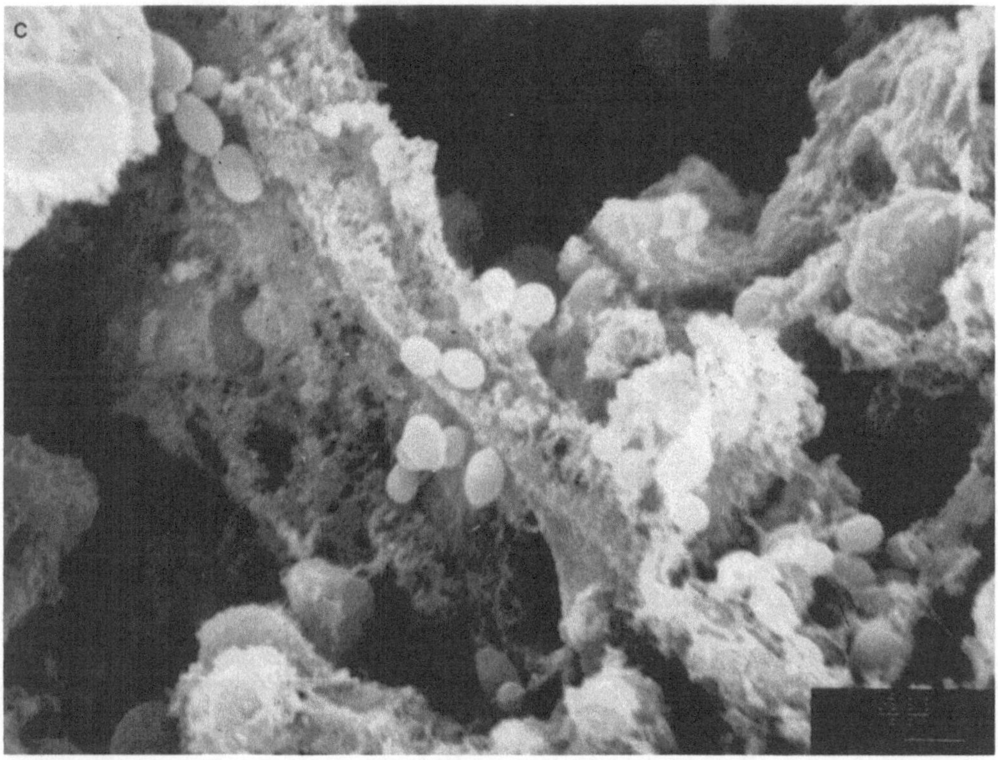

FIGURE 2C.

quantitative dilution plate counting. Animals were sacrificed and dissected under aseptic conditions. Body organs were removed to sterile Ystral x 10/20 glass homogenizers (Ystral GmbH, Gottingen, Germany), in the following order: both kidneys, spleen, liver, stomach, and whole intestine. The stomach and whole intestine were then gently washed in three changes of sterile PBS to remove digesta and nonadherent microorganisms. Tissues were homogenized in 5 ml sterile PBS and serial ten-fold dilutions in sterile PBS were prepared from these homogenates. One ml samples of appropriate dilutions were plated in triplicate in Sabouraud dextrose agar (Difco) containing chloramphenicol (50 mg l^{-1}); the plates were left to dry and incubated at 37°C. Yeast colonies were counted after 2 to 3 d and the results were expressed as the mean CFU per gram weight of tissue.

3. Effect of Inhibitors on the Systemic Spread from the GI Tract of Mice

After oral challenge with 10^8 CFU per mouse, animals were sacrificed at various times during a 72 h period postinoculation, and body organs of saccharide-treated and nontreated control animals were scored for the presence or absence of *C. albicans* (Table 2). The intestines were almost uniformly infected at each of the five test periods employed. Viable *C. albicans* were recovered from all the stomachs cultured irrespective of whether the animals were saccharide treated or not. The only exception was for animals treated with NAG where the number of

TABLE 2
Body Organs Positive For *Candida Albicans*

Organ	Time post-infect. (h)	% of samples positive for *C. albicans* Control	2% Mannose	2% NAG[a]	2%CSE[b]
Kidneys	3	45.5 (5/11)	33.3 (2/6)	0.0 (0/6)	0.0 (0/6)
	6	36.3 (4/11)	33.3 (2/6)	0.0 (0/6)	0.0 (0/6)
	24	36.3 (4/11)	0.0 (0/6)	0.0 (0/6)	0.0 (0/6)
	48	27.2 (3/11)	0.0 (0/6)	0.0 (0/6)	0.0 (0/6)
	72	27.2 (3/11)	0.0 (0/6)	0.0 (0/6)	0.0 (0/6)
Spleen	3	45.5 (5/11)	16.6 (1/6)	0.0 (0/6)	0.0 (0/6)
	6	45.5 (5/11)	0.0 (0/6)	0.0 (0/6)	0.0 (0/6)
	24	36.3 (4/11)	0.0 (0/6)	0.0 (0/6)	0.0 (0/6)
	48	36.3 (4/11)	0.0 (0/6)	0.0 (0/6)	0.0 (0/6)
	72	27.2 (3/11)	0.0 (0/6)	0.0 (0/6)	0.0 (0/6)
Liver	3	72.7 (8/11)	66.6 (4/6)	50 (3/6)	16.6 (1/6)
	6	72.7 (8/11)	50 (3/6)	50 (3/6)	16.6 (1/6)
	24	72.7 (8/11)	33.3 (2/6)	50 (3/6)	16.6 (1/6)
	48	63.6 (7/11)	16.6 (1/6)	50 (3/6)	0.0 (0/6)
	72	54.5 (6/11)	16.6 (1/6)	50 (3/6)	0.0 (0/6)
Stomach	3	100 (11/11)	100 (6/6)	100 (6/6)	100 (6/6)
	6	100 (11/11)	100 (6/6)	83.3 (5/6)	100 (6/6)
	24	100 (11/11)	100 (6/6)	83.3 (5/6)	100 (6/6)
	48	100 (11/11)	100 (6/6)	66.6 (4/6)	100 (6/6)
	72	100 (11/11)	100 (6/6)	66.6 (4/6)	66.6 (4/6)
Whole intestine	3	100 (11/11)	100 (6/6)	100 (6/6)	100 (6/6)
	6	100 (11/11)	100 (6/6)	100 (6/6)	100 (6/6)
	24	100 (11/11)	100 (6/6)	100 (6/6)	100 (6/6)
	48	100 (11/11)	100 (6/6)	100 (6/6)	100 (6/6)
	72	100 (11/11)	100 (6/6)	100 (6/6)	100 (6/6)

Note: Numbers in parentheses are number positive per total number examined.

[a] N-Acetylglucosamine.
[b] Chitin soluble extract.

infected stomachs was 100% at 3 h postinfection but declined thereafter.

C. albicans organisms spread systemically from the gut to visceral organs in the control animals with the number of infected livers, spleens, and kidneys highest at between 3 and 6 h postinfection. In contrast, systemic spread of *C. albicans* from the gut in saccharide-treated animals was drastically reduced with no yeast recovery from the kidneys and spleens of NAG- and CSE-treated animals (Table 2). However, *C. albicans* was recovered from the kidneys of mannose-treated animals at 3 h and 6 h postinfection and spleens were transiently colonized at 3 h postinfection only. Although the liver of most saccharide-treated animals showed infectivity, the percentage of livers positive for yeast was much lower than the control animals. This was particularly true for CSE-treated mice.

TABLE 3
Population Levels of *Candida Albicans* in the GI Tract and Visceral Organs of Untreated and Saccharide-Treated Mice

Organ	Time (h)	Mean CFU per gm organ (× 10^4) Control	2% Mannose	2% NAG[a]	2% CSE[b]
Kidneys	3	1.0 ± 0.03	0.35 ± 0.02	ND	ND
	6	0.95 ± 0.03	0.09 ± 0.007	ND	ND
	24	1.2 ± 0.05	ND	ND	ND
	48	0.75 ± 0.02	ND	ND	ND
	72	0.3 ± 0.01	ND	ND	ND
Spleen	3	75 ± 3.2	60 ± 1.8	ND	ND
	6	150 ± 6.0	ND	ND	ND
	24	22.5 ± 0.38	ND	ND	ND
	48	32.5 ± 1.0	ND	ND	ND
	72	46 ± 1.25	ND	ND	ND
Liver	3	2.9 ± 0.08	1.25 ± 0.048	0.125 ± 0.04	0.35 ± 0.02
	6	0.6 ± 0.0024	0.8 ± 0.02	0.8 ± 0.02	0.3 ± 0.01
	24	1.65 ± 0.05	0.125 ± 0.006	0.08 ± 0.003	0.05 ± 0.005
	48	2.1 ± 0.06	0.075 ± 0.005	0.08 ± 0.004	ND
	72	4.1 ± 0.012	0.2 ± 0.013	0.065 ± 0.002	ND
Stomach	3	13500 ± 80.0	4435 ± 40.3	610 ± 7.1	835 ± 11.4
	6	6750 ± 50.2	2000 ± 24.0	478.5 ± 6.7	9100 ± 60.8
	24	1015 ± 20.0	1000 ± 11.95	411.5 ± 5.9	3140 ± 45.2
	48	1012.5 ± 20.1	50 ± 0.98	310 ± 4.2	118 ± 3.1
	72	1010 ± 19.2	67.5 ± 1.25	208 ± 2.8	23.5 ± 0.38
Whole Intestine	3	3300 ± 49.3	1470 ± 18.7	1615 ± 21.1	585 ± 6.9
	6	1115 ± 20.8	1500 ± 22.8	1216 ± 17.5	477 ± 7.0
	24	160 ± 4.2	477 ± 7.3	915 ± 13.1	220 ± 3.6
	48	155 ± 4.0	33 ± 0.95	27.5 ± 0.45	18.5 ± 0.4
	72	150 ± 4.1	66 ± 1.2	7 ± 0.2	7.5 ± 0.3

Note: ND, Not detected.

[a] N-Acetylglucosamine. Values given are average count from six animals ± SD × 10^4.
[b] Chitin soluble extract.

4. Colonization of the GI Tract and Visceral Organs.

Heavy colonization of the GI tract and visceral organs of infant mice follows oral inoculation with *C. albicans*.[47] The GI infection persists in survivors to an age at which the animals are normally resistant. Mice with persistent infections, therefore, were used to study the effects of saccharide treatment on the colonization and dissemination of *C. albicans*.

Treatment of mice with saccharides resulted in markedly reduced levels of *C. albicans* in the visceral organs (Table 3). This was particularly pronounced in the kidneys and spleen, where *Candida* cells were not cultured from these organs of most of the animals. As shown in Table 3, control animals had high gut population which decreased with time. Similarly, saccharide-treated mice showed high GI tract colonization. The levels of stomach colonization by *C. albicans* in mannose- and NAG-treated animals were lower than that obtained in the control animals throughout the study period. Counts were between 2.5-fold to 22-fold lower than the control values. CSE-treated animals initially showed a slightly higher infection than the control,

TABLE 4
Population Levels of *Candida Albicans* in the GI Tract and Visceral Organs of Untreated and Pretreated Mice with 20 μl Saccharide

Organ	Time (h)	Mean CFU per gm organ ± SD (× 10^4)[a] Control	2% Mannose	2% NAG	2% CSE
Kidneys	3	0.56 ± 0.03	0.31 ± 0.01	0.13 ± 0.02	0.16 ± 0.05
	6	0.98 ± 0.06	0.75 ± 0.03	0.24 ± 0.03	0.26 ± 0.02
	24	1.5 ± 0.04	4.8 ± 0.6	0.63 ± 0.05	0.92 ± 0.06
	48	0.66 ± 0.02	2.1 ± 0.08	0.33 ± 0.002	ND
	72	0.41 ± 0.01	ND	ND	ND
Spleen	3	26 ± 0.38	5.3 ± 0.7	1.2 ± 0.04	0.8 ± 0.02
	6	75 ± 0.62	22.3 ± 1.1	2.3 ± 0.6	1.9 ± 0.04
	24	24 ± 0.46	36.5 ± 3.2	3.8 ± 0.7	2.3 ± 0.06
	48	26 ± 0.81	ND	ND	ND
	72	32 ± 0.2	ND	ND	ND
Liver	3	1.3 ± 0.01	12 ± 0.9	1.1 ± 0.8	0.4 ± 0.02
	6	2.1 ± 0.04	10.3 ± 0.8	1.0 ± 0.3	0.7 ± 0.03
	24	1.9 ± 0.03	9.5 ± 0.4	0.2 ± 0.05	0.06 ± 0.001
	48	2.6 ± 0.06	5.1 ± 0.2	0.9 ± 0.02	ND
	72	1.1 ± 0.01	2.4 ± 0.1	0.5 ± 0.001	ND
Stomach	3	417 ± 2.9	4120 ± 3.4	815 ± 4.6	7773 ± 10.3
	6	636 ± 2.3	536 ± 2.2	99140 ± 12.3	9414 ± 9.3
	24	149 ± 0.9	198 ± 0.6	416 ± 6.5	8482 ± 8.3
	48	512 ± 3.6	6437 ± 14.4	901 ± 7.4	7379 ± 12.2
	72	468 ± 2.9	580 ± 2.2	1460 ± 9.3	5530 ± 11.4
Whole Intestine	3	9734 ± 16.2	7411 ± 13.4	7699 ± 11.2	5760 ± 11.6
	6	1940 ± 14.1	4120 ± 14.0	8603 ± 12.1	3133 ± 7.3
	24	796 ± 4.6	1952 ± 20.3	3122 ± 6.7	615 ± 4.6
	48	188 ± 1.3	87.0 ± 6.7	1297 ± 4.6	595 ± 2.4
	72	168 ± 1.1	5.8 ± 0.9	25 ± 2.3	377 ± 2.1

Note: ND, not detected.

[a] Values given are average count from six animals ± SD × 104.

particularly at 6 h and 24 h (Table 3). This was followed by lower levels of colonization at 48 h and 72 h (counts between 9-fold and 43-fold lower than in control animals). In general, the intestines of saccharide-treated animals had lower counts than the control animals at 3 h post-infection (between 2-fold and 3.7-fold lower than the control). At 6 h and 24 h postinfection the number of CFU in saccharide-treated animals were either similar to controls or slightly higher. This situation reversed thereafter (Table 3).

Treatment of mice with a single dose of 2% solution of NAG or CSE, 30 min prior to yeast inoculation led to a reduction in the levels of *Candida* in the kidneys, spleen, and liver (Table 4). In contrast, the transient colonization of the stomach and intestine, exposed to the same treatment, increased significantly. Similar treatment with mannose gave inconsistent results (Table 4).

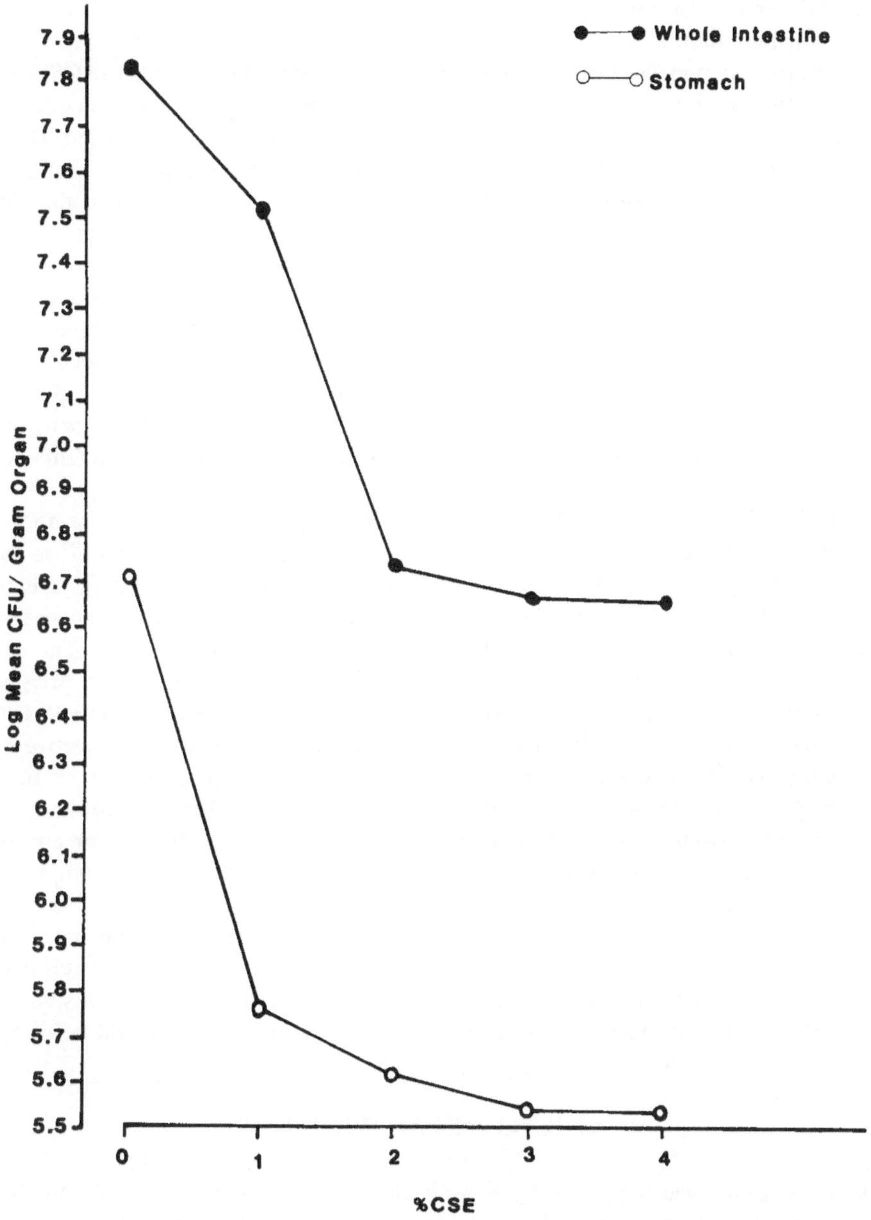

FIGURE 3. Dose-response effect of CSE in both whole intestine and stomach.

Treatment with the inhibitors post-*Candida* inoculation did not produce any significant differences in transient colonization of the GI tract or systemic spread between the control and treated mice and are, therefore, not shown.

Concentration dependence of the inhibitory activity of CSE was tested at a single time point (48 h). Figure 3 shows that inhibition by this saccharide derivative of transient yeast colonization of the intestine and stomach of mice is dose dependent, with higher concentrations resulting in lower colonization. At 1% concentration, CSE inhibited yeast colonization of the intestine only slightly (4% reduction as compared with the control), while inhibition of stomach colonization was higher at this concentration (14% inhibition as compared to the control). However, no significant difference was observed in the inhibition level between 3 and 4% CSE solution. In both cases the inhibition was around 14% for the intestine and 17% for the stomach.

Our data provides evidence that treatment of infant mice with mannose, NAG, or CSE, which are components of the *Candida* cell wall,[48,49] leads to a remarkable reduction in systemic spread of *C. albicans* from the gut to visceral organs, particularly to the kidneys and spleen. This was true for number of positive organs infected (Table 2) and for levels of colonization (Table 3). In addition, the levels of yeast colonization of the stomach and intestine were generally lower than those obtained for the control. This inhibition was most significant when the treatment was carried out 2 d prior to yeast colonization, with a daily saccharide dose throughout the study period. Although a single dose of saccharides administered 30 min prior to yeast inoculation did reduce systemic spread of *Candida* to visceral organs, such a dosing regimen increased the transient colonization of the stomach and intestine. When mice were treated postinoculation, no inhibitory effect was observed. These results indicate that the dosing regimen employed for saccharide treatment is important. Treatment for a long period prior to yeast inoculation is preferable to a single dose treatment. In addition, post-*Candida* inoculation treatment will not prevent colonization and dissemination.

The mechanism(s) of how saccharide treatment interfered with the GI colonization and dissemination of *Candida* is not clear. However, these substances could bind to the GI epithelial mucosa and thereby block the attachment of *C. albicans*, preventing or reducing the level of GI colonization and subsequent systemic spread. Another possiblity is the immune response of the GI mucosa to saccharides. It is known that chitin and its derivatives,[50] NAG,[51,52] and mannans[53,54] exert immunopotentiating effects on animals challenged with tumor cells or with pathogenic microbes. Also Suzuki et al.[50] have shown that both chitin and chitosan are able to exhibit immunopotentiating action for lethal challenge of *C. albicans* thus exhibiting a protective effect on mice administered with these polysaccharides. Possibly, saccharides stimulate the mucosal immune response, triggering protective immunological reactions which reduce the level of colonization and dissemination. Whether one or both of these possibilities, or any other mechanism(s), are involved in the protection process is yet to be elucidated. The complexity of this problem is only starting to emerge.

In conclusion, our investigations provide evidence that saccharides known to inhibit adherence of yeast *in vitro* interfere with the colonization and dissemination of the GI tract by *C. albicans* in experimental animals. They also stress the fact that more *in vivo* studies are required to reveal the mechanism(s) involved so that such an approach could be of future practical value in prophylaxis.

ACKNOWLEDGMENT

This investigation was supported by Kuwait University, Research Council, and Kuwait Foundation for the Advancement of Science, Grant number SO 040/87-07-27.

REFERENCES

1. **De Villez, R. L. and Lewis, C. W.**, Candidiasis seminar, *Cutis*, 19, 69, 1977
2. **Odds, F. C.**, *Candida and Candidosis, A Review and Bibliography*, Bailliere-Tindall, London, 1988.
3. **Parker, J. C., Mc Closkey, J. J., and Knauer, A. K.**, Pathobiologic features of human candidiasis. A common deep mycosis of the brain, heart and kidney in the altered host, *Am. J. Clin. Pathol.*, 65, 991, 1976.
4. **Drouhet, E. and Dupont, B.**, Evolution of antifungal agents: past, present, and future, *Rev. Infect. Dis.*, 9, S4, 1987.
5. **Allen, J. C.**, Infection complicating neoplastic disease and cytotoxic therapy, in *Infection and the Compromised Host*, Allen, J. C., Ed., Williams & Wilkins, Baltimore, 1976, 151.

6. **Tattersall, M. H.**, Aggressive cancer treatment and its role in predisposing to infection, *Eur. J. Cancer*, 11(Suppl.), 9, 1975.
7. **Armstrong, D., Chmel, H., Singer, C., Tapper, M., and Rosen, P.**, Non-bacterial infections associated with neoplastic disease, *Eur. J. Cancer*, 11(Suppl.), 79, 1975.
8. **Ofek, I. and Beachey, E. H.**, Bacterial adherence, in *Advances in Internal Medicine*, Vol. 25, Stollerman, G. H., Ed., Year Book Medical, Chicago, 1980, 503.
9. **Costerton, J. W., Marrie, T. J., and Cheng, K. J.**, Phenomena of bacterial adhesion, in *Bacterial Adhesion, Mechanisms and Physiological Significance*, Savage, D. C. and Fletcher, M., Eds., Plenum, New York, 1985, chap. 1.
10. **Beachey, E. H.**, Bacterial adherence: adhesin-receptor interactions mediating the attachment of bacteria to mucosal surfaces, *J. Infect. Dis.*, 143, 325, 1981.
11. **Freter, R. and Jones, G. W.**, Models for studing the role of bacterial attachment in virulence and pathogenesis, *Rev. Infect. Dis.*, 5(Suppl. 4), S647, 1983.
12. **Sparling, P. F.**, Bacterial virulence and pathogenesis: an overview, *Rev. Infect. Dis.*, 5(Suppl. 4), S637, 1983.
13. **Beachey, E. H.**, *Bacterial Adherence, Receptors and Recognition*, Series B, Vol. 6, Chapman and Hall, London, 1980.
14. **Ofek, I. and Beachey, E. H.**, General concepts and principles of bacterial adherence, in *Bacterial Adherence, Receptor and Recognition*, Beachey, E. H., Ed., Series B, Vol. 6, Chapman and Hall, London, 1980, 1.
15. **Ghannoum, M. A.**, Mechanisms potentiating *Candida* infections. A review, *Mycoses*, 1988, 31, 543, 1988.
16. **Douglas, L. J.**, Adhesion to surfaces, in *The Yeasts*, Vol. 2, 2nd ed., Rose, A. H. and Harrison, J. S., Eds., Academic Press, London, 1987, 239.
17. **Savage, D. C. and Fletcher, M.**, *Bacterial Adhesion, Mechanisms and Physiological Significance*, Plenum Press, New York, 1985.
18. **Beachey, E. H., Eisenstein, B. I., and Ofek, I.**, *Bacterial Adherence in Infectious Diseases*, Upjohn Company, Michigan, 1982.
19. **Ciba Foundation Symposium 80**, *Adhesion and Microorganism Pathogenicity*, Pitman Medical, London, 1981.
20. **Segal, E.**, Pathogenesis of human mycoses: role of adhesion to host surfaces, *Microbiol. Sci.*, 4, 344, 1987.
21. **Aronson, M., Medalia, O., Schori, L., Mirelman, D., Sharon, N., and Ofek, I.**, Prevention of colonization of the urinary tract of mice with *Escherichia coli* by blocking of bacterial adherence with methyl α-D-mannopyrnoside, *J. Infect. Dis.*, 139, 329, 1979.
22. **Ryley, J. F.**, Pathogenicity of *Candida albicans* with particular reference to the vagina, review article, *J. Med. Vet. Mycol.*, 24, 5, 1986.
23. **Ray, T. L., Digre, K. B., and Payne, C. D.**, Adherence of *Candida* species to human epidermal corneocytes and buccal mucosal cells: correlation with cutaneous pathogenicity, *J. Invest. Dermatol.*, 83, 37, 1984.
24. **McCourtie, J. and Douglas, L. J.**, Relationship between cell surface composition, adherence, and virulence of *Candida albicans*, *Infect. Immun.*, 45, 6, 1984.
25. **Ghannoum, M. and Abu ElTeen, K.**, Correlative relationship between proteinase production, adherence and pathogenicity of various strains of *Candida albicans*, *J. Med. Vet. Mycol.*, 24, 407, 1986.
26. **Lehrer, N., Segal, E., and Barr-Nea, L.**, *In vitro* and *in vivo* adherence of *Candida albicans* to mucosal surfaces, *Ann. Microbiol. (Inst. Pasteur)*, 134B, 293, 1983.
27. **Segal, E., Gottfried, L., and Lehrer, N.**, Candidal vaginitis in hormone-treated mice: prevention by a chitin extract, *Mycopathologia*, 102, 157, 1988.
28. **Lehrer, N., Segal, E., Lis, H., and Gov, Y.**, Effect of *Candida albicans* cell wall components on the adherence of the fungus to human and murine vaginal mucosa, *Mycopathologia*, 102, 115, 1988.
29. **Krause, W., Mathesis, H., and Wulf, K.**, Fungaemia and funguaria after oral administration of *Candida albicans*, *Lancet*, 1, 598, 1969.
30. **Stone, H., Geheber, C., Kolb, L., and Kitchens, W.**, Alimentary tract colonization by *Candida albicans*, *J. Surg. Res.*, 14, 273, 1973.
31. **Stone, H., Kolb, L., Currie, C., Geheber, C., and Cuzzell, J.**, *Candida* sepsis: Pathogenesis and principles of treatment, *Ann. Surg.*, 179, 697, 1974.
32. **Kennedy, M. J. and Volz, P. A.**, Effect of various antibiotics on gastrointestinal colonization and dissemination by *Candida albicans*, *J. Med. Vet. Mycol.*, 23, 265, 1985.
33. **Kimura, L. and Pearsall, N.**, Adherence of *Candida albicans* to human buccal epithelial cells, *Infect. Immun.*, 21, 64, 1978.
34. **King, R., Lee, J., and Morris, A.**, Adherence of *Candida albicans* and other *Candida* species to mucosal epithelial cells, *Infect Immun.*, 27, 667, 1980.
35. **Sandin, R., Rogers, A., Patterson, R., and Beneke, E.**, Evidence for mannose-mediated adherence of *Candida albicans* to human buccal cells *in vitro*, *Infect. Immun.*, 35, 79, 1982.
36. **Sobel, J., Myers, P., Kaye, D., and Levison, M.**, Adherence of *Candida albicans* to human vaginal and buccal epithelial cells, *J. Infect Dis.*, 143, 76, 1981.

37. **Segal, E., Soroka, A., and Lehrer, N.**, Attachment of *Candida* to mammalian tissues—clinical and experimental studies, *Zbl. Bakt. Hyg.*, 257 A, 257, 1984.
38. **Lee, J. and King, R.**, Characterization of *Candida albicans* adherence to human vaginal epithelial cells *in vitro*, *Infect. Immun.*, 41, 1024, 1983.
39. **Maisch, P. and Calderone, R.**, Role of surface mannan in the adherence of *Candida albicans* to fibrin-platelet clots formed *in vitro*, *Infect. Immun.*, 32, 92, 1981.
40. **Clancy, J. and Savage, D. C.**, Another colicin V phenotype *in vitro* adhesion of *Escherichia coli* to mouse intestinal epithelium, *Infect. Immun.*, 32, 343, 1981.
41. **Ghannoum, M., Motawy, M., Mubarak, A., and Al-Awadi, H.**, *Candida albicans* strain differentiation in cancer patients undergoing therapy, *Mykosen*, 28, 388, 1985.
42. **Ghannoum, M., Motawy, M., Mubarak, A., and Al-Awadi, H.**, Incidence and susceptibility of yeasts isolated from cancer patients in Kuwait, *Mykosen*, 28, 619, 1985.
43. **Ghannoum, M., Khatter, M., Kayed, F., and Shridar, P.**, The influence of antineoplastic agents on experimental candidosis in mice, *Mykosen*, 28, 369, 1985.
44. **Segal, E. and Savage, D. C.**, Adhesion of *Candida albicans* to mouse intestinal mucosa *in vitro:* development of the assay and test of inhibitors, *J. Med. Vet. Mycol.*, 24, 477, 1986.
45. **Pope, L., Cole, G., Guentzel, M., and Berry, L.**, Systemic and gastrointestinal candidiasis of infant mice after intragastric challenge, *Infect. Immun.*, 25, 702, 1979.
46. **Field, L., Pope, L., Cole, G., Guentzel, M., and Berry, L.**, Persistence and spread of *Candida albicans* after intragastric inoculation of infant mice, *Infect. Immun.*, 31, 783, 1981.
47. **Guentzel, M. and Herrera, C.**, Effects of compromising agents on candidosis in mice with persistent infections initiated in infancy, *Infect. Immun.*, 35, 222, 1982.
48. **Farkas, V.**, Biosynthesis of cell wall of fungi, *Microbiol. Rev.*, 43, 117, 1979.
49. **Poulain, D. C., Trouchin, G., Cubremetz, J. F., and Biguet, J.**, Ultrastructure of the cell wall of *Candida albicans* blastospores: study of its constitutive layers by the use of a cytochemical technique revealing polysaccharides, *Ann. Microbiol.*, 129, 141, 1978.
50. **Suzuki, K., Okawa, Y., Hashimoto, K., Suzuki, S., and Suzuki, M.**, Protecting effect of chitin and chitosan on experimentally induced murine candidiasis, *Microbiol. Immunol.*, 28, 903, 1984.
51. **Azuma, I., Sugimura, K., Taniyama, T., Yamawaki, M., Yamamura, Y., Kusumoto, S., Okada, S., and Shiba, T.**, Adjuvant activity of mycobacterial franctions: adjuvant activity of synthetic N-acetyl muramyldipeptide and the related compounds, *Infect. Immun.*, 14, 18, 1976.
52. **Azuma, I., Taniyama, T., Hirao, F., and Yamamura, Y.**, Anti-tumor activity of cell wall skeletons and peptidoglycolipids of mycobacteria and related microorganisms in mice and rabbit, *GANN*, 65, 493, 1974.
53. **Hashimoto, K., Okawa, Y., Suzuki, K., Okura, Y., Suzuki, S., and Suzuki, M.**, Antitumor activity of acidic mannan fraction from bakers yeast, *J. Pharm. Dyn.*, 6, 668. 1983.
54. **Okawa, Y., Okura, Y., Hashimoto, K., Matsumoto, T., Suzuki, S., and Suzuki, M.**, Protective effect of D-mannan of baker's yeast against *staphylococcus aureus* infection in mice, *Carbohydrate Res.*, 108, 328, 1982.
55. **Cole, G.T., Seshan, K.R., Pope, L.M., and Yancey, R.J.**, Morphological aspects of gastrointestinal tract invasion by *Candida albicans* in the infant mouse, *J. Med. Vet. Mycol.*, 26, 173, 1988.

INDEX

D

E

F

S

T